THE PRENTICE HALL ILLUSTRATED HANDBOOK OF ADVANCED MANUFACTURING METHODS

Jerome H. Fuchs, CMC

PRENTICE HALL
Englewood Cliffs, New Jersey 07632

Prentice-Hall International (UK) Limited, *London*
Prentice-Hall of Australia Pty. Limited, *Sydney*
Prentice-Hall Canada, Inc., *Toronto*
Prentice-Hall Hispanoamericana, S.A., *Mexico*
Prentice-Hall of India Private Limited, *New Delhi*
Prentice-Hall of Japan, Inc., *Tokyo*
Simon & Schuster Asia Pte. Ltd., *Singapore*
Editora Prentice-Hall do Brasil, Ltda., *Rio de Janeiro*

10 9 8 7 6 5 4 3 2 1

Library of Congress Cataloging-in-Publication Data

Fuchs, Jerome H.
The Prentice-Hall illustrated handbook of advanced manufacturing methods / by Jerome H. Fuchs.
p. cm.
Includes index.
ISBN 0-13-698318-9
1. Manufacturing processes—Handbooks, manuals, etc. I. Title.
TS183.F83 1988
658.5—dc19 87-32285
CIP

ISBN 0-13-698318-8

PRENTICE HALL
BUSINESS & PROFESSIONAL DIVISION
A division of Simon & Schuster
Englewood Cliffs, New Jersey 07632

PRINTED IN THE UNITED STATES OF AMERICA

With experience comes wisdom
With wisdom comes sound judgment
With sound judgment comes proper decisions
With proper decisions comes appropriate action
With appropriate action comes solutions to problems leading to . . .

CONSTRUCTIVE CHANGE
AND ACCELERATED GROWTH AND PROFITS

About the Author

Jerome H. Fuchs is senior partner of Fuchs Associates, an independent management consulting firm founded in 1960 and based in the New York City area. The authoritative industry publication, *Consultants News,* ranks Fuchs Associates as one of the one hundred Leading Management consulting firms in the United States.

During his career in industry, Mr. Fuchs has been executive vice president of Business Supplies Corporation of America, executive assistant to the president of Rockbestos Wire and Cable, and vice president of Manufacturing at United Aircraft Products. While in these positions, he had administrative responsibility over manufacturing, production planning and control, materials management, and human resources. In this capacity, he was instrumental in increasing labor productivity, reducing cost of goods sold, and contributing to improved corporate earnings. Prior to this, Mr. Fuchs was manager of systems and methods at Pennwalt, Lukens Steel, and AMAX where he engaged in various projects involved in the development of computerized management information systems, factory and office automation, and direct labor, administrative, and clerical work methods.

Since forming Fuchs Associates, he has served over one hundred industrial and commercial firms, government agencies, and private institutions. This service has involved numerous in-depth studies for many prominent concerns, including R.J. Reynolds Nabisco, Scovill, The Allen Group, Brunswick, Cerro (The Marmon Group), Certain-Teed, Curtiss-Wright, Asarco, Kennecott (Standard Oil), American Electric Power (AEP), American District Telegraph (ADT), BASF, Sulzer Brothers, St. Joe Minerals (division of Fluor), Air Products & Chemicals, Dun & Bradstreet, Continental Can (division of Keiwit Murdoch), Getty Oil (Texaco), Primamerica, AMAX, Shearson/Lehman American Express, Hudson Pulp & Paper (Georgia Pacific), ILC (Rapid American), and Kayser-Roth (Subsidiary of Gulf + Western Industries), Wagner (division of Cooper Industries), and A.T. Massey (subsidiary of Scallop/Royal Dutch Petroleum).

Mr. Fuchs is presently on the Board of Directors of Del Electronics and Extended Technologies.

As a member of the staff (evening graduate school), he has lectured in courses on the graduate level at Syracuse University, McCoy College of Johns Hopkins, Drexel Institute, Queens College of the City of New York, and the State University of New York at Stoney Brook. These courses were in production and operations management, quality management, organization planning and control, human relations, personnel administration, problems in supervision, labor economics, and finance.

Mr. Fuchs is the author of *Computerized Inventory Control Systems, Computerized Cost Control Systems, Administering the Quality Control Function, Making the Most of Management Consulting Services,* and *Management Consultants in Action,* as well as numerous articles on various business management subjects.

He holds a B.S. in sales management and an M.B.A. in production management from Syracuse University. While obtaining his advanced degree, Mr. Fuchs also had an industrial research assistantship in personnel administration.

Mr. Fuchs is a charter member and past President of the Society of Professional Management Consultants, and is presently on the SPMC Board of Directors. He also is a founding member of the Institute of Management Consultants, a Certified Management Consultant, and is included in *Who's Who in Finance and Industry, Who's Who in the East,* and *Contemporary Authors.*

Special Attributions

Throughout the preparation of this handbook individuals in virtually every segment of manufacturing provided me with valuable source material. To all these people and their fine organizations, I extend a most heartfelt thanks.

I also would like to mention with great respect and admiration, some of my good friends and colleagues in the consulting profession for their support in providing me with industry contacts and a range of subject matter taken from their extensive technical libraries: H. William Nordyke, President, H. William Nordyke & Associates, Inc., Albuquerque, N.M.; Jack Davis, President, Distributed Consultancy International, Inc., Flushing, N.Y.; George Gross, President, J. George Gross & Associates, Woodbridge, N.J.; Harmon Houghton, President, Extended Technologies Corporation, Weehawken, N.J.

I also am greatly indebted to Randy R. Boldosser, Member of the Technical Staff, AT&T Bell Laboratories, Holmdel, N.J. and Rod Griffith, Manager, Supervisory Control Applications Marketing, Digital Equipment Corporation, Marlboro, MA., for reviewing the text in rough draft form and offering many constructive suggestions in content and substance.

I would like to thank Dianne Kaplan, Word Works Secretarial Services, Seaford, N.Y., and Karen O. Basch, Superior Typing Service, Massapequa, N.Y., for their outstanding editing, word processing, and typing skills in converting my handwritten material into professionally prepared final copy.

It is very difficult to put into words how important illustrations can be in enriching the reader's interest in and understanding of the author's text. This important task was performed admirably by Susan Decker, Windsor, N.Y., whose outstanding artistic talent brought to life scores of graphs, charts, and other visual renderings.

I also was most fortunate to have the services of Olivia Lane, Acquisitions Editor of Prentice Hall who possesses outstanding professional editorial capability.

Most of all, I am grateful to my wife Eleanor for her constant encouragement and understanding during the many months of concentrated effort required to write this extensive handbook.

Acknowledgments

I would like to express my sincere appreciation for the wholehearted cooperation of these individuals who provided invaluable material that has been included in this book.

Jervis C. Webb, President and Chairman of the Board, Jarvis B. Webb Company, Farmington Hills, MI.

David Harrigan, National Marketing Manager, Integrated Automation, Alameda, CA.

D. Scott Sink, Ph.D., P.E., Director, Virginia Productivity Center, Virginia Polytechnic Institute and State University, Blacksburg, VA.

Barry Fox, Vice President, Technical Educational Consultants, Hicksville, NY.

John T. Blackney, President, American Supplier Institute, Inc., Dearborn, MI.

W. W. Scherkenbach, Corporate Director Statistical Methods, and P.T. Jessup, Statistical Methods Associate, Statistical Methods Office, Ford Motor Company, Dearborn, MI.

Guy W. Potok, Vice President and General Manager, International and Automotive Assembly Division, GM Fanuc Robotics Corporation, Troy, MI.

Robert Carper, Vice President CIM Group, American Cimflex Corporation, Pittsburgh, PA.

Phillip A. Angelillo, Manager, Modeling and Simulation, and Michael C. Dunn, Programmer, Grumman Data Systems Corporation, Woodbury, NY.

Luis B. Boza, Charles D. Chavous, Jay C. Hsu, Robert D. Lake, and Michael K. Pratt, Operations Engineering, AT&T Bell Labs, Short Mills, NJ.

Thomas A. Zack, Vice President, Nu-Tec Corporation, S. Portland, ME.

Jon M. Baldwin, Division Manager, Geomet Systems Division, Helmel Engineering Products, Inc.

Andres V. Karolin, Systems Analyst, United McGill Corporation, Groveport, OH.

Society of Manufacturing Engineers, Dearborn, MI.

Gregory Coticcia, Product Manager, Vision, American Industrial Vision, an American Cimflex Company, Pittsburgh, PA.

Hanno Behm, Project Manager, Beumer Corporation, Scarsdale, NY.

Training magazine, Lakewood Publications, subsidiary of Maclean Hunter Publishing Company, Minneapolis, MN.

Ronald L. Reimink, Sales Manager, Ann Arbor Computer, division of the Jervis B. Webb Company, Ann Arbor, MI.

Roger N. Nagel, Director, MSE, Institute for Robotics, Lehigh University, Bethlehem, PA.

Michael G. Semmens, P.E., Group Vice President, The BDM Corporation, Albuquerque, NM.

William Tandler, President, Multi Metrics, Inc., Redwood City, CA.

William Phelps, President, and David Billheimer, Engineer, Product Services, Polhemus Navigation Sciences Division, McDonnell Douglas Corporation, Colchester, VT.

David B. Wortman, Executive Vice President, Pritsker & Associates, West Lafayette, IN.

Charles A. Fried, Executive Vice President, Creative Output, Milford, CT.

Emerson Schmidt, Marketing Services Manager, Sheffield Measurement Division, Warner & Swassey/A Cross & Trecker Company, Dayton, OH.

Ron Adams, Manager, Marketing/Sales, Ingersoll-Rand Waterjet Cutting Systems, Baxter Springs, KS.

Paul DiBono, Vice President Marketing, American Software, Inc., Atlanta, GA.

James Leitch. C. Mfg. T, FMC Project Manager, LTV Aerospace and Defense Company, Vought Aero Products Division, Dallas, TX.

Dr. R. Hubach, Chairman, View Engineering, Inc., Simi Valley, CA.

Richard Thau, President, and David A. Meyer, CPIM, Executive Vice President, Micro-MRP, Inc., Foster City, CA.

George Gugliotta, President, Automation Systems, Inc., Oxford, CT.

Daniel Jackson, Sales Engineer, LDJ Electronics, Inc., Troy, MI.

Philip Rhodes, President, Systems Specialists, Inc., Dallas, TX.

M. R. Hawk, Manager of Industrial Technology Modernization Program, and Lisa A. Kennedy, Supervisor, Industrial Engineering, General Dynamics Corporation, Fort Worth Division, Fort Worth, TX.

Richard C. Dunne, Director of Communications, AIL Division of Eaton Corporation, Deer Park, NY.

Brian Salisbury, Manager Product Publicity, Grumman Corporation, Bethpage, NY.

John Foggiato, Director, Technical Marketing, Flexible Manufacturing Systems, Inc., Los Gatos, CA.

Michael Urban, Vice President of Marketing, Production Machinery Corporation, Mentor AGVS, Inc., Mentor, OH.

James Jones, President and CEO, Apogee Robotics, Inc., Ft. Collins, CO.

Charles Rosbeck, Section Head of Engineering/Manufacturing Integration, Fairchild Republic Corporation, Farmingdale, NY.

Michael Hitchcock, Project Engineer, Manufacturing Technology Division of the Air Force Wright Aeronautical Laboratories, Wright-Patterson AFB, OK.

Daniel Simpson, Sales Manager, and John E. Chilton, Vice President Engineering, Escort Memory Systems, Inc., Scotts Valley, CA.

Laura Hanson, Publisher, Bar Code News, Peterborough, NH.

David G. Campolo, Business Manager, National Productivity Review, New York, NY.

Jerry Gross, Press Relations Manager, Hewlett Packard, Information Networks Division, Cupertino, CA.

Kathy Kimbell, Publicity Coordinator, Hewlett Packard, Fort Collins Systems Division, Fort Collins, CO.

William P. Hakanson, Executive Director, Automatic Identification Manufacturers (AIM), Pittsburgh, PA.

James M. Doherty, Manager, Engineering Systems and Standards, U.S. Electrical Motors, Division of Emerson Electric Company, Milford, CT.

Thomas V. Sobczak, Ph.D., President, Little People's Productivity Center, Baldwin, NY.

Harry G. Smart, Director of Management Systems, VSI Aerospace Products Division, Culver City, CA.

Jim Feenstra, Director of Automotive Marketing, Computervision Corporation, Bedford, MA.

Carl G. Thor, Senior Vice President, American Productivity Center, Houston, TX.

Tracy McAllister, Publications Specialist, Intel Corporation, Santa Clara, CA.

Martin C. Woodhouse, Vice President, CEO, Planring U.S.A., Inc., Farmington Hills, MI.

Victor Mandrillo, Manager Communications, Eaton-Kenway, Subsidiary of Eaton Corporation, Salt Lake City, UT.

E. J. De Gayner, President, Catalyst USA, Inc., Grafton, WI.

Lewis E. Lachter, Director Marketing and Public Relations, Staff Marketing Services, 3M Corporation, St. Paul, MI.

Harry Apkarian, Chairman and CEO, Mechanical Technology, Inc., Latham, NY.

Donald Kiser, Communications Specialist, International Business Machines Corporation, Lexington, KY.

Robert Price, President, E.S.I., Albany, NY.

Luann T. Ebert, Manager, Marketing Communications, OIR/Organization for Industrial Research, Bedford, MA.

Control Data, David S. Lindsey, Communications Program Manager, Minneapolis, MN.

Robert Callahan, President, Ingersoll Engineers Incorporated, Rockford, IL.

Judy Doelling, Marketing Communications Specialist, McDonnell Douglas, St. Louis, MO.

Ron Keil, Chairman, GM MAP Task Force, General Motors Corporation, Warren, MI.

Kurt T. Hertzog, Supervisor, Advanced Process Development, Rochester Products Division of General Motors Corporation, Rochester, NY.

F. Hide, Manager, Accusembler Robotic Systems, Kanematsu-Gosho (USA) Inc., Somerset, NJ.

William J. Higgins, National Sales Manager, Mannesmann DeMag Material Handling Division of Mannesmann DeMag Corporation, Grand Rapids, MI.

John Simontacci, Sales Manager, Westinghouse Electric Corporation, Advanced Production Technology Systems, Pittsburg, PA.

Gary Wagner, Director of Sales and Marketing, Penn Video, a subsidiary of Ball Corporation, Akron, OH.

Mathew L. Montforte, President and CEO, Montforte Robotics, Inc., a subsidiary of Multivisions Corporation, West Trenton, NJ.

Ernie Martinez, Jr., Marketing Telecommunications Specialist, Unimation, a Westinghouse Company, Danbury, CT.

Ralph J. Olson, President, Interlake Material Handling Division of Interlake/Conco-Tellus AGVS, Burr Ridge, IL.

MacNeal-Schwendler Corporation, Los Angeles, CA.

Stephen L. Kirkwood, Vice President and General Manager, Continental Systems Technology, Inc., Chicago, IL.

Robert Metzger, Engineering Group Leader, Robotic Vision Systems, Inc., Hauppauge, NY.

Carlos R. Bejarano, Numerical Control Systems Engineer, Fairchild Republic Corporation, Farmingdale, NY.

Kathy Klatman, Marketing Communications Specialist, Data General, Westboro, MA.

Contributor Organizations

The author wishes to thank the 260 companies listed below for providing valuable research material during the development of this book. In particular:

Autocad
Amtech
Albuquerque Technical–Vocational Institute
Adept Technology
American CIMFLEX
Applicon
The Austin Company
Allen Industries
AVX Corp.
American Software
AT&T (AT&T Bell Laboratories)
Applied Scanning Technology
American Industrial Vision
Air Technical Industries
Automatic Systems
American Productivity Center
Arizona State University
The American University
AGS Management Systems
Automatic Identification Manufacturers (AIM)
Amano
Advanced Manufacturing Systems
APOGEE Robotics
ASEA Robotics
American Supplier Institute
American Industrial Vision
Automation Systems
Allen Translift, a unit of Wilson Machine Division

Angus & Rudin
Autonetics
Ashton–Tate
ASK Computer Systems
Allen-Bradley, a Rockwell International Company
The BDM Corp.
Barrett
Brigham Young University
Beumer
Bar Code News
Boeing Computer Services
Banner Engineering
Baker
Brown & Sharpe
Creative Output
Control Data
Computervision
Carnegie-Mellon Univ.
CI Compunetics
Center Development, Inc.
Cincinnati Milicron
COMSERV
Clark Equipment
Continental Systems Technology
CMW, Inc./Advanced Manufacturing Systems, Inc.
Cross & Treaker
Cullinet Software
Cahners Exposition Group
Catalyst USA
Clark
Control Engineering
CIMLINC
Columbia Labeling Machinery
Control Engineering
CAERE Corp.
Digital Equipment Corp.
Data General
DocuGraphix
Del Electronics
Design Technology Corp.
Daner–Hayes
Dynex
Distribution Concepts and Issues
Distribution Centers, Inc.
DAI
Dennison Industrial Systems
Eaton Corporation—AIL Division
Eaton–Kenway, a subsidiary of Eaton Corporation
Evans & Sutherland
Extended Technologies

ESAB North America, Inc.
Escort Memory Systems (EMS)
E.S.I.
Franson & Associates
Floating Point Systems
Fairchild Republic
Ford Motor Co.
Flexible Manufacturing Systems
Frost Controls
FGL Management Group
Farmingdale State University of New York Agricultural and Technical College, School of Engineering Technologies
Grumman Data Systems
Grumman Aerospace
Gulton Industries—Data Systems Division
General Dynamics—Fort Worth Division
Gould Electronics
GMF Robotics Corp.
General Motors—MAP/TASK Force
General Electric
Gerber Scientific Instrument Co.
Georgia Tech. Eng. Experiment Station, Georgia Productivity Center
Geomet Systems Division of Helmel Engineering Products
George Gross & Associates
Grumman Corporation
Hewlett Packard—Manufacturing Systems Group
Hewlett Packard—Design Systems Group
Honeywell
Hazeltine
Harvard Production Technology, Work, and Character
Hartford Graduate Center—Productivity Group
Hand-Held Products
Hudson Robotics
Harnischfeger Engineers
Infolink
ICC International Cybernetics Corp.
IBM
ILC Data Device
INTEL Corp.
Intro-Logic
Institute of Industrial Engineers
Intech Systems
International Logistics Corp.
Institute for Productivity
IIT Center—Manufacturing Productivity Center
Instrument Systems Corp.
Ingersoll Engineers
Integrated Automation
Ingersoll Rand
IFS Ltd.

Intergraph
Infra Pak
Interlake Material Handling Division of Interlake/Conco-Tellus AGVS
INTERMEC
Journal of Manufacturing Systems
Japan Productivity Center
Jervis B. Webb
Kennametal, Inc.
Key Image Systems
Kanematsu-Gosho (USA)
Little People's Productivity Center (LPPC)
Los Alamos National Laboratory
Litton Industrial Automation Systems
LDJ Electronics
Lehigh University, Institute for Robotics
LTV Aerospace & Defense Co., Vought Aero Products Division
Learning International
Lakewood Publications—*Training* magazine
LXE, a division of Electromagnetic Sciences, Inc.
L-Tron Associates Corp.
Micro—MRP
McDonnell Douglas Manufacturing Industry Systems Co.
Microcad
Mechanical Technology, Inc.
The MacNeal-Schwendler Corp. (MSC)
Motorola
Micross Manufacturing Systems
Monforte Robotics, a subsidiary of Multivisions Corp.
Machine Vision International
M.I.T.—Laboratory for Manufacturing and Productivity
Microsoft
Microcad
Material Handling Institute
Mentor AGVS, Inc., a subsidiary of Production Machinery Corp.
Multi Metrics
Menasha
Manufacturing Productivity Center
Mannesmann De Mag Material Handling Systems Division of Mannesmann Demag Corporation
Measurex Automation Systems
Martin Marietta Data Systems
MEDIA-NET
Nastec
Newsday—Technical Business Section
National Productivity Review
National Industrial Security Association
North Carolina State Univ., Productivity Research & Extension Program
National Technical Information Service, Productivity Information Center
NU-TEC Corp.
William Nordyke & Assoc.

Northrup
NBI
NCR
Ohio State Univ.
Oklahoma State Univ.
Oregon State Univ.
Organization for Industrial Research (OIR)
On-Line Computer Systems
Pritsker & Associates
Prime Computer
Professional Computer Resources
Polytechnical Institute of New York
The Pennsylvania State Univ., Pennsylvania Technical Assistance Program
Purdue University—Computer Integrated Design Manufacturing and Automation Center
Positech
Penn Video
Prab Robotic Systems and Automation Division
Productivity, Inc.
Planring, U.S.A.
Portec
Polhemus Navigation Sciences Division, McDonnell Douglas Corp.
PCI
Republic Storage Systems
Robotic Vision Systems
Robotics Age
Robotics Industry Directory
Rensselaer Polytechnic Institute—Center for Manufacturing Productivity and Technical Transfer
Rapistran, a Lear Siegler Company
Rochester Products Division of General Motors
Raymond
Sandia National Laboratories
Spacesaver
Sensing Devices, Inc.
Starcut Sales, Inc.
Swedcon AB
Skan-A-Matic
Society of Manufacturing Engineers (SME)
Siemens Components, Inc.
Stonybrook University
Sulzer
SPS Technologies—Hartman Systems
Symbol Technologies
State Univ. of Northern Virginia, George Mason University
SAI—Sparacino Associates
SYSSCAN
SPECTRON Technology
SMI
Systems Specialist, Inc.

Sheffield Measurement Division, Warner & Swassey, A Cross & Trecker Co.
Systems Modeling Corp.
SI Handling Systems
Schlage Electronics
Trax Instrument Corp.
Tandem Computers
Telephonics
Towa Corporation of America
Texas Technical University, Texas Center for Productivity and Quality of Life
3M
Technical Educational Consultants
Technology Transfer Institute
Tritek
Televideo Systems
Texas Instruments
University of New Mexico, College of Engineering
U.S. Electrical Motors Div., Emerson Electric
University of Maryland, Maryland Center for Productivity and Quality of Working Life
Univ. of Miami, Laboratory Artificial Intelligence
Unimation, a Westinghouse Company
Uniking
Visi-Tech
Viatech Systems
V.P.I. and State University, Dept. of Industrial Engineering and Operations Research
Virginia Technical University, Virginia Productivity Center
View Engineering
VSI Aerospace Products Division of VSI Corp.
Visi-Trak
Volvo Automated Systems
Westinghouse Electric—Advanced Production Technology Systems
Work in America Institute
Westinghouse Electronics Systems Division
Westinghouse Electric—Voice Systems
Webb-Triax
Xerox Computer Service

What This Book Will Do For You

The Prentice Hall Illustrated Handbook of Advanced Manufacturing Methods provides a complete understanding of the latest techniques and innovations in factory automation. Through comprehensive discussion of all elements of the manufacturing cycle and a generous sample of interesting, leading-edge case examples, you can learn how to design the most applicable automation plan for your company.

Topics in this book flow in a logical sequence from one chapter to the next, which enables you to more fully comprehend the *Total Integration of all Manufacturing Elements* (TIME as shown in figure 0–1). Physically integrating all of these components would result in the ultimate competitive technology of the so called "Factory of the Future." This handbook covers these advanced technologies in a straightforward language that includes discussions of how they can fit into your modernization program. This information can enable you to slash unit labor costs, improve product quality, provide more responsive customer service, significantly reduce overhead, and gain a competitive marketing edge.

Here's a sample of the many concepts this book can enable you to implement successfully:

Planning and Resource Allocation

- The seven critical characteristics that can affect your plant modernization strategic planning process
- Twenty-two key concerns to remember when allocating resources to fund your company's modernization program
- Ten important issues to resolve before deciding to continue or cancel a proposed modernization project
- Recognized accounting methods and measuring devices that you should use to determine project profitability and risk

Laser Technology

- The advantages of laser systems, types of applications, types of laser-based systems, and the future of laser-based technology

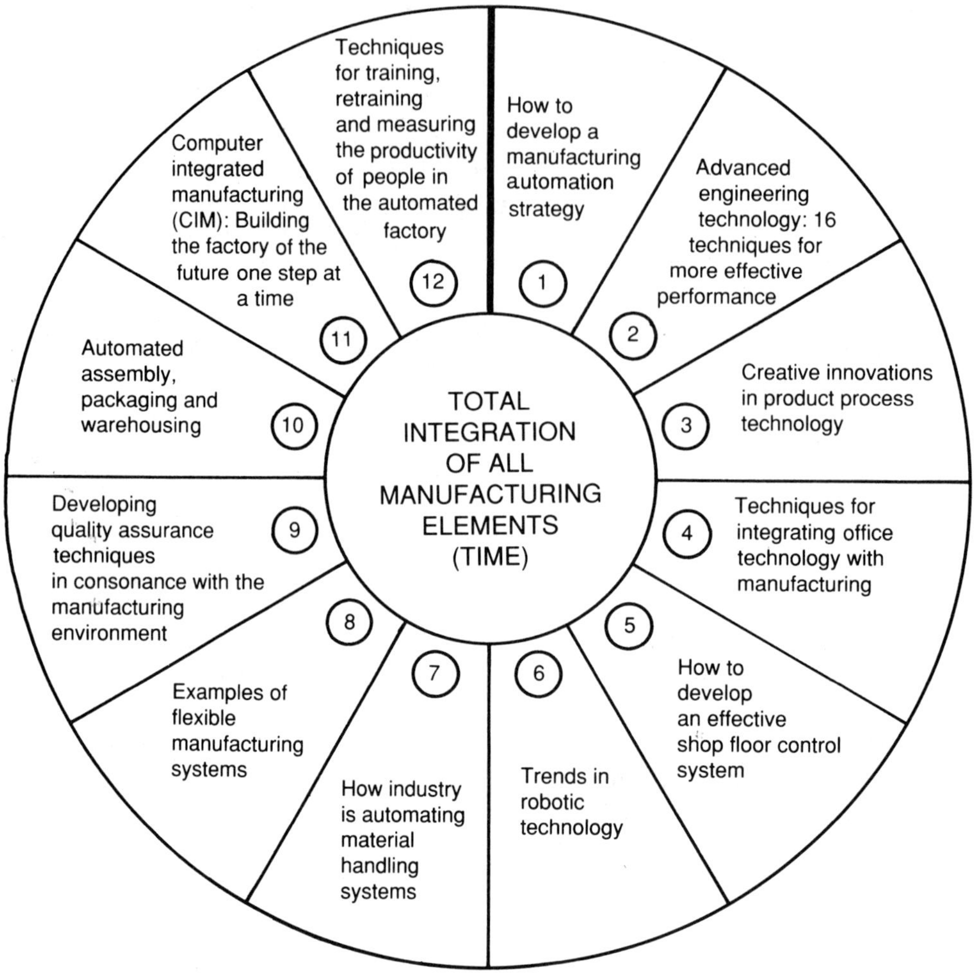

FIGURE 0–1 Total Integration of All Manufacturing Elements.

Quality Assurance

- Quality assurance computerized systems, integrated dimensional metrology analysis system (IMS); computer-aided tolerance analysis (CATA); quality information management system (QIMS); quality operations planning (QOP)

Automated Assembly

- Nine key factors to consider when developing automated assembly operations

Packaging

- A complete checklist of all the factors you should consider when selecting automatic packaging equipment, systems, and materials

Warehousing

- Five primary benefits of a computerized warehouse and shipping control system (WCS)

- Four modules that make up an integrated warehouse control system
- Seven productivity measurement formulas for use in controlling your warehousing costs
- Five steps involved in developing a warehouse simulation model
- How you can develop strategic and planning techniques for installing an integrated warehouse automation system

Material Handling

- Twenty principles of material handling
- Thirteen types of material handling equipment and methods
- Twelve guidelines for determining the correct level of material-handling automation you should employ

Picking Systems

- What are "paperless picking systems" and pick director systems? How can they speed up the picking operation and reduce direct labor costs?

Computer-Integrated Manufacturing

- Eight specific goals of a computer-integrated manufacturing (CIM) program
- How to apply strategic planning methodology toward implementation of CIM
- Steps involved in CIM simulation modeling
- What you should know about various approaches to installing a CIM system—modular or vertical approach?
- The seven components of a CIM system and how they are interrelated

Automated Guided Vehicle Systems

- Thirteen benefits of automated guided vehicle systems (AGVS)
- Eleven major categories of AGVS and how to determine which types best suit your needs
- Seven types of AGV technology, with advantages and disadvantages of each type
- The criteria to use in selecting an AGV system

Manufacturing Resource Planning

- How you can compare and evaluate the "Just-in-Time" (JIT) manufacturing system versus manufacturing resource planning (MRP II)

Shop Floor Control

- The seven unforeseen circumstances to guard against in shop floor control
- The ten features of an effective shop floor control system

Tactical Ratio and Monitoring System

- How to control work-in-process lots using the tactical ratio and monitoring system (TRAMS)

Identification Technologies

- When and how to use automatic and other identification technologies such as: bar code scanners, optical mark sensing devices, optical character recognition (OCR), surface acoustical wave (SAW) systems, machine vision, voice recognition, speech technology, and radio frequency identification systems

Training

- How much is the United States presently spending on training and what is the outlook for meeting our country's needs in training and retraining the work force?
- Ten goals of an effective training program
- Four basic training methods: (1) small semi-autonomous team; (2) productivity action team (PAT); (3) participant action planning (PAPA); (4) participant management
- How to address the critical question of displaced worker retraining, the need for retraining, and government, corporate, and union involvement

Measuring Productivity

- Four primary techniques used in industry for measuring productivity

What To Expect in Automation

- What do knowledgeable people envision as the seven areas of attainable computer technology in the near future?
- Seventeen predictions of future innovations using existing technologies as a benchmark

The case examples interwoven throughout this handbook illustrate how companies have successfully implemented advanced technology. Here are several examples:

Integrated Robotic Cell

A major aerospace company realized that its light machining of the complex-contour sheetmetal parts for aircraft skins, panels, fairings, and doors were fabricated using tedious labor-intensive manual operations. After considerable study, corporate management approved the development and installation of an integrated robotic cell (IRC) that performs trimming and drilling of complex parts. This determination was made based on finding a robot capable of both highly efficient and economical operations. Based on the experience acquired by the company since the IRC has been in operation, it is estimated that it saves approximately 7000 direct worker hours per year.

The fixturing has been designed for maximum flexibility, and the robot can be reprogrammed easily to handle virtually any part within its machining capability. The ability of the robot to exchange one powered tool for another enables ease of programming. When all these features are considered, this company's IRC is one of the most flexible machining cells in the aerospace industry.

Computer-Integrated Organizational Automation

A company specializing in the manufacture of high quality aluminum forging dies realized that it needed to modernize if it was to stay competitive and survive. This meant investigating ways to speed the design and manufacture of forging dies with an eye to cutting lead times and costs, and ways to improve accuracy and repeatability. That investigation led to installing computer-aided design and computer-aided manufacturing (CAD/CAM), and then to a sophisticated system of graphic terminals used to design forging dies; computerized numerical control (CNC) machines that have automated die production; accounting and word processing terminals that handle the business operations of the plant; a complete energy management system; personal computers that access programmable logic controllers (PLCs) on production equipment; and super minicomputers that manage the entire system. All of these are connected by a state-of-the-art, high-speed broadband network compatible with the manufacturing automation protocol (MAP), endorsed by General Motors as a communications standard for the automation of manufacturing organizations.

As this case history illustrates, as these steps fall in place the company is gradually instituting true computer-integrated organizational automation, with every group and department linked together in a completely integrated network system.

Flexible Manufacturing System Simulation

A company manufacturing castings was interested in designing and installing a flexible manufacturing system (FMS) to perform its machining operations. But before moving ahead, it decided to employ simulation methodology to minimize unforeseen problems that may occur during the various physical stages of the project. This was a logical approach since experience has shown that excessive cost and time overruns in installing an FMS are primarily due to inadequate planning and lack of simulated test procedures. The objectives of the simulation were to:

- Evaluate system balance and productivity
- Determine the need for additional equipment
- Determine which underutilized resources can be eliminated
- Evaluate machine mix (i.e., number of dedicated and flexible machines)

A MAP/1 simulation model was used to evaluate alternate milling machine center configurations. With a configuration consisting of nine dedicated machines and one flexible machine, simulation experimentation revealed that several pieces of equipment—at a total cost of well over $200,000—could be eliminated while still maintaining the desired production goal of 157,000 finished parts per year.

Flexible Manufacturing Cell

An aerospace and defense company, after exhaustive analysis, decided to replace its conventional machining processes and handling operations with a completely integrated flexible manufacturing cell (FMC). In the evolutionary process of developing the FMC, a holistic systems engineering approach was first used to establish a baseline of aerospace machining tasks.

On the basis of issues and concerns defined through analysis of the *as-is* model, a streamlined, flexible model was constructed to represent future machining cell functions in the FMC. The resultant FMC installed by this aerospace and defense company consists of a number of machining centers (fixtures, tooling, and cutters), coordinate measuring machines (CMMS), a wash station, automated wire-guided vehicles (AWGVs), complete computer control capable of machining more than 500 different parts with an economic order quantity of one. As a result of installing this advanced FMC, the company was able to achieve its objectives of maximizing production throughout, reducing labor costs, improving machine utilization, improving operational control, and reducing inventory.

Automated Inspection System

A can manufacturing company employed a quality control method that involved a large staff of full-time technicians using contact sensors to perform measurements at several workstations. These techniques resulted in high labor costs, unavoidable inaccuracies in measurement, and delays of up to 12 hours from the time a part leaves the production line to the time out-of-tolerance can ends are identified.

In seeking to improve efficiency and accuracy in measuring can ends, the company considered systems that use contact or noncontact gauging. Based on this study the company decided to install an automated inspection system. This precision measurement system uses five video cameras, structured light sources, a vision processor, and a host computer to measure thirteen critical dimensions on each can end in less than 3 minutes. A robotic loader and a precision motorized stage are used to provide fully automated handling of parts during the inspection process.

The system performs noncontact measurement with an accuracy of up to 100 microinches, completes the inspection process at a single site, and records and analyzes data via a host computer. This automated inspection system is a unique example of specialized vision techniques applied to large-volume inspection of a precision part. It illustrates how various equipment and techniques such as precision positioning, robotics, structured light, machine vision, and image processing can be combined to address a specific functional need in an industrial quality control environment.

Automated Packaging and Shipping System

A major manufacturer of office supplies realized that to meet the challenge of aggressive foreign competition without losing market share, it would have to materially improve its packing and shipping of spare parts and scheduled maintenance kits to field service personnel. As a result, the company decided to institute a modernization program in its distribution center with heavy emphasis on mechanization of its packaging operation.

After analyzing the characteristics of its distribution center's activities, the company embarked on a multiphase modernization program, involving:

- Installation of five new packaging lines
- Integration of the packaging cells that were created in the pre-pack and overpack areas
- Physically integrating the packaging cells by fifteen automatic guided vehicles
- Installation of automatic identification and data transmission technology

- Introduction of robots and automatic order picking devices for assembling scheduled maintenance kits
- Installation of robotic palletizers to pick small cartons from the twin packaging lines into large corrugated containers

After implementation, the company was able to identify the following dramatic results due to mechanization of its packaging operations:

- Cycle times about three times faster than they were before mechanization
- 25% reduction in personnel engaged in the packaging operation
- 37% overall reduction in personnel directly and indirectly supporting the packaging operations
- A reduction in the number of packaging workstations from twenty-five to eight with no reduction in throughout capability.

Computer-Integrated Manufacturing System

A major appliance manufacturer undertook a comprehensive market research effort to determine precisely what the consumer wanted in its dishwasher product line. The result of this study indicated that the customers demanded performance, reliability, durability, capacity, convenience, and quiet operation. Based on these findings, the company developed a new line of dishwashers and its supporting manufacturing system for an investment of $60 million, with some of this sum attributable to increased product volume.

The computer integrated manufacturing system (CIM) incorporates the use of exotic technology and modern manufacturing approaches. Technology that includes using automation at selected critical stages in the production process to ensure high levels of consistency and precision, and manufacturing approaches like just-in-time inventory control and "management by sight" to help assure that the proper part is in the right place at the right time, and that it's right in every respect—to achieve a high standard of quality.

After installation, this advanced CIM system achieved the following:

- Reduction by 85 percent the manual manipulations—the "handlings"—of tubs and door liners during the manufacturing process. Point-of-use manufacturing does away with long-term storage and inefficient delivery of structural components.
- Computerized, on-the-spot-checking of every automated operation substantially reduces the on-line repair of finished goods and eliminates incomplete units from progressing on through the factory.
- A "worker-paced" assembly line allows the product to move on only if the worker is satisfied that the work has been done properly.
- Assembly-line changes and automation together have reduced total manufacturing time—from raw materials to finished product—to less than 8 hours from three days.
- Productivity has improved 18 percent.
- A computer-based plant quality information system collects test and inspection measurement data for each unit at key points in real time, signaling process control engineers if unfavorable trends develop.

- Prior to installing this computer integrated manufacturing system, market share was about 28 percent. Afterwards the company achieved "first place" in this highly competitive market with an *increase* of more than 33 percent in market share.

Factory of the Future

A major manufacturer of computers and peripheral equipment decided to design and build the most advanced automated operation of its kind in the world to manufacture proprinters—the so-called "Factory of the Future."

There are eight separate assembly lines that can collectively turn out one million products a year. The eight assembly lines are independent and can be used to simultaneously manufacture different products. The key to the efficiency of this production operation lies in the flexible automated systems that control the movement of everything that is to be assembled.

There are seven robotic systems in each automated line. The products move through the series of automated stations while robotic systems add parts from the base up in an efficient "layered" building process. Each proprinter is automatically tested and then automatically shuttled by conveyor to the packaging area.

The automated factory control system (AFCS), a series of programs developed specifically for this product, control the entire production operation. The programs specify to the injection molding presses (that make about 70 percent of the Proprinter's parts), when and in what quantity to produce parts. They also act as traffic controllers, ensuring that parts and finished products move—at 60 feet a minute—over the production operation's more than one mile of conveyors. Another piece of software handles the manufacturing, production, and inventory control system.

This fully automated Factory of the Future that begins with raw materials in the shape of plastic pellets—and produces the Proprinter as its end product—is not designed just to assemble only Proprinters. It is a generic production operation since any product within a certain size and number of parts also can be assembled. Since each assembly line is independent, this automated system can be used to assemble several products at the same time. And, this advanced technology can be readily expanded, should additional production be needed.

As these examples clearly indicate, American industry is moving aggressively to modernize its industrial base. Certainly, much of this emphasis is due to international pressures in the marketplace. But the primary catalyst behind this surge of activity is the availability of proven technologies in the manufacturing environment. Through proper selection of these varied methods, a company can satisfy virtually any requirement—from phased upgrade to complete automation—with almost unlimited ability to integrate these disparate elements to suit each company's unique operational needs.

From 1980 to 1987, investments in factory automated systems have grown from $7 to $24 billion. It is expected that by 1990 this figure will reach $38 billion. What this means is that corporate management now realizes that strategic business planning must be heavily oriented toward capital investment in manufacturing technological modernization. *Low-cost production has now emerged as the key element in a company's ability to compete on a global scale.* Considering how rapidly innovative change occurs in manufacturing, a company must master its production processes to remain at the cutting-edge of the latest technological advances.

Today, opportunity for successful industrial modernization is within our technological grasp—and this book can show you how to attain it. As we move ahead, the following factors are worth noting:

- Modernization requires a considerable investment commitment over a number of years.
- Modernization should result in a reduction in both direct and indirect overhead costs.
- Modernization should lower costs (breakeven point) and generate higher profit margins.
- Modernization should enhance a company's skills base and improve worker quality of life.
- Modernization should result in a more competitive stance in the marketplace.

Contents

CHAPTER 3 Creative Innovations in Product Process Technology 55

CHAPTER 7 How Industry is Automating Material-Handling Systems 225

chapter 1

How to Develop a Manufacturing Automation Strategy

WHY STRATEGY IS THE KEY PREREQUISITE TO AUTOMATION

Any company dedicated to modernizing its manufacturing operation begins with the realization that the wave of the future is computer-automated systems coupled with robotic technology. Meeting this challenge requires a plan, a plan for implementing advanced technology with adequate capital investment based on a long-term payback. The time horizon for modernization must be sufficient to allow for selection of hardware and software, vendor lead time, acceptance by employees, training and retraining programs, development of methods to measure results, and complete top-management support.

New concepts in factory automation and robotics rapidly become industrial applications, which makes a decision to undergo a specific change analogous to shooting at a moving target. Today's leading edge technology may be supplanted with an improved approach at any time. Or a proven technology may become uneconomical as new state-of-the-art advances hit the marketplace. This brings us to one central point: How can a company move along methodically in achieving sound results in a modernization program without making mistakes in timing and cost?

Experience has shown that many manufacturing resource planning (MRP) systems and flexible manufacturing systems (FMS) have not lived up to their expectations. Although there are a number of installed robotic "islands of automation," few companies have achieved complete implementation of computer integrated systems (CIM), let alone the ultimate "Factory of the Future." The concept of "just in time" has only been marginally successful in American industry, while in Japan numerous companies have successfully installed JIT and, by comparison, Japan has already established a substantial base of computer integrated robotic technology.

This lack of implementation in the United States suggests that FMS, CIM, and all the other advanced manufacturing technologies may be at the threshold of fulfillment but they are still in their infancy in terms of *proven applied accomplishments.* It clearly points to a need to recognize the situation and move ahead in a systematic manner so that results *can* be attainable, measurable, and an integral part of a company's long-range strategic plan.

Sizing up the Status Quo: Where Do You Fit In?*

Basically, manufacturing companies fall into one of four categories:

1. Those that do nothing or very little to automate production.
2. Those that attempt to install advanced technology.
3. Those that aggressively move ahead with automation with sufficient discretionary cash flow to invest.
4. Those that have top executives who decide to spend money because they foresee a fundamental change in their markets, usually brought about by new international competition or overcapacity, or both. Thus, they decided quantum improvements were needed in cost, quality, and delivery if they were to stay in business. They reasoned that technology was the required ingredient for survival.

CASE-IN-POINT:

TECHNOLOGY SAVES OIL PUMP FIRM MILLIONS

From a positive viewpoint, the last two categories in the list above are the prime movers in today's industrial climate. These companies are ready to lead the way toward upgrading plant operations. A typical example of gains that can be achieved by an aggressive, well conceived program is an oil pump equipment company that obtained the impressive results shown below in figure 1–1.

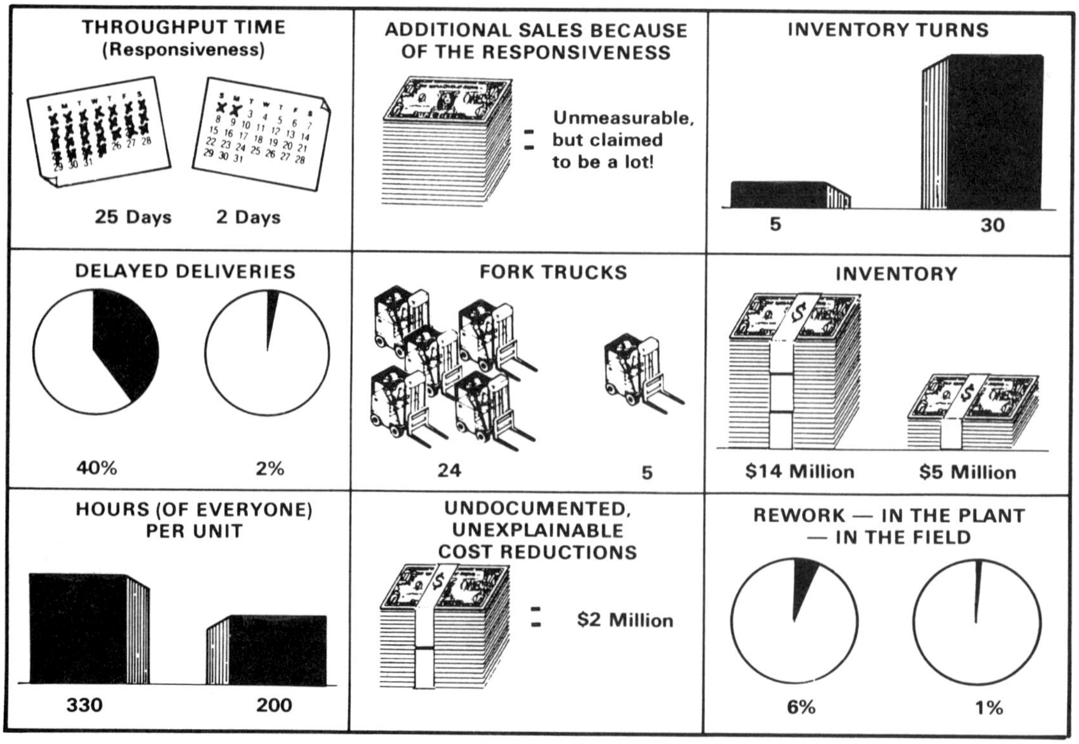

FIGURE 1–1 An Oil Pump Manufacturer (Courtesy of Ingersoll Engineers Inc.).

*Permission granted by William Arrott Associates, Inc. for Ingersall Engineers.

Area of Benefit	Before Technology	After Technology
Throughput time (responsiveness)	25 days	2 days
Additional sales because of responsiveness	0	Unmeasureable but claimed to be considerable
Inventory turns	5	30
Delayed deliveries	40%	2%
Fork trucks	24	5
Inventory	$14 million	$5 million
Hours (of everyone) per unit	330	200
Undocumented, unexplainable cost reductions	—	$12 million
Rework: plant and field	6%	1%

FIGURE 1–1 *(cont'd)*

The oil pump equipment company's annual value of the improvements is an impressive $10 million to $15 million. The cost for achievement was $7 million. Although in balance sheet terms the cost was minus $2 million because of the one-time inventory reduction of $9 million, other savings and benefits more than paid for the whole project.

Why Some Companies Succeed while Others Fail

Basically, successful companies all have the following characteristics in common:

- Technology was implemented in stages that allowed employees a reasonable time span to absorb change before moving up to the next plateau.
- No one got hung up on traditional justification procedures.
- The companies did not have strong functional lines to hamper people from working as a team in crossing over normal territorial boundaries.
- Firms had an idea of where they wanted to be several years into the future. In short, they *all* had some form of strategy or plan.
- Most importantly, all of the people who could make or break the technology were involved and sold on it. Many were retrained.
- Each firm had a champion of automation to constantly move the program along with total commitment.

By contrast, some companies have failed in their attempt to introduce advanced technology for these reasons:

- Some wanted to buy a solution before defining the problem.
- Some were convinced by persuasive technical salespersons representing systems houses, computer companies, software houses, machine tool companies or robot manufacturers to install a technology that did not meet the needs of the company.
- Some decided to become too esoteric and unrealistic in their modification program.

- Some moved ahead without proper planning because of management edict, either real or implied.
- Their employees were not deeply involved in advancing the technology.
- The attempted technology leap was too long.
- The technology was not part of an overall strategy or plan.

The most important fact for implementing a successful program is that *most, if not all, technology is cross-functional.* This means that many people from a *number* of functions have to understand the why, what, and how of the technology from each functional perspective.

In every case of failure, nonmanufacturing "people" issues were not addressed vigorously. The necessary changes to each function *must* be identified. Retraining *must* occur, to some extent, in almost every function. This takes time and money.

The single most apparent reason for mishaps is the failure to achieve this level of involvement. Another factor that has been validated time and again is that no manufacturing technology has worked up to its capability where the people on the shop floor did not feel a measure of "ownership" in it. Union people also need deep involvement. The most successful companies have involved worker-level employees and their union representatives in the development and implementation of the program.

Tips for Successful Automation: Transition to Complete Automation

To move from no automation or moderate automation to complete automation, you should do the following:

- Make certain that process flows are synchronized.
- Develop a cohesive floor layout plan that fully integrates all elements of production.
- Develop and institute basic elements that lead to production efficiencies, such as part families, process families, or group technology.
- Standardize tooling and fixturing methods.
- Make certain that product design is done with manufacturing fit in mind.
- Have already installed a CAD/CAM capability.

To avoid failure, a transition to some manufacturing cells can point out problems connected with redesigning parts, tools, and fixtures, and thus correct any deficiencies prior to going to an FMS mode.

Also, you should recognize that sophisticated automated technology requires almost complete mastery of the considerable hardware involved.

To introduce "high" technology into a plant, you must first start with a master plan for the whole plant and its support functions. This plan must be built around a computer-based system that will enable management to capture, coordinate, and analyze data in a real-time mode to control the two critical flows in manufacturing: information and material.

WHY THE COMPUTER IS THE CORE OF FACTORY AUTOMATION

Computerized systems are now being designed to replace the rigid boundaries of discrete functional departments with an integrated flow based on a logical sequence of events. This series of steps flows from product design to robotic instructions, selection of optimum production processes, automatic creation of computer numerical control (CNC) tapes, modeling simulation of the complete process, flexible manufacturing and assembly, computer-based planning, scheduling and shop floor control, automated materials handling, warehousing and distribution, cost control, and worker–productivity measurement techniques.

In this environment, corporate strategies have to be developed based on long-term commitment to change in the context of the whole enterprise rather than as a conglomerate of individual organizational components.

Today, computers are much more powerful and much less costly than their predecessors. Input devices have achieved faster response time, and with the advent of voice recognition, radio frequency systems, laser technology and optical scanning devices, they are more attuned to the needs of plant operations. Conventional keyboards, which are cumbersome as a worker input station, are being replaced by audio command (particularly in process inspection), and touch-sensitive screens.

Common data bases are being developed to enable storage and retrieval of pertinent data by design and manufacturing engineers; production planners; manufacturing, purchasing, and MRP personnel; inspectors; cost accountants; and others. These data bases are being designed as both distributed and integrated networks. The user may have either selective or complete access to these common data bases, often without knowing where the data physically resides. The information comes from many sources and can be available in virtually any format desired.

Flexibility Is the Challenge Today in Manufacturing

A key goal in factory automation is to provide operation machining flexibility to significantly reduce manufacturing throughput time. This change from conventional planning, scheduling, and machine loading of lot size production with "wait time" at each operation is taking hold with fully automated systems. Responsiveness at the shop floor level is increasing because of advances such as on-line systems, visual monitor graphics, direct numerical control, programmable controllers, data collection devices, fiber optics, and voice recognition.

All of these improvements enable management to focus on future considerations rather than historical events. Software is being developed to help companies anticipate, plan for, and predict future conditions. Such planning enhances the firm's ability to mitigate unforeseen events and stay on target.

Innovations to Expect in the Future

Using existing technologies as a benchmark, observers predict numerous future innovations, such as:

- *CAD/CAM*—portable graphic terminals will provide immediate access to all drawings stored on the mainframe computer.
- *Solid Modeling*—computers will automatically code and search for similar parts by merely sketching the general appearance of the part with overall dimensions and material information.
- *Voice Input and Output*—workstations, data collection devices, and office automation devices will accept voice input and be capable of talking back.
- *Workstations*—all existing devices would be combined into a single management workstation.
- *Communications*—digital, computer-enhanced, global communications using satellite, microwave, or fiber optics will be used universally.

- *Software Applications*—reliance on purchased software packages will increase. Higher level language will be used for unique or proprietary software.
- *Manufacturing*—flexible manufacturing systems will enable companies to produce small lot batches economically.
- *Product Processes*—least costly, optimum sequence of manufacturing processes, will be automatically generated.
- *Simulation Modeling*—will enable complete simulation of part design and process prior to production.
- *Material Handling*—automation will extend throughout the entire production cycle without any manual intervention.
- *Machine Tools*—tool wear and breakage will be substantially reduced by such techniques as energy beams and electrical discharge.
- *Plant Utilization*—the unmanned factory of the future will operate continuously to achieve maximum plant utilization and to recover capital investment.
- *Inspection*—automated, real-time, in-process control and inspection using laser and/or vision systems, and image enhancement and signal/sensor processing.
- *Materials*—composites, plastics, ceramics, and adhesives will be used as performance characteristics improve.
- *Robotics*—robots will become more mobile, gain vision, use audio instructions, have improved sensors and more intelligence, be networked and more fully integrated into the entire manufacturing cycle. A "standard programming language" will likely emerge, allowing universal programs that will be understood by various robots from various manufacturers. This in turn will allow the quicker integration of off-line robotic programming systems.
- *Flexible Manufacturing Systems (FMS)*—FMS will be the logical progression from numerical control (NC) to computer numerical control (CNC), and the last step leading to the unmanned factory of the future.
- *Artificial Intelligence and Expert Systems*—will become the software technology used to run the factory of the future.

HOW TO BEGIN AN AUTOMATION PROJECT

With this overview of future events, you can now begin to develop sound approaches that can capitalize on current state-of-the-art. Your plans for upgrading your manufacturing capability should be based on a matrix of time and cost. How much can you afford to invest in computerized factory automation and over what period of time? Does this modernization program fit into the company's overall strategic plan?

The benefits to be derived from an industrial technological modernization program can be substantial. For example, benefits can be achieved through the following: parts designed more easily, common family of parts and production processes, scores of machines replaced with FMS, sharply reduced inventories, faster response in product delivery, direct labor cost containment, elimination of manual materials handling, improved materials flow, computer integration of engineering and manufacturing, better product quality, improved bottom line profits, and higher return on invested capital.

Tips for Successful Automation: The Long-Range Technological Program

To implement a long-range technological program you should follow these guidelines:

- Objectively evaluate technologies already in place to determine whether they have achieved targeted results and what was learned during their use.
- Investigate the availability and status of new technologies.
- Develop a logical sequence of implementation.
- Determine the rate of change that the company can tolerate.
- Establish the real cost of the program and its potential benefits.
- Carefully select the key person who will be responsible to spearhead the program.
- Evaluate the status of competition as it relates to the need and cost of the program.

Why You Need A Strategic Plan for Modernization

The rewards for total integration of all manufacturing elements can be measured in tangible savings, improved quality, and better delivery performance. On the downside, however, risk of inappropriate or misguided expenditures can have a most detrimental effect on a company's financial well being. Development of advanced technology involves a substantial commitment in human and capital resources. It may require building new plants, paying much more for power and light, and incurring expenses for indirect labor such as programmers, methods engineers, maintenance personnel, and others.

For optimum results you must undertake strategic planning not merely as a thought process with a generalized concept of the future, but as a clearly stated strategy of where the company is headed and at what point you expect to achieve certain growth objectives. This strategic assessment must be quantified in terms of definitive markets, product lines, and products as well as whether growth will be nurtured internally by mergers and acquisitions or a combination of both. These strategies must be translated into resource allocation to fund research and development efforts, office and factory automation, and investments in external ventures.

How to Begin the Strategic Planning Process

The key to strategic planning is the quality of strategic development and problem solving. Every company has a strategy, be it formal or informal, but without an explicit, understandable plan to which all members of the organization adhere, chances of success are diminished.

Strategic planning is much more than forecasting. Strategic planning is an attempt to evaluate a company's business environment, compare it with management's objectives, and determine how to achieve stated goals. An organization must first assess its strengths and weaknesses and determine what business it is in. These corporate attributes are then compared with the competitive environment. A firm's internal strength must be assessed on the basis of financial, technological, human, marketing, and manufacturing resources. All of these elements must be weighed to achieve optimum long-term results commensurate with management's objectives. This appraisal allows management to formulate its strategic purpose and direction.

Any strategic plan will ultimately involve modifications or changes in corporate direction and introduction of a constant stream of new technologies. Thus, those involved in a factory automation program must be able to shape, nurture, promote, and reinforce the proposition that through technological innovation, changes will occur. This must be expressed in positive terms so that all individuals fully embrace it as corporate philosophy.

The very nature of the strategic planning process can provide consensus direction for establishing the framework to improve productivity. If done properly, it can build commitment to the productivity effort and shape the internal forces needed to tackle long-term goals and objectives.

Seven Key Components of Effective Strategic Planning

An effective strategic planning process has certain proven, time-tested characteristics that work. These characteristics are as follows:

1. The plan is structured but not as formal or rigid as in the past.
2. The plan focuses on the process as a means to an end, rather than the other way around.
3. The plan drives the budget so that it is more flexible and dynamic.
4. All levels of management (and workers) are involved with the planning process.
5. The strategic plan balances the often conflicting desires of various corporate functions.
6. The planning horizon is sufficiently long (depending on the industry) to enable management to develop needed allocation of capital resources in a well modulated fashion.
7. The planning process is directed and coordinated by senior management with a free-flow of information on progress and results (both positive and negative) flowing upward from various levels of the organization.

Tips for Successful Automation: How to Translate Plans into Tangible Results

To achieve tangible productivity gains it is important to:

1. Focus on fundamental management principles and practices.
2. Develop a workable productivity measurement and evaluation system at individual, group, and organizational levels.
3. Integrate productivity planning with strategic planning to keep them in synch.
4. Constantly search for changes and innovative ideas that will upgrade existing methods and techniques in a dynamic environment.

Why Organizational Planning Is Critical in Developing A Strategy

Organizational planning is an integral component of the strategic planning process because a company must translate its program into sequential organizational designs that identify the anticipated changes in the organizational structure in relation to the planned corporate strategies. Change must be managed since the growth process will involve changes in a company's size, complexity, products, markets, and other internal and external factors.

Whether change is sudden or gradual, it can have a decided impact on how advanced technology is introduced into the manufacturing environment.

Many companies employ the strategic planning matrix (invest/grow, earn/protect, harvest/divest) to allocate capital and technological and human resources. The matrix also serves as a usable tool for developing more detailed planning at the operating levels.

The strategic planning matrix is based on two assumptions:

1. Different strategic business units (SBUs) have varying degrees of maturity and opportunity, depending on their strengths, characteristics, and life-cycle profile.
2. An SBU should be evaluated within the context of its products, markets, and competition. The position of the SBU within the matrix determines management strategies and objectives.

Strategic planning evaluates all of the business units within a corporation to determine what impact various external and internal factors will have on them five, ten, or more years in the future. What will the needs of the marketplace be in the future and how can a company institute technology to satisfy these expected demands in terms of cost, quality, and delivery?

This planning process can compel a company to make major decisions today that, at the very least, will have an impact in three to five years. Most will go beyond that time-frame.

Companies able to develop or maintain proprietary or technological advantages will continue to grow rapidly. Those that can predict with some degree of certainty what the world of tomorrow will be like will also prosper. Those able to move quickly and adjust to a changing environment also will succeed. Others, concentrating only on immediate expediency and short-term solutions, may not survive.

FROM STRATEGY TO RESULTS: HOW TO PLAN FOR PROFIT AND ALLOCATE RESOURCES EFFECTIVELY

After a company clearly enunciates its strategic plan, it must shift to short-range planning. A company does this by formulating growth and profit objectives for the planning period, and by programming its resources to ensure attainment of these objectives. Through this systematic approach, a company can develop programs that fit into the company's overall planning horizon.

The heart of a business is not its products, which change in time, but how executive management identifies the right incremental investment opportunities on a continuing and consistent basis. After undergoing this critical analysis, the profit plan should be developed so that the firm invests monies in selected projects that will continue the company's successful growth path. This decision-making process should result in the allocation of resources in areas where results can be measured over a projected period of time. When allocating resources to fund the company's technological modernization program, management must resolve the following questions:

1. What are the technical risks in terms of chances of success, and the technical problems and difficulties in carrying out the program?
2. What are the technological advantages or the competitive superiorities of the program? How long can the program maintain its technological advantage? What are the current or prospective technological trends? What is the competition doing?
3. Will the program result in increased utilization of existing facilities? Improved production rates? What is the current capacity and rate of production? On what equipment operating rate is the program based?

4. Will additional facilities be required? If capacity is increased, can it be used profitably based on current projections?
5. What are the maintenance requirements of new equipment?
6. What is the firm's experience factor on this or similar applications?
7. Is there sufficient engineering talent, technical competence, and process know-how to complete the program?
8. What are the major components of capital investment—plant capacity, cost, special design features, manufacturer support, long-lead time delivery requirements?
9. Is the new equipment standard, or must specially designed equipment be purchased?
10. What are the general plant/equipment layout requirements for production flow?
11. What additional personnel is required? Number and type? If labor savings will result from this program, what will be done for displaced people, and how will this affect estimated savings?
12. Will this program require retraining of displaced personnel? If so, what type of training and at what cost and duration? Will retrained people be placed, and if so, approximately when?
13. What pre-operating problems are anticipated? What will be incurred in start-up and other nonrecurring expenses? What is the basis for the estimate?
14. How long will it take to achieve breakeven operation after the program is implemented?
15. What are the current and future uses of the equipment? Does it satisfy projected requirements of the strategic plan?
16. Does the planned facility allow for future expansion? Flexibility for periods of heavy demand? What portion of the proposed investment represents provision for future expansion? When will this expansion take place? What are the relative costs of providing for expansion now versus addition of facilities later?
17. What are the assumptions supporting economic life and relationship to physical life of facility, tax depreciable life, and other concerns involving the life of facility?
18. Will the program use surplus utilities? At premium or penalty rates? Will approval of this program result in additional current investment in utilities or off-site facilities planned for future periods? What pro rata share of the investment in these facilities applies to this program? What are the operating and utility requirements for power, steam, or electricity? What is the basis for requirements? What is the source of utilities, and are all necessary utilities available?
19. Has a simulation been performed? If so, what were the results? Did any problems occur?
20. What alternative investments or alternative means of technological development are available? What are the comparative yields and recoveries?
21. Is production seasonal? Will it complement existing seasonal patterns? Can production be leveled?
22. What degree of integration will this program accomplish? What time element is involved? How many phases will be required to implement the funded program?

Ten Key Factors for Determining the Economic Justification for Automation

After addressing the questions about operations, you must then determine the economic justification of the program. Developing and analyzing the information described below helps management determine whether to proceed or cancel the proposed modernization program.

1. The investment yield on the program versus the present situation, based on the relationship of additional cash income or savings to the net new cash outlay required.
2. Income and/or cost analyses, as applicable, which compare operating results before and after completion of the program.
3. How long will it take to recover initial new investment? Total investment? What is the yield on a total investment basis, including amounts expended to date? How will the program affect the return on the operation or the business?
4. The cause and nature of specific cost reductions and/or increases in income such as higher production rates, lower maintenance, reduced fuel consumption, and higher quality resulting in price increase. Compare specific cost improvements by major element of cost, department, process or cost center, and show the impact of changed production rates separately. Separate the effect on profits of higher volume versus improvement resulting from lower cost. Determine the effect of volume on manufacturing, distribution, selling, and administrative expense.
5. The cost of major items of equipment, additional facilities, and installation. Include the source and basis for cost estimates. Note the resources needed for start-up dismantling, transfer, nonrecurring losses or other special expenses anticipated—nature and amount.
6. Breakeven analysis. Total unit sales volume at anticipated price levels required to cover all costs. Determine breakeven capacity rate-production in dollars.
7. Volume and margins over the life of the program. Will they deteriorate or decline due to loss of technological advantage?
8. Basis for net working capital requirements.
9. Any long-term financing possibilities or arrangements which are unique to this program—amount, term, source, and interest rate.
10. Tax problems and advantages affecting the program and the best means of maximizing return in view of tax regulations. Determine method and rate of tax depreciation.

It is imperative that each stage of the modernization program have some supportable justification. The following section describes various accounting methods and types of measuring devices used to determine program profitability and risk.

How to Monitor Results Using Any of Four Accounting Methods

Most capital expenditures are intended to generate a cash flow in excess of the cost of the venture and yield a return on investment. In general, two types of measurement devices are used to show a percentage return: the time-adjusted and the nontime-adjusted rates of return.

The use of a time-adjusted rate of return method provides a measurement over the full life of the proposed investment. Realistic comparisons can be made between the present use of funds and the proposed investment or investments. Even though various stages of the program under consideration may differ widely in original cost, method of financing, and economic life, quantitative evaluation is still possible as the end result is always a percent.

Investment Yield (Discounted Cash Flow)

This method is a time-adjusted device which derives an investment yield on a program. The program's cost equals the interest rate at which the sum of the after-tax cash earnings is discounted yearly to its present value. This method of computing investment yield assumes:

1. An investment is a trade of present cash outlay for a future inflow of cash.
2. Cash in hand is worth more than cash in the future, and cash in the near future is worth more than cash in the far future.
3. The relationship between the outlay and future inflow is an actual compound interest rate.

Given these assumptions, the inflows and the outflows of cash are set down in a time sequence. Reference is then made to a discount table (the inverse of a compound interest table) to determine at what rate of interest the future flow of cash can be discounted to present worth.

The principal advantages of the discounted cash flow method are:

- Recognition that time is money; hence, near money is more valuable than distant money.
- Recognition that cash flow is more important than book income, hence, capitalization policy and resulting book costs are irrelevant for capital decisions except to the extent they affect taxes.
- Recognition of the divergent time patterns of the investment outlay and the cash returns to reflect the important differences in the value of near and distant cash flows.
- The ability to compare cost of capital ratios so that decisions may be considered on the relationship between the indicated yield and the value of money to the company.

The formula for discounted cash flow (DCF) rate of return considering a lower value of future money compared with the same amount of money today, can be calculated as:

$$CO = \sum \frac{NCI}{(1 + r)^n}$$

where

CO = cost of financing to pay for the investment
NCI = future net cash inflows in different years
n = expected life of the investment
r = internal rate of return on the investment

With an expected return on the investment (savings, net cash inflow) over five years the expanded formula would be:

$$CO = \frac{NCI_1}{(1+r)^1} \quad \frac{NCI_2}{(1+r)^2} \quad \frac{NCI_3}{(1+r)^3} \quad \frac{NCI_4}{(1+r)^4} \quad \frac{NCI_5}{(1+r)^5}$$

NCI_1 is the net cash inflow in Year 1, NCI_2 is Year 2, and so forth. Using this method you can then determine the rate of return, if positive, compared with the cost of financing, as an annual interest rate. Thus, the margin by which the internal rate of return exceeds the cost of financing, the greater the expected profitability of the investment.

> ***Key Point:*** To do a meaningful discounted cash flow analysis, it is necessary to obtain figures for the predicted net cash flow generated by the project in each year of the life of the project.

Payout

The calculated period required for the cash investment in a project to be recovered through cash flow is another measurement method. It is computed by accumulating, from operational start-up, the cash return by the project until the total equals the cash investment.

The cash investment in a project is the total cash outlay for plant and equipment, for other capitalized costs—that is, preproduction interest, and for that working capital required to maintain normal operations.

Payout is a key factor in risk measurement. In theory, depending on the nature of the project, the shorter the payback time, the less the risk. This method serves as a coarse screen to pick out the high-profit projects that are so clearly desirable they do not require refined investment yield and reject quickly those projects that show no financial merit. It also can be useful for appraising high-risk investments where the economic life is particularly hard to predict.

For most purposes, however, payout used as a sole criterion is inadequate. It is a cash concept designed to answer the single question, "How soon will the cash outlay be returned to the company's treasury?"

As such, it ignores the economic life of the project, the extent and shape of its cash flow beyond the payback point, disposal value and the time value of money.

The formula for payout is:

$$P = \frac{I}{O}$$

where

P = payout period in years
I = initial extra amount invested
O = uniform annual extra cash inflow from operations

Production Rate Payout Period

This formula is used to account for either a faster or slower rate of production generated by an automated operation compared with the human operator. The formula is expressed as

$$\text{(P) Production Rate Payout Period} = \frac{I}{L - E + / - q(L + Z)}$$

where

P = payout period in years
I = total capital invested in the project
L = total annual labor savings
E = annual expense of automated operation upkeep
Z = annual depreciation costs of associated equipment
q = production rate coefficient (the percentage in which the automated operation is faster (+) or slower (−) than the human operator

$$L - E + / - q(L + Z) = \text{annual incremental cash flow}$$

Return on Investment

Return on investment (ROI) is the average annual book income before taxes over the economic life of the project divided by the original investment, including the working capital required to attain normal operations. The ROI formula is:

$$ROI = \frac{S \times 100}{I} \text{ (percent)}$$

where

I = capital cost of investment in automated equipment and accessories
S = annual savings resulting from automated operation

A CHECKLIST FOR IMPLEMENTING A SUCCESSFUL MODERNIZATION PROGRAM

Before committing substantial resources to a major factory automation effort, it may be beneficial to gain insight into what lies ahead, by drawing from the experiences of those who have already implemented an industrial modernization program.

The following points clearly demonstrate the need for a complete systems approach to automation within a company's strategic plan. *Technology alone will not suffice.* The key to success is to fit products and technologies into the overall strategic direction of the corporate body.

- Plan from top down and implement from bottom up.
- Analyze all manufacturing processes with a long-term view.
- Install sound information systems and material flow procedures before obtaining any automated equipment.
- Design a system that will enable unbuffered continuous production flow.
- Remove dependence on paperwork by use of electronic transactions only.
- Prioritize the modernization program agenda so that changes can be implemented in a sequence that will cover automation costs on an on-going basis.
- Recognize that in addition to direct labor, significant savings can be made in material, indirect labor, utilities, and other areas of expense.
- Integrate product design, product processes, production scheduling, material requirements planning (MRP), shop floor control, and other related activities with factory automation to gain maximum benefit.
- Gain the complete support of all plant employees, and set attainments for each person.

- Maintain open channels of communication during and after implementation of the program.

This chapter stresses the need for a company to integrate its industrial modernization program with its strategic business planning process. Because a company can reasonably sustain many gradations of technology, management must recognize the matrix components of time and cost in this process.

Today, most informed people realize that industrial gain must be measured and evaluated on a long-term basis. This healthy change in attitude may bring about the desired results needed to retain our country's preeminence as the world's leading industrial power.

However, from a pragmatic standpoint, it still will be necessary to evaluate individual projects using recognized accounting methods. Otherwise, companies may become overextended with expectations outstripping bottom line numbers, which could eventually affect a company's well being in terms of profit and loss statements (P & L), balance sheets, and cash flow. Wisdom and common sense must dictate how much to invest in long-range payback and what demands must be placed on management for short-term and intermediate accountability.

The chapters following focus on trends, innovative techniques, and cutting-edge issues that are of the greatest concern in manufacturing management.

chapter 2

Advanced Engineering Technology: 16 Techniques for More Effective Performance

Engineering has rapidly moved to the forefront in computerized systems with *computer-aided design* and *computer-aided manufacturing* (CAD/CAM), *computer-aided design and drafting* (CADD), *computer-aided engineering* (CAE); and many other engineering software packages. These systems contain extensive data bases to enable the user to engage in virtually any task required. This chapter contains detailed information on the types of systems/software that have been developed to aid engineering and manufacturing people perform more effectively. The advanced methods and techniques include:

- Integrated computer-aided design and computer-aided manufacturing
- Engineering data library/configuration manager
- Data-base management software
- Computer graphics
- Solid modeling
- Geometric modeling
- Mechanical modeling
- Graphic–finite element modeling
- Integrated–finite element analysis and design
- Automatic digitizer
- High-resolution tracker
- Low-cost tracker

- Human-computer interaction through use of body language
- Group technology
- Relational data-base system
- File management systems

COMPUTER-AIDED DESIGN AND COMPUTER-AIDED MANUFACTURING: THE KEY TO FULLY AUTOMATED FACTORY

One of the major advances taking place in industry is the interface between CAD/CAM. When all elements are in place, a company will have many of the basic requirements for factory automation (figure 2–1). To accomplish this objective, all of these activities must be integrated:

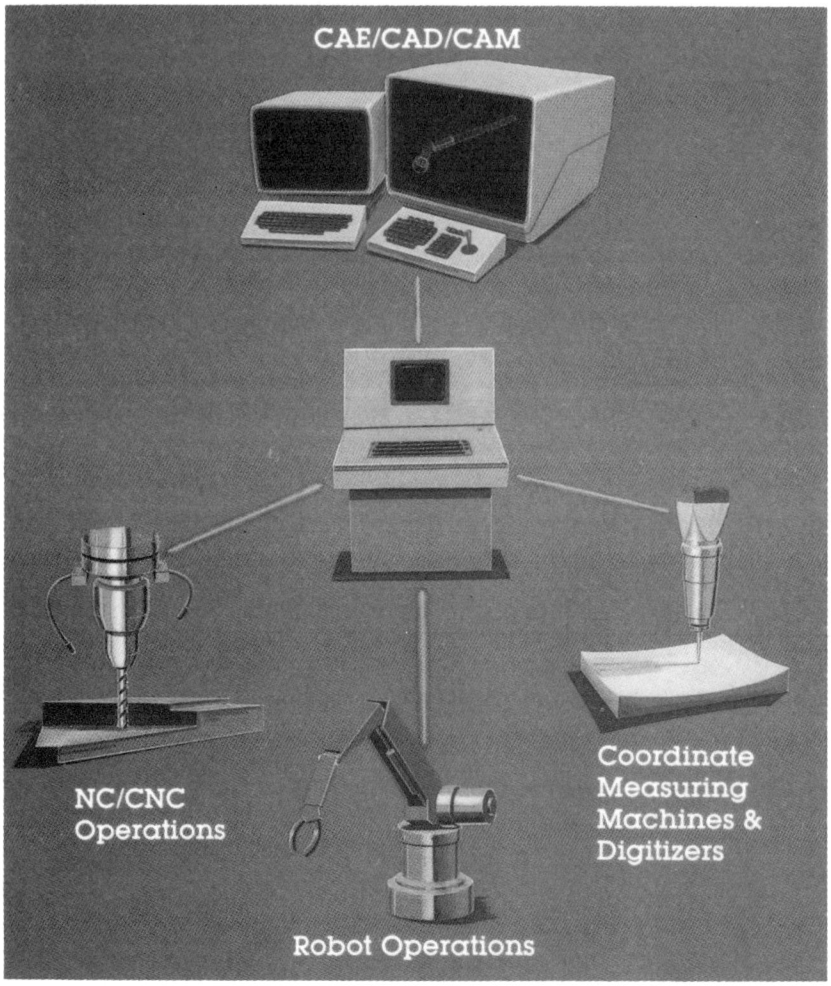

FIGURE 2–1 Interface between CAE/CAD/CAM (Courtesy of McDonnell Douglas Company).

- Device programming
- Numerical control programming
- Robot programming
- Group technology
- Process planning
- Geometric modeling
- Sheetmetal fabrication
- Automated quality control
- Facility design

How to Assess the Costs and Benefits of CAD/CAM

As discussed in Chapter 1, the amount of capital investment must be determined both in terms of cost and when these sums will be expended. After these decisions have been made, a more detailed analysis is required to cost justify individual projects. This analysis requires studying the following:

- Desired return on investment
- Whether to develop software in-house or purchase software packages
- What the system will achieve, and will the result be measurable in terms of tangible cost savings
- Will the payback be short-term, or will it be a long-range "building block" approach that relies on additive value accruing from complete automation

Based on this analysis, integration of CAD/CAM should provide a company with these benefits:

- Ability to accelerate the product development cycle.
- Ability to speed up introduction of new products into the marketplace, thereby gaining market share, increased market penetration, and added sales revenue.
- Standardization, consistency, and uniformity which will improve quality and reduce scrap and rework.
- Ability to capture all information in the CAD data base for multiple retrieval purposes.

Five Objectives Critical to the Success of CAD/CAM

Maximum impact of installing CAD/CAM can be achieved by seeking optimization of automation technology in consonance with strategic business plan objectives. This requires a commitment to accomplish the following goals:

1. Control all software involved in the program which consists of setting up a standard library that contains simple formats of drawings, plots, and screens and continues on to structured procedures for managing, transporting, sending, converting, and controlling all pertinent data.

2. Establish a total distributed data communications network to enable all individuals access and use of required data bases.
3. Train people in CAD/CAM so that they become skilled at handling DNC programming, robotic programming, flexible automation, automated materials handling, and other integrated skills.
4. Indoctrinate management into understanding that work skills are now radically different. (The tool designer has become the machine tool programmer; the methods engineer now integrates product processes; and manufacturing people now execute these commands.)
5. Instill in all people involved in the program the need to be selective in assessing these upgrades so that they remain in consonant with goals, missions, and corporate objectives.

A Five-Step Plan for Installing CAD/CAM

Once management has committed necessary resources to this program, specific actions must be taken to implement the program. This involves putting in place all of the "building block" activities mentioned previously.

1. *Geometric Modeling*—part definition consisting of the part model designed in wire frame, solids, or surfaces. These data, when broken down into points, lines, circles, and arcs, are then stored in a data base. When manipulated or arranged on the display screen, the data shows the geometric construction of part designs, textual information for dimensioning, and notes representing the geometric model stored in the computer. The part model is the first element in building a data base for automation.
2. *Process Planning*—determining which machine tools, operational sequence, materials, setup and run times, and any special fixturing are required to manufacture a product based on an engineering drawing.
3. *Group Technology*—development of a classification and coding system to standardize and group similar parts and assemblies into families of like items. This methodology utilizes the geometric features of the product's design. Since group technology develops the optimum design and manufacturing of a product, it should be regarded as a key to the data base. In addition, group technology affects the manufacturing process by enabling quicker development of process plans as well as new product design.
4. *Facility Design*—design and layout of a plant facility involving automated machine tools, automated materials handling, and robot work cells. CAD/CAM technology and graphics are of considerable value since layout drawings and their associated logic represent the two key elements of facility design.
5. *Device Programming*—provision of the operating instructions needed to drive the automation equipment's control devices. This involves interpretation of engineering drawings, manufacturing methods, and the working environment. CAD/CAM technologies enable the geometry of the part to be stored in data bases so that it then becomes available for programming specific devices such as NC machines.

CASE-IN-POINT

MAJOR APPLIANCE MANUFACTURER

Situation

A major appliance manufacturer wanted to reduce engineering costs by computerizing the detailed engineering drawing activity and developing numerical control (NC) part programs. Previously, a graphic terminal was used as an input device, which to a certain extent was analogous to using a conventional drawing board. Therefore, the company reasoned, by eliminating the graphic scope as the main input device, it could realize some very significant productivity increases.

Solution

The new system utilizes a data base to store each terminal session. Hence, any part created through this system can be called up and edited. The output from the system is a part model stored in the company's CAD/CAM system, and an NC part program, ready to feed into a machine tool.

When an individual wants to produce an engineering drawing, the operation is performed by accessing the computer via a CRT. The program displays a series of questions about the geometry of the part. The program executes that process and passes the data to the CAD/CAM system in the form of a composite part model and simultaneously updates a geometric data base outside of the CAD/CAM system. The computer part model that is received and stored in the CAD/CAM system can be accessed and displayed on the graphic scope and plotted or changed like any other CAD/CAM model.

The data in the geometric external data base, not the CAD/CAM model, is the data used for creating the NC part program. When a new NC part program is required, the engineer selects a series of options from the interactive screen on a CRT which determines the specific machine tool and machining parameters. When a program is executed, it uses the machining parameters, part geometry, and the family of part programs in the computer-assisted NC language to create the final NC part program for use on the machine tool.

The time required to input the data and run the program required to create the drawing is only a few minutes. To create the NC part program requires only another few minutes.

This approach proved to be excellent for families of parts. It provided the company with the ability to start the actual machining of a part within minutes of completion of the design requirements.

From this methodology, the company is able to move closer to designing a computer integrated manufacturing system by developing a program to display CAD/CAM drawings on a personal computer on the shop floor. The personal computer (PC) displays data to a machine operator (including a part print and a tool set-up sheet), and acts as a DNC terminal for downloading part programs to an NC machine tool.

Results

As a result of developing and implementing this CAD/CAM system the company was able to:

1. Create computer models and NC part programs from a common data base input from a CRT.
2. Reduce engineering time creating detailed engineering drawings.
3. Provide shop people with faster and better information on engineering drawings and machine tool requirements.

4. Install group technology, which resulted in significant cost savings through standardization and consolidation of families of like parts.

HOW TO USE THE ENGINEERING DATA LIBRARY AND CONFIGURATION MANAGEMENT TO INTEGRATE A COMPUTERIZED SYSTEM*

The core of an integrated computerized engineering/manufacturing system is the *engineering data library* (EDL). The EDL integrates all design, engineering, and manufacturing functions. It handles data storage and retrieval, allowing parts geometry and other data to be contained in or referenced through a single data base. Once the data is entered for an initial application, EDL automatically transfers them among all succeeding applications thus, eliminating the necessity to re-enter basic data over and over again. Inputted information can be shared, retrieved, added to, and deleted in either interactive or batch mode. As a result, this technique creates a huge information resource for designers, engineers, and manufacturing personnel.

A software *configuration management* interface into the EDL, collects, coordinates, formats, stores, updates, and distributes product and project data wherever needed, online, and in real time. This configuration management software expands and compliments the capabilities of the EDL to provide total product control throughout the product life cycle. Thus, this entire software configuration is an automated data base application that manages information pertaining to the planning, identification, control, and accounting, with status and interrelationships of product components. It provides information on the total life cycle of the product in a faster, more effective manner, at less expense than a system that is not fully integrated.

Key Features of the Engineering Data Library/Configuration Management

The integrated computer-aided engineering and manufacturing system, and configuration management (ICEM CM), entail the following features:

- Accesses all information immediately about any aspect of a product, including drawings, part usage, changes, work-in progress, software, and program administration
- Tracks the complete history of all revisions
- Manages the custody and location of all engineering and product documentation
- Standardizes data entry format throughout the organization
- Automatically applies all related information once information is entered into the system
- Controls access for viewing, changing, or approving information
- Automates generation of reports and queries, both internal and external
- Provides simplified data entry through preformatted terminal screens

How the System Works

The ICEM CM works as follows: Data enters the system in either of two ways. First, through the EDL, which transfers all design/drafting information automatically throughout

*Permission granted by Control Data.

the ICEM CM system, or second, through manual input of data using preformatted terminal screens. Selection may be made from a variety of screens for entering, changing, and viewing data. Numerous screens provide information on such topics as engineering specifications, related drawings, related parts, engineering changes, work-in-progress, and product configurations.

Figure 2–2 is representative of terminal screen information contained on the integrated computer electronic management (ICEM) configuration manager master menu (lower half of screen) having been accessed through the EDL displayed on the top half of the screen. Users can then select whatever area of ICEM CM they would like to access, change, or update. This may involve processing data on various components of the system such as:

- *Part Information Screen* which contains part specifications, part title, and material description.
- *Engineering Change Order Screen* which contains all relevant data pertaining to a particular part.
- *Parts Usage Inquiry Screen* which contains a complete listing of all next higher assemblies that use this particular part.

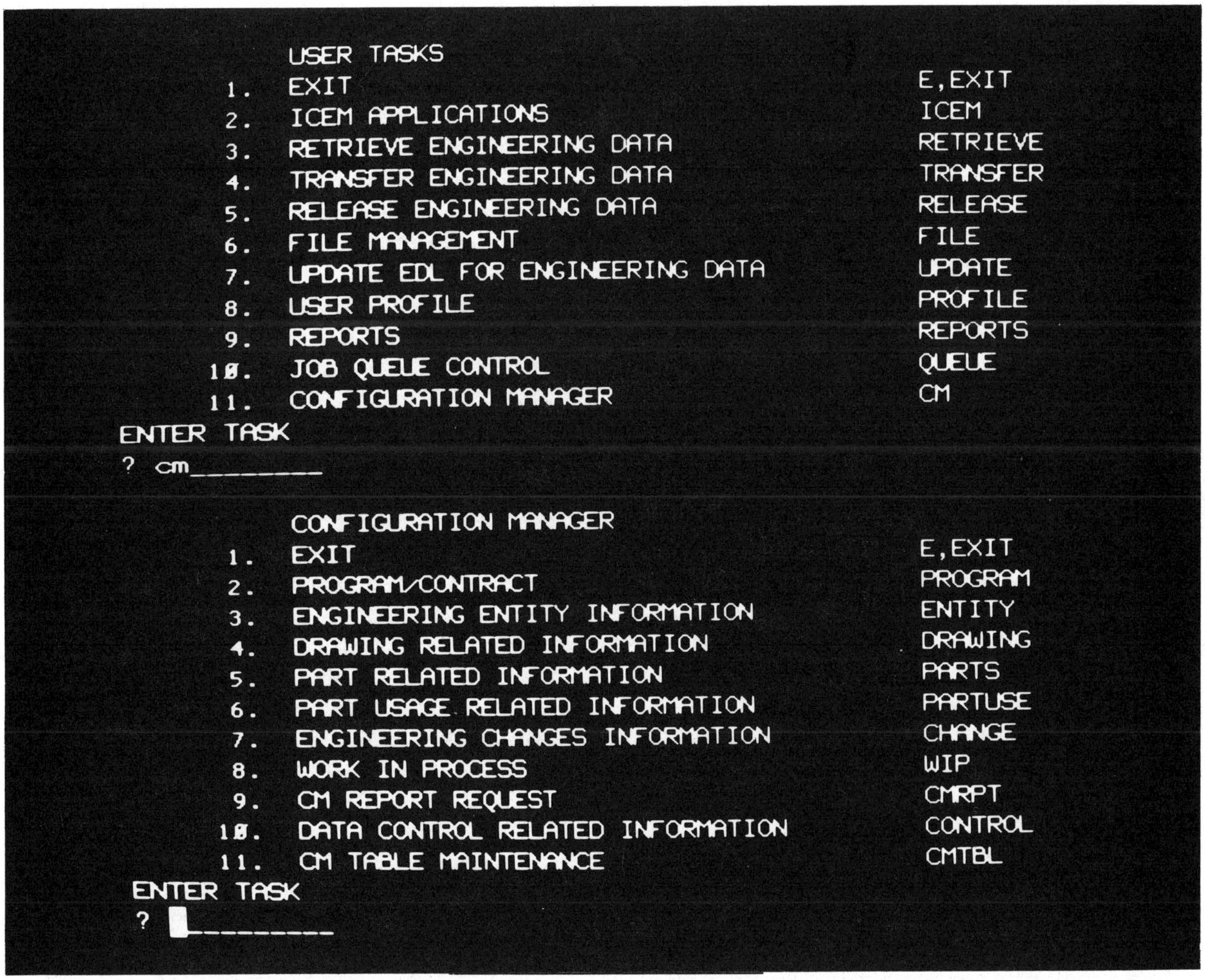

FIGURE 2–2 Engineering Data Library/Configuration Manager (Courtesy of Control Data).

A substantial number of different reports can be generated from data stored in this system.

WHY DATA-BASE MANAGEMENT SOFTWARE CAN IMPROVE PRODUCTIVITY*

To manage large CAD/CAM data bases it is necessary to obtain *data-base management software.* This software enables engineers and drafting and manufacturing personnel to make better decisions and work more productively by giving them quick access to up-to-date, accurate project information from a central source. At the same time this software can protect data integrity and security.

Through a modular structure data-base management software can provide:

- Efficient on-line storage capabilities for large amounts of data
- Ability to integrate and manage data bases from several systems
- Fast data access and distribution mechanisms
- Data-base security, through controls for project-information release and for revision, archiving, and recovery of information
- Interaction query function
- Interface and structure tailored to CADDs users

In addition, this software can offer the following:

- Improved monitoring of engineering change orders
- Release control to enable users to track the progress of product information from concept to final release
- Message handling that allows users to view, print, and delete electronic messages, as well as to create, delete, and view lists of users to receive messages
- File maintenance mechanisms that allow users to modify file attributes to meet individual needs
- User table maintenance that helps users define tables to take advantage of the software's data security capability for nongraphic data

CREATIVE IDEAS FOR APPLYING COMPUTER GRAPHICS SOFTWARE

Computer graphics technology has advanced very rapidly and now offers many unique features because a user can interact with a common data base via a graphic terminal to design and manufacture a part from start to finish (figure 2–3). As a result, the display feature of the terminal has taken on a role of its own. This section will cover many of the innovations developed for graphic display.

Six Ways to Apply Graphic Machining Modules

Software vendors have developed *graphic machining modules* which provide an interactive graphics alternative to conventional NC programming techniques. For instance, software developed for machining applications (figure 2–4) may be used for:

*Permission granted by Grumman Corporation.

FIGURE 2-3 Computer Graphics (Courtesy of Grumman Corporation).

- 2-axis lathe motion through interactive programming
- Two $2^1/_2$ axis regional milling
- Creating cutter paths for the machining of any surface
- Containing cutter movements
- Optimizing sheetmetal movements
- Generating and unfolding triangulated flat patterns

How Interactive Graphics and Imaging Can Enhance Design*

Graphics software can enhance the work of the design engineer by:

- Enabling the user to make design changes as needed, on-line.

*Permission granted by the Eaton Corporation.

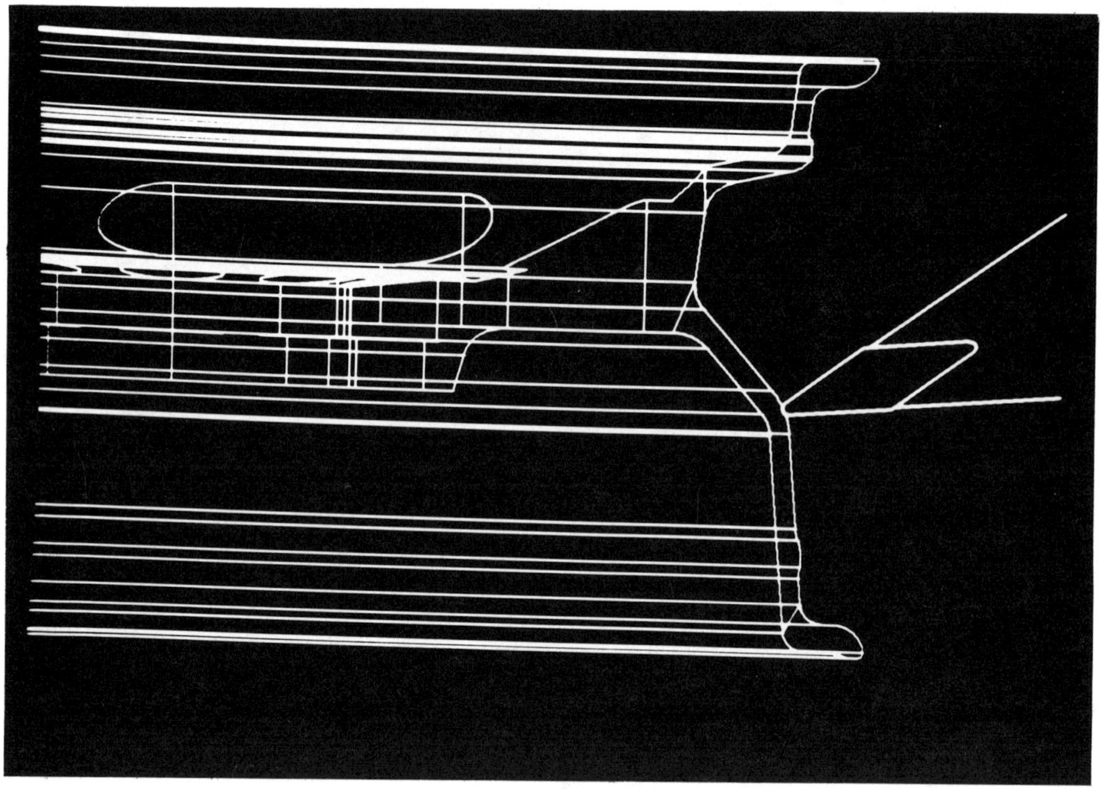

FIGURE 2-4 Tool Path Generation (Courtesy of Computervision Corporation).

- Providing manipulation functions, including zooming, magnification and rotation, translation and scaling in several different coordinate systems.
- Providing user-editable interactive graphic commands with on-line documentation.
- Creating, storing and recalling standard drawing formats.
- Containing a data-base library to support the creation of symbols, parts, textfiles, and other graphic and nongraphic information.
- Providing viewing functions for both perspective and orthographic projection, shrunken elements, hidden surface removal, and three-dimensional clipping.
- Providing post-processing facilities which include the specifications of data output in printed form, in tabular form or model plots, surface contour plots, or X–Y plots. In addition, selected output data can be plotted in combinations of displacement, velocity, acceleration, applied loads, forces of constraint, element forces and stresses, element, or grid ID numbers, time or frequency, and super elements or subcases.
- Providing continuous tone shaded images and hidden line removal. This feature is especially helpful in presentation graphics, aesthetic evaluation of parts, and illustrations for technical publications.
- Providing imaging capabilities, including flat or Gourand shading, production of shadows, translucency, control of background color, independent light sources, and surface textures.

- Providing construction aids such as
  - a) font selection and definition
  - b) blinking and highlighting
  - c) rectangular and radial grids
  - d) geometric identification via system or user-defined tags

Mechanisms Module: A Tool for Detailed Analysis

This graphics software is used for design and analysis of mechanisms or linkages in products such as automobiles, aircraft, and construction equipment. It allows the engineer to perform more detailed analysis of the product.

Key features of the mechanisms module include interactive synthesis of 4-bar linkages, multi-element linkage design, analysis and animation, and interfaces to other software modules.

Mechanisms synthesis is normally difficult and impractical, requiring laborious mathematical calculations. A mechanisms module's graphics representations give the designer a complete, yet easily understood, solution. The engineer can specify from two to five positions that the mechanism must pass through at the proper orientation. As it explores possible solutions, the module provides continuous feedback to the engineer.

Multi-element linkage system design, analysis, and animation can be performed with the module. To create the mechanical system in part geometry, the designer may use actual parts already stored in the data base, develop the schematic or skeletal representation of the system, or create skeletal geometry as part of the 4-bar synthesis process previously discussed.

Once the motion of the mechanism has been defined, the designer can animate the device through its cycle of motion. This helps identify problems without building expensive prototypes.

The designer can conduct static analysis, kinematic analysis (motion without reference to the forces causing motion), and dynamic analysis (motion in relation to the forces causing them) of the mechanism.

The mechanism module serves as a powerful graphics-based, pre- and post-processor to these tools. The output from these analysis packages is easily formatted into tabular listings or graphics.

SOLID MODELING: A QUICK, EFFECTIVE TOOL FOR DESIGN AND MANUFACTURING*

Solid modeling software enables the user to quickly create and display solid models, either as color-shaded images or as views with hidden lines removed automatically (figure 2–5). Solid modeling calculates comprehensive mass properties for solid parts of assemblies and produces boundaries (edge) files which may be used as a point of departure for finite element modeling and NC tool path generation applications.

Interactive model creation is facilitated by prompting sequences and on-screen menus which guide the user through the steps of the construction process. Choosing from a set of volumetric shapes called features, the user combines, intersects, and subtracts features to create parts or more complex assemblies. Features range from simple shapes, such as a box and cylinder, to complex volumes, such as free-form sculptured solids and solids of revolution.

*Permission granted by Hewlett Packard.

FIGURE 2–5 Solid Modeling Design Visualization (Courtesy of Hewlett-Packard).

At this point, solid models may be viewed as coursely-defined color-shaded images, or approximate wireframe representations of the solid model with hidden lines dashed or removed.

After all editing operations have been performed, mass properties, output for display and plotting of high-resolution color-shaded images, precisely-defined views with hidden lines removed, and boundary (edge) files are obtained.

In addition to the construction of complete parts and assemblies, solid modeling software provides the user with facilities for quickly creating full, half, offset, revolved, auxiliary, and thin section cuts in any orientation. Furthermore, solid modeling produces a geometrically complete, accurate, three-dimensional data base which can be used as a point of departure for the creation of engineering drawings and technical illustrations.

What Solid Modeling Can Do and Why It Is Effective

Solid modeling software offers a user the ability to:

- Generate geometrically, complete, three-dimensional solid models of parts and assemblies.
- Provide facilities for solid model creation, editing, and display.
- Perform combination, intersection, or subtraction operations among solid features, parts, and assemblies.

- Automatically remove hidden lines to facilitate drawing production.
- Calculate comprehensive mass properties for both individual components and assemblies.
- Provide user control over model display resolution.
- Enable the user to derive complex surface intersections.
- Automatically generate full, half, offset, revolved, auxiliary, and thin section cuts.
- Provide for creating high-resolution, large-format, full-color plots on a color plotter.

Solid modeling also offers many other benefits such as increasing the productivity of designers, drafters, and manufacturing personnel by enabling them to quickly view and analyze realistic, three-dimensional, color-shaded representations of their designs. A user can easily generate complex section cuts in any orientation. These features permit users to check clearances and interferences early in the design cycle. The automatic generation of mass properties for complex parts and assemblies facilitates analysis. Drafting and drawing production are also facilitated by automatic hidden-line removal and profile-line addition capabilities.

Solid modeling permits design concepts to be examined with greater comprehension than that provided by traditional engineering drawings. The ability to create full-color hardcopy using a color plotter permits solid models to be used in presentations, proposals, or as additional documentation for manufacturing applications such as metalworking, mold design, assembly, and process planning.

Five Ways to Apply Solid Modeling in Manufacturing

Solid modeling can be extended beyond design and into the manufacturing environment. Although still in the early stages of this extension in use of solid modeling methodology, the following list consists of some of the applications where it may apply:

- Automatic NC output for machining a part represented as a solid model
- Representation and analysis of tolerances using solid modeling
- Die cavity modeling by subtracting the model geometry from the die block geometry
- Interference checking among NC tools, fixtures, and workpieces
- Simulation of flexible manufacturing cells as an aggregate simulation of the cell elements that are modeled as solids

CASE-IN-POINT

*COMPUTER-AIDED DESIGN AND SOLID MODELING PENSKE RACE CAR: FROM SCREEN TO TRACK**

Situation

Previously Penske Racing, Inc. obtained its race car chassis from one produced by outside manufacturers. Instead, the company purchased a CDS 4000 from Computervision to analyze the design of the race car and determine if the design could be improved upon. Penske hoped that by creating an improved design, they would capture a competitive edge that

*Permission granted by Computervision Corporation.

would consistently put its drivers in the winner's circle during the 1985 CART/PPG Indy Car World Series. At that time, owner Roger Penske had not decided which chassis to race in 1985—his own or one produced by manufacturers.

Solution

Penske obtained a CDS 4000 from Computervision to be used to analyze the existing chassis and determine what improvements in design could be made. Engineers digitized all three chassis into the system, created 3D models, and analyzed them. They added tension, torque, and suspension systems to the models and were able to rotate and view the components from various angles (figure 2–6).

Designers discovered that the exhaust system in a new version of the Penske chassis actually interfered with the chassis design. Shading of the 3D solid model revealed the underbody work did not allow enough room for the exhaust to clear. Analysis also demonstrated that the March chassis had been manufactured differently from its design. Based on these findings, Penske opted to modify the March chassis and race it in the 1985 season.

Through use of this analytical tool, Penske has redesigned the flaps, wings, and airfoils on the car as well as the underchassis. The CDS 4000 has enabled designers to take the geometry of the concept car and produce an accurate one-third scale model for wind-tunnel testing. By so doing, Penske feels he has made considerable progress since he has been able to perform more analysis and testing before actually constructing a model. Based on this approach, he does not plan to use any outside manufacturer's components on future cars. By the next year, engineers will design a Penske chassis and engine completely on the CAD system.

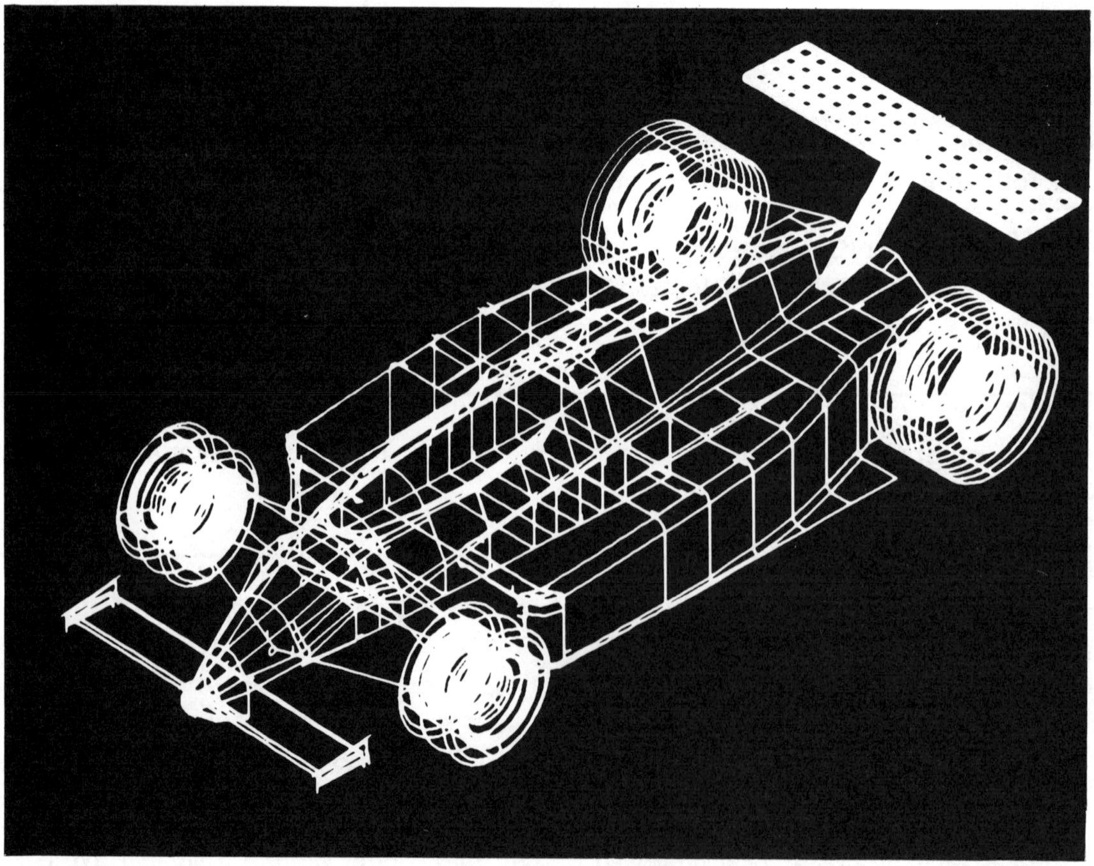

FIGURE 2–6 Wireframe Model of Penske Race Car (Courtesy of Computervision Corporation).

Results

Work on a new engine is being done through a joint venture among Penske, Chevrolet Division of General Motors, and an English firm called Ilmor Engineering. Ilmor designers have completed proprietary work on fuel intake ports using CAD. They have performed analysis on the structural geometry of the engine, taking only three days with CAD compared with an estimated six weeks manually.

Penske and Ilmor are sharing a common data base, which allows Penske to reduce lead time on development significantly. Since Ilmor transmits data directly to Penske designers, this enables the designers to explore concepts more quickly without manufacturing the parts. The same holds true for the Chevrolet Division of General Motors. An engine may evolve from this approach which will be the new technology for the Indy Car racers.

Two critical factors in race car performance are speed and endurance. CAD contributes to optimizing both. In terms of speed, CAD has increased precision design of manufactured parts. Components fitted correctly eliminate drag resulting from poorly fitted parts. The system also has enabled designers to re-engineer components quickly to meet aerodynamic demands. In terms of endurance, CAD equipment lets the Penske team forecast conditions under which components on the car will fail. Engineers input material, car design, and operating conditions such as force and temperature in the model and then analyze the data. In addition, Penske can track similar information on the competition.

The Penske CAD system consists of a CAD 4000, based on 32-bit architecture. Features include 32-bit RAM with associated control and error-correcting logic, 64-bit processor for operations requiring double-precision calculations, and 16-kilobyte cache memory. Other components include a high-resolution color workstation and a P1000 pen plotter. Also included are a tape punch and a line printer. Software consists of the CADDs 4X design and drafting packages in addition to third-party analysis programs.

Both Penske and Ilmor are obtaining Computervision's new CADD station products based on Sun Microsystem 68020 workstations. Each faciity will have four CADD stations for design and analysis as well as having a direct link to the shop floor.

GEOMETRIC MODELING: A BASIC TOOL FOR COMPUTER-AIDED DESIGN

The geometric model is the baseline of the model design within the computer-aided environment. The model may represent a piece part, a sub-assembly, or an assembly. Other nongeometric data is normally needed to complete the definition of a model at the design stage (figure 2–7).

Geometric models that interface to CAM provide basic controlling data such as raw material, part, tool and fixture, and the user–input data. Supporting data is provided by the geometric modeler and the various technology libraries. The input data are those from other CAM applications, and the output data are additional CAM related geometric and nongeometric data.

A CAD/CAM data management system will contain a program for mechanical design applications using geometric modeling. Through geometric modeling a designer defines the geometry of a part using basic geometric entities and geometric transformations. The system provides construction functions for the following geometric entities:

- Point
- Arc
- Curves: line, circular arc, general conic, and parametric spline
- Solids

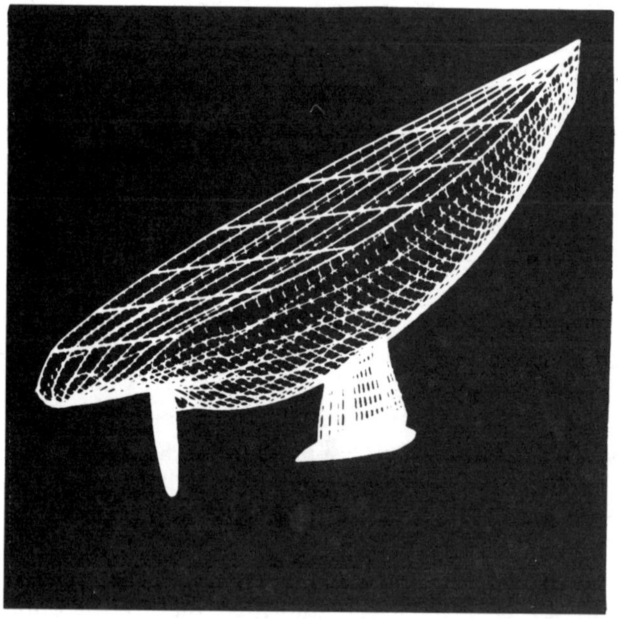

FIGURE 2–7 Geometric Modeling—Structural Analysis (Courtesy of McDonnell Douglas Company).

- Surfaces: plane, ruled surface, surface of revolution, and bi-cubic surface patch (sculptured surface)

Geometric and graphic utility functions allow viewing the model from any angle, blanking or deleting entities, editing and verifying geometric properties, and transforming and duplicating groups of entities. Groups of entities can be saved in a file to use later as sub-assemblies or symbols.

The geometric model supports associativity through structure entities such as tree or list.

MECHANICAL MODELING: A CONVENIENT, TIME-SAVING MEANS OF MECHANICAL ANALYSES

The design of mechanical linkages is a complex, mathematical puzzle. The positions of interconnected, moving parts are too diverse to determine accurately; torque and forces are time-consuming to calculate. In addition, specifications for size, weight, and cost may further complicate the design of a mechanical assembly.

Because the mathematics required to calculate the effects of all these factors on a chosen design are so complex, designers have traditionally resorted to "pin-and-cardboard" models. Through trial-and-error manipulation of these hardware prototypes, designers arrive at a final design, but the process is tedious, and at times, inaccurate. In fact, the constraints of time and labor may make it difficult to perform comparisons between design alternatives, or to optimize the final design. Computers have been used to solve portions of the mechanical analysis puzzle. However, because of cost and complexity, their use has been limited to a small set of applications.

Software has emerged that allows for kinematic analysis of mechanical linkages. This software contains the following features:

- Relative and absolute velocity
- Acceleration
- Reaction of forces
- Displacement

How Mechanical Modeling Works

Geometry created during the conceptual design phase can be accessed for analysis in a fully integrated software system. Individual parts or complete assemblies can be stored and recalled from the data base to facilitate comparative analysis. Alternatively, a rudimentary representation of a mechanical assembly can be created, and the analysis results used to optimize the final configuration.

After the mechanical components have been specified, the mechanisms software prompts the user to specify the direction of motion, the contacts between moving parts, and the internal and external forces which operate on the mechanism. In most cases, the inputs are defined graphically, so the possibility of creating an incorrect model is minimized.

The mechanism can be assembled automatically into a user-specified initial position and viewed on the workstation screen. This mode of analysis is useful for checking whether a mechanism can be, in fact, assembled while still conforming to the topological restraints imposed by joints and defined motion.

After the input is defined, designers can choose to analyze the kinematic, static, or dynamic behavior of the mechanical linkage assembly; take the input data and "connect" the parts together with the prescribed joints; apply the given motions and/or forces; and calculate all the required equations to numerically solve the problem. This analytical process can be done very quickly when done in an on-line mode.

Results of the analysis can be presented graphically on a workstation screen. These graphical displays provide a way to visually determine whether interferences occur and whether the linkage performs as required. Fan diagrams superimpose the image of the mechanism at various points in the cycle and present a step-by-step evaluation of the dynamic behavior.

To better understand the results of dynamic analysis, designers can, through animation capability, obtain an additional visual confirmation that cannot be achieved when viewing static, single-frame output. Animated displays show the complete motion of the mechanical linkage and provide the designer with a "feel" for such characteristics as velocity or acceleration.

What Mechanical Modeling Can Do for Engineers

Mechanism software can produce tabular reports which list numerical data on velocity, acceleration, and load versus time. Using mass properties data generated with solids modeling software, it is then possible to simulate the effects of inertia and gravity. The results of these analyses can then be inputted into software that perform finite element modeling and analysis, and then continue the design analysis and optimization process. Should the mechanism analysis reveal problems or show where the assembly can be optimized, the designer can access the systems design data base to make modifications to the original model and submit it for re-analysis.

In summary, mechanisms software enables design engineers to perform analysis quickly and easily, as an integrated part of the ongoing analysis capability, and thus can prove and optimize designs at a time and cost savings well below the more conventional build-and-test method.

WHO CAN BENEFIT FROM GRAPHIC FINITE ELEMENT MODELING AND WHY

Graphic finite element modeling software provides a set of comprehensive tools for creating, editing, and viewing finite element models and subsequent analysis results.

This software's automated capabilities and user-friendly environment are ideally suited to the design engineer who may perform only preliminary analysis. However, this software also meets the needs of the dedicated analyst who requires a wide range of model creation, mesh generation, and editing capabilities to define very complex fine element models.

How Graphic Finite Element Modeling Works

With this software users can generate finite element meshes over model geometry created with basic solid modeling software. The user has complete control over the distribution of nodes and elements. After the user interactively specifies boundary node locations, the software automatically generates interior nodes and elements to define a mesh. The user may select from a variety of mesh types, including planar, surfaced, or volumetric. In addition, it automatically numbers nodes and elements and optimizes the sequence to minimize array bandwidths. This feature provides for extremely efficient analysis processing (figure 2–8).

The user then interactively assigns attributes (loads, restraints, boundary conditions, and element and material properties) to any individual or group of nodes and elements. These attributes are used as direct input to an integrated finite element analysis and design software package.

During model preparation, this software provides tools to prevent the creation of faulty meshes. It enables the user to automatically check for ill-formed elements, duplicate nodes, and unusual nodes. Additionally, a shrinking feature permits detection of missing elements. The user also may specify minimum spacing between automatically generated nodes; the software will merge those nodes within the minimum spacing limits into one node and graphically identify the merged nodes on the display. These capabilities eliminate the possibility of submitting an incorrect model for analysis, and reduce the time and cost typically associated with finite element methods.

Optimization of node placement also is facilitated with the software's three-dimensional viewing capabilities which permit the user to view the model from any direction in multiple windows.

As a post processor, this software enables the user to display the results of analysis as deflected (deformed) shapes; graphic deformation may be magnified or exaggerated for clarity. Stress and strain contours can be displayed as color contour lines. A range of colors is assigned to denote degrees of stress from minimum to maximum values. Additionally, this software supports display of dynamic node shapes.

Features and Benefits of Graphic Finite Element Modeling

Graphic finite element modeling offers a user the ability to:

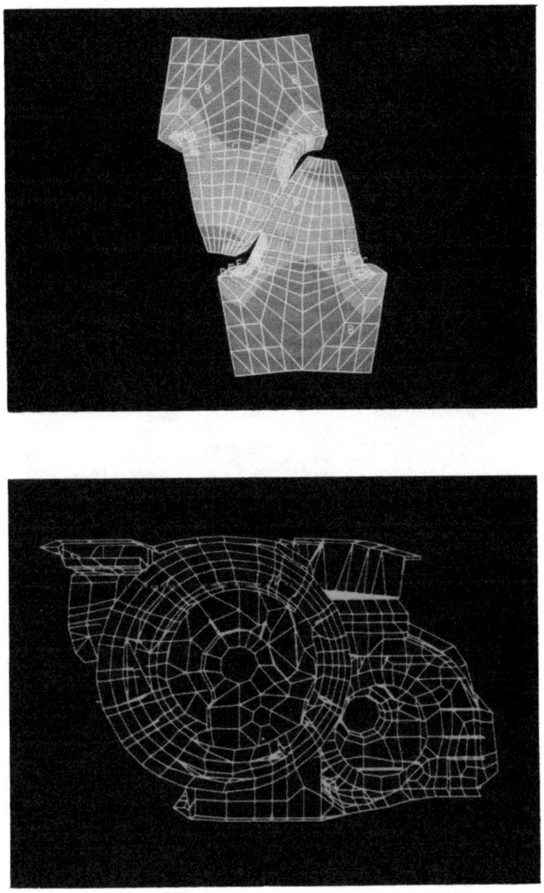

FIGURE 2-8 Graphic Finite Element Modeling (Courtesy of MacNeal-Schwendler).

- Facilitate rapid construction, editing, and display of 2D or 3D finite element models from geometry created with various software packages.
- Automatically generate 2D-planar, 3D-surface, and 3D-volumetric meshes.
- Provide comprehensive editing capabilities for loads, boundary conditions, and element properties.
- Allow full user control over node and element distribution; automatically number nodes and elements.
- Provide automatic facilities for model checking and verification.
- Automatically optimize node and element numbering to provide for efficient processing.
- Perform geometric creation, manipulation, and display.
- Display results of analysis as defected shapes, node shapes, and stress/strain contours highlighted by color contour plots.

Graphic finite element modeling offers a user the following benefits: Facilitates the rapid construction of finite element models from existing geometry. Automatic meshing generation reduces the time required to build models for analysis. User control over attribute assignment, mesh density, and boundary parameters provides the analyst with the flexibility needed for the most complex modeling requirements.

Three-dimensional viewing enables users to check models from any orientation in multiple views and to optimize node/element configurations and verify mesh placement.

These automated capabilities minimize the possibility for errors and reduce the time and costs associated with analysis.

Moreover, the graphical display of analysis results greatly simplifies the task of analysis interpretation and eliminates the tedious review of lengthy numeric computer printouts. Model deformation can be displayed alone or superimposed over the undeflected model. Stresses and strains are highlighted graphically as a color contour plot, enabling the designer or analyst to quickly determine design flaws or weaknesses. Following is a listing of typical element types:

- One-, two-, and three-dimensional
- Rigid
- Mass
- Heat transfer
- Fluid
- Rigid constraint
- Solid
- Curved-beam
- Shear panel
- Linear-strain
- Damping
- Axisymmetric
- Scalar
- Shell
- Quadrilateral plate

In addition, these systems enable the user to supply their own element descriptions for non-standard analysis.

INTEGRATED FINITE ELEMENT ANALYSIS AND DESIGN: A COMPONENT IN STRESS ANALYSIS

Integrated finite element analysis and design performs linear static and dynamic analyses on finite element models created with graphic finite element modeling software. Together these two software packages provide an integrated easy-to-use environment which facilitates the entire mechanical stress analysis process.

This software employs algorithms to perform analysis and requires no input deck preparation or data formatting since it uses model data created from graphic finite element modeling software.

This software contains a library of finite elements, including beam, spring, shell, planes, and solid elements, and accepts a variety of loadings and constraints. Output generated by the software can be transferred directly to graphic finite element modeling software for post-processing and results display. Integrated finite element analysis and design output consists of:

- Model displacements
- Beam and bar end forces and moments

- Finite element stresses and strains
- Reactions at support points
- Nodal strain energy density

Integrated finite element analysis and design offers a user the ability to:

- Provide an integrated, easy-to-use facility for finite element analysis.
- Perform linear static analysis on finite element models created with graphic finite element modeling software.
- Perform dynamic model analysis on linear structures.
- Contain a library of finite elements, including bar, beam, spring, shell, planar, and solid elements.
- Output directly to graphic finite element modeling software for results interrogation and display.

Integrated finite element analysis and design software offers a user the following benefits: The application of computer analysis and simulation techniques to design concepts enables engineers and designers to develop better, more reliable products in less time than it would take using conventional prototyping techniques.

Graphic finite element modeling software and integrated finite element analysis and design software enable both the designer and the finite element analyst to implement analysis techniques at the earliest stages in product design. These software aides compress the development cycle by allowing designers and engineers to evaluate a product's performance and reliability during the preliminary design phases, rather than as a physical prototype. With integrated finite element analysis and design, design models can be analyzed for their reaction to various real-world environment conditions; alternate designs can be compared on the basis of performance and reliability, until an optimum configuration is determined.

Graphic finite element modeling simplifies model preparation and helps ensure that only complete and accurate models are made available for analysis. Fully integrated with the graphics data base, integrated finite element analysis and design directly accesses model data created by graphic finite element modeling thus eliminating the formatting and preparation of input decks and the errors that typically result from hand coding.

Integrated finite element analysis and design utilizes powerful, state-of-the-art algorithms and advanced processing techniques to solve linear static and dynamic model analysis problems.

WHY THE AUTOMATIC DIGITIZER IS USEFUL IN COMPUTER-GENERATED IMAGERY SYSTEMS*

The process of *digitizing* begins when an operator enters XYZ coordinate data of a complex geometry into the system's host computer by means of a hand-held digitizer stylus. The 3 SPACE Digitizer (manufactured by Polhemus Navigation Sciences, a division of McDonnell Douglas Electronics company) returns not only position X, Y, Z data, but also orientation (azimuth, elevation, roll) data on its stylus. This provides a unique capability to digitize any nonmetallic object, whether mock-up, prototype, or component part, and synthesizes it in a computer. Once there, the 3 SPACE Digitizer's six degrees of freedom capability

*Permission granted by McDonnell Douglas Corporation.

allows users to animate computer-generated representations, direct their movement through space, and create artistic designs in a mode that is sometimes called "video paint-brushing." The system can be used to build quickly vast data bases of models for computer-generated imagery systems, and to aid designers in sketching mechanical parts and visualizing complex geometries (figure 2–9).

A software interface provides the direct link between the digitizer and the CAD/CAM system. The digitizer interface allows an operator to create 3D entities in an off-line mode. A new part can be created or data can be added to an existing part. Basic capabilities include controlling the orientation, depth, scale (each axis), units (English or metric), entity layer, and color within the part. With the digitizer, an operator can digitize a three-dimensional object to create graphic entities—points, lines, arcs, splines, conics, strings, polygons, and groups.

Digitizers are able to accept parts up to 20 × 20 × 10 inches with full accuracy. Larger objects can be digitized in sections using a meshing function to relate precisely one digitized section to the next (figure 2–10).

HIGH-RESOLUTION TRACKER: A TOOL FOR MONITORING UP TO FOUR OBJECTS

Polhemus Navigation Sciences also manufactures a high-resolution tracker that continuously updates the position and orientation—that is, six degrees-of-freedom, of one to

FIGURE 2–9 Automatic Digitizing Demonstration (Courtesy of Polhemus Navigation Sciences, a division of McDonnell Douglas Electronics Company).

FIGURE 2–10 Automatic Digitizing System (Courtesy of Polhemus Navigation Sciences, a division of McDonnell Douglas Electronics Company).

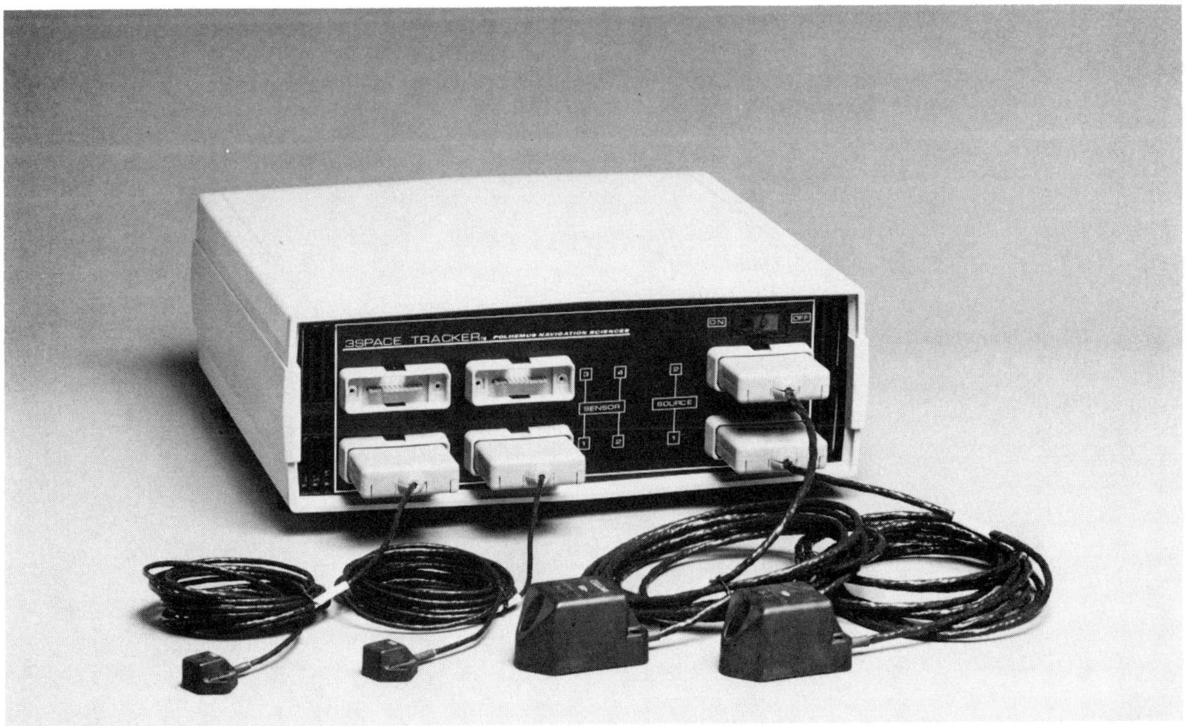

FIGURE 2–11 High-Resolution Tracker (Courtesy of Polhemus Navigation Sciences, a division of McDonnell Douglas Electronics Company).

four small, freely movable electromagnetic sensors with respect to one or two fixed electromagnetic sources. Their 3 SPACE Tracker's sensor can be unobtrusively mounted on any nonmetallic object and will continuously update its location as it moves through free space. The tracker transmits data to a host on command or at a rate of up to sixty measurements a second (figure 2–11).

THE LOW-COST TRACKER: AN AFFORDABLE ALTERNATIVE

Another product of Polhemus Navigation Sciences is a low-cost tracker called the 3 SPACE Isotrak. This instrument has measurement capabilities identical to the 3 SPACE Tracker but is not quite as accurate. The Isotrak is capable of supporting one sensor and one source (figure 2–12).

The 3 SPACE Isotrak offers users a natural and simple-to-use peripheral. It is compatible with personal computers and is capable of replacing multiple joysticks as well as 2D digitizers. With a sweep of the hand, the Isotrak can be used to trace out an animation path, program a robot's motion, or quantify human movement.

The Isotrak's multiple degree of freedom can be used to control special displays and 3D cursors and to interact with spatially oriented menus and images. Its electromagnetic technology handles applications requiring medium resolution and low cost. The system employs

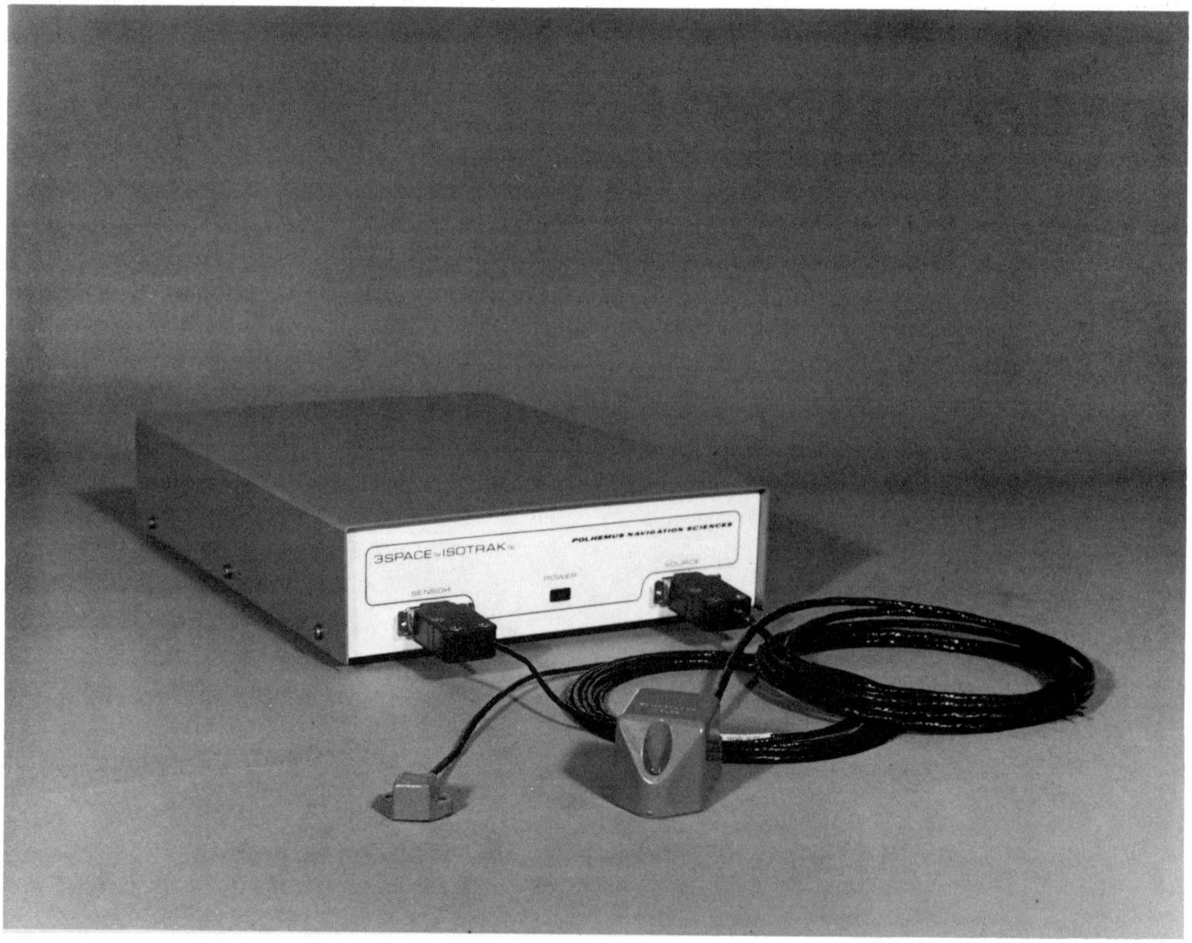

FIGURE 2–12 Low-Cost Tracker—3 SPACE Isotrak (Courtesy of Polhemus Navigation Sciences, a division of McDonnell Douglas Electronics Company).

two encapsulated modules, each consisting of three small coils that form an orthogonal triad. One of the modules acts as a fixed source (transmitter) to generate a long frequency electromagnetic field, while the other module acts as a movable sensor (receiver) to sample the field. In addition, an electronics unit comprising analog circuitry with digital processing and a power supply, performs all the necessary computations and interfaces with the outside world.

Working from a complex algorithm, the electronics unit sorts out the signals it receives from the sensor and computes the sensor's position in rectangular coordinates and the sensor's orientation in angular coordinates thus providing a full six degree-of-freedom solution.

THE NEXT INNOVATION IN HUMAN-COMPUTER INTERACTION: USE OF BODY LANGUAGE

The development of a magnetic position and orientation sensing technology enables the human body to be used to control sophisticated computer processes. This means that body language can be considered another way of interacting with computers.

Computers are becoming truly conversant in our language forms including written, spoken, and body forms. "User friendly" written languages are now commonplace, and spoken language is now used with the advent of voice recognition and synthesis. The next step is to execute computer commands in the third mode of human communications: body language.

Software is now available which provides a natural means of human–computer interaction. It continually computes both the position and orientation of its sensor within a defined reference frame. It's freedom of movement, responsiveness, and high accuracy evoke a progression of implementations for coupling natural body movements with computerized systems. These essential components comprise this type of system:

- A source which generates an electromagnetic field.
- A sensor which samples the field.
- A computer which performs all required control, calculation, and interface tasks. (These components interconnect.)

The system operates by generating three orthogonal fields that are well defined in the area of interest. The fields are sampled by the sensor and the resultant data is computer processed to determine position and orientation of the sensor.

HOW GROUP TECHNOLOGY CAN INCREASE PRODUCTIVITY BY REDUCING DESIGN DUPLICATION*

Group technology is a soundly conceived methodology used to integrate computerized systems in design and manufacturing. It performs the functions of computer-assisted process planning, design retrieval, design and manufacturing standardization, machine tool selection, and long-term planning and analysis. Through the total integration of automated systems, it is possible to create a data base of company technology and to retrieve, analyze, and use information from that data base for a wide range of design and manufacturing applications.

Experience has shown that companies tend to create designs and processes for similar parts over and over again. With the advent of group technology, design engineers can retrieve drawings using a classification and coding system, thus significantly reducing design

*Permission granted by Organization for Industrial Research.

proliferation. By creating families of similar parts, manufacturing benefits with fewer setups, longer production runs, and less redundant parts to control and maintain in inventory.

Defining Similarities: A Key Task of Retrieval Systems

One of the most important elements in developing a workable retrieval system is how to define similarities of parts. The design engineer tends to think in terms of shape, dimension, and related attribution. Although the manufacturing engineer may recognize similarities in shape, he or she may only be concerned with the type of material involved and the manufacturing process.

Since each company has its own historical precedence, funding limitations, human resource availability, and other factors to consider, companies have resorted to a variety of approaches for analyzing how parts should be grouped, such as:

- *Occular.* Occular grouping, or "eye balling," is a popular way of grouping parts. Unfortunately, its limitation is that a company will usually have thousands of parts, so making judgments through this method becomes impractical in terms of time and dimension.
- *Nomenclature.* Part names can be misleading since designers may call one part by one name and another similar part by a different name.
- *Production Flow Analysis.* Another approach is to group all of the parts that are produced by a given group of machine tools. Such an approach may be excellent from the manufacturing engineer's point of view, however, it does not meet the design engineer's needs.
- *Classification and Coding.* With a classification and coding system, the parts are grouped by technological characteristics—design, manufacturing, or whatever seems most appropriate. The system is developed to meet specific criteria, and parts are dropped into "buckets" placed in the classification scheme. A code number is assigned to each part for retrieval purposes. Classification puts parts into buckets, and code numbers help to get them out. Classification and coding is a technique of capturing the relevant design *and* manufacturing characteristics of parts.

The key to the successful use of classification and coding systems is the capability to analyze the parts once they have been classified. Through analysis, it is then possible to standardize design features, manufacturing approaches, and other related factors.

Why Group Technology Can Double as a Means of Systems Integration

A *classification and coding system* also is a communication tool since it provides the user with a consistent vocabulary for part identification because it can involve both design and manufacturing characteristics. Also, because it is a consistent identification method, it can be the key to integrating computer-aided design and computer-aided manufacturing.

You cannot access similar parts by part number, but code numbers do allow for ease of retrieval of parts which are similar in design, manufacturing, quality control, or other characteristics.

The technology data base, built with inputs from the classification and coding system and other related systems, enables a user to store data on shapes, dimensions, tolerances, lot sizes, materials, functions, parts, assemblies, purchased parts, and anything else needed or wanted.

A user can group parts or other components and assemblies by common characteristics to take advantage of important similarities. By grouping parts in this manner, a user can reduce dramatically the number of retrieval search items. Classes of parts that are retrieved can be analyzed for design and/or manufacturing standardization and for other purposes.

CASE-IN-POINT

DESIGN APPLICATION

A company had a small standard part which had several design characteristics important to its function, and several more that made no difference at all. An analysis found that the company was producing twenty-two versions of this same basic part. In each variation, the differences were unimportant in terms of use—they had been arbitrarily created by design engineers over a period of time. While the parts seemed cheap to make, it turned out that each required a special cutting tool, and the average tool cost per part was $148.60. By standardizing this part so that it could be made with one standard tool, with a tool cost of only $12.25 per part, the result is a reduction of $136.35 per manufactured part.

CASE-IN-POINT

MANUFACTURING STANDARDIZATION

A company had 523 different gears in production. There were 477 different process plans for the manufacturing of these gears. Since each plan averaged about three pages, there were nearly 1,500 pages of process plans in the files. Rather than search through all these process plans, the methods engineer would simply prepare a new plan for each new gear. A study was made which resulted in the elimination of 400 process plans. Since each plan took two to three hours to prepare, considerable cost savings were achieved through standardization of process plans for the manufacture of gears.

The Organization of Industrial Research Multi-II Group Technology System

The Organization of Industrial Research (OIR) has developed Multi-II, which is an advancement in Group Technology since it contains multiple retrieval options, enhanced mass updating, and other important features. Multi-II is a powerful management and integration tool that provides advanced reporting capabilities of the industry standard relational data base and user-configurable bridges to other CAD/CAM systems. It includes the following task-oriented modules:

- MultiTreeve II Group Technology classification and retrieval
- MultiGroup II Group Technology analysis
- MultiCapp II computer-assisted process planning
- MultiCats II automated time standards
- MultiTrack II tool-tracking and inventory control

All of these modules utilize a single data base—the industry standard in relational data base technology (Oracle). Multi-II enables a user to build a comprehensive data base of all design and manufacturing information, and to use that data base for Group Technology applications as well as for virtually all other CAD/CAM applications including technologies such as MRP, FMS, and robotics.

Design Retrieval: OIR MultiTreeve II

Effective automation throughout the factory begins with the creation of a data base of design and manufacturing information. To be useful, the data base must be organized for the types of retrieval and analysis that are important to design and manufacturing engineers and to management. Information must be entered with a consistent and disciplined system.

MultiTreeve II makes it possible to enter part descriptions quickly and easily, in a tightly organized framework that supports retrieval and analysis effectively. It classifies parts so they can be grouped by relevant common attributes.

MultiTreeve II also can be used to retrieve information and to form part families based on those common attributes. Thus it provides the foundation for many Group Technology applications.

The user can retrieve parts by keywords, features, individual characteristics, part numbers, classification numbers, or other data. The system uses common terms, tailored to individual company practices. With keyword retrieval, for example, the user can ask for a part by a name, such as "gear." A menu then appears on the CRT with a list of types of gears, such as bevel, Gleason, helical, spur, hypoid, and so on. When the specific type is chosen, simple questions about diameter, length, and material appear on the screen. Once the questions have been answered, the system searches the data base and retrieves the relevant part or parts.

The user can search for an exact match or for ranges of similarities.

MultiTreeve II works with any type of component, including machined parts, electronics, purchased parts, tooling, and assemblies.

Analysis, Standardization, and Decision Support: OIR MultiGroup II

The OIR MultiGroup II module makes it possible to analyze the data in the Multi-II data base and to produce a wide range of reports and planning aids for applications such as the standardization of designs and manufacturing methods: the development of part families and work cells or flexible machining centers: the analysis of machine tool requirements, machine shop layout, and workload and product mix analyses.

It makes possible the analysis of productivity and identification of strengths and weaknesses; evaluation of the potential impacts of other CAD/CAM technologies; planning of future production facilities, and effective modification of existing facilities. It is an extremely useful management tool which provides unique insights into production processes and production capacities.

Reports and analysis results are presented in clear and easy-to-understand graphic form with bar graphs, pie charts, line graphs, and highlighted tables. Thus they are ideal for management review and decision making as well as for day-to-day observation and control of design and manufacturing activities.

Users also can take full advantage of the Multi-II relational data base to customize reports and for ad hoc reports on issues of timely importance.

Computer-Assisted Process Planning: OIR MultiCapp II

The MultiCapp II computer-assisted process planning module enables planners to use the computer to create, edit, store, and retrieve process plans efficiently and effectively.

Optimized manufacturing methods and standardized operation descriptions can be stored in the data base and used to create new process plans. Existing plans can be retrieved by any header field, including name, classification number, part number, planner, or any other descriptors specified by the user. Information stored in the relational data base also includes all data associated with the process plan, such as time standards, tooling, gaging, and bill of materials.

MultiCapp II also can be integrated with graphics so that text and illustrations can be merged to provide shop floor personnel with detailed instructions in pictures as well as words.

When integrated with the Group Technology retrieval features of MultiTreeve II, MultiCapp II allows for the development and use of standardized process plans related to families of parts. These standards incorporate the company's most efficient and cost effective manufacturing methods, so that MultiCapp II cannot only significantly increase planning productivity, but also can impact upon overall manufacturing quality and productivity.

Automated Time Standards: OIR MultiCats II

MultiCats II is an automated time standards module which can be used as an integrated component of the OIR MultiCapp II package.

With MultiCats II, it is possible to automatically calculate machining times and set up and handling times. Machining calculations can be based on the company's experience or on machining parameters from the *Machining Data Handbook* developed by Metcut Research Associates. The user can modify the Metcut standards to reflect variations in company practices, if and where, necessary. Set-up and handling time calculations are based on user standards.

MultiCats II can make use of all relevant data in the header and manufacturing sections of a MultiCapp II process plan when calculating machining times. The planner needs only to input data which is not in the process plan. The calculations are carried out and the results are automatically inserted in the process plan.

Although MultiCats II can be used for individual parts, it is unique in its capability to also take advantage of Group Technology benefits and be used with families of parts. The cost and effort involved in the development of company-specific standards are greatly reduced when the standards are developed for families rather than individual parts.

Through the Multi-II relational data base, these standards can be related to optimized design and manufacturing methods, thus helping to assure maximum efficiency and high quality.

Tool Tracking and Inventory Control: OIR MultiTrack II

The OIR MultiTrack II module eliminates waste in tooling costs, assures cost effective purchasing, and increases productivity because tools are available when and where needed and the best tools can be identified for each job.

MultiTrack II is an on-line system that tracks the use, movement, and status of perishable and durable tooling, including fixtures, gages, and other items of concern to tooling departments.

It also generates reports tailored to company requirements to help minimize tooling costs and to aid management decisions regarding resource allocation. These reports can deal with issues such as check-in and check-out, tool usage and consumption, inventory status, where-used, and kitting.

When integrated with the classification and retrieval features of MultiTreeve II, it can also be used to standardize tooling and to identify and retrieve correct tooling for specific operations.

MultiTrack II can be fully integrated with MultiCapp II computer-assisted process planning. When preparing a process plan, the user can access tool information and automatically insert pertinent tool data in the plan.

How Group Technology Can Serve as the Integrating Link for CAD/CAM

The use of the computer and the capturing of both design and manufacturing information in the code number have made it possible to set up a data base which contains both design and manufacturing information. With a common data base, Group Technology can become a hub for the integration of CAD/CAM. Essentially, group technology has become a valuable technique because of the need for standardization in both design and manufacturing in order to reduce costs and automate manufacturing processes.

In a fully automated system, the designer works at a computer graphics terminal linked to a common data base. Working from a sketch, the designer codes the part, searches the file, and locates the optimum design. In a similar mode, the manufacturing engineer can search the data base and locate the best manufacturing method and detailed process plans and routings. With standardization, the system can automatically computer-generate process plans, thus integrating CAD/CAM with computer-assisted process planning (CAPP).

By integrating all of these capabilities, you can group machine tools with common tooling to produce families of parts, and generate the detailed process plans and routings required to produce these parts. The end result is a completely computerized system of scheduling work in the factory.

Purpose of Code Structures in Group Technology

There are three basic types of code structures:

- Hierarchal or monocode
- Attribute or polycode
- Hybrid

Hierarchal. In a hierarchal code, each position relates to the previous position. For example, if the first digit defines a main shape, such as rotational, the second digit in the code will define a feature related to a rotational part, and the next digit a feature related to the feature defined in the second digit. This is the methodology used to develop a hierarchal code structure.

Although this method can satisfy the needs of design engineers when retrieving stored data, it cannot be used in a total systems integration because, for other applications, the same digit in the same position may have different meanings at different times depending on the preceding digit. As a result, this method has limited computerization potential, and hence is of minimal value as an analytical tool.

Attribute. An attribute code structure is based on the total population of different attributes. Each digit and each position (value) of each digit has a specific meaning, hence the meanings of digit values for any given part would be consistent in the system. Compared with the limitation of hierarchal codes, this method enables a user to make a detailed computer analysis that can identify specific machining operations required to produce parts within a family tree.

Since there are so many possible attributes when all parts are included in a typical manufacturing environment, a third method referred to as a hybrid code structure is most commonly used. This coding system combines the advantages of hierarchal and attributes.

Hybrid. In the hybrid coding structure, the first digit refers to the main configuration, and an attribute matrix is then linked to each main shape. Thus, this method is more compact than the attribute coding system, and hence more amenable to computer analysis. As a result most classification systems use the hybrid coding structure.

Six Universal Attributes of Classification and Coding Systems

Regardless of which classification and coding system is used, certain attributes are found universally in every company and every industry. These are:

- Main shape—rotational, box form, sheet
- Shape elements on the main shape—cones, holes, slots
- Element position—where the shape elements are located
- Dimension—main dimension, auxiliary dimension, and ratio of dimensions
- Accuracy—how critical the tolerances are
- Material

In addition to these basic attributes, certain specific attributes such as lot size, set-up time, run time, and others, may be required depending on the needs of a particular industry, factory, or machinery operation.

A Seven-Step Method for Developing a Classification and Coding System

1. Take a representative sample of all active parts in the existing parts data base.
2. Assemble the drawings obtained from this sampling into like families of parts, as defined by different discrete features.
3. Refer to industry standards as a guide to verify that groupings appear to be proper.
4. Identify the frequency of occurrence of the chosen features in the drawings contained in the selected sample.
5. Develop a hierarchy of these features based on the samples, and link them to the machine tool capabilities of your company.
6. Verify that the system is satisfying your needs, or, if not, make whatever adjustments deemed necessary.
7. Analyze the data to reduce costs through standardization of design and product processes, and determine the optimum method to manufacture a family of parts.

How to Determine Load Balancing

When developing optimum methods of manufacturing like parts, it is essential that machine loads be balanced to maximize machine utilization and the plant throughout. Thus, the coding system must include such factors as lot size and numbers of releases per year. This data can then be used to determine load demand of each family and each part within each family over a period of time. After this information has been captured, it is then necessary to determine which machines, or groups of common machines, will have sufficient capacity to produce these families of parts. From the information on parts and machine tools available to manufacture them, you can develop matrices. This code number matrix identifies the features of a part by code numbers, and uses the same code numbers to identify the attributes that can be produced on a particular machine tool.

RELATIONAL DATA-BASE SYSTEMS: A FLEXIBLE APPROACH TO ON-LINE ACCESS

Advanced CAD/CAM systems require on-line capability to have fast access to all the stored drawings, documentation, material information, and models needed in designing a product. CAD/CAM systems must link to commercial data processing systems and be compatible with systems from different vendors. In addition, CAD users need computer assistance to locate information stored off-line in archival storage.

Most CAD/CAM systems are incapable of analyzing the engineering specification and data about its application to know whether a part already exists. This is because the data structures they use make it difficult or impossible to associate more than a small number of different data types with a specific part, or to arrange data associations hierarchically. In addition to group technology discussed earlier, introduction of a *relational database system* into CAD/CAM helps engineers determine if a part has been designed previously.

Relational data management and control software tracks what parts go with a certain assembly, tracks information associated with new designs, and tracks the many parts of an overall activity.

A relational data base has superior flexibility over network and hierarchal approaches. Engineering and manufacturing data bases are characterized by change, while relational methods do not impose structural restrictions on data. An integrated shared data base incorporates three principal differences compared with a conventional CAD system:

1. Engineering applications requiring an ability to store very large data elements, such as pictures of scan drawings or test results that require long data elements. They also must be integrated with data that provides attributes describing these elements. Business data bases are not good at supporting this requirement.
2. Engineers need a large collection of records to describe all the data and all aspects of the object. In a relational system a series of records are linked together logically and physically to provide added functions useful for engineering applications.
3. Business transactions touch upon only a few records in the data base. However, when an engineer makes a change to a part, many records are involved. Since these records are more complex, it requires considerable time to complete the transaction.

Relational data-base systems enable engineers to access multiple data bases and thereby gain the full benefit of an integrated CAD/CAM system. Each data base is optimized for a particular function, with each carrying a common subset of information. Thus there is no need for one monolithic data base, which would be unwieldy and impractical.

FILE MANAGEMENT SYSTEMS: A COMPARATIVE GUIDE*

Companies use either of two types of systems for electronic document management:

- Micrographic computer-assisted retrieval system (CAR)
- Optical disk document management system

Each of these systems performs the same basic functions: capture, storage, retrieval, transmission, and duplication of document images. The system configurations of hardware

*Permission granted by 3M.

may even appear to be similar in both systems, but these two systems can be distinguished by the extent to which a company uses one or the other in a particular application.

The micrographic computer-assisted retrieval system is generally thought to provide better cost performance where documents are organized in chronological sequence. In these applications the cost of optical disks is difficult to cost justify.

On the other hand, when files are not in chronological order requiring random filing, then even though the system is a large volume application, the relatively slow retrieval speed of CAR tilts the cost equation toward an optical disk document management system. What normally occurs, is that the data is stored on more than one roll of microfilm, and these multiple rolls must be located and inserted into the reader/printer for retrieval of the requested images. Because of this, retrieval speed is slower with a CAR system than with optical disk. These differences demonstrate that although the functions of the two electronic systems are the same, they go about performing those functions differently, and therefore require an analysis of a company's unique applicational requirements in the selection process.

The following sections of this chapter compare the primary functions of these two types of electronic document management systems.

Capture: Filming versus Scanning

Capture in an automated micrographics system is the process of filming a document and capturing its image on microfilm. The procedure is very fast, usually requiring less than one second for each document. However, most CAR systems require an interim period during which the film is processed or developed. This process may take anywhere from two hours if done internally, to as much as a day or two if sent to an outside film developer. During that time, the documents are not available for retrieval.

By comparison, in an automated optical disk system, capture is accomplished by a scanner which scans the document (placed face down on the scanning flat-bed) and then converts the document's image to minute dots called *pixels* (figure 2–13). Scanning, using this process, may take three seconds compared with one second for filming, but its advantage is that it offers real-time access of documents. No further processing is required, and after storage the document image is available immediately for viewing or printing.

Storage: Cost and Storage Life Are Primary Issues

Today, the CAR system enjoys a lower cost-per-unit storage ratio than that of optical disk storage. However, this cost advantage may be overcome in the next few years because of increased storage capacities of optical disks and the reduced cost of the disks. In this case, both media will be nearly equal in their cost-per-unit storage ratios, so storage cost is not a major factor. Both systems also are equally efficient in their conservation of space for stored documents.

Although micrographics is already established as a recognized method of archival storage with an ability to exceed seven years—which covers the vast majority of documents—optical disk manufacturers already project a lifetime of ten to twenty years for the integrity of the images stored on optical disk. Thus, integrity of acceptable storage life to satisfy any document archival classification should soon be acceptable on an optical disk system as well as micrographics.

Micrographics is an accepted storage medium for evidentiary documentation in the legal process. The optical disk storage medium has yet to be tested. Yet, given the accurate duplication of the document's image on an optical disk, and the security measures of the system to protect that image from modification once captured, the images stored on optical disks will soon be regarded as legally admissible evidence.

FIGURE 2–13 File Management System Display Screen (Courtesy of 3M Corporation).

Retrieval: How Important Is Speed?

A strong case can be made that optical disk systems will emerge as the champions over micrographics systems in retrieval speed capabilities. Magnetic tape and microfilm rolls are, by their very constructions, serial media. This means the system's machinery must scroll through rolls of film to find the requested document image. The result is multiple searches for multiple documents and multiple winding and rewinding. Depending on the size of the document, the retrieval process may take twenty to forty seconds. With optical disk, the input operator designates the location of the image on the optical disk at the time of indexing. Upon retrieval, the system goes straight to the location and retrieves the image—in most cases in less than six seconds (figure 2–14).

However, even this circumstance is not without its limitations. Input (or more accurately, indexing and capturing) has always been the counterbalance to output (retrieval and transmission/duplication). Consider it a matter of trade-offs. Does the quick input capability of a micrographics system (one second capture/indexing document) for a particular application compensate for the slower retrieval speed? Or is the three-second capture/indexing speed of the optical disk system permissible to obtain its faster retrieval speed? Front-end

FIGURE 2-14 File Management System—Indexing and Retrieving Documents (Courtesy of 3M Corporation).

processing speed versus back-end retrieval speed: The trade-offs must be weighed in the balance of the application.

Transmission: A Possibility for Each System

Traditionally, electronic transmission of microfilmed images has not been practiced. However, from a technological point of view, there is no reason why it can't be electronically transmitted. If an image can be digitized, then it can be transmitted. However, because of the retrieval trade-off discussed earlier, the hybrid type of system is essentially off-line. Even with a robotic roll film retriever, there is still the inherent weakness of roll film: serial recording and multiple roll selection.

Printing: What Quality at What Cost?

Image printing is achieved at a remote workstation in a micrographics environment with a reader–printer. In an electronic optical disk environment, it is achieved with a laser printer. Both cost and quality are issues to be considered when comparing the two technologies.

Both systems can print images on plain paper and on coated paper. However, a micrographics system does so with a reader–printer, a combination unit. The optical disk system requires a separate display and laser printer, which are costly components of this alternative system.

Yet, in some cases, the additional cost of the laser printer may be justified by the flexibility afforded by its capabilities. An optical disk laser printer can actually improve the quality of the printed image by sharpening the edges of lines within the image and by increasing the contrast between the lines and the document's background. The result is an enhanced printed image, which may, in the case of recording old birth records or similarly poor quality records, be warranted or even required for the application.

Conclusion: A Combination Approach Is Possible

There is a strong temptation to view optical disk technology as the panacea of automated document management systems. However, not all applications will find optical disk technology the answer to their document management needs. Nor will all applications find a perfect match in micrographics automated systems. But it is a fair assumption that between the two, a pretty good choice can be made.

To make this choice, the key is efficient, cost-effective document management through software that controls the indexing, capturing, storing, retrieving, transmitting, and printing or reproducing of the documents regardless of the storage medium used. For example, 3M has what is called Document Management Software, or DMS, which enables the micrographics system user to manage documents in the same manner as the optical disk system user. To bridge the gap between the two technologies, DMS will make the transition from one technology to the other, or the integration of both optical disk and micrographics technologies in a multi-media application.

CASE-IN-POINT

*ENGINEERING DOCUMENT TRACKING**

Situation

A large midwestern utility controls a central file of over 750,000 of their engineering drawings. When engineering change notices call out revisions to an original drawing, the drawing is removed from the central repository and routed to the engineering department for modification. Each day over 800 drawings are moved in or out of the central file and tracked by a computerized system. When drawings are moved to or from the file, a drawing identification number (averaging 16 characters) is keyed into the computer terminal, along with a six-character access control code for employee identification. Because of stringent nuclear regulatory requirements, more accurate drawing records, as well as speeding up the drawing check-in/check-out process, are needed.

Solution

A bar code printer is located immediately outside the entrance to the file room and is used to print a label for each drawing. The label is then applied to the lower right corner of the original drawing (figure 2–15).

As each drawing is removed from or returned to the file area, the label is scanned into a portable bar code reader. The access control code is scanned from the bar coded *drawing loan access control list,* which has been produced from the computer printer or electronic copier.

Twice a day, three control labels are scanned into the bar code reader to identify the "end" of data, block length, and cursor tab control. The reader is connected to the CRT

*Permission granted by Automatic Identification Manufacturers (AIM).

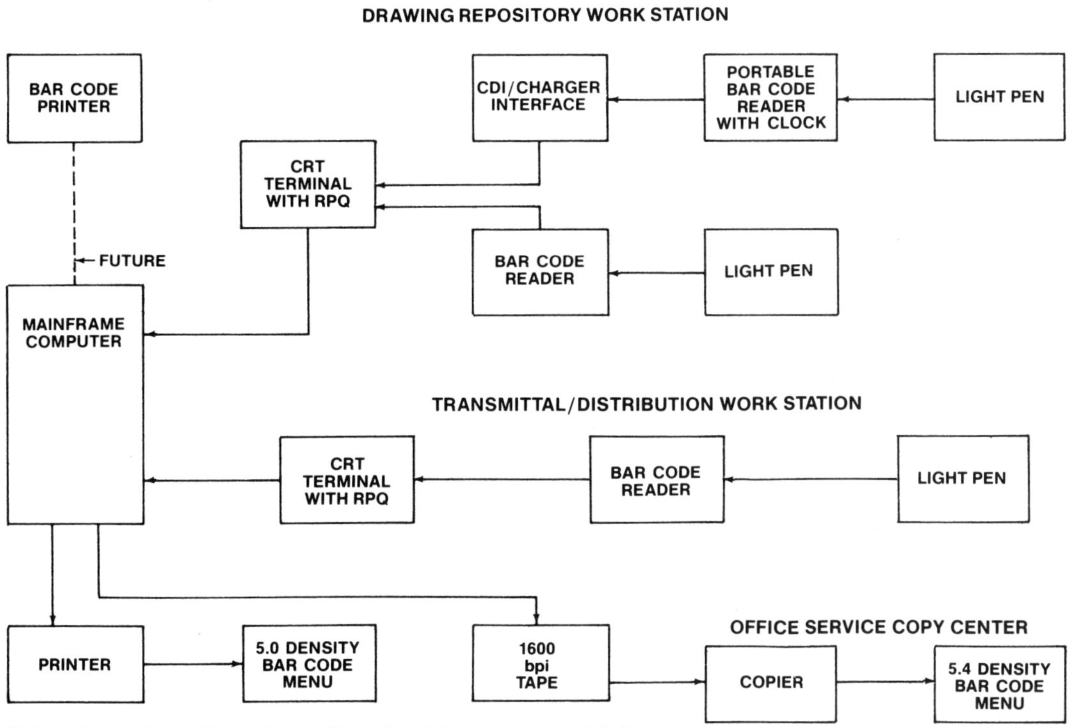

The bar code system diagram illustrates the layout for creating labels, drawing access control, drawing removal and return and computer index updating.

FIGURE 2–15 Engineering Document Tracking (Courtesy of Automatic Identification Manufacturers AIM).

terminal located in the engineering document services area, and the data is transmitted to the terminal. Program control reads the ASCII to EBCDIC converted character strings, creates appropriate transactions, and posts the drawing status to the drawing loan system through the main computer.

Each label contains a human-readable drawing identification code from seven to twenty-three characters in length containing the following: drawing prefix (equivalent to drawing size), division (applicable to a manufacturer's drawings only), series (a unique numeric value of a specific drawing), dash and sheet (further uniqueness of drawing series for multiple sheets).

Results

Prior to installing the bar code system, data entry took two hours daily for an average of 800 drawings yielding a 2 percent error factor.

The bar code system, by allowing the utility to accurately identify drawing status, has reduced entry time by 95 percent. The savings in re-entry of data in error, lost drawings, and turnaround contribute to greater annual savings. In addition, improved accuracy and drawing accountability save time in drafting, engineering, construction, and operation disciplines, resulting in higher corporate productivity and response time.

chapter 3

Creative Innovations in Product Process Technology

HOW MANUFACTURING ENGINEERING FITS INTO THE PRODUCTION SYSTEM: AN OVERVIEW

The next phase in the *total integration of all manufacturing elements* (TIME) is the transition from design engineering to manufacturing engineering. Manufacturing costs are not only determined by the design of a part, but also by the tooling, material, methods, and machining processes involved in production and fabrication. Engineering design and drafting supplies blueprints, specifications, and other technology requirements. Manufacturing engineering provides manufacturing producibility cost data and equipment in-process capability. This involves development of advanced manufacturing and materials technology in addition to the support of production shop tooling, operation sheets, equipment and processes, tool prove-out, and updating of the latest technology. Of equal importance is that manufacturing engineering is the focal point for feedback of improvements identified by shop personnel, which result in product, tooling, and processing enhancements.

WHY COMPUTER-AIDED PROCESS PLANNING IS A KEY TO PRODUCTIVITY

In many companies, process planners use computer-aided process planning software. This computerized system can significantly increase process planning productivity by eliminating manual labor associated with writing or typing process plans. A computer-aided process planning system is characterized by its ability to quickly produce accurate process plans that are both consistent and free of error.

How the Computer Simplifies the Process Plan

The process plan is an important part of the manufacturing activity. It provides workers on the shop floor with vital information, such as sequence of operations to make a part; machine tools to use; perishable tooling requirements; jigs, fixtures and dies; cut parameters; part tolerances and dimensions; inspection criteria; and other items. Sound planning can reduce manufacturing costs, while poor planning can result in a higher incidence of scrap and rework, and longer delivery lead time.

Often, process plans consist of a substantial number of operations, which in certain instances are very lengthy, complex, and meticulous. As a result, companies usually attempt to reduce the complexity of the process plan by incorporating group technology (discussed earlier in Chapter 2) into the computer-aided process planning system. By developing "families of parts" on the CAPP system, considerable cost savings can accrue since it reduces significantly the amount of search time to develop the complete process plan, and also aids the shop floor in facilitating production.

By incorporating group technology into the system utilizing classification and coding, code numbers can be used for retrieval of significant manufacturing information. The storage, retrieval, and editing functions are associated with the process planning efficiencies. This system also allows planners to standardize process plans and develop preferred routings for part families. The retrieval is accomplished by code number or code number matrix. The system then generates the preferred routing—or optimal routing of the part—based on the stored experience of process planners and the tools available to do the job.

How Users Benefit from CAPP

Stored data is developed by the process planner who gathers all applicable process plans, and then prepares a detailed production flow analysis to arrive at families of common parts. Next, a matrix is constructed which consists of a vertical listing of the routing in descending order of each machine tool used in the processing of the part. The horizontal axis of this matrix consists of the respective part numbers of the family. This procedure results in the identification of common operations and the assignment of operation numbers wherever a part number and machine tool intersected. The vertical column on the matrix is used to record a machine class or process type. At this point, an experienced process planner will review each routing in detail to create a standard routing. This standard routing is then inputted into the CAPP system.

Cost information is another component of a computer-aided process planning system. Cost–calculation modules assist the process planner to realize the cost implications of a decision and also make it possible to predict costs. This can be done most effectively when an *automated time standard* (ATS) system is linked to the process planning system. Companies using this system have been able to predict the manufacturing cost of a finished part within 5 to 10 percent, based on a selected code number. Expressed another way, code numbers can be used as a means of arriving at a cost estimate.

A computer-aided process planning system (CAPP), linked with ATS and MRP (we'll discuss this further in Chapter 4), can be very effective in improving shop floor productivity.

HOW PICTORIAL PROCESS PLANNING ENHANCES CAPP

A process plan is often lengthy, with many steps and considerable detail accompanying each step. Interpreting lengthy instructions can be difficult, especially when most people are becoming more familiar with visual communications. Advancements now make it possible to link computer-aided process planning and computer graphics to create pictorial process plans which contain not only verbal instructions, but illustrations as well.

The text of the process plan is developed on an alphanumeric or graphics terminal, and then formatted on the terminal. For example, all text may appear on the left or right side, and the graphics information, such as setup, machining details, jigs, and fixtures can be created on the remaining part of the graphics screen. Furthermore, the tool path can be visualized to prove out the operation before going to the shop floor.

The finished process plan, with verbal and visual instructions, is then sent to the shop floor for machine operation use. It is becoming more prevalent to use common data bases for both graphics and process planning, thus improving the interface of these complementary activities.

PREPARED PLANS VERSUS ARTIFICIAL INTELLIGENCE: HOW PROCESS PLANNING SYSTEMS DIFFER

There are two types of computer-aided process planning systems: variant and generative. In a variant system, a process plan is retrieved by part numbers, code numbers, or some other key. If the retrieved plan is for exactly the same part entering into production, it can be used without change. If the part is in any way different, the retrieved plan can be varied to meet its specific requirements. While in a generative system, the part under consideration is described to the computer, and the computer produces the plan to manufacture the part exactly, based on manufacturing logic.

In the variant mode, standard or prepared process plans are stored for future use. By contrast, in a generative mode, manufacturing logic is stored to create process plans as needed. For the most part, the majority of systems currently in use is of the variant mode, although work is progressing to develop generative process planning systems.

Artificial intelligence is one approach to generative process planning. By means of artificial intelligence, the computer can look at those attributes of the part in question and use decision rules programmed into it to decide exactly how to manufacture the part. However, this advanced technology will only be applied through use of artificial intelligence after appropriate manufacturing rules have been established.

Another approach available more immediately is the use of multiple classification and coding systems and group technology as discussed previously. This approach begins with the development of a manufacturing data base that can be accessed by part number, nomenclature, or code number. Using group technology, optimized processes can be developed for each part attribute and part family. Design and manufacturing considerations can be optimized to the point where all factors can be considered thus providing a plan with the best design and manufacturing technology for the specific facility in which the part is to be produced.

Using a multiple classification and coding system, these optimized process plans can be fine-tuned to a high degree. Matrices can be developed for both parts and the tools available to manufacture them. Tight parameters can be developed for the specific facility to meet exacting needs. Essentially, by going from variant process planning to design and manufacturing optimizing using classification and coding group technology, you arrive gradually at a manufacturing rationale (logic) for individual part families. These rationales are geared specifically to an individual production facility.

The system can be refined continually to the point where it is possible to ask the computer to generate a plan for such things as a gear or a shaft. The system asks the user several questions to define the part further, and then searches through its matrices and generates a process plan which describes the optimal method to manufacture it. The group technology multiple classification and coding approach to generative process planning then leads to the rationale to generate process plans (and designs) automatically.

UNDERSTANDING THE PROBLEM-SOLVING NATURE OF MANUFACTURING ENGINEERING

Manufacturing engineering is responsible for developing methods, tooling requirements, machine processes, and material requirements for a broad range of manufacturing operations. This activity may extend from the development of small basic tooling and fixturing to large tools used for fabrication of major assemblies.

The following case-in-point examples illustrate the problem-solving methodology employed by manufacturing engineers.*

CASE-IN-POINT

SOLID RUBBER PRESS

Situation

The company's previous solid rubber press operations were performed by five presses in four separate work centers. The individual presses were restricted in capability to size and pressure application. Although the presses are rubber pad (Guerin type) equipment and essentially operate the same way as the newer type, the manner of loading the tools and material into the press differs. Three of the presses had a shuttle platen that required the operator to "wait out" the press cycle time. One press had a Lazy Susan arrangement feeding a dedicated press loader/operator. The last press had an automatic platen system that fed the press and returned the platens to stations for reloading. The press required a part-time operator to activate the press upon correcting a misalignment problem inherent in the design of the system. These operations constituted a major portion of the forming operations producing approximately 500,000 pieces per year.

Solution

After analyzing the existing method, manufacturing engineering decided to obtain a new solid rubber press with an integrated pallet handling system to replace the five existing presses. The state-of-the-art press with a larger platen size and variable pressure provides more flexible capability in the forming area. The individual pallet loading/unloading stations (six in all) now have the ability to select the required pressure for their work by a terminal input. The press senses the dedicated pallet assigned to that station and applies the appropriate pressure. The ability to vary pressure within the 20,000 PSI limitation enables manufacturing to selectively increase pressure or group parts to improve part definition thereby reducing the need for further hand refinement.

Results

The solid rubber press acquisition benefited the company by:

1. Avoiding the cost required to rebuild and relocate five antiquated presses.
2. Reducing production costs in forming parts on the automatic press.
3. Providing increased capacity and capability in the formation area.
4. Providing back-up capability for another new type press.

*Permission granted by Grumman Corporation.

CASE-IN-POINT

SUNSTRAND 4-AXIS NC PARTSMAKER

Situation

Previously, extrusion stock was drawn and sawed to an overall length (total length required). Extrusions were then transported to a queue area adjacent to the router room where a set of templates were joined. These templates represented the part at various stages of profiling. First, the extrusion was sawed to size, legs ripped, saw mitred or chamfered. Each cut required a new setup of stops and guides on the table saw. Holes were then either drilled via drill press or punched one piece at a time. If the part had more than straight cuts to it, the Marwin Miller was employed. The particular model used in the router room had manual turn wheels to control the movements of the cutter. An experienced operator had to control the movement of a stylus, tracing the part contours. Each side required a separate setup. To blend the edges where legs met, each piece required a flush milling after profiling.

Solution

To automate this operation, manufacturing engineering obtained a White Sunstrand 4-axis NC Partsmaker, which had been designed specifically for fabricating detail parts for the aerospace industry.

Although initially limiting the machine's workload of parts to under 15 inches in length fabricated from extruded stock, as the company gained additional knowledge, the complexity and lengths of parts increased considerably. Experience with feeds and speeds has allowed stock shifts to be made with no compromise of tolerance requirements.

Unique jaw design developed by methods people has extended its use to 39-inch parts, and even a 130-inch test piece (with twelve indexes of the workpiece). This new jaw design also has allowed production to run a wide range of extrusion types and sizes with minimal change of jaws and inserts.

Results

The Sunstrand 4-Axis NC Partsmaker is a numerically controlled computer machining center that has provided the company with the capability to produce complete parts from extruded shapes or bar stock from a variety of materials suitable for high speeds, low torque milling, and drilling operations.

As a result of this advancement in manufacturing automation, the company has substantially reduced setup and operator labor costs, gained machining flexibility, improved job scheduling performance, and expanded its range of part manufacturing capabilities.

AN INNOVATION TO WATCH: MACHINING-TO-FUNCTION

Manufacturing engineers are currently involved in the development of a machining approach which considers the part function rather than conventional manufacturing drawing tolerances. This technology consists of an optically based, microcomputer-controller metrology system that monitors the orifice area as the part is being machined, compares the progress of the machining against a standard, and adjusts the tool path accordingly. While this operation is proceeding, the computer creates a final inspection record of the part.

Why This Technique Saves Time

This technique reduces machining and inspection time while obtaining an acceptable machined part. Processing time is increased by using a high-speed spindle and coated carbide tools, and by inspecting the machine's edges with an optical gauge while the part is still in the machine.

The system reduces the number of rejected parts by monitoring the coordinates of the edge produced by the machine tool and correcting the tool path when necessary. Adjustments are made automatically when the trend of the monitored edge coordinates indicate that the cut is drifting toward the tolerance limits. Another approach to this technique known as *adoptive control* prevents damage to the part from unwanted changes in cutting force by altering the cutting tool feed rate appropriately.

How Well Does It Perform?

The concept of machining-to-function is based on developing all necessary control features so that accurate dimensional measurements can be made while the part is being machined. The overriding criterion for the acceptance of a part should be *how well it performs its intended function.* This concept can have a considerable impact on current manufacturing practices that decide machined part acceptability based on tolerance criteria rather than function. As a result of this approach, it is possible to relax tightly controlled tolerance dimensions without compromising the function of the part. This change can be made provided that the sum of the total orifice flow area is within the area tolerance at any point along the part path. Thus, if one of the orifices is out of tolerance because a tool broke or some other event occurred unexpectedly, the dimensions of the next orifice would be adjusted to bring the total orifice area within tolerance. As you can readily see, application of the machine-to-function technique can result in a considerable reduction in the number of rejected parts.

Four Conditions to Avoid

There are many common concerns intrinsic in any machining operation that the manufacturing engineer must recognize and overcome. Four of these are:

- The distance between the spindle center line and the instantaneous line on the part generated by the machining operation is neither constant nor known with sufficient precision.
- Tools wear out and have to be reground or replaced by new tools.
- New tools may differ in diameter.
- Tools can deflect in operation.

Any of these conditions may occur at one time or another, thereby causing uncertainty in determining the distance between the spindle and the cut. The hardness of the material also can be a factor because it affects tool deflection during the machining operation. Together, these conditions have a bearing on the tolerance that can be held on certain types of machine tools and production processors. Therefore, you should recognize that some modifications will be needed on any standard machine tool used for an in-process tool path control application to allow for these various contingencies.

Why Use of On-Machine Inspection Is Expanding

The in-process gauge system must be able to control the tool path and make the final inspection of the part. The critical requirements for the gauge are speed and accuracy. There are a number of different types of gauges that can be used, such as tactile sensor, capacitance principle, and optical principle.

Based on progress-to-date, the concept of machining-to-function is a cost-saving approach to manufacturing that can be applied to a number of different parts. In most cases, the manufacturing process for these parts are, to some extent, dictated by the requirements of the instrumentation that drives the system.

Optically based instrumentation has the speed, accuracy, and reliability that is necessary for in-process control of machining. The technical and economic feasibility for closed loop machine control electro-optical based systems appears to be expanding. As the application of electro-optical sensors to machine tools becomes more developed, "on-machine" inspection will become feasible for more types of parts. In addition to reducing manufacturing time and cost, on-machine inspection has the potential for substantially increasing product quality through 100 percent inspection, especially when combined with in-process control of machining.

Machines that use electro-optical control systems must be very stiff. They must be free from structural vibrations and stick-slip motions of the slideways that could interfere with taking accurate measurements. As these machine-by-function systems begin to expand into the manufacturing environment, they will create change in maintenance, operations, design of machines, and other areas, all of which could result in a more productive, higher quality machining capability.

TECHNIQUES THAT ARE CHANGING METHODS AND PROCESSES FOR DIFFERENT TYPES OF MATERIALS

Manufacturing engineers also are involved in the development of material specifications and material properties as well as the means for testing these materials.

For example, a company that required fabrication of large formed parts, developed a super plastic form and diffusion bonding process which combined materials of unique characteristics with customized tooling. This was accomplished by laying two sheets of titanium discretely on a welding material, then putting the two sheets in a die and pumping air and gas between the two sheets. With the application of high temperature and high pressure, materials elasticity formed itself into a new titanium structure shape.

In a number of companies, manufacturing engineers are developing new methods and processes using composite materials. For example, Grumman Aerospace Corporation recognized the need for more fully automated fabrication of gently contoured composite structures from unidirectional tape. This led to the development and implementation of a facility dedicated to composites production (Integrated Laminating Center shown in figure 3-1). One application produced by the center is the 56-ply covers of the F-14A horizontal stabilizer from three inch-wide boron-graphic/epoxy demonstration articles and for the inboard end of the B-1 horizontal stabilizer.

Grumman also has developed mechanized equipment to fabricate severely contoured, integrally stiffened, complex structures for high-performance aircraft such as the ATS. The prime material is graphite/epoxy. This automated integrated manufacturing system (AIMS) consists of three modules: a contour ply handler, a broadgoods prepreg dispenser, and a translaminar stitcher (figure 3-2).

FIGURE 3–1 Integrated Laminating Center (Courtesy of Grumman Corporation).

Contour Ply Handler. The *contour ply handler* automatically locates, drapes, and forms larger-area plies to highly contoured tools at a rate of 200 square inches per second. This module interfaces with the ILC flip table so that it can automatically receive large-area plies prepared on deformable carrier film. Plies are formed to contour by rotary brush action in a programmable cycle that controls tool rise, split-table separation, and brush arbor motion. Brush rotational speed, bristle density and contour, and programmable cycle configuring provide flexibility to meet specific requirements. Use of this automated system eliminates manual tailoring of strips to contour and permits direct use of automatically prepared flat plies for contoured configurations.

Broadgoods Prepreg Dispenser. Another application developed by Grumman manufacturing engineers is the Broadgoods Prepreg Dispenser.

The Broadgoods Prepreg Dispenser automatically dispenses and trims either of two selected prepregs (kevlar, fiberglass, or graphite/epoxy) in widths of up to 80 inches. The dispenser interfaces with the ILC via the transfer gantry, which pulls out lengths of material, and the laser trimming station, which cuts off the dispensed lengths. This module features quick-lead chucks, a roll-away freezer storage capability, and optional dewrinkler roll action.

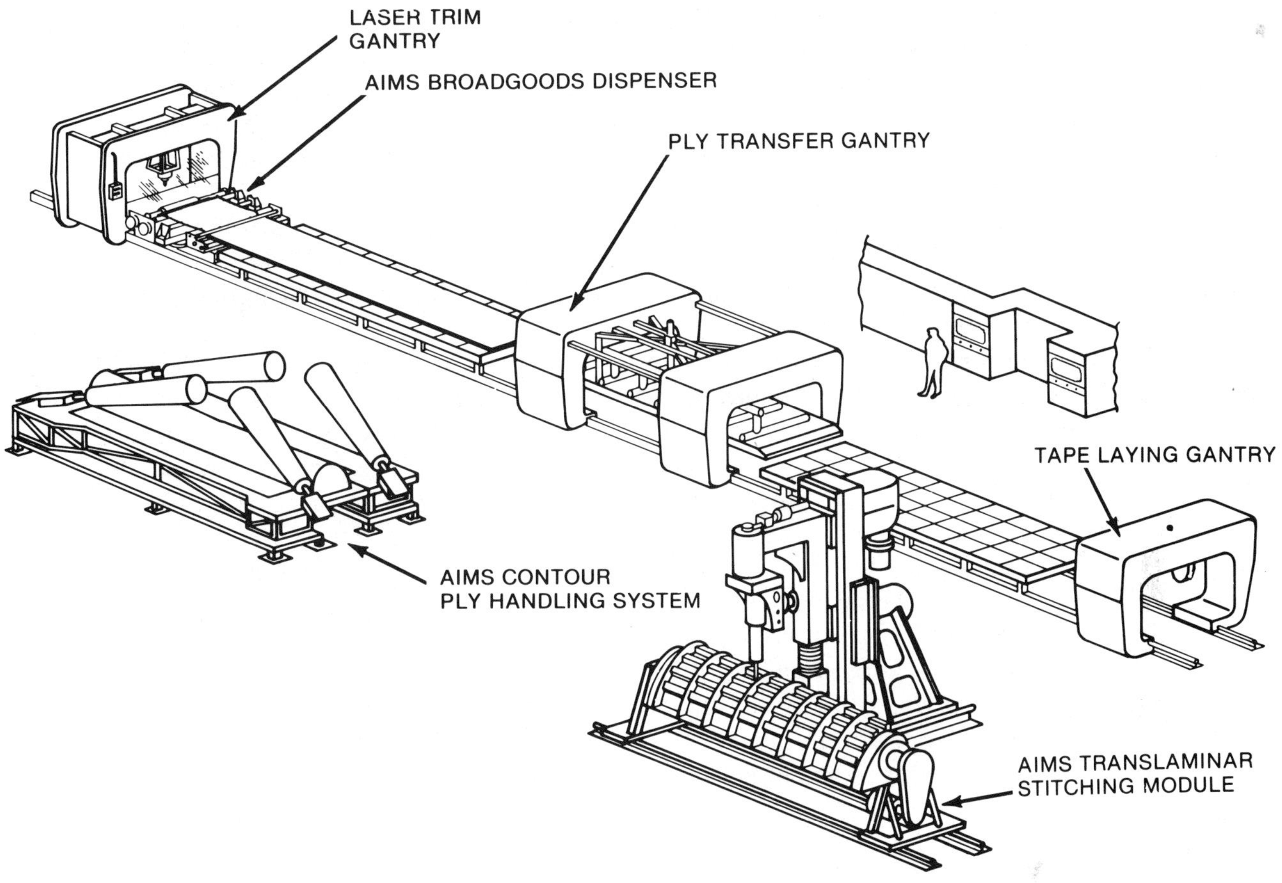

FIGURE 3–2 Automated Integrated Manufacturing System (AIMS) (Courtesy of Grumman Corporation).

Use of nested computerized geometry data provides efficient laser trimming of contoured details with appropriate flange foldouts and ply orientation control. The dispenser can produce more than 300 graphite/epoxy, contoured frame-station plies with computer-generated contours, flanges and hat stiffener foldouts, together with 150 splice strips and fiberglass edge buildups. The laser trimming station also prepares bleeder and glass cloth details, using the same nested detailed numerical control data.

Translaminar Stitcher. Another advanced method developed by Grumman manufacturing engineers is the Translaminar Stitcher. This self-digitizing, multi-axis module can stitch along straight, bowed, twisted, and highly contoured (circumferential) paths. Its computer interpolates between selected point inputs and inserts the required stitch pitch. Using Kevlar thread with selected pitch, edge-distance and gauge, translaminar stitching assures the out-of-plane load-carrying capability of integrally cured flange-to-skin joints. For example, over 1600 mechanical fasteners were eliminated by translaminar stitching—the first accomplishment of this kind in the aerospace industry.

ADVANTAGES OF USING ULTRASONIC WELDING

Ultrasonic welding is an improved process developed by manufacturing engineers and aimed at applications in joining materials and assembly. This technology is in use in aerospace, automotive, and general manufacturing applications.

Ultrasonics are vibrational waves at frequencies of 10 to 75 KHz. Ultrasonic welding uses this high-frequency energy to disrupt and plasticize the surfaces being bonded. This causes interatomic diffusion of the metals at temperatures below their melting points.

Ultrasonic weld-bonding uses similar techniques and adds an adhesive bonding material that requires oven curing between the parts to be joined for added strength. Ultrasonic weld-bonding can utilize either film or paste adhesives.

The ultrasonic welding process can replace rivets and fasteners, resulting in considerable cost savings and reduction in weight of the end product. It can further reduce weight by using a thinner gauge metal since the welded panel would not have to be designed to support holes drilled into the metal for fasteners.

Further fine-tuning improvements in this process have resulted in changes in weld-tip geometry, a coaxial ring clamping system, and welding controls that regulate power. Clamping pressure and time all lessen the chances for weld distortion, especially on aluminum components. These refinements have enabled companies using this process to isolate the generation of the ultrasonically welded spot and not affect the surrounding structure.

Other features of this process that can be helpful in a production environment include automation of the ultrasonic welding machine to a single pedal operation, methods to prevent the welding tip from sticking to the workpiece, longer welding periods, and features for easy weld tip and anvil changes for various shapes and sizes. Control of these parameters improves the consistency of ultrasonic welding applications.

The potential for this process may include all types of alloys and material combinations, such as aluminum-to-aluminum weld-bonds, steel, titanium, plastics, and combining ultrasonic components with superplastic formed aluminum.

SEVEN REASONS THE GRUMMAN CORPORATION USES AUTOMATED INTEGRATED MANUFACTURING SYSTEM

In designing AIMS, Grumman Manufacturing engineers included many advantages into its operation, such as:

- Computerized design/fabrication data base
- High-speed mechanized fabrication of large-area, high-contoured plies
- Computer-controlled, multi-axis translaminar stitching
- Automated dispensing and nested trimming of details from broadgoods
- Mechanized layup of prepared details
- Innovative tooling materials and methods
- Application of conventional part-handling techniques to preformed composite details

CAD/CAM plays an important part in AIMS operations because it provides:

- Computerized air-passage contour data bases
- Computerized detail design of flat ply patterns, templates, and curing tools
- Computer-generated nesting pattern for automatic laser trimming

To fabricate the integrally stiffened component required for the program, innovative tooling concepts had to be developed in several areas by manufacturing engineers. These included integration of the final curing tool as a moving mandrel in the ply handler, an orthogonally slotted stitching rack, CAD/CAM-generated templates, and rapidly installed locating tools for skin buildups. Also, Teflon mandrels were made to be re-usable for bowed/twisted 100 inch long, closed stiffener sections.

THE INTEGRATED ROBOTIC CELL THAT SAVED 7,000 WORK HOURS

Manufacturing engineers in many companies are heavily involved in robotic development. One application of this advanced technology is the integrated robotic cell (IRC). A typical robotic integrated cell consists of a:

- Shear clip dispenser
- Liquid shim dispenser
- Drill/countersink unit
- Hole-bore probe
- Fastener installer
- Fastener head shaver

CASE-IN-POINT

INTEGRATED ROBOTIC CELL

Situation

The Grumman Corporation realized that its light machining of the complex-contour sheetmetal parts for aircraft skins, panels, fairings, and doors were fabricated using tedious labor-intensive manual operations. In manual routing and drilling of formed sheetmetal parts, both a base holding fixture and a hand routing jig were used to support the parts and establish the geometry/location of edges and holes (figure 3-3).

More than 3500 sets of jigs and fixtures were used to produce sheetmetal detail parts for its various aircraft. This operation required more than 3000 hours per month on parts ranging in size from less than 1 square foot to 18 square feet.

Corporate management decided to install a robotic system to drive direct labor costs down, improve quality, and remain competitive with other firms in the industry that were already using robots or contemplating their use.

Solution

After a detailed analysis of machining processes and methods by manufacturing engineers, the company installed an IRC which is centered around an ASEA Model IRB-60 with 60-kg capacity, that performs trimming and drilling of complex parts (figure 3-4).

This determination was made based on finding a robot capable both of highly efficient and economical operation. The robot selected also had to perform trimming and drilling without the aid of templates. This requirement dictated that the candidate robot would have to be accurate to within ± 0.030 inches under dynamic loads. Based on these specifications and an estimated end-of-arm tooling weight, a minimum accuracy of ± 0.030 inches at a 100-pound load was required. The IRB-60 was able to meet these specified accuracy and load requirements. Six axes of programmable motion are available, and all drives are electric for clean and quiet operation.

Moreover, the close-tolerance unit permits drill and trim operations without guide templates and bushings. Additionally, the ability to mount the robot on precision tracks adds

FIGURE 3-3 Manual Drill/Trim Tooling for Contoured Aluminum Parts (Courtesy of Grumman Corporation).

FIGURE 3-4 Integrated Robotic Cell (IRC) (Courtesy of Grumman Corporation).

the range necessary to allow continual operation of the IRB-60 at other stations while parts or fixtures are being set up.

In addition to the robot, the IRC comprises five other basic components or modules, including linear track, workstations, tool-exchange module, part tooling, and cell controller. The components are integrated under the supervisory control of the cell host computer.

The track system has been designed to provide 20 feet of linear motion under program control. This provides the robot with a tremendously increased work envelope.

Each workstation was designed to provide a modular precision platform for location and clamp-down of part-holding fixtures. Its 90° pivot capability provides a convenient platform for manual part loading and unloading. Most important, the workstation establishes mechanical coordination between part fixturing and the robot coordinates.

Drilling and routing are automated to eliminate the manual intervention required to change power units and cutters. To accomplish this, an interface was designed for the end of the manipulator to allow the robot to grasp the tool required for the operation.

A minicomputer-based, custom-designed control system with supporting peripheral equipment supervises cell functions and also provides communication with factory-center processors. Concurrent coordination, scheduling, and communication of the cell's control task are handled by an RSX-11M, Digital Equipment PDP-11 with a real-time, multiprogramming, event-driven executive.

Dual 10-megabyte disk drives provide on-line mass program storage and local database capability. A floating point processor speeds mathematical operations.

A serial communication link provides the computer interface to the ASEA robot controller. Robot-program transmission to and from the computer, as well as robot-status monitoring are provided by this link.

A CRT terminal is used during program development for file creation, editing, and system commands. During trim and drill operations, the terminal serves as the primary operator communication/interface device.

A line printer is used to generate program listings and status reports, and a sufficient number of general-purpose digital input/output ports are included to facilitate discrete control–function outputs and monitoring operations.

For production operations, a program for each part must first be taught to the robot and then stored on the controller's disk system. Robotic fabrication of parts is accompanied by recalling the appropriate part programs. The parts are then automatically drilled and/or routed under cell control. Any part taught to the system can be trimmed robotically and drilled on any of the four workstations.

The IRC system operates in the following manner (figure 3-5):

1. The host computer has all necessary part programs contained on a disk.
2. Fixtures for parts are located at workstations, with associated parts to be drilled and routed.
3. The robot-caddy is loaded with necessary cutting tools and equipment to accomplish planned jobs.
4. The operator notifies the host computer of the identity of parts planned for fabrication at various workstations in the cell.
5. The operator loads a part on a fixture and presses a button. The trunnion workstation pitches 90° from the horizontal setup (or "load") position to the vertical work (or "ready") position. When the fixture is vertical, the host computer is signaled that the station is ready for the robot.
6. The robot with tool required to perform the task, moves along tracks to the ready station at which the computer has stored the part geometry for drill and trim operations. The host computer sends this data to the robot microprocessor, and the robot

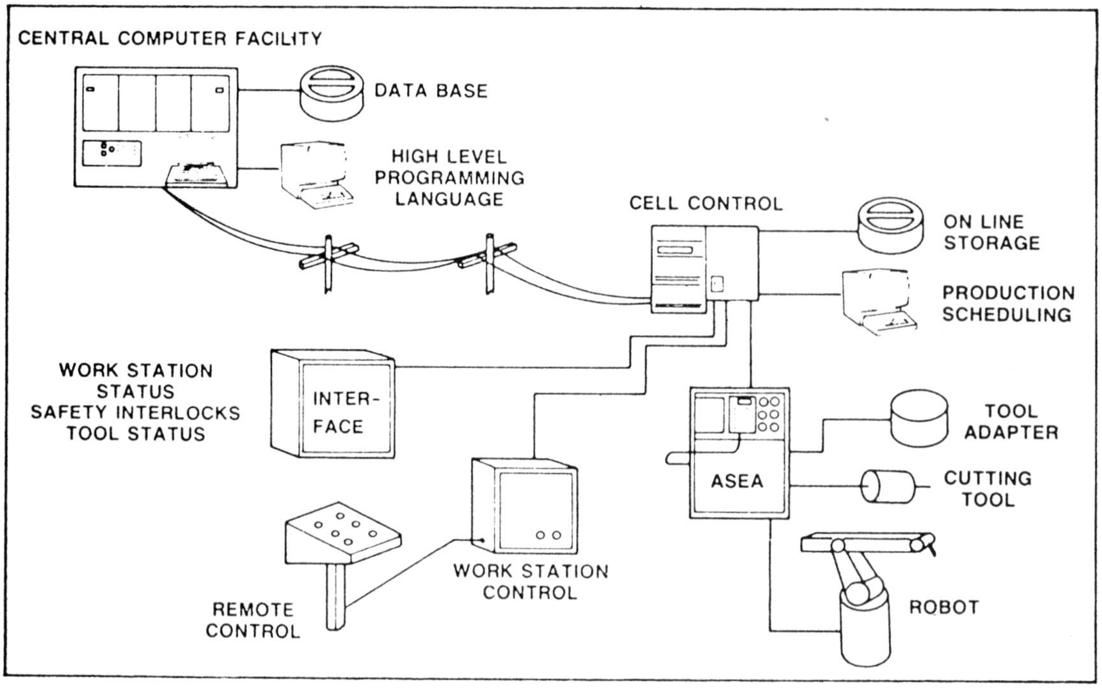

FIGURE 3-5 Integrated Robotic Cell (IRC) System Structure (Courtesy of Grumman Corporation).

automatically picks up the necessary tools and performs programmed drill and trim operations.

7. When the robot has completed its task, it notifies the host computer. The computer then sends the robot to the next ready workstation and rotates the completed workstation to a horizontal position to enable the operator to unload the finished part and reload the next part.
8. When the last part is completed at a particular workstation, a new fixture is set up, and the host computer is notified of a part change at the station.

Results

Based on the experience acquired by Grumman since the IRC has been in operation, the company estimates that it saves approximately 7000 direct worker-hours per year.

The fixturing has been designed for maximum flexibility, and the robot can be reprogrammed easily to handle virtually any part within its machining capability. The ability of the robot to exchange one powered tool for another enables ease of programming.

When all these features are considered, Grumman's IRC is one of the most flexible machining cells in the aerospace industry.

COST-SAVING INNOVATIONS IN PRODUCIBILITY ENGINEERING AND PLANNING

In addition to advanced materials development and fabrication of tools, manufacturing engineers also engage in producibility engineering. This involves the application of engineering and manufacturing expertise to hardware design and manufacturing planning, which results in the satisfaction of quality and function requirements at least cost. The purpose is to develop alternative approaches that achieve the optimum method of manufacturing. This methodology evaluates a product as it becomes more advanced in the design engineering stage before manufacturing and establishes a baseline on what is going to be built, its target cost, and performance criteria. This then becomes the final evaluation of the configuration to be produced. The tasks and procedural steps involved in producibility engineering and planning are listed below.

Task	Procedure
Evaluate concept–identify concerns	Review conceptual sketches, applicable specifications, and customer requirements
Prepare producibility program plan	Follow task outline for producibility plan defined by customer
Establish evaluation criteria	Cost/schedule impact to assure compliance with customer requirements
Establish program baseline	Establish engineering baseline configuration; establish baseline manufacturing plan; establish baseline cost
Evaluate configuration	Review design alternatives; prepare trade-off studies; participate in design and mock-up reviews

There are three phases in a producibility engineering and planning program:

- Phase I—Preliminary design engineering
- Phase II—Detail design
- Phase III—Drawing release/production development

Phase I: The Best Place to Avoid Unnecessary Expense

Phase I involves design engineering concerns in regard to the preliminary design proposal. Producibility influences this phase by creating detailed assembly flows, automation concepts, and applicable manufacturing approaches. The reason for becoming involved with a product in the preliminary design phase is that experience has shown that the greatest cost avoidance potential is in the front end of the program where expenditures amount to only about 5–10 percent of total program cost. Once in production, the major portion of nonrecurring expenditures, engineering configuration, and manufacturing tooling has already been spent and it then becomes very expensive to make any adjustments without affecting program profitability. The rationale used for trade-offs of various approaches is to try to satisfy both product performance requirements and cost/schedule requirements.

On a typical product, the first step is to break down the product into various major assembly components from deliverable end items, and then further break these down to identify how to manufacture detailed parts that go into subassemblies. Information developed through manufacturing planning, product producibility planning, and design engineering results in establishing a sound baseline for use in tracking costs. As the design becomes more mature various cost drivers are identified. A typical example is the assembly cost drivers shown in figure 3-6. In this illustration, fastening accounts for 35 percent of the cost, and drilling 45 percent. Hence, these two operations must be analyzed for other processes, methods, or equipment such as automating the drilling operation.

Next, a producibility plan is developed for the baseline configuration so that future trade-offs can be assessed as the design matures, with particular attention given to the high cost of such operations as fastening shown on the pie graph on figure 3-6. This analysis enables company personnel to maintain a tracking system on specific items against the cost baseline.

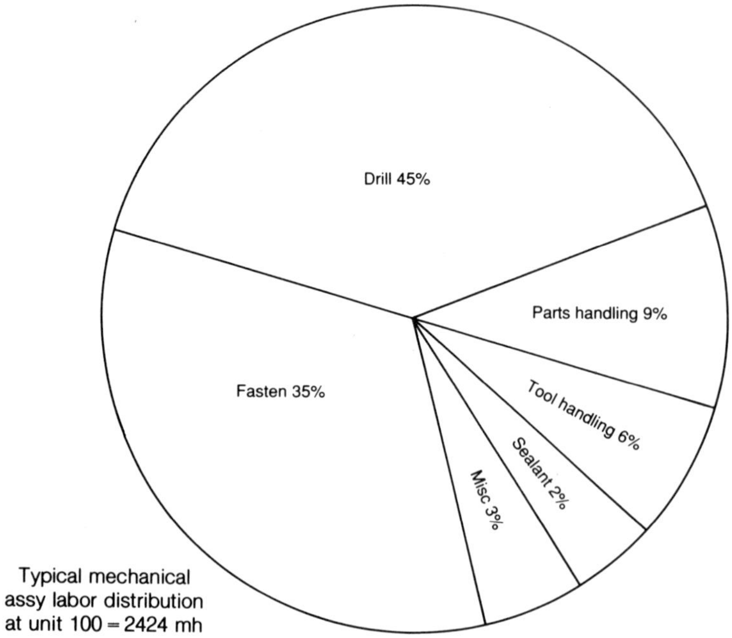

FIGURE 3-6 Identify Assembly Cost Drivers (Courtesy of Grumman Corporation).

In addition, the company can establish a *producibility priority guide* which breaks down the product into areas that will receive the most concentrated effort from an automation standpoint. As noted in figure 3-4, to reduce drilling costs, the company must automate assembly drilling by employing a robot cell for drilling, fastening, and as an unload station, which automatically cycles back to reload again. Thus, manufacturing engineers would have accomplished one step in their primary Phase I (preliminary design) tasks: shaking out high cost areas while satisfying product performance criteria.

Phase II: A Ten-Step Path Toward Automation

Phase II introduces detail design blueprints, specifications, and other documentation into the manufacturing environment. This requires expertise of methods engineers engaged in advanced tool design, development of operation sheets, identifying equipment processes, and other concerns of a more definitive nature. For example, method engineers were able to simplify the method of producing an assembly without affecting its function, and thereby eliminate fasteners, machine parts, details, and an assembly fixture.

The sequence of events involved in automation implementation, resulting in optimum product cost/performance, consist of the following:

- Requirements analysis
- Conceptual design
- Systems/process analysis
- Proof of concept
- ROI analysis
- Acquisition/development
- User training
- Technical/cost demonstration

- Production implementation
- Validation

Requirements Analysis, which is performed to justify conceptual design is the first step in implementation. In the previous example, process analysis not only involved development of a robot drilling cell, but also integration of it to associated systems. Validation and proof of concept involves queries such as: How risky is it? Is the process outdated? Can the automated system be bought off the shelf? If not, can the part be purchased and the remainder done in-house? After the desired approach is selected, ROI analysis is made to cost justify the capital investment (refer to Chapter 1). Upon conceptual design acceptance, acquisition/development proceeds as planned. This involves the remaining steps of: user training of the new technology; technical/cost demonstration; production implementation; and validation of achieved results.

A typical automation implementation evaluation was illustrated in figures 3-3 and 3-4. The after-form sheetmetal trimming operation was a manual operation, hazardous, and unfriendly to the worker. The robotic cell (workstation) that replaced this manual operation, is a fully automated integrated system that also is user friendly in terms of noise and other worker discomforts.

Phase III: The Four-Step Transition from Planning to Production

The last phase of this cycle consists of release and production development. As discussed in Chapter 2, this step involves computer-aided design and computer-aided manufacturing, encompassing three-dimensional math modeling, which is used to create NC tapes for programming machine tools to produce various component parts. A typical sequence of events performed in this activity and comparative cost savings of a Wing Skin Mold Form (MOF) are shown in figure 3-7.

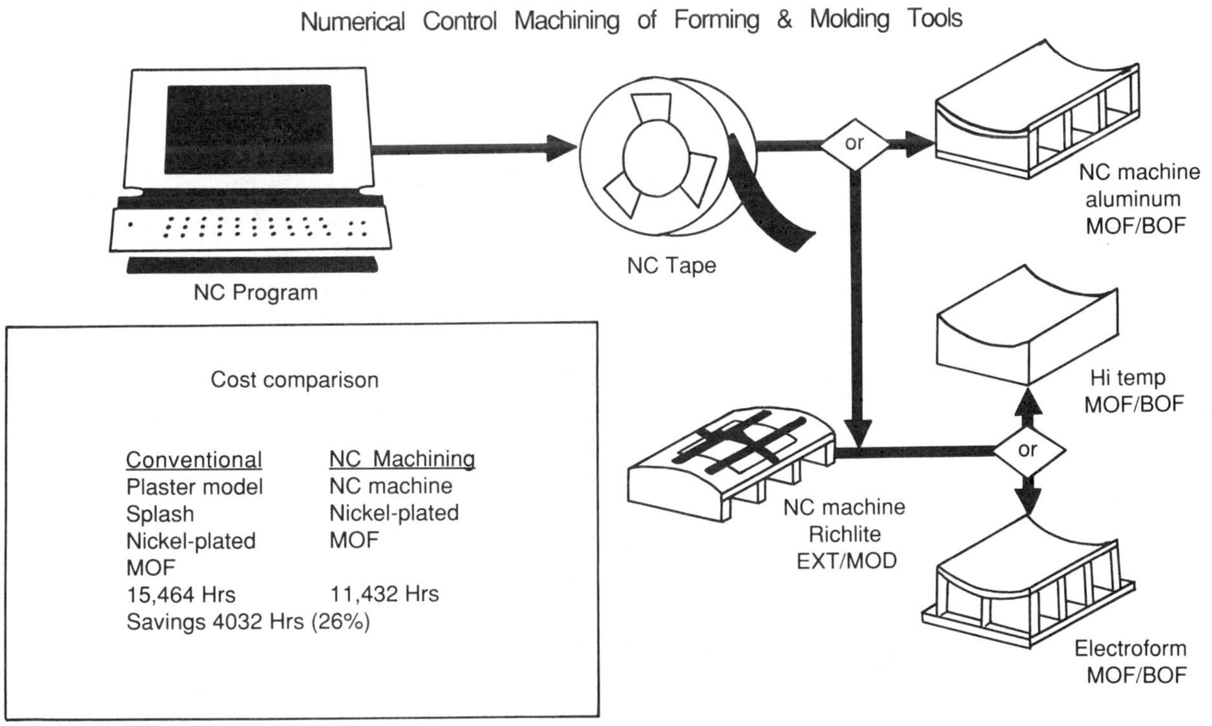

FIGURE 3-7 Computer-Aided Manufacturing (Courtesy of Grumman Corporation).

In addition, CAD/CAM can automatically create all drawings and documentation using computerized software.

CASE-IN-POINT

TRUMPF CNC DRILL AND ROUTE MACHINE

Situation

Previously, sheet stock was drawn and sheared to a rough dimension larger than the actual part. These blanks were then transported to a queue area adjacent to the router room where the routing fixture and drill template were joined. Utilizing drills with plywood table tops, the template was placed on top of a stack of blanks and nailed down. Each hole was drilled as defined by the template. When complete, the nails were removed, the blanks rejoined with the routing tool and moved into a routing area. Part configuration was created by pushing and pulling the router bit through the stacked material tracing the pattern as defined by a path in the router board or the template edge. When routing was achieved by tracing the template, a radial arm router was used and the material and template were once again nailed into plywood.

Each part had at least one tool that had to be maintained and stored. In the event of a breakdown, new tools had to be fabricated.

Solution

After manufacturing engineers, analyzed existing methods, the company decided to obtain a Trumpf CNC Drill and Router machine to perform this machining operation. The company used the following determining factors in its vendor selection process:

- Material types 2024 and 7075 aluminum
- Part sizes
- Material thickness (nest no greater than .500″)

In the new process, prior to machining, sheets are loaded onto an aluminum base plate and fixed with clamps or plastic straps. This preparation takes place away from the machine.

Following the loading of the base plate into the coordinate guide workpiece clamps, the sheet metal stack is attached to the base plate with a few rivets. The clamps are removed and the machining cycle begins.

Maximum savings are achieved by nesting several different parts on the stack. Each individual part, however, is riveted to the base plate prior to drilling and routing operations. The conventional method of attachment of screws was eliminated. An automated riveting system is used. After rivet holes are drilled through the stack and base plate, rivets are automatically inserted and expanded.

The cycle time for drilling small holes averages about one second, depending on rivet size. Up to five different drills can be used without any tool change. Since all of the drilling spindles run continuously at the selected speeds, no time is lost when the drill is programmed to operate. During each drilling operation the sheetmetal stack is held together tightly by a hydraulic presser foot. The result is burr-free round holes.

The main machining operation is routing. When completed, parts and scrap remain fixed to the base plate with rivets. These are then removed in a secondary operation.

In the last operation, the rivets are drilled out by the same drill that placed the initial hole for the rivet. After the finished parts are removed, the empty base plate is rinsed and prepared for the next stack of material.

Results

Results of post-operative audit revealed a 50 percent savings in labor, and 28 percent material savings when compared with conventional fabrication methods. This indicated that the company's CNC stack routing method was an economical alternative to a previous, more conventional approach.

As the program progressed, other benefits were achieved. For example, as the company gained additional experience with the Trumpf CNC Drill and Router Machine, and software was modified, it converted flat patterns into the NC data base. Other software developments enabled the company, after engineering the detail parts, to obtain tooling information without prototyping or hard tooling.

In addition, the routing cell includes applicable deburring and part identification areas to complete the profiling process.

Compliance Verification: An Important Checkpoint

This step in the sequence of events involving the manufacturing engineering technical plan consists of compliance verification through the control mechanism of producibility engineering sign-off. This step is employed to verify adherence to agreements made with engineering on how the part is to be fabricated and what processes are going to be used to accommodate engineering drawings.

As discussed previously in this chapter, computer-aided process planning is an on-line/real-time system that determines how to methodize the part to be fabricated. When engineering drawings are received, manufacturing engineering establishes a part code, and inputs its pertinent features on a CRT terminal. The computer software program has stored manufacturing operations and alternative approaches to producing the part. Manufacturing engineering then either accepts the logic as a function of the program, or edits the operational sequence to arrive at a finalized selected production method. This approved version is then displayed automatically, stored on disk, and available for hard-copy generation, as required.

Another program implemented by manufacturing engineers in collaboration with design engineering and production people, is group technology, discussed previously in Chapter 2. Figure 3-8 is an illustration of classification codes developed by part shape employing a computer-aided process planning system. The computer program then "asks" the manufacturing engineer various questions as the manufacturing engineer progresses through the decision tree (shown in figure 3-9) starting with the part shape and progressing down to flat or bent; straight bend or curved bend; parallel or nonparallel; flanges, same or different direction; holes or no holes; and hole tolerance, greater or less than .005 inches.

How the Standards and Routing System Contributes to Production

After the operational sequence is determined, the computer-aided process planning (CAPP) generated classification code is automatically computer interfaced to a *standards and routing system* (STAR). STAR is an on-line real-time system that stores manufacturing operation descriptions and instructions that are extracted to create a computer-printed routing. STAR applies the operational sequence, plant departments, work centers, operation codes, workstations, machine number, tool code, operation text, and quantity to be set up and run. Once the parts are classified, generation of the operation/route sheet can be virtually instantaneous in a real-time mode.

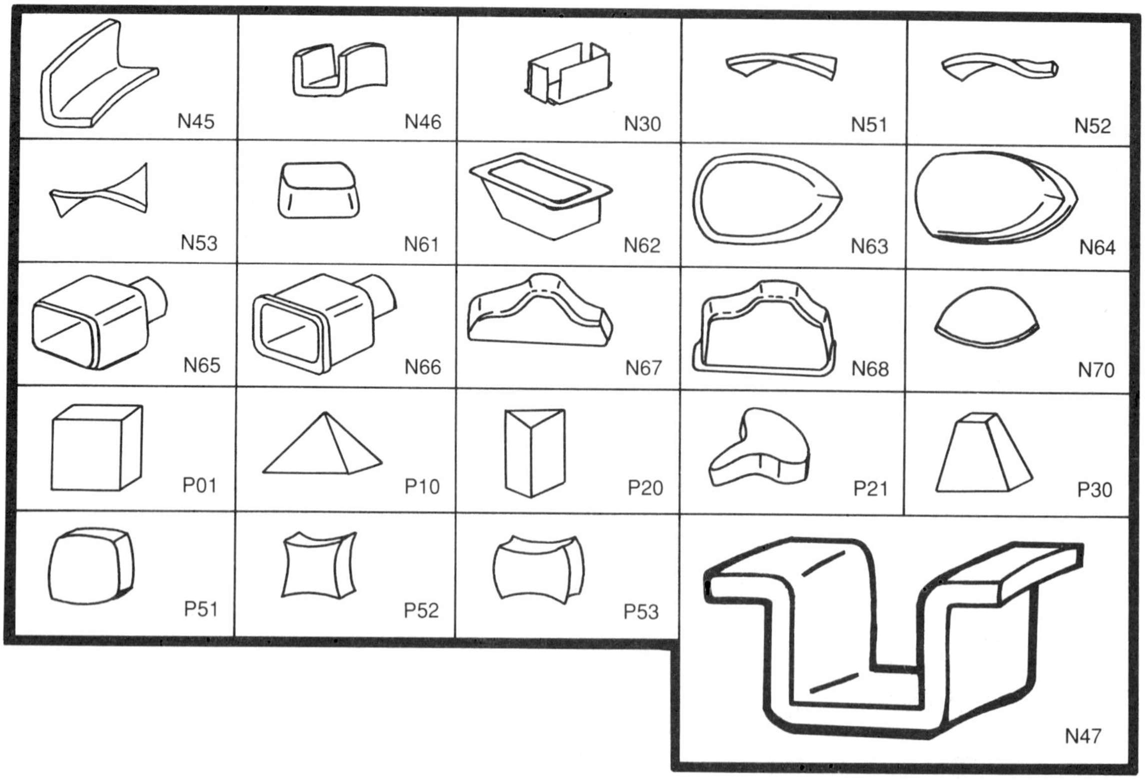

FIGURE 3–8 Part Classification Codes (Courtesy of Grumman Corporation).

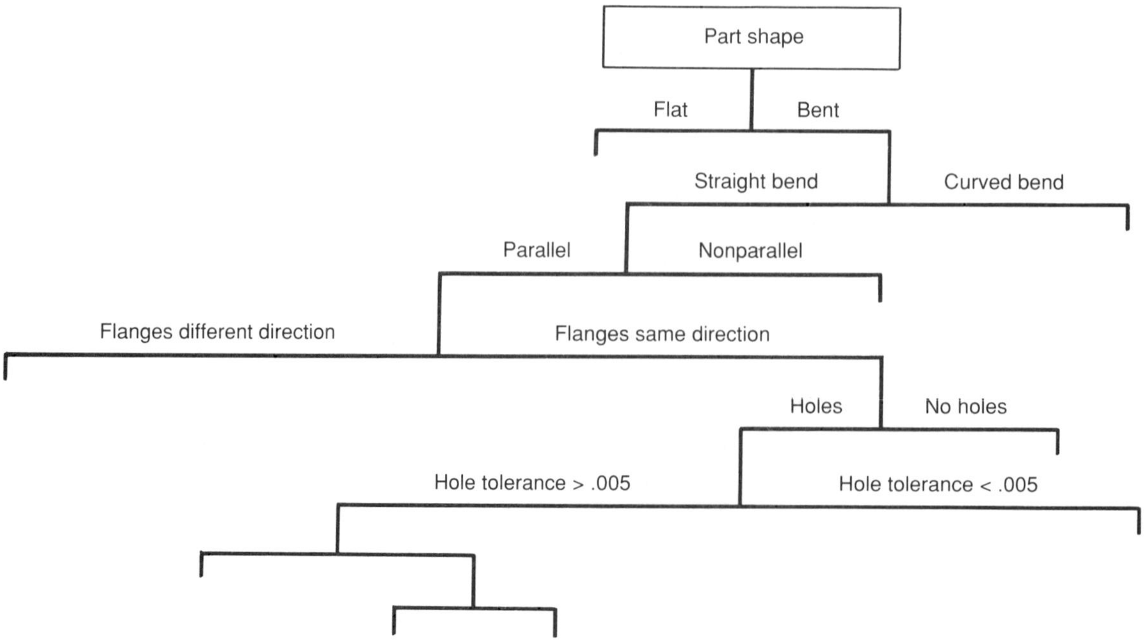

FIGURE 3–9 Decision Tree (Courtesy of Grumman Corporation).

Five Considerations in a Producibility Evaluation Criteria Review

As the Phase III sequence of events continues, manufacturing engineering engages in a producibility evaluation criteria review. This evaluation consists of:

- Technical assessment
- Cost factors
- Return on investment
- Customer impact
- Schedule

Producibility Improvement Protocol System: A Method of Evaluating Changes

Manufacturing engineers review the tool design and tool fabrication processes and material specifications on the developed units. In addition, manufacturing engineers identify as a function, improvements that were made which may further benefit the company in economic and productivity terms. This is accomplished through the *producibility improvement proposal system* (PIP). PIP makes a technical assessment of any manufacturing changes, evaluates cost, returns on investments, customer impact, and scheduling. As a result, a program review team evaluates any changes that appear necessary for technical cost and customer impact. Changes, if approved, are implemented in an orderly fashion.

Tips for Successful Automation: Why Adhere to Sequence?

A good rule to follow is that manufacturing engineering should take great care not to impose on the production line any type of changes through imposition, or in a haphazard fashion, that forces the production line to start to work out of sequence rather than the agreed upon sequence. Estimates of forcing changes out of sequence run as much as four times smooth introduction.

In summary, this complete three-phase cycle follows manufacturing engineering's area of activity from design engineering output through production. It shows how manufacturing engineers use various skills and resources to make this transition as smooth as possible by:

- Identifying future needs (new program to put in place)
- Automation (robotics technology)
- Advanced processes
- Solving manufacturing problems (feedback from production)
- Using orderly discipline to approve the product from an economic standpoint—design, development, economics

A NEW COST-BENEFIT ANALYSIS TECHNIQUE THAT SUPPLEMENTS TRADITIONAL ACCOUNTING METHODS*

As a greater number of companies implement industrial modernization programs, manufacturing engineers are developing more definitive approaches to evaluating project results that supplement established accounting methods.

*Permission granted by General Dynamics Corp., Fort Worth Division.

One of these techniques was developed by General Dynamics under Air Force Contract F33657-80-G-0007 Order 0002. This cost-benefit analysis methodology is applicable both before project implementation, as well as after measuring actual benefits after the project is on-line in production (figures 3-10 and 3-11). This analysis function is intended to play a major role in optimizing the enabling technologies and measuring the cost-benefit of facility items on the shop floor.

Three Reasons to Conduct a Cost-Benefit Analysis

The cost-benefit analysis methodology consists of three distinct but closely related functions. They are:

- *Economic Analysis*—estimates savings prior to a project's implementation.
- *Performance Assessment*—documents actual savings of on-line facility items after implementation. It provides the means to accurately measure actual savings of a number of facility items, both for incentive purposes and to provide feedback to management.
- *Program Savings Forecast*—provides the ability to estimate savings resulting from an entire technology modernization program, and for use in updating the overall project periodically.

> **Tip for Successful Automation:**
> **Why Engineers Use the Micro-Matic Methods Measurement System**
>
> The cornerstone of technology analysis has been the use of the *micro-matic methods and measurement* (4-M) system. The 4-M system is a predetermined time standard system with all standard elements stored in a data base, which General Dynamics is using to meet MIL-STD 1567 tracebility requirements. The 4-M system is most suitable for providing estimates of proposed projects since a detailed description of how the system will work enables the manufacturing engineer to apply the 4-M system to the new project or method, even if it does not yet physically exist. The micro-matic methods and measurement system was obtained from the MTM Association for Standards and Research.

A Nine-Step Procedure for Economic Analysis

The Economic Analysis function consists of a nine-step procedure:

1. Gather data on a project.
2. Reconcile parts list.
3. Sample parts list, if necessary.
4. Apply standards to parts list for both old and new methods.
5. Infer savings over entire part universe, if sampling technique was applied.
6. Project labor-hour savings.
7. Apply dollar rates.
8. Calculate return on investment (ROI) and payback.
9. Review with appropriate personnel.

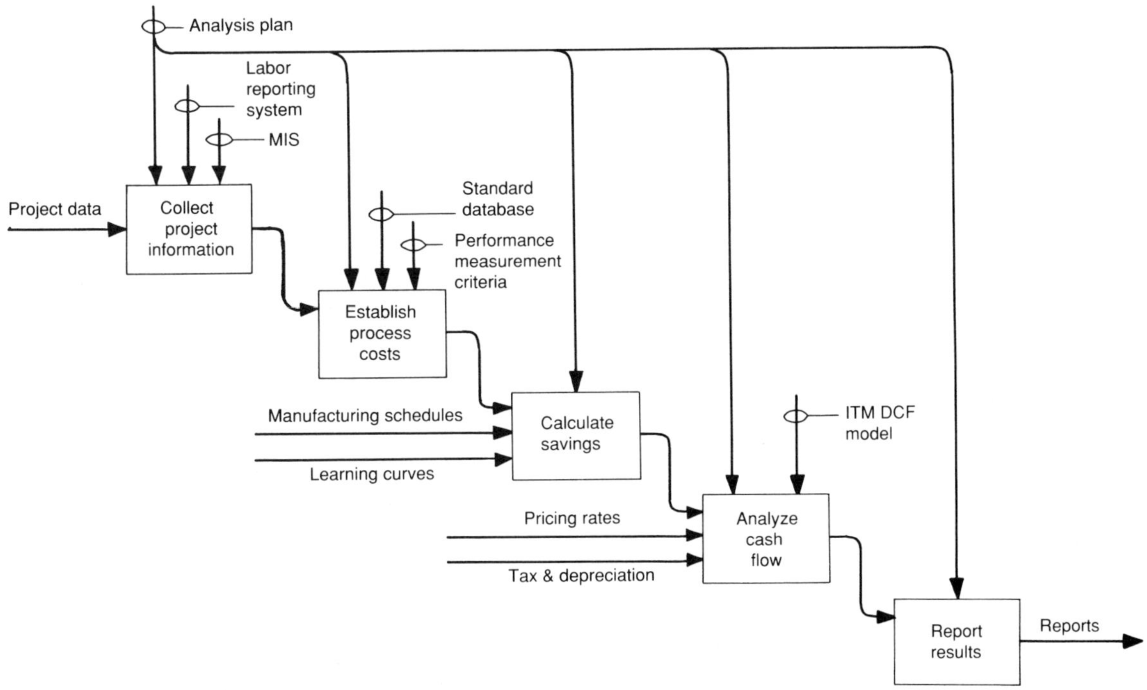

FIGURE 3-10 Cost-Benefit Analysis Methodology—Sequence of Activities #1 (Courtesy of General Dynamics Fort Worth Division).

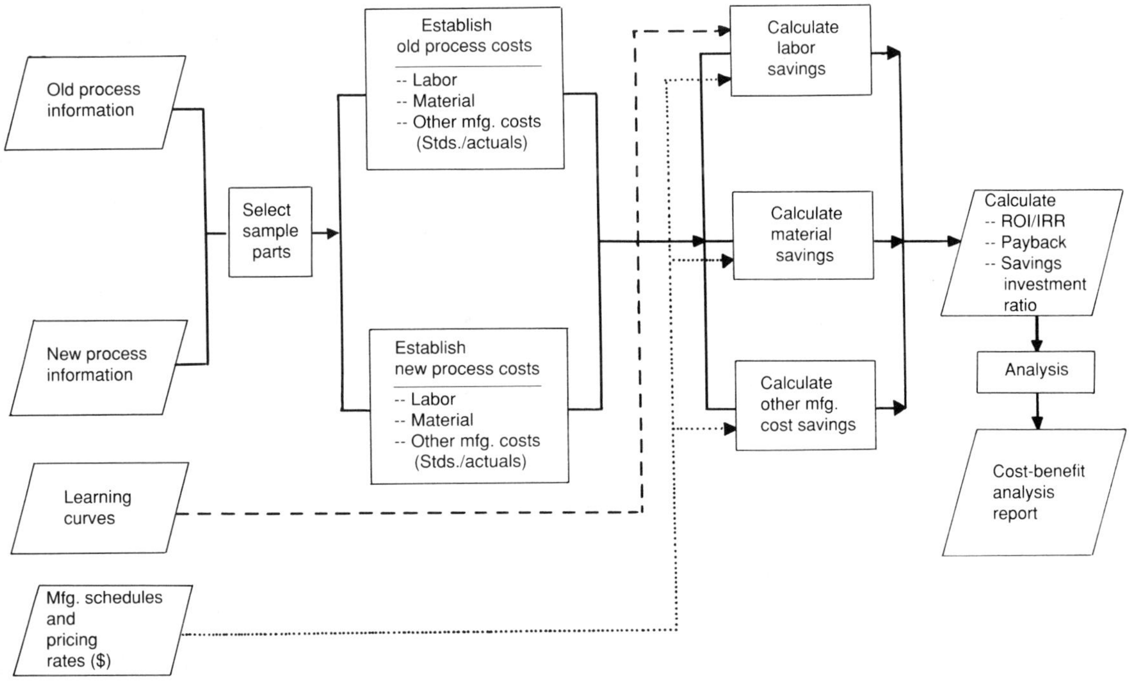

FIGURE 3-11 Cost-Benefit Analysis Methodology—Sequence of Activities #2 (Courtesy of General Dynamics Fort Worth Division).

1. Gather Data. Gathering data is the most important step for obtaining an accurate analysis. It also is the step most dependent on human judgment. The manufacturing engineer must obtain descriptions of the new process as it will operate, as well as data on the existing processes that will be replaced. A list of the part numbers affected must be obtained along with the implementation plan or schedule. The implementation costs must also be acquired. (*Note:* In actual practice, all of these parameters may become subject to change, as the project progresses.)

2. Reconcile Parts List. The parts list should be validated as to its current updated status before use. The most common reason for performing this validation is to spot parts that have become obsolete by being replaced due to an engineering change.

3. Sample Parts List. A random sample of the entire parts universe is made to determine which parts will undergo the detailed 4-M analysis. The size of the sample is selected to permit a 90 percent confidence level. (*Note:* To simplify the selection process, a computer program was developed to randomly select a statistically significant sample of part numbers for 4-M standards development.)

4. Apply Standards. Once a sample parts list is obtained, the list and all pertinent data are forwarded to a special standards group. Planning data and existing 4-M application studies provide most of the baseline standards, other than new methods standards which are derived straight from the new process description obtained in the first step of gathering data. Other refinements have occurred where the standard is a unit of measure other than parts, such as rolls of adhesive.

5. Infer Savings. Once standards are applied to the sample part family, it is necessary to infer the results of the standards study over the entire parts universe. The sample parts are then dispersed to the various *scheduled work breakdown structures* (SWBSs), and the total SWBS universe inferred by a simple ratio of number of total parts to number of sample parts. The total 4-M standards are then calculated for both the baseline and the new method.

6. Project Labor-Hour Savings. Composite learning curves are used for many economic analysis estimates. However, customized learning curves, tailored to an individual project or process, such as NC machining or punch press operation, are becoming more common. These tailored learning curves bring the economic analysis directly in line with the performance assessment methodology, where individual learning curves are derived.

7. Apply Dollar Rates. The application of labor rates and overheads to labor hours is computerized at General Dynamic (as with most companies). Regardless of which system a user may have, each company would specify its own dollar rates and overheads, and in turn, obtain summary data on total dollar savings.

8. Calculate Return on Investment and Payback. The calculations for return on investment (ROI) and payback were discussed in Chapter 1. General Dynamics uses a proprietary computer financial model, known as LARCAR for these calculations. Inputs to the program include:

- Investment types and installation dates
- Years of savings
- Investment costs

- Depreciation rates and tax credits
- Savings by year

LARCAR Version I, calculates a discounted cash flow ROI and a payback period, as well. Once LARCAR has been run, the analysis portion of the cost-benefit task is complete.

9. Review with Appropriate Personnel. Key General Dynamics personnel review the results of the economic analysis portion of the project before publication. Sometimes the outcome is a revision in the project's scope, or even its technical approach. Economic benefits are an essential issue since the objective is to optimize a project so that it is fully developed.

The following case histories are indicative of the type of tangible savings that have been accrued to both General Dynamics and the Air Force from this industrial modernization program.

CASE-IN-POINT

AUTOMATIC WING DRILLING

Situation

In the previous method the wing drilling task was a completely manual operation. Approximately 6200 holes were drilled in the upper and lower wing skins and wing substructure for each ship set. Nineteen different pneumatic drill units were positioned manually into the drill templates to drill and countersink the various sized holes. This method was extremely slow and fatiguing to the operator. During operation, drilling efficiency often dropped due to a lack of tool condition monitoring. This sometimes affected hole quality.

Solution

The method implemented consists of four industrial T^3 robots (Cincinnati Milacron) arranged to fabricate upper and lower F-16 wing skins. The robotic end effectors are equipped with spacematic units capable of drilling and countersinking in one operation. In addition, the improved software and special drilling bits ensure an increase in the feed and speed rate. These two features offer an attractive improvement in productivity and consistency of hole quality.

Parts handling and fixture clamp-up were improved. The setup bolts for each drilling operation are performed while the robot is removing metal on the other side. Also, a specialized fixture clamp-up operation featuring vacuum pressure, is incorporated thus replacing many of the manual type clamps.

Results

The following data contains the results of this project:

Total labor-hour savings (total aircraft)	42,493
Total labor-hour savings (USAF aircraft)	39,814
Internal rate of return	27.9%
Payback (in years)	2.5%
GD capital expenditures	$555,000
USAF expenditures	$850,000
Savings to investment ratio (GD)	2.78/1
Savings to investment ratio (USAF)	1.72/1

CASE-IN-POINT

AUTOMATED TRIM CELL

Situation

Under the old methods employed by the company, approximately 430 F-16 discrete aluminum sheetmetal parts, including interchangeable/replaceable (I/R) panels and doors and fuselage skins, were identified as being processed in the conventional drill-and-rout area. The parts flow into the area with excess material on their perimeters and require rough trimming to remove the excess material before the primary operations of drilling and routing can be accomplished. Depending on the size and shape of the parts, the rough trimming is performed manually with a band saw, a pneumatic hand saw, or a router. After rough trimming, the parts are loaded into hard tooling (clamshell type) and then installed on positioners or clamped to work benches for manual drilling and routing. After these operations have been accomplished, the parts are removed from the hard tooling for the secondary operation of deburring or of countersinking/chamfering and then deburring.

Solution

The proposed method uses four industrial robots, two-part presentation systems, and drilling and routing end effectors with integrated computer controls. The production environment is patterned after the work cells concept.

Robotic end effectors for routing sheetmetal are adapted and designed to rout each I/R panel EOP (net). The resultant trim operation eliminates a setup and run operation for sawing off excess material. An increase in feed and speed rates over the manual drilling and routing units will enhance productivity and hole quality.

The existing tooling shells have been adapted to be used with rotatable positioners. Each positioner can be loaded with two parts. While one part is being processed, the operator on the other side is setting up for the next operation. This manufacturing method eliminates manual setups for drilling and routing. In addition, much of the parts handling is being done at the same time as robotic processing thus increasing production further.

Results

The following data contains the results of this project:

Total labor-hour savings (total aircraft)	40,463
Total labor-hour savings (USAF aircraft)	35,881
Internal rate of return	29.0%
Payback (in years)	2.4%
GD capital expenditures	$554,000
GD capital expenditures (tooling)	$396,000
USAF expenditures	$810,000
Savings to investment ratio (GD)	1.62/1
Savings to investment ratio (USAF)	1.72/1

How General Dynamics Uses Performance Assessment

The assessment of performance is one of the major functions of a cost-benefit analysis. General Dynamics developed a methodology to measure and document the cost savings generated by the technology modernization program.

Methodology Development. Performance assessments include the development of the old method baseline, new method actual cost, and comparison of the two. This methodology was refined by using various statistical approaches and by the development of an automated cost tracking system. The flowchart in figure 3-12 identifies the information required to generate labor-hour savings or loss in a performance assessment.

Assessment Procedures. An assessment begins with the collection of certain key information to establish the old method (baseline), the new method cost, and to make a comparison between the two. This key information is described in the following paragraphs.

Task Center Information. For all new projects installed on the production floor, a unique task center number that represents the new process is assigned. The task center connected with the process to be replaced is identified as the "old process." In many cases, however, the new process will replace several old processes and their associated task centers must all be identified as "old processes." The new method task center is identified in the industrial engineering cost center bulletin while the identification of the old task centers involves a more extensive evaluation of the new process planning versus old process planning.

Part Number Information. The assessment process is designed to calculate the project's savings on a part/operation basis. It is therefore necessary to identify all part numbers affected by a new process. To facilitate determination of the old process, the process planning data for the parts list are obtained for both old and new processes. Also needed is the related SWBS in which each part belongs and the quantity of parts needed per ship set.

Actual On-Line Date. The actual on-line date indicates when the new process was implemented. This date serves as a reference to determine the date the performance assessment should begin.

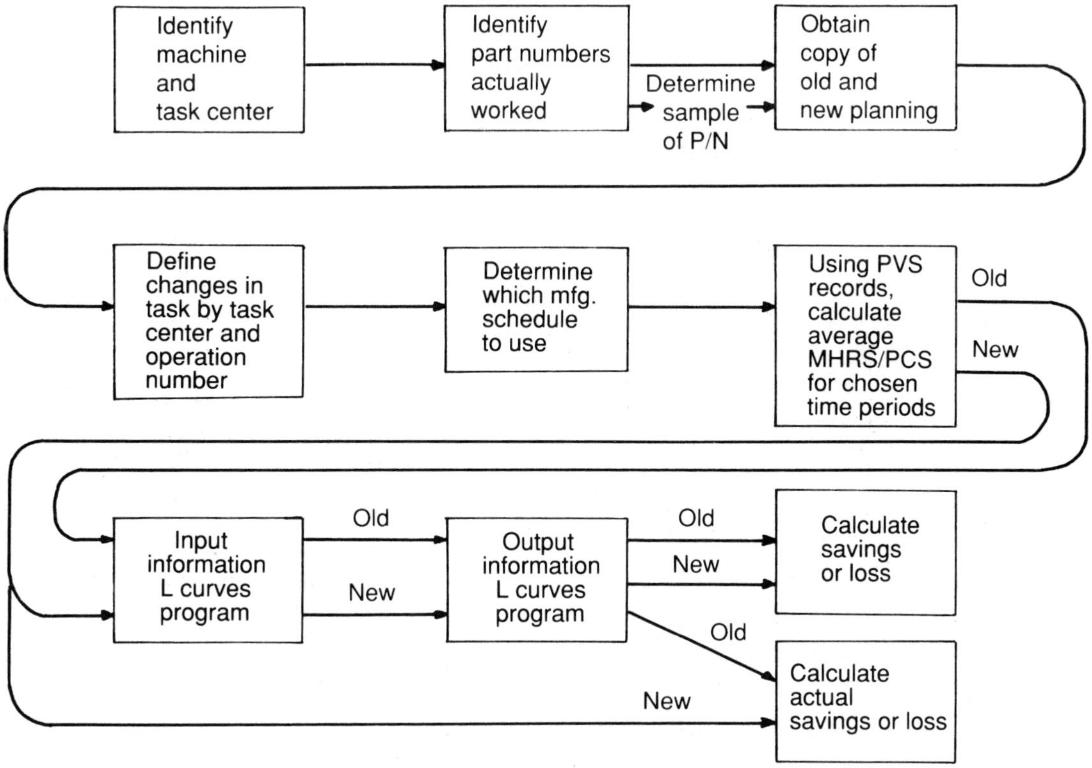

FIGURE 3-12 Methodology for Performance Assessment (Courtesy of General Dynamics Fort Worth Division).

Baseline Development. After the previous data has been accumulated, the baseline for the old method costs must be developed. The labor hours needed to accomplish a given SWBS task are compared with estimated hours for the new process or equipment. The average hour differential between old and new process shows actual hours either saved or lost by the new process. This evaluation only can be accomplished if accurate part time for the old method process has been accumulated.

Preliminary Baseline Tasks. The preliminary baseline tasks include the assessment procedures discussed above as well as the tasks described below.

1. Identify new process machine(s) and the affected task centers.
2. Obtain a list of parts planned for the new process machines.
3. Review the planning on selected part numbers and determine which task centers and operations were replaced by the new process. (Task centers that are replaced by the new processes are designated as old process task centers.)
4. Obtain a list of parts that have actually run on the new process machine(s).
5. Record the SWBS number, aircraft effectivity, quantity of pieces per aircraft, and part family.
6. Develop a list of part numbers acceptable for analysis. Eliminate non-USAF part numbers and part numbers which have run on the new process but are not planned from the list.
7. Classify part numbers into the nine component groups by means of the SWBS identifiers. Sum up the total number of parts planned for each component.
8. Identify, for analysis only, those components which contain at least 10 percent of the total parts that are planned for the new machine(s).
9. Revise the list of part numbers to be assessed by eliminating part numbers, which although planned for the new task center, have not yet run in the new task area; part numbers that have a limited effectivity; and part numbers that have insufficient baseline data.

Upon completion of these preliminary tasks, the development of the baseline may begin.

How Baseline Data Is Developed

Baseline data is developed in the following ways:

1. Review and record the four-cycle averages and cumulative good parts quantity from the PVS-target data base.
2. Group part numbers by component and four-cycle averages by lot.
3. Calculate a weighted average hour-per-piece for each lot across all part numbers. Sum up the averages across all part numbers for each lot in each aircraft component.
4. Replace erratic values with estimated values and document and justify all changes.

The baseline is calculated by plotting the labor actuals on a log-log scale by SWBS and manufacturing lot. The line is plotted by SWBS to incorporate the differences in both lot quantities and manufacturing schedules into the savings calculations. The labor actuals are the sum of the average lot hours calculated for each part in the SWBS.

To develop the old method cost line, the lot labor actuals for the old method are analyzed, and the lot cost is "best fitted" to establish actual demonstrated performance up to the point of implementation. Factors influencing the old method cost line include variables such as delivery rules, remaining development impact, and machine-paced versus human-paced operations.

How Much Does the New Process Cost?

The development of the new cost line is, in many respects, like that of the old method baseline development. The labor hours needed to accomplish the original task are now tracked for the new machine. The basic difference is that the automated cost tracking system can be utilized to replace the manual operations and reduce the time required to extract data for analysis.

Development of the new cost trends begins with the creation of computer files containing assessable part numbers and old and new task centers. The automated cost tracking system then extracts labor hours by part number and component from General Dynamic's Fab WIP data base, and produces a report containing average labor hours by part number, per piece, per ship set, and by manufacturing lot for each component.

A new cost line is not projected through the new process labor actuals until an accurate trend is apparent. In many cases, the designated parts break in gradually on the new process, with portions being run on the old and new process concurrently. Therefore, in some cases, the new process labor hours are greater at the start, at least until debugging and start-up problems are resolved.

For the new process, the labor actuals are plotted only for the lots completed as of the assessment date. The difference between the baseline labor hours and the new method labor hours for each new process lot is then calculated. The difference represents the labor hour savings per ship set.

A Seven-Step Technique to Calculate Costs or Savings

The savings for an individual project results from the difference between the old method and the new method costs. The calculations used to derive these savings are as follows:

1. Calculate what the old method process hours would have been (had the new process not been implemented) by measuring the delta difference from each new method lot midpoint to where it intersects the old method (baseline).
2. Calculate the average savings per aircraft for each lot by subtracting the new process average hours per aircraft from the baseline average hours per aircraft.
3. Calculate the total savings for all aircraft in the lot by multiplying the average savings per aircraft by the number of aircraft produced in each lot.
4. Calculate the total savings per components (for the assessed parts) by summing up the savings for each lot.
5. Calculate the total savings for all assessed components by summing up the savings of all components evaluated.
6. Calculate a total savings factor for the new process by dividing the total number of parts run on the new process machine(s) by the number of parts assessed.

7. Calculate the total savings to date for the new process machine(s) by multiplying the total savings for all components by the total savings factor.

The savings are summarized by fiscal year for each CAR item that has met the criteria for assessment. These savings are then documented and presented to the Air Force in the form of incentive billing.

Assessing Labor-Hour Savings

The labor-hour savings data generated from performance assessments are compiled semiannually and reported to USAF for incentive payment. The incentive payment is based on the ability of General Dynamics to achieve the total estimated labor-hour savings for each fiscal year.

The calculation of incentives earned is derived from the results of items assessed. The percent of savings assessed, when compared with the estimated labor-hour savings for the assessed items, yields a factor that is applied to the total unmeasured labor-hour savings to produce figures on the factual savings. Thus, the amount of earned incentive is calculated for assessed items and is factored for all unmeasurable items.

Examples of Time-Saving Projects

Auto Drilling Vertical Fin. Vertical fin skin holes are drilled by a Cincinnati Milacron T^3 Industrial Arm Robot. The holes are drilled undersized and later enlarged to full size on the assembly line, because they are close tolerance holes. This machine has saved a total of 3,720 labor hours when last reported.

OKK Milling Machine with MDI. The milling machines with MDI provide for automatic machining (milling) from data entered in the microprocessor memory/control. Once the data is entered, the cycle can be repeated automatically for each part. All data can be recorded on a cassette tape for machining the next lot. Parts are held in a hydragrip chuck and the machine is programmed to run parts consecutively with few operation setups and at fast cutting speeds. This assessment project had savings of 33,512 labor hours upon last observation.

Turret Drill. The NC Burgmaster turret machine incorporates drilling and its various derivatives efficiently without numerous task center visits. Computer-controlled automatic cutter changing, cutting performances, and holding fixture indexing are cost saving enhancements. Precision metal removal is augmented with computerized sensing to maintain tolerance requirements. This contributes significantly in maintaining or enhancing overall part quality. Last recorded savings were 1,946 labor hours.

Forecasting Labor-Hour Savings

As a part of the F-16 Technology Modernization Program, a savings forecast methodology was established with a capability to project the labor-hour savings for facilities purchased by the Fort Worth division. The purpose of the technology modernization savings forecast is to project potential savings to establish yearly incentives and determine savings for the total technology modernization program.

The technology modernization savings forecast originated with the facilities projects at the inception of the program. From that point, the savings calculations broadened to include facility capital authorization requests (CARS), avionics intermediate shop (AIS), alternate mission equipment (AME) and spares.

The savings methodology consists of sophisticated computer learning curve models, the pricing system (TPS) utilization, and automated computer pricing.

Contract incentives described in the performance assessment methodology are based on direct labor-hour savings. The labor hours needed to accomplish a given WBS task entail a comparison of existing processes or equipment and estimated hours with the new process or equipment. Differential hours (average hour difference between old and new method) for each piece of equipment within each CAR are applied to the applicable bid learning curve at the midpoint of the lot in accordance with the scheduled on-line dates. The differential costs are projected to a synthetic Number 1 cost, which provides the means for forecasting and projecting by component, country, and fiscal year in accordance with the multiyear-buy business base. These forecasts and/or projections improve the decision-making process by providing management with accurate and objective analyses of savings.

Savings are estimated for manufacturing, engineering, quality assurance, tooling, direct and indirect labor. Airframe, spares, avionics intermediate shop (AIS), and alternate mission equipment (AME) are data sources included in the estimate. The labor-hour savings forecast provides the data base for generated dollar savings as priced by TPS. Labor-hour savings forecasts also afford the advantage of a common unit because of their noninflationary nature. Therefore, all savings forecasts are based on labor-hour savings.

chapter 4

Techniques for Integrating Office Technology with Manufacturing

THE FIVE-MODULE, COMPUTERIZED LINK BETWEEN PLANNING AND MANUFACTURING*

The computer interface between production planning and manufacturing is critical because it enables all systems elements to be fully integrated and thus achieve optimum results.

The various systems modules that provide this linkage are:

- Bill of materials
- Master scheduling
- Material requirements planning
- Inventory control
- Purchasing control

Twelve Key Functions of a Bill of Materials Module

The bill of materials module is the cornerstone for automating a company's manufacturing and related functions. This module, in conjunction with the other systems modules, enables the user to gain complete control over purchased parts and items produced on the shop floor. Various systems menus and data entry screens can provide the user with access to the entire manufacturing data base in an on-line mode.

The successful implementation of "closed-loop" manufacturing resources planning (MRP) requires accurate bills of materials. The bill of materials module contains all the parts and resources used in any assembly, at all levels of the product (family) tree. This enables the user to determine how an assembly is built or what assemblies use a part.

*Permission granted by Micro-MRP, Inc.

In a bill of materials module, product trees are defined by parent part, component part, and effectivity date "relationships." Other data may include quantity per assembly, part or assembly descriptions, standard costs, and references.

A bill of materials module can provide these features:

- Add, change, delete, and review parent–component relationships for multilevel product structures up to ninety-nine levels.
- Maintain data instantly on-line.
- Display or print parent–component relationships in any combination of single or multilevel explosions or implosions.
- Maintain separate units of measurement for costing and manufacturing activities.
- Maintain multiple bill of material types (planning, engineering, resource, etc.).
- Maintain multiple part types (purchased, manufactured, bulk, phantom, etc.).
- Maintain decimal or percentage quantities per assembly.
- Specify optional scrap factors to be planned by MRP separately for each component.
- Establish resource bills of materials for rough cut capacity planning.
- Change product structures with "start date effectivity" logic and maintain part revision levels.
- Calculate standard costs using material, labor, material burden, and labor burden cost elements and "cost rollup" logic.
- Automatically copy entire bills of materials onto new parts.

Figure 4–1 illustrates a multilevel explosion report by low-level sequence, which will:

- Sort by low level or by part.
- Print indented or alternate costed bills.
- Review entire bill of materials, or explode any lower-level assembly.

Figure 4–2 shows a recalculate cost rollups report. Component scrap and part yield are both factored to obtain accurate standard cost for manufactured assemblies and finished goods. Software programming automatically calculates costs in both manufacturing and cost units of measure.

Figure 4–3 shows an engineering parts list by part identifier report. This report can be used to verify data entry such as part numbers, costs, part types, commodity codes, drawing numbers, and bill of materials units of measure.

Figure 4–4 shows a single-level explosion by low-level sequence report. This report enables a user to verify the product structures by reviewing on-line displays of single or multilevel explosions (seeing how a product is made) and single or multilevel implosions ("where used" or "goes into").

At this stage, rough cut capacity planning allows a user to construct resource bills of materials to review demands on critical constraining influences before committing the company's resources to a detailed production plan (see Chapter 5, "Shop Floor Control," for more information on employing advanced techniques for handling bottleneck constraints).

11-04-1985 10:55:57 MICRO-MRP, INC. PAGE NO. 1

MAX * BILL OF MATERIALS * MULTI-LEVEL EXPLOSION BY LOW LEVEL SEQUENCE

LOW LEVEL	PART IDENTIFIER	DESCRIPTION	EFF DATE	QTY/ASSY	Q C	UOM	LT OFF	T C	SCRAP	ECN NUMBER	ALTERNATE PART
0	11000	PC - 2 FLOPPY - 128K				EA					
1	13000	PC KEYBOARD	830101	1.00	U	EA	0	N	0		
.2	50001	PC KEYBOARD CASE	830101	1.00	U	EA	0	N	0		
..3	70001	KEYBOARD CASE TOP	830101	1.00	U	EA	0	N	0		
...4	70016	METAL, 4 X 4, SHEET	830101	0.75	U	SH	0	N	0		
..3	70002	KEYBOARD CASE BOTTOM	830101	1.00	U	EA	0	N	0		
...4	70016	METAL, 4 X 4, SHEET	830101	1.50	U	SH	0	N	0		
..3	70003	KEYBOARD FEET	830101	4.00	U	EA	0	N	0		
..3	70004	PAINT, PC GREY	830101	0.33	U	OZ	0	N	0		
.2	70005	PC KEYBOARD KEY SET	830101	1.00	U	EA	0	N	0		
.2	70006	POWER CABLE, 6 FEET	830101	1.00	U	EA	0	N	0		
.2	70007	SCREWS, 1/4" X #4	830101	4.00	U	EA	0	N	0		
1	14000	MONITOR	830101	1.00	U	EA	0	N	0		
.2	50002	MONITOR CASE	830101	1.00	U	EA	0	N	0		
..3	70009	MONITOR FRONT	830101	1.00	U	EA	0	N	0		
...4	70016	METAL, 4 X 4, SHEET	830101	2.00	U	SH	0	N	0		
..3	70010	MONITOR BACK	830101	1.00	U	EA	0	N	0		
...4	70016	METAL, 4 X 4, SHEET	830101	1.75	U	SH	0	N	0		
.2	70007	SCREWS, 1/4" X #4	830101	4.00	U	EA	0	N	0		
.2	70008	CRT, 10", GREEN	830101	1.00	U	EA	0	N	0		
.2	70011	MONITOR CORD, 24"	830101	1.00	U	EA	0	N	0		
.2	70012	KEYBOARD FEET, 1/2"	830101	4.00	U	EA	0	N	0		
1	15000	PC, SYSTEM UNIT - 128K	830101	1.00	U	EA	0	N	0		
.2	50003	PC MOTHERBOARD - REV A	830101	1.00	U	EA	0	N	0		
..3	70013	PC BOARD	830101	1.00	U	EA	0	N	0		
..3	70014	16K RAM CHIP	830101	8.00	U	EA	0	N	0		
..3	70015	8088 PROCESSOR	830101	1.00	U	EA	0	N	0		
.2	50004	CASE ASSEMBLY	830101	1.00	U	EA	0	N	0		
..3	50005	SYSTEM UNIT BASE	830101	1.00	U	EA	0	N	0		
...4	70016	METAL, 4 X 4, SHEET	830101	1.25	U	SH	0	N	0		
..3	50006	SYSTEM UNIT COVER	830101	1.00	U	EA	0	N	0		
...4	70004	PAINT, PC GREY	830101	0.35	U	OZ	0	N	0		
...4	70016	METAL, 4 X 4, SHEET	830101	2.75	U	SH	0	N	0		
..3	70003	KEYBOARD FEET	830101	1.00	U	EA	0	N	0		
..3	70007	SCREWS, 1/4" X #4	830101	4.00	U	EA	0	N	0		
.2	70017	FLOPPY DRIVE, 320K, DSDD	830101	2.00	U	EA	0	N	0		
.2	70018	SPEAKER, 2", 4 OZ	830101	1.00	U	EA	0	N	0		
.2	70019	POWER SUPPLY, 120 VOLT	830101	1.00	U	EA	0	N	0		
1	LABOR	FINAL ASSY LABOR	830101	1.00	U	DL	0	N	0		

FIGURE 4–1 B/M-Multi-Level Explosion Report by Low Level Sequence (Courtesy of Micro-MRP, Inc.).

How the Master Scheduling Module Facilitates Planning

The master scheduling module allows a user to plan production of finished goods, create forecast orders, and maintain customer demand. This enables a user to obtain consistent, coordinated information necessary for establishing a feasible production plan that "drives" the manufacturing and purchasing operations.

This module provides current information on orders (both actual and forecasted), availability of materials, resource capacities, and current backlog. It becomes the plan that balances the company's resources and demand for its products in a way that satisfies the company's policies on customer service, production efficiency, and inventory investment. Once an order for finished goods is inputted into the master schedule, this module will automatically create demands for the component parts and resources necessary for building that product, whether the parts are to be purchased or manufactured in-house.

11-04-1985 11:00:07 MICRO-MRP, INC. PAGE NO. 3

MAX * BILL OF MATERIALS * RECALCULATE COST ROLLUPS

PART IDENTIFIER	DESCRIPTION	LL	QTY/ASSY	UOM	C T	PERCENT SCRAP:YIELD	COSTS PER UNIT MATERIAL	LABOR	MATL BUR	LABOR BUR	TOTAL
TOTAL IN COST UNITS ** MANUALLY COSTED AT: 60.50				HD			55.00	0.00	5.50	0.00	60.50
70017	FLOPPY DRIVE, 320K, DSDD	2	1.00	EA	M		50.00	0.00	5.00	0.00	55.00
SUBTOTAL IN BOM UNITS				EA			50.00	0.00	5.00	0.00	55.00
YIELDED TOTAL IN BOM UNITS				EA		:100.00	50.00	0.00	5.00	0.00	55.00
TOTAL IN COST UNITS ** MANUALLY COSTED AT: 55.00				EA			50.00	0.00	5.00	0.00	55.00
70012	KEYBOARD FEET, 1/2"	2	1.00	BX	M		0.08	0.00	0.01	0.00	0.09
SUBTOTAL IN BOM UNITS				EA			0.08	0.00	0.01	0.00	0.09
YIELDED TOTAL IN BOM UNITS				EA		:100.00	0.08	0.00	0.01	0.00	0.09
TOTAL IN COST UNITS ** MANUALLY COSTED AT: 13.20				BX			12.00	0.00	1.20	0.00	13.20
70011	MONITOR CORD, 24"	2	1.00	HD	M		0.20	0.00	0.02	0.00	0.22
SUBTOTAL IN BOM UNITS				EA			0.20	0.00	0.02	0.00	0.22
YIELDED TOTAL IN BOM UNITS				EA		:100.00	0.20	0.00	0.02	0.00	0.22
TOTAL IN COST UNITS ** MANUALLY COSTED AT: 22.00				HD			20.00	0.00	2.00	0.00	22.00
70008	CRT, 10", GREEN	2	1.00	EA	M		75.00	0.00	7.50	0.00	82.50
SUBTOTAL IN BOM UNITS				EA			75.00	0.00	7.50	0.00	82.50
YIELDED TOTAL IN BOM UNITS				EA		:100.00	75.00	0.00	7.50	0.00	82.50
TOTAL IN COST UNITS ** MANUALLY COSTED AT: 82.50				EA			75.00	0.00	7.50	0.00	82.50
70006	POWER CABLE, 6 FEET	2	1.00	HD	M		0.50	0.00	0.05	0.00	0.55
SUBTOTAL IN BOM UNITS				EA			0.50	0.00	0.05	0.00	0.55
YIELDED TOTAL IN BOM UNITS				EA		:100.00	0.50	0.00	0.05	0.00	0.55
TOTAL IN COST UNITS ** MANUALLY COSTED AT: 55.00				HD			50.00	0.00	5.00	0.00	55.00
70005	PC KEYBOARD KEY SET	2	1.00	HD	M		0.10	0.00	0.01	0.00	0.11
SUBTOTAL IN BOM UNITS				EA			0.10	0.00	0.01	0.00	0.11
YIELDED TOTAL IN BOM UNITS				EA		:100.00	0.10	0.00	0.01	0.00	0.11
TOTAL IN COST UNITS ** MANUALLY COSTED AT: 15.84				HD			14.40	0.00	1.44	0.00	15.84
50004	CASE ASSEMBLY	2	1.00	EA	A		0.00	3.60	0.00	3.60	7.20
50005	SYSTEM UNIT BASE	3	1.00	EA	A	0.00:	13.65	0.00	0.00	0.00	13.65
50006	SYSTEM UNIT COVER	3	1.00	EA	A	0.00:	24.51	0.00	0.00	0.00	24.51
70003	KEYBOARD FEET	3	1.00	HD	M	0.00:	0.11	0.00	0.00	0.00	0.11
70007	SCREWS, 1/4" X #4	3	4.00	BX	M	0.00:	0.44	0.00	0.00	0.00	0.44
SUBTOTAL IN BOM UNITS				EA			38.71	3.60	0.00	3.60	45.91
YIELDED TOTAL IN BOM UNITS				EA		:100.00	38.71	3.60	0.00	3.60	45.91
TOTAL IN COST UNITS				EA			38.71	3.60	0.00	3.60	45.91
50003	PC MOTHERBOARD - REV A	2	1.00	EA	A		0.00	4.50	0.00	4.50	9.00

FIGURE 4–2 B/M-Recalculate Cost Rollups Report (Courtesy of Micro-MRP, Inc.).

This module will store scheduling information in a detailed "bucketless" format, and enable the user to call up the schedule in either display or print-out mode. This data generally consists of availability to forecast and availability to promise in either a detailed (vertical) format or in a weekly bucketed (horizontal) format.

In essence, the master schedule is a statement of what a company plans to manufacture. A well structured and defined master schedule module provides a user with greater assurance that the plan is based on the most accurate and timely information available; that

11-04-1985 10:53:47 MICRO-MRP, INC. PAGE NO. 1

MAX * BILL OF MATERIALS * ENGINEERING PARTS LIST BY PART IDENTIFIER

PART IDENTIFIER	DESCRIPTION	T C	LL	BOM UOM	E S	PROD DATE	OBSL DATE	COMMODITY CODE	DRAWING NUMBER	YIELD	ACTIVE DATE	COST/UNIT
11000	PC - 2 FLOPPY - 128K	M	0	EA	2	831217		FIN GOOD	A-7895235	100	851025	412.3386
12000	PC - 10 MEG HD - 128K	M	0	EA	2	831217		FIN GOOD	B-5230798	100	831218	912.3386
13000	PC KEYBOARD	A	1	EA	2	831217		FIN GOOD	A-9649925	100	851010	30.4468
14000	MONITOR	A	1	EA	2	831217		FIN GOOD	A-8647825	100	831218	134.2766
140836135	REAR FILLER	M	0	EA	2	851023		FIN GOOD	135	93	000000	0
140836135-	REAR FILLER	B	0	EA	2	851023		RAW	135	93	000000	0
15000	PC, SYSTEM UNIT - 128K	A	1	EA	2	831217		FIN GOOD	A-2363475	100	831218	227.6152
16000	PC-HD, SYSTEM UNIT - 128K	A	1	EA	2	831217		FIN GOOD	A-6552340	100	831218	723.6152
50001	PC KEYBOARD CASE	A	2	EA	2	831217		MFG ASSY	B-9435576	100	851010	24.3468
50002	MONITOR CASE	A	2	EA	2	831217		MFG ASSY	B-3638745	100	831218	40.75
50003	PC MOTHERBOARD - REV A	A	2	EA	2	831217		MFG ASSY	B-7354787	100	831218	26.5
50004	CASE ASSEMBLY	A	2	EA	2	831217		MFG ASSY	A-3283923	100	851010	48.0102
50005	SYSTEM UNIT BASE	A	3	EA	2	831217		MFG ASSY	A-6536974	100	831218	14.25
50006	SYSTEM UNIT COVER	A	3	EA	2	831217		MFG ASSY	A-6893566	100	831218	25.2102
70001	KEYBOARD CASE TOP	A	3	EA	2	831217		FAB ASSY	A-4893429	100	831218	6.95
70002	KEYBOARD CASE BOTTOM	A	3	EA	2	831217		FAB ASSY	B-6758769	100	831218	12.9
70003	KEYBOARD FEET	B	3	EA	2	831217		PUR PART	B-6777550	100	851023	11
70004	PAINT, PC GREY	B	4	OZ	2	831217		PUR PART	A-8766486	100	831218	5.5
70005	PC KEYBOARD KEY SET	B	2	EA	2	831217		PUR PART	A-2575789	100	831218	15.84
70006	POWER CABLE, 6 FEET	B	2	EA	2	831217		PUR PART	A-8478294	100	851010	55
70007	SCREWS, 1/4" X #4	B	3	EA	2	831217		PUR PART	B-6536755	100	831218	110
70008	CRT, 10", GREEN	B	2	EA	2	831217		PUR ASSY	A-0086475	100	831218	82.5
70009	MONITOR FRONT	A	3	EA	2	831217		FAB ASSY	A-7312652	100	831218	17.2
70010	MONITOR BACK	A	3	EA	2	831217		FAB ASSY	B-7367586	100	831218	17.55
70011	MONITOR CORD, 24"	B	2	EA	2	831217		PUR PART	B-3645822	100	831218	22
70012	KEYBOARD FEET, 1/2"	B	2	EA	2	831217		PUR PART	A-6748148	100	831218	13.2
70013	PC BOARD	B	3	EA	2	831217		PUR ASSY	A-5665471	100	831218	2.2
70014	16K RAM CHIP	B	3	EA	2	831217		PUR PART	B-1496906	100	831218	110
70015	8088 PROCESSOR	B	3	EA	2	831217		PUR PART	A-3867462	100	831218	5.5
70016	METAL, 4 X 4, SHEET	B	4	SH	2	831217		PUR PART	A-5482321	100	831218	6.6
70017	FLOPPY DRIVE, 320K, DSDD	B	2	EA	2	831217		PUR ASSY	A-6362718	100	831218	55
70018	SPEAKER, 2", 4 OZ	B	2	EA	2	831217		PUR PART	A-6746917	100	831218	60.5
70019	POWER SUPPLY, 120 VOLT	B	2	EA	2	831217		PUR ASSY	B-5236833	100	831218	27.5
70020	HARD DISK, 10 MEG	B	2	EA	2	831217		PUR ASSY	A-7563124	100	831218	550
LABOR	FINAL ASSY LABOR	B	1	DL	2	831217		RESOURCE		100	000000	0
PHANTOM		P	0	EA	2	851021				100	000000	0
TOLYAL	SOLVENT	B	0	EA	2	851023		RAW		100	851023	0
WA3967	PAINT WHITE	B	0	EA	2	851023		RAW		30	000000	0

FIGURE 4–3 B/M-Engineering Parts List by Part Identifier Report (Courtesy of Micro-MRP, Inc.).

11-04-1985 10:55:13 MICRO-MRP, INC. PAGE NO. 1

MAX * BILL OF MATERIALS * SINGLE-LEVEL EXPLOSION BY LOW LEVEL SEQUENCE

LOW LEVEL	PART IDENTIFIER	DESCRIPTION	EFF DATE	QTY/ASSY	Q C	UOM	LT OFF	T C	SCRAP	ECN NUMBER	ALTERNATE PART
0	11000	PC - 2 FLOPPY - 128K				EA					
1	13000	PC KEYBOARD	830101	1.00	U	EA	0	N	0		
1	14000	MONITOR	830101	1.00	U	EA	0	N	0		
1	15000	PC, SYSTEM UNIT - 128K	830101	1.00	U	EA	0	N	0		
1	LABOR	FINAL ASSY LABOR	830101	1.00	U	DL	0	N	0		

FIGURE 4–4 B/M-Single-Level Explosion by Low Level Sequence Report (Courtesy of Micro-MRP, Inc.).

all involved personnel are working on the same plan; and that the master schedule that has been developed will achieve the company's overall objectives.

A master scheduling module can provide these features:

- Maintains customers forecast and master schedule data on-line instantly
- Maintains master schedule at any level or multiple levels of the product structure
- Stores demand and supply data in an unlimited planning horizon
- Integrates customer demand with forecast, and schedule data for accurate scheduling by part using "availability to promise" and "availability to forecast" logic
- Accomplishes time-phased rough cut capacity planning with user-defined resources to plan work hours, capital, or work center requirements
- Approves master schedule orders for release to the shop floor
- Prints routings and pick lists for master scheduled orders that have been released for manufacturing

Figure 4–5 shows a master schedule summary report by part identifier. This report enables a user to analyze the production schedule in a weekly, summarized format up to fifty-two weeks, and to compare supply against forecast demand (available to forecast) and against actual customer and dependent demand (available to promise) to optimize inventory levels.

The formulas used in this segment of master scheduling are:

Available to Forecast = on-hand inventory + master schedule orders – forecast demand
Available to Promise = on-hand inventory + master schedule orders – customer orders – dependent demand

11-04-1985 14:37:12 MICRO-MRP, INC. PAGE NO. 1

MAX * MASTER SCHEDULE * SUMMARY REPORT BY PART IDENTIFIER

PART : 11000 DESCRIPTION 1 : PC - 2 FLOPPY - 128K DESCRIPTION 2 : FCST : 0

COMMODITY : FIN GOOD TYPE : M POLICY : D BOM UOM : EA COST UOM : EA COST CONV : 1

MFG LEADTIME : 5 PUR LEADTIME : 0 LOW LEVEL : 0 QUM HOURS : 2.50

PLANNER : SAM BUYER : 000 SS : 0 MIN : 0 MAX : 0 MULT : 0

ISSUES MTD : 0 ISSUES YTD : 0 SALES MTD : 0 SALES YTD : 0 ON-HAND : 62

PERIOD ENDING	OVERDUE	11/08/85	11/15/85	11/22/85	11/29/85	12/06/85	12/13/85	12/20/85	12/27/85
FORECAST DEMAND	150.00	100.00	200.00	150.00	300.00	0.00	0.00	0.00	0.00
CUSTOMER DEMAND	300.00	150.00	175.00	0.00	0.00	275.00	0.00	0.00	0.00
DEPENDENT DEMAND	0.00	0.00	0.00	0.00	0.00	0.00	0.00	0.00	0.00
TOTAL DEMAND	300.00	150.00	175.00	0.00	0.00	275.00	0.00	0.00	0.00
MASTER SCHEDULE	600.00	200.00	150.00	250.00	0.00	200.00	0.00	0.00	0.00
AVAILABLE TO FORECAST	512.00	612.00	562.00	662.00	362.00	562.00	562.00	562.00	562.00
AVAILABLE TO PROMISE	362.00	412.00	387.00	637.00	637.00	562.00	562.00	562.00	562.00

FIGURE 4–5 Master Schedule-Summary Report by Part Identifier (Courtesy of Micro-MRP, Inc.).

Figure 4–6 shows a master schedule detail report by part identifier. This report enables a user to review day-to-day detail, including order life-cycle (status), order due dates and quantities, along with available to forecast and available to promise logic. This module can sort by part, ABC class code, commodity code, or part type.

Figure 4–7 shows a master schedule resource detail report by part identifier. This report is used for rough cut capacity planning to analyze and smooth demands on critical resources, and to identify future over-capacity situations before implementing the production plan.

Approved master schedule orders are released to the shop floor to start the production process. This takes the form of an automatically generated pick list and route sheet based on the standard bill of materials and part routing (figures 4–8 and 4–9).

As discussed in subsequent chapters, systems also are in use for pickless paperwork, and magnetic tape and bar coded route sheets. However, this discussion covers the more basic on-line method of automatically generated documents to handle releases to the manufacturing environment.

Material Requirements Planning Module: Achieving Balance between Supply and Demand

The material requirements planning (MRP) module balances the demand and supply for all component parts needed to meet the requirements of the master schedule. It provides a user with an automated, accurate planning and replanning vehicle that can respond to changes on the shop floor.

11-04-1985 14:13:37 MICRO-MRP, INC. PAGE NO. 1

MAX * MASTER SCHEDULE * DETAIL REPORT BY PART IDENTIFIER

PART : 11000 DESCRIPTION 1 : PC - 2 FLOPPY - 128K DESCRIPTION 2 :

COMMODITY : FIN GOOD TYPE : M POLICY : D BOM UOM : EA COST UOM : EA COST CONV : 1

MFG LEADTIME : 5 PUR LEADTIME : 0 LOW LEVEL : 0 QUM HOURS : 2.50

PLANNER : SAM BUYER : 000 SS : 0 MIN : 0 MAX : 0 MULT : 0

ISSUES MTD : 0 ISSUES YTD : 0 SALES MTD : 0 SALES YTD : 0 ON-HAND : 62

TYPE	ORDER NUM	STAT	FP	CUR. DUE	ORG. QTY	CUR. QTY	INDEP DEM	DEP DEM	TOTL DEM	SCHED REC	AVL FCST	AVL PROM
CU	2000030000	3		10/15/85	100.00	100.00	100.00	100.00	100.00	0.00	62.00	-38.00
MS	3000140000	1	Y	10/18/85	200.00	200.00	0.00	100.00	100.00	200.00	262.00	162.00
FC	1000100000	3		10/19/85	150.00	150.00	250.00	100.00	250.00	200.00	112.00	162.00
CU	2000040000	3		10/25/85	200.00	200.00	450.00	300.00	450.00	200.00	112.00	-38.00
MS	3000090000	1	Y	10/25/85	100.00	100.00	150.00	300.00	450.00	300.00	212.00	62.00
MS	3000150000	1	Y	11/01/85	300.00	300.00	150.00	300.00	450.00	600.00	512.00	362.00
CU	2000050000	3		11/03/85	150.00	150.00	600.00	450.00	600.00	600.00	512.00	212.00
MS	3000100000	1	Y	11/03/85	200.00	200.00	150.00	450.00	600.00	800.00	712.00	412.00
FC	1000110000	3		11/05/85	100.00	100.00	700.00	450.00	700.00	800.00	612.00	412.00
CU	2000060000	3		11/10/85	175.00	175.00	875.00	625.00	875.00	800.00	612.00	237.00
FC	1000120000	3		11/10/85	200.00	200.00	1075.00	625.00	1075.00	800.00	412.00	237.00
MS	3000110000	1	Y	11/10/85	150.00	150.00	450.00	625.00	1075.00	950.00	562.00	387.00
FC	1000130000	3		11/18/85	150.00	150.00	1225.00	625.00	1225.00	950.00	412.00	387.00
MS	3000120000	1	Y	11/20/85	250.00	250.00	600.00	625.00	1225.00	1200.00	662.00	637.00
FC	1000140000	3		11/25/85	300.00	300.00	1525.00	625.00	1525.00	1200.00	362.00	637.00
CU	2000070000	3		12/01/85	275.00	275.00	1800.00	900.00	1800.00	1200.00	362.00	362.00
MS	3000130000	1	Y	12/01/85	200.00	200.00	900.00	900.00	1800.00	1400.00	562.00	562.00

FIGURE 4–6 Master Schedule—Detail Report by Part Identifier (Courtesy of Micro-MRP, Inc.).

11-04-1985 14:21:20 MICRO-MRP, INC. PAGE NO. 1

MAX * MASTER SCHEDULE * RESOURCE DETAIL REPORT BY PART IDENTIFIER

PART : LABOR DESCRIPTION 1 : FINAL ASSY LABOR DESCRIPTION 2 : PLANNER : 000

COMMODITY : RESOURCE TYPE : R BOM UOM : DL COST UOM : DL COST CONV : 1 ON-HAND : 0

TYPE	ORDER NUM	STAT	CUR. DUE	ORG. QTY	CUR. QTY	FC DEM	TOTL DEM	AVL DEM
RQ	3000050000	1	02/22/85	120.00	120.00	0.00	120.00	-120.00
PL	4001920000	1	02/22/85	120.00	120.00	0.00	120.00	0.00
RQ	3000060000	1	05/24/85	144.00	144.00	0.00	264.00	-144.00
PL	4001930000	1	05/24/85	144.00	144.00	0.00	264.00	0.00
RQ	3000070000	1	08/23/85	156.00	156.00	0.00	420.00	-156.00
PL	4001940000	1	08/23/85	156.00	156.00	0.00	420.00	0.00
RQ	3000140000	1	10/11/85	200.00	200.00	0.00	620.00	-200.00
PL	4001950000	1	10/11/85	200.00	200.00	0.00	620.00	0.00
RQ	3000090000	1	10/18/85	100.00	100.00	0.00	720.00	-100.00
PL	4001960000	1	10/18/85	100.00	100.00	0.00	720.00	0.00
RQ	3000100000	1	10/25/85	200.00	200.00	0.00	920.00	-200.00
RQ	3000150000	1	10/25/85	300.00	300.00	0.00	1220.00	-500.00
PL	4001970000	1	10/25/85	200.00	200.00	0.00	1220.00	-300.00
PL	4001980000	1	10/25/85	300.00	300.00	0.00	1220.00	0.00
RQ	3000110000	1	11/01/85	150.00	150.00	0.00	1370.00	-150.00
PL	4001990000	1	11/01/85	150.00	150.00	0.00	1370.00	0.00
RQ	3000120000	1	11/13/85	250.00	250.00	0.00	1620.00	-250.00
PL	4002000000	1	11/13/85	250.00	250.00	0.00	1620.00	0.00
RQ	3000080000	1	11/22/85	168.00	168.00	0.00	1788.00	-168.00
RQ	3000130000	1	11/22/85	200.00	200.00	0.00	1988.00	-368.00
PL	4002010000	1	11/22/85	168.00	168.00	0.00	1988.00	-200.00
PL	4002020000	1	11/22/85	200.00	200.00	0.00	1988.00	0.00

FIGURE 4-7 Master Schedule—Resource Detail Report by Part Identifier (Courtesy of Micro-MRP, Inc.).

PART IDENTIFIER : 13000

* * PICK LIST * *

DESCRIPTION 1 : PC KEYBOARD

SHOP ORDER : 500003

ORDER QUANTITY: 200.00

COMPONENT PART	DESCRIPTION	UOM	DUE QUANTITY	STOCK LOCATION	---- PICK ---- QTY	LOC
50001	PC KEYBOARD CASE	EA	200.00	STOCK-A	______	______
70005	PC KEYBOARD KEY SET	EA	200.00	STOCK-B	______	______
70006	POWER CABLE, 6 FEET	EA	200.00	STOCK-B	______	______
70007	SCREWS, 1/4" X #4	EA	800.00	STOCK-B	______	______

FIGURE 4-8 Pick List (Courtesy of Micro-MRP, Inc.).

PART IDENTIFIER : 13000

* * * ROUTING * * *

DESCRIPTION 1 : PC KEYBOARD

SHOP ORDER : 500003

ORDER QUANTITY: 200.00

DUE DATE : 11/13/85

SEQ#	DESCRIPTION	WRKCT	RUN TIME	TYPE	LOAD QUANTITY	-- COMPLETION -- QUANTITY	INIT
0010	PICK & STAGE	PICK	0.0000	U	200.00	_________	____
0020	FINAL ASSEMBLY	ASSY	1000.0000	U	200.00	_________	____
	TOOL REF :JIG 53467						
0030	INSPECTION	INSP	400.0000	U	200.00	_________	____
	TOOL REF :JIG 77777						

FIGURE 4-9 Route Sheet (Courtesy of Micro-MRP, Inc.).

This module should enable a user to obtain up-to-the-minute (supply and demand) information in weekly summary buckets or in complete detail. The MRP module reschedules existing orders to meet changes in the current demand and create new orders, if required to increase supply to meet new demand. It examines all the details of a component before suggesting new orders, such as the quantity on-hand, existing scheduled receipts, the manufacturing or purchasing lead time, and the due date of the order. "Order qualifiers" such as minimum, maximum, multiple order quantities, and safety stock can be used to control the process. Due dates for component parts are calculated automatically using "backward scheduling" so that the right part gets to the required operation at the right time, but not before it can build up work-in-process inventory carrying costs. The system should enable a user to have the option of running MRP on either an exception (net change) basis, or on a complete (regen) basis for all parts.

A material requirements planning module provides these features:

- Calculates component requirements with net change or regen explosion logic
- Reviews demand and supply in detail (bucketless) or summarize (bucketed) format
- Reviews time-phased shortage reports, demands, and scheduled receipts for any level of the bill of materials
- Approves planned orders to either manufacture or purchase any part (make/buy decision)
- Plans parts using standard MRP order policies, reorder point logic, or weekly order policy
- Reschedules order due dates automatically based on demand-supply balance, but also is able to "freeze" (firm plan) the date of any order
- Creates exception messages for order release, rescheduling, and identification of parts that need further attention
- Defines a firm planned stability window to control orders within a user-defined planning horizon

Figure 4–10 shows a part list by part identifier report. This report can be stored by planner, part, part types, commodity code, or ABC class code.

Figure 4–11 shows an order signal report by part type code. This report contains all of the currently unapproved planned orders generated by the MRP explosion. Order due and start dates are shown, and any start date prior to the current date is flagged with an asterisk (*). It can be sorted by planner, part, part type, commodity code, or ABC class code.

Figure 4–12 shows an explosion shortage report. This report can be used to predict component shortages up to thirty-nine weeks in the future, based on all current demands and approved scheduled receipts.

Figure 4–13 shows an implosion shortage report. This report can be used to analyze the demands created by every parent part that uses critical components, and can make allocation decisions with the knowledge of how that will affect finished goods.

Figure 4–14 shows an MRP explosion exception message report. The net change explosion shown on this report replans only those component parts whose supply and demand is no longer in balance. Exception messages (identifying parts that need immediate attention) are either sent directly to the printer or can be saved for later printing. Variations in exception messages can be developed depending on user requirements. This illustration contains ten different types of exceptions.

11-04-1985 14:31:07 MICRO-MRP, INC. PAGE NO. 1

MAX * MRP * PART LIST BY PART IDENTIFIER

PART IDENTIFIER	DESCRIPTION	TYP CDE	CLS CDE	POL CDE	COMMODITY	PLANR	SAFETY STOCK	MIN QTY	MAX QTY	MULT QTY	AVE QTY	LAST TRANS DATE
11000	PC - 2 FLOPPY - 128K	M	A	D	FIN GOOD	SAM	0.00	0.00	0.00	0.00	0.00	10/10/85
12000	PC - 10 MEG HD - 128K	M	A	D	FIN GOOD	SAM	0.00	0.00	0.00	0.00	0.00	12/18/83
13000	PC KEYBOARD	A	A	L	FIN GOOD	SAM	0.00	100.00	1000.00	100.00	0.00	10/10/85
14000	MONITOR	A	A	L	FIN GOOD	SAM	0.00	100.00	1000.00	100.00	0.00	12/18/83
15000	PC, SYSTEM UNIT - 128K	A	A	L	FIN GOOD	SAM	0.00	100.00	1000.00	1000.00	0.00	12/18/83
16000	PC-HD, SYSTEM UNIT - 128K	A	A	L	FIN GOOD	SAM	0.00	100.00	1000.00	100.00	0.00	12/18/83
50001	PC KEYBOARD CASE	A	B	L	MFG ASSY	SUE	0.00	50.00	500.00	50.00	0.00	10/10/85
50002	MONITOR CASE	A	B	L	MFG ASSY	SUE	0.00	50.00	500.00	50.00	0.00	12/18/83
50003	PC MOTHERBOARD - REV A	A	B	L	MFG ASSY	JOE	50.00	100.00	1000.00	100.00	0.00	12/18/83
50004	CASE ASSEMBLY	A	B	L	MFG ASSY	JOE	0.00	100.00	500.00	50.00	0.00	10/10/85
50005	SYSTEM UNIT BASE	A	B	L	MFG ASSY	JOE	0.00	50.00	500.00	50.00	0.00	12/18/83
50006	SYSTEM UNIT COVER	A	B	L	MFG ASSY	SUE	0.00	50.00	500.00	50.00	0.00	12/18/83
70001	KEYBOARD CASE TOP	A	B	L	FAB ASSY	SUE	0.00	50.00	500.00	50.00	0.00	12/18/83
70002	KEYBOARD CASE BOTTOM	A	B	R	FAB ASSY	SUE	0.00	50.00	500.00	50.00	0.00	12/18/83
70003	KEYBOARD FEET	B	C	R	PUR PART	JAN	0.00	0.00	0.00	0.00	0.00	12/18/83
70004	PAINT, PC GREY	B	C	R	PUR PART	JAN	5.00	0.00	0.00	0.00	0.00	12/18/83
70005	PC KEYBOARD KEY SET	B	C	L	PUR PART	JAN	0.00	100.00	1000.00	100.00	0.00	12/18/83
70006	POWER CABLE, 6 FEET	B	C	L	PUR PART	JAN	0.00	100.00	2000.00	100.00	0.00	10/10/85
70007	SCREWS, 1/4" X #4	B	C	R	PUR PART	JAN	1000.00	0.00	0.00	0.00	0.00	12/18/83
70008	CRT, 10", GREEN	B	B	L	PUR ASSY	JAN	10.00	100.00	500.00	10.00	0.00	12/18/83
70009	MONITOR FRONT	A	C	L	FAB ASSY	SUE	0.00	100.00	500.00	50.00	0.00	12/18/83
70010	MONITOR BACK	A	C	L	FAB ASSY	SUE	0.00	100.00	500.00	10.00	0.00	12/18/83
70011	MONITOR CORD, 24"	B	C	R	PUR PART	JAN	25.00	0.00	0.00	0.00	0.00	12/18/83
70012	KEYBOARD FEET, 1/2"	B	C	R	PUR PART	JAN	100.00	0.00	0.00	0.00	0.00	12/18/83
70013	PC BOARD	B	C	L	PUR ASSY	JAN	0.00	200.00	1000.00	10.00	0.00	12/18/83
70014	16K RAM CHIP	B	B	R	PUR PART	JAN	100.00	0.00	0.00	0.00	0.00	12/18/83
70015	8088 PROCESSOR	B	B	R	PUR PART	JAN	100.00	0.00	0.00	0.00	0.00	12/18/83
70016	METAL, 4 X 4, SHEET	B	C	L	PUR PART	JAN	0.00	100.00	1000.00	100.00	0.00	12/18/83
70017	FLOPPY DRIVE, 320K, DSDD	B	A	L	PUR ASSY	JAN	0.00	100.00	1000.00	100.00	0.00	12/18/83
70018	SPEAKER, 2", 4 OZ	B	C	R	PUR PART	JAN	25.00	0.00	0.00	0.00	0.00	12/18/83
70019	POWER SUPPLY, 120 VOLT	B	A	L	PUR ASSY	JAN	0.00	100.00	600.00	100.00	0.00	12/18/83
70020	HARD DISK, 10 MEG	B	A	L	PUR ASSY	JAN	25.00	100.00	800.00	10.00	0.00	12/18/83
LABOR	FINAL ASSY LABOR	R	D	D	RESOURCE	000	0.00	0.00	0.00	0.00	0.00	00/00/00

FIGURE 4–10 Part List by Part Identifier Report (Courtesy of Micro-MRP, Inc.).

Inventory Control Module Tells What Happened and When

The *inventory control module* tracks and maintains inventory balances for each stocked part, including finished goods, assemblies, component parts, and raw materials. It provides the user with an ability to kit an order or to transfer parts between stockrooms.

This module keeps a history of all inventory transactions as an audit trail of what happened and when. It lets the user analyze surpluses to eliminate obsolete inventory and reduce inventory costs.

An inventory control module has these features:

- Instant on-line entry of receipts, issues, and cycle counts
- Automatically generates inventory value reports at standard cost
- Maintains multiple stockrooms per part and multiple parts per stockroom
- Creates user-defined stockrooms that are either nettable or nonnettable for MRP
- Transfers among stockrooms with one entry
- Issues all components for an order with one transaction
- Maintains separate units of measure for costing and stocking activities

MAX * MRP * ALL ORDER SIGNAL BY PART TYPE CODE

PART IDENTIFIER	DESCRIPTION	TYP CDE	CLS CDE	POL CDE	COMMODITY	PLANR	BUYER	ORDER	LT	ORIG DUE	START DT	CUR DUE	ORD. QTY
13000	PC KEYBOARD	A	A	L	FIN GOOD	SAM	000	400168	5	03/01/85	02/22/85	03/01/85	100.00
								400169	5	05/24/85	05/17/85	05/24/85	100.00
								400170	5	08/23/85	08/16/85	08/23/85	100.00
								400171	5	10/11/85	10/04/85	10/11/85	200.00
								400172	5	10/18/85	10/11/85	10/18/85	100.00
								400173	5	10/25/85	10/18/85	10/25/85	500.00
								400175	5	11/13/85	11/06/85	11/13/85	200.00
								400176	5	11/22/85	11/15/85	11/22/85	400.00
								400044	5	12/06/85	11/29/85	12/06/85	200.00
14000	MONITOR	A	A	L	FIN GOOD	SAM	000	400177	5	02/22/85	02/15/85	02/22/85	100.00
								400178	5	05/24/85	05/17/85	05/24/85	100.00
								400179	5	08/23/85	08/16/85	08/23/85	100.00
								400181	5	10/18/85	10/11/85	10/18/85	100.00
								400182	5	10/25/85	10/18/85	10/25/85	500.00
								400183	5	11/01/85	10/25/85	11/01/85	200.00
								400184	5	11/13/85	11/06/85	11/13/85	200.00
								400185	5	11/22/85	11/15/85	11/22/85	400.00
								400045	5	12/06/85	11/29/85	12/06/85	200.00
15000	PC, SYSTEM UNIT - 128K	A	A	L	FIN GOOD	SAM	000	400186	5	10/11/85	10/04/85	10/11/85	1000.00
								400187	5	11/13/85	11/06/85	11/13/85	1000.00
16000	PC-HD, SYSTEM UNIT - 128K	A	A	L	FIN GOOD	SAM	000	400188	5	02/22/85	02/15/85	02/22/85	100.00
								400189	5	05/24/85	05/17/85	05/24/85	100.00
								400190	5	08/23/85	08/16/85	08/23/85	100.00
								400191	5	11/22/85	11/15/85	11/22/85	200.00
50001	PC KEYBOARD CASE	A	B	L	MFG ASSY	SUE	000	400203	5	02/22/85	02/15/85	02/22/85	50.00
								400204	5	05/17/85	05/10/85	05/17/85	100.00
								400205	5	08/16/85	08/09/85	08/16/85	100.00
								400206	5	10/04/85	09/27/85	10/04/85	200.00
								400207	5	10/11/85	10/04/85	10/11/85	100.00
								400208	5	10/18/85	10/11/85	10/18/85	500.00
								400209	5	10/25/85	10/18/85	10/25/85	200.00
								400210	5	11/06/85	10/30/85	11/06/85	200.00
								400211	5	11/15/85	11/08/85	11/15/85	400.00
								400047	5	11/29/85	11/22/85	11/29/85	200.00
50002	MONITOR CASE	A	B	L	MFG ASSY	SUE	000	400212	5	02/15/85	02/08/85	02/15/85	50.00
								400213	5	05/17/85	05/10/85	05/17/85	100.00
								400214	5	08/16/85	08/09/85	08/16/85	100.00
								400215	5	10/04/85	09/27/85	10/04/85	200.00
								400216	5	10/11/85	10/04/85	10/11/85	100.00
								400217	5	10/18/85	10/11/85	10/18/85	500.00
								400218	5	10/25/85	10/18/85	10/25/85	200.00
								400219	5	11/06/85	10/30/85	11/06/85	200.00
								400220	5	11/15/85	11/08/85	11/15/85	400.00
								400048	5	11/29/85	11/22/85	11/29/85	200.00
50003	PC MOTHERBOARD - REV A	A	B	L	MFG ASSY	JOE	000	400221	5	05/17/85	05/10/85	05/17/85	100.00
								400222	5	08/16/85	08/09/85	08/16/85	100.00
								400223	5	10/04/85	09/27/85	10/04/85	1000.00
								400224	5	11/06/85	10/30/85	11/06/85	1000.00
								400225	5	11/15/85	11/08/85	11/15/85	200.00
50004	CASE ASSEMBLY	A	B	L	MFG ASSY	JOE	000	400226	5	02/15/85	02/08/85	02/15/85	100.00

FIGURE 4–11 Order Signal Report by Part Type Code (Courtesy of Micro-MRP, Inc.).

11-04-1985 16:16:34 MICRO-MRP, INC. PAGE NO. 1

MAX * MRP * EXPLOSION SHORTAGE REPORT

PARENT PART ID : 11000 DESCRIPTION : PC - 2 FLOPPY - 128K LEADTIME : 5

WEEK ENDING	QTYPER	ONHAND	11/08/85	11/15/85	11/22/85	11/29/85	12/06/85	12/13/85	12/20/85	12/27/85	01/03/86	01/10/86	01/17/86	01/24/86
CUM DEMAND														
11000		62	800	950	1200	1200	1400	1400	1600	1600	1600	1600	1600	1600
NET AVAILABLE														
13000	1	151	-949	-1199	-1539	-1539	-1739	-1739	-1739	-1739	-1739	-1739	-1739	-1739
14000	1	78	-1022	-1272	-1612	-1612	-1812	-1812	-1812	-1812	-1812	-1812	-1812	-1812
15000	1	59	-891	-1141	-1341	-1341	-1541	-1541	-1541	-1541	-1541	-1541	-1541	-1541
LABOR	1	0	-1370	-1620	-1988	-1988	-2188	-2188	-2188	-2188	-2188	-2188	-2188	-2188

FIGURE 4–12 Explosion Shortage Report (Courtesy of Micro-MRP, Inc.).

11-04-1985 16:18:12 MICRO-MRP, INC. PAGE NO. 1

MAX * MRP * IMPLOSION SHORTAGE REPORT

COMPONENT PART ID: 13000
ON-HAND BALANCE : 151.00

DESCRIPTION: PC KEYBOARD LEADTIME: 5

WEEK ENDING	QTYPER	11/08/85	11/15/85	11/22/85	11/29/85	12/06/85	12/13/85	12/20/85	12/27/85	01/03/86	01/10/86	01/17/86	01/24/86	01/31/86
SUPPLY														
13000	1	200	0	0	0	0	0	0	0	0	0	0	0	0
LEVEL REQUIREMENTS														
1 11000	1	950	250	200	0	200	0	0	0	0	0	0	0	0
1 12000	1	350	0	140	0	0	0	0	0	0	0	0	0	0
TOTAL DEMAND		1300	250	340	0	200	0	0	0	0	0	0	0	0
NET AVAILABLE		-949	-250	-340	0	-200	0	0	0	0	0	0	0	0
CUMULATIVE NET AVAILABLE		-949	-1199	-1539	-1539	-1739	-1739	-1739	-1739	-1739	-1739	-1739	-1739	-1739

FIGURE 4–13 Implosion Shortage Report (Courtesy of Micro-MRP, Inc.).

Message Description	#	CODE
Below Reorder Point	01	BR
Order was De-expedited	02	DE
Order would have been De-expedited but it was firm-planned	03	DF
No Activity Index Record	04	NI
Order would have been Rescheduled but it was firm-planned	05	RF
Order was rescheduled	06	RE
Order created in less than leadtime	07	LL
Order rescheduled to a past due date	08	PD
Positive available at end of window	09	PA
Negative available at end of window	10	NA
Excess order coverage	11	EO
Reserved for future use	12	

FIGURE 4–14 MRP Explosion Exception Message (Courtesy of Micro-MRP, Inc.).

- Receives or issues in decimal quantities
- Uses cycle count logic for inventory accuracy and control
- Defines cycle count tolerances in both dollar amounts and in percentages for accurate inventory tracking
- Displays or prints inventory transaction histories for all transactions, by part and by order number
- Displays all the parts in a stockroom or all the stockrooms where a part is stored

Figure 4–15 shows an inventory master parts list report by class code. This report indicates the stocking position of parts by commodity class.

Figure 4–16 shows an inventory control cycle count report by stockroom ID sequence. This exception report is used to zero-in on parts that have not been counted during their specified cycle count period. It identifies how many times a part was counted and how many

11-04-1985 11:21:41 MICRO-MRP, INC. PAGE NO. 1

MAX * INVENTORY CONTROL * INVENTORY MASTER LIST BY CLASS CODE

PART IDENTIFIER	DESCRIPTION	T C	C C	DELIVER TO STOCK	REORDER POINT	REORDER QTY	SAFETY STOCK	MIN QTY	MAX QTY	MULT QTY	AVE QTY
11000	PC - 2 FLOPPY - 128K	M	A	STOCK-A	0	0	0	0	0	0	0
12000	PC - 10 MEG HD - 128K	M	A	STOCK-A	0	0	0	0	0	0	0
13000	PC KEYBOARD	A	A	STOCK-A	0	0	0	100	1000	100	0
14000	MONITOR	A	A	STOCK-A	0	0	0	100	1000	100	0
15000	PC, SYSTEM UNIT - 128K	A	A	STOCK-A	0	0	0	100	1000	1000	0
16000	PC-HD, SYSTEM UNIT - 128K	A	A	STOCK-A	0	0	0	100	1000	100	0
70017	FLOPPY DRIVE, 320K, DSDD	B	A	STOCK-B	0	0	0	100	1000	100	0
70019	POWER SUPPLY, 120 VOLT	B	A	STOCK-B	0	0	0	100	600	100	0
70020	HARD DISK, 10 MEG	B	A	STOCK-B	0	0	25	100	800	10	0
50001	PC KEYBOARD CASE	A	B	STOCK-A	0	0	0	50	500	50	0
50002	MONITOR CASE	A	B	STOCK-A	0	0	0	50	500	50	0
50003	PC MOTHERBOARD - REV A	A	B	STOCK-B	0	0	50	100	1000	100	0
50004	CASE ASSEMBLY	A	B	STOCK-A	0	0	0	100	500	50	0
50005	SYSTEM UNIT BASE	A	B	STOCK-A	0	0	0	50	500	50	0
50006	SYSTEM UNIT COVER	A	B	STOCK-A	0	0	0	50	500	50	0
70001	KEYBOARD CASE TOP	A	B	STOCK-A	0	0	0	50	500	50	0
70002	KEYBOARD CASE BOTTOM	A	B	STOCK-A	0	0	0	50	500	50	0
70008	CRT, 10", GREEN	B	B	STOCK-B	0	0	10	100	500	10	0
70014	16K RAM CHIP	B	B	STOCK-B	1000	5000	100	0	0	0	0
70015	8088 PROCESSOR	B	B	STOCK-B	200	500	100	0	0	0	0
70003	KEYBOARD FEET	B	C	STOCK-B	500	2000	0	0	0	0	0
70004	PAINT, PC GREY	B	C	STOCK-B	10	100	5	0	0	0	0
70005	PC KEYBOARD KEY SET	B	C	STOCK-B	0	0	0	100	1000	100	0
70006	POWER CABLE, 6 FEET	B	C	STOCK-B	0	0	0	100	2000	100	0
70007	SCREWS, 1/4" X #4	B	C	STOCK-B	1000	10000	1000	0	0	0	0
70009	MONITOR FRONT	A	C	STOCK-A	0	0	0	100	500	50	0
70010	MONITOR BACK	A	C	STOCK-A	0	0	0	100	500	10	0
70011	MONITOR CORD, 24"	B	C	STOCK-B	200	600	25	0	0	0	0
70012	KEYBOARD FEET, 1/2"	B	C	STOCK-B	1000	5000	100	0	0	0	0
70013	PC BOARD	B	C	STOCK-B	0	0	0	200	1000	10	0
70016	METAL, 4 X 4, SHEET	B	C	STOCK-B	0	0	0	100	1000	100	0
70018	SPEAKER, 2", 4 OZ	B	C	STOCK-B	200	600	25	0	0	0	0
PHANTOM		P	C		0	0	0	0	0	0	0
140836135	REAR FILLER	M	C		0	0	8900	0	0	0	0
140836135-	REAR FILLER	B	C		0	0	0	0	0	0	0
TOLYAL	SOLVENT	B	C		30	120	15	0	0	0	0
WA3967	PAINT WHITE	B	C		30	120	15	0	0	0	0
LABOR	FINAL ASSY LABOR	B	D		0	0	0	0	0	0	0

FIGURE 4–15 Inventory Master Parts List Report by Class Code (Courtesy of Micro-MRP, Inc.).

11-04-1985 11:36:27 MICRO-MRP, INC. PAGE NO. 1

MAX * INVENTORY CONTROL * CYCLE COUNT REPORT BY STOCKROOM ID

PART IDENTIFIER	DESCRIPTION	T C	C C	COMMODITY CODE	STOCK ID	ON-HAND QTY	ON-HAND VALUE	C C	LAST CYC DATE	YTD CYC	TOL %	TOL $	CYC OOT
TOLYAL	SOLVENT	B	C	RAW	PAINT	100.00	0.00	W	00/00/00	0	8	55	0
11000	PC - 2 FLOPPY - 128K	M	A	FIN GOOD	STOCK-A	52.00	21358.40	W	10/25/85	1	0	0	1
12000	PC - 10 MEG HD - 128K	M	A	FIN GOOD	STOCK-A	53.00	48338.04	W	00/00/00	0	0	0	1
13000	PC KEYBOARD	A	A	FIN GOOD	STOCK-A	151.00	4386.05	W	00/00/00	0	0	100	1
14000	MONITOR	A	A	FIN GOOD	STOCK-A	78.00	10270.78	W	00/00/00	0	0	0	1
15000	PC, SYSTEM UNIT - 128K	A	A	FIN GOOD	STOCK-A	59.00	13157.89	W	00/00/00	0	0	100	1
16000	PC-HD, SYSTEM UNIT - 128K	A	A	FIN GOOD	STOCK-A	82.00	58951.04	W	00/00/00	0	0	100	1
50001	PC KEYBOARD CASE	A	B	MFG ASSY	STOCK-A	57.00	1336.46	M	00/00/00	0	0	100	1
50002	MONITOR CASE	A	B	MFG ASSY	STOCK-A	62.00	2427.30	M	00/00/00	0	0	100	1
50004	CASE ASSEMBLY	A	B	MFG ASSY	STOCK-A	63.00	2832.34	M	00/00/00	0	0	100	1
50005	SYSTEM UNIT BASE	A	B	MFG ASSY	STOCK-A	123.00	1678.95	M	00/00/00	0	0	100	1
50006	SYSTEM UNIT COVER	A	B	MFG ASSY	STOCK-A	17.00	416.67	M	00/00/00	0	0	100	1
70001	KEYBOARD CASE TOP	A	B	FAB ASSY	STOCK-A	49.00	330.75	M	00/00/00	0	0	100	1
70002	KEYBOARD CASE BOTTOM	A	B	FAB ASSY	STOCK-A	271.00	3414.60	M	00/00/00	0	0	100	1
70009	MONITOR FRONT	A	C	FAB ASSY	STOCK-A	79.00	1327.20	Q	00/00/00	0	0	100	1
70010	MONITOR BACK	A	C	FAB ASSY	STOCK-A	159.00	2695.05	Q	00/00/00	0	0	100	1
70006	POWER CABLE, 6 FEET	B	C	PUR PART	STOCK-A	30.00	16.50	Q	00/00/00	0	0	100	1
50003	PC MOTHERBOARD - REV A	A	B	MFG ASSY	STOCK-B	179.00	4564.50	M	00/00/00	0	0	100	1
70003	KEYBOARD FEET	B	C	PUR PART	STOCK-B	1984.00	218.24	Q	00/00/00	0	0	100	1
70004	PAINT, PC GREY	B	C	PUR PART	STOCK-B	28.00	4.81	Q	00/00/00	0	0	100	0
70005	PC KEYBOARD KEY SET	B	C	PUR PART	STOCK-B	62.00	6.82	Q	00/00/00	0	0	100	0
70006	POWER CABLE, 6 FEET	B	C	PUR PART	STOCK-B	379.00	208.45	Q	00/00/00	0	0	100	1
70007	SCREWS, 1/4" X #4	B	C	PUR PART	STOCK-B	4500.00	495.00	Q	00/00/00	0	0	100	1
70008	CRT, 10", GREEN	B	B	PUR ASSY	STOCK-B	257.00	21202.50	Q	00/00/00	0	0	200	1
70011	MONITOR CORD, 24"	B	C	PUR PART	STOCK-B	391.00	86.02	Q	00/00/00	0	0	100	0
70012	KEYBOARD FEET, 1/2"	B	C	PUR PART	STOCK-B	1044.00	95.70	Q	00/00/00	0	0	100	0
70013	PC BOARD	B	C	PUR ASSY	STOCK-B	295.00	649.00	Q	00/00/00	0	0	100	1
70014	16K RAM CHIP	B	B	PUR PART	STOCK-B	1259.00	1384.90	M	00/00/00	0	0	100	1
70015	8088 PROCESSOR	B	B	PUR PART	STOCK-B	251.00	1380.50	M	00/00/00	0	0	100	1
70016	METAL, 4 X 4, SHEET	B	C	PUR PART	STOCK-B	316.00	2085.60	Q	00/00/00	0	0	100	1
70017	FLOPPY DRIVE, 320K, DSDD	B	A	PUR ASSY	STOCK-B	235.00	12925.00	W	00/00/00	0	0	200	1
70018	SPEAKER, 2", 4 OZ	B	C	PUR PART	STOCK-B	169.00	102.25	Q	00/00/00	0	0	100	1
70019	POWER SUPPLY, 120 VOLT	B	A	PUR ASSY	STOCK-B	216.00	5940.00	W	00/00/00	0	0	100	1
70020	HARD DISK, 10 MEG	B	A	PUR ASSY	STOCK-B	79.00	43450.00	W	00/00/00	0	0	600	1
11000	PC - 2 FLOPPY - 128K	M	A	FIN GOOD	STOCK-B	10.00	4107.39	W	10/25/85	1	0	0	1
TOTALS							271904.72						

FIGURE 4–16 Inventory Control-Cycle Count Report by Stockroom Identifier Sequence (Courtesy of Micro-MRP, Inc.).

times the part was out of defined tolerances. This report sorts by part, part type, commodity code, planner, or ABC class code.

Figure 4–17 shows an inventory value report by class code. This report "dollarizes" inventory using computer-calculated standard costs. It sorts data by part, planner, part type, commodity code, or ABC class code for nettable, nonnettable, or all stock locations.

Figure 4–18 shows a physical inventory report by stockroom identifier. This report is generated as an aid in the physical inventory process, and contains information on the current on-hand inventory balance, or excluding the on-hand, at the user's discretion. It can be sorted by part or stockroom.

Figure 4–19 shows an obsolescence report by part type code. This report tracks the date of the last transaction and year-to-date usage to quickly identify parts to be considered for obsolescence. It can be sorted by planner, commodity code, part, ABC class code, and part type.

11-04-1985 11:27:16 MICRO-MRP, INC. PAGE NO. 1

MAX * INVENTORY CONTROL * INVENTORY VALUE REPORT BY CLASS CODE

PART IDENTIFIER	DESCRIPTION	T C	C C	COMMODITY CODE	ON-HAND QTY	ON-HAND VALUE $	ISSUES YTD $	SALES YTD $	LAST TNX DATE
11000	PC - 2 FLOPPY - 128K	M	A	FIN GOOD	62.00	25465.79	0.00	0.00	10/25/85
12000	PC - 10 MEG HD - 128K	M	A	FIN GOOD	53.00	48338.04	0.00	0.00	12/18/83
13000	PC KEYBOARD	A	A	FIN GOOD	151.00	4386.05	0.00	0.00	10/10/85
14000	MONITOR	A	A	FIN GOOD	78.00	10270.78	0.00	0.00	12/18/83
15000	PC, SYSTEM UNIT - 128K	A	A	FIN GOOD	59.00	13157.89	0.00	0.00	12/18/83
16000	PC-HD, SYSTEM UNIT - 128K	A	A	FIN GOOD	82.00	58951.04	0.00	0.00	12/18/83
70017	FLOPPY DRIVE, 320K, DSDD	B	A	PUR ASSY	235.00	12925.00	0.00	0.00	12/18/83
70019	POWER SUPPLY, 120 VOLT	B	A	PUR ASSY	216.00	5940.00	0.00	0.00	12/18/83
70020	HARD DISK, 10 MEG	B	A	PUR ASSY	79.00	43450.00	0.00	0.00	12/18/83
NON-NETTABLE AND ONHAND SUB TOTALS						222884.59	0.00	0.00	

FIGURE 4–17 Inventory Control—Inventory Value Report by Class Code (Courtesy of Micro-MRP, Inc.).

11-04-1985 11:41:24 MICRO-MRP, INC. PAGE NO. 1

MAX * INVENTORY CONTROL * PHYSICAL INVENTORY REPORT BY STOCKROOM ID

PART IDENTIFIER	DESCRIPTION	T C	C C	COMMODITY CODE	STOCK ID	ON-HAND QTY	PHYSICAL COUNT	COUNTED BY	VERIFIED COUNT	VERIFIED BY
TOLYAL	SOLVENT	B	C	RAW	PAINT	100.00				
11000	PC - 2 FLOPPY - 128K	M	A	FIN GOOD	STOCK-A	52.00				
12000	PC - 10 MEG HD - 128K	M	A	FIN GOOD	STOCK-A	53.00				
13000	PC KEYBOARD	A	A	FIN GOOD	STOCK-A	151.00				
14000	MONITOR	A	A	FIN GOOD	STOCK-A	78.00				
15000	PC, SYSTEM UNIT - 128K	A	A	FIN GOOD	STOCK-A	59.00				
16000	PC-HD, SYSTEM UNIT - 128K	A	A	FIN GOOD	STOCK-A	82.00				
50001	PC KEYBOARD CASE	A	B	MFG ASSY	STOCK-A	57.00				
50002	MONITOR CASE	A	B	MFG ASSY	STOCK-A	62.00				
50004	CASE ASSEMBLY	A	B	MFG ASSY	STOCK-A	63.00				
50005	SYSTEM UNIT BASE	A	B	MFG ASSY	STOCK-A	123.00				
50006	SYSTEM UNIT COVER	A	B	MFG ASSY	STOCK-A	17.00				
70001	KEYBOARD CASE TOP	A	B	FAB ASSY	STOCK-A	49.00				
70002	KEYBOARD CASE BOTTOM	A	B	FAB ASSY	STOCK-A	271.00				
70009	MONITOR FRONT	A	C	FAB ASSY	STOCK-A	79.00				
70010	MONITOR BACK	A	C	FAB ASSY	STOCK-A	159.00				
70006	POWER CABLE, 6 FEET	B	C	PUR PART	STOCK-A	30.00				
50003	PC MOTHERBOARD - REV A	A	B	MFG ASSY	STOCK-B	179.00				
70003	KEYBOARD FEET	B	C	PUR PART	STOCK-B	1984.00				
70004	PAINT, PC GREY	B	C	PUR PART	STOCK-B	28.00				
70005	PC KEYBOARD KEY SET	B	C	PUR PART	STOCK-B	62.00				
70006	POWER CABLE, 6 FEET	B	C	PUR PART	STOCK-B	379.00				
70007	SCREWS, 1/4" X #4	B	C	PUR PART	STOCK-B	4500.00				
70008	CRT, 10", GREEN	B	B	PUR ASSY	STOCK-B	257.00				
70011	MONITOR CORD, 24"	B	C	PUR PART	STOCK-B	391.00				
70012	KEYBOARD FEET, 1/2"	B	C	PUR PART	STOCK-B	1044.00				
70013	PC BOARD	B	C	PUR ASSY	STOCK-B	295.00				
70014	16K RAM CHIP	B	B	PUR PART	STOCK-B	1259.00				
70015	8088 PROCESSOR	B	B	PUR PART	STOCK-B	251.00				
70016	METAL, 4 X 4, SHEET	B	C	PUR PART	STOCK-B	316.00				
70017	FLOPPY DRIVE, 320K, DSDD	B	A	PUR ASSY	STOCK-B	235.00				
70018	SPEAKER, 2", 4 OZ	B	C	PUR PART	STOCK-B	169.00				
70019	POWER SUPPLY, 120 VOLT	B	A	PUR ASSY	STOCK-B	216.00				
70020	HARD DISK, 10 MEG	B	A	PUR ASSY	STOCK-B	79.00				
11000	PC - 2 FLOPPY - 128K	M	A	FIN GOOD	STOCK-B	10.00				

FIGURE 4–18 Inventory Control—Physical Inventory Report by Stockroom Identifier (Courtesy of Micro-MRP, Inc.).

11-04-1985 11:31:03 MICRO-MRP, INC. PAGE NO. 1

MAX * INVENTORY CONTROL * OBSOLESCENCE REPORT BY PART TYPE CODE PRIOR TO 850601

PART IDENTIFIER	DESCRIPTION	T C	C C	COMMODITY CODE	E S	ON-HAND VALUE	ON-HAND QTY	YTD ISSUE VALUE	YTD ISS QTY	LAST TNX DATE	DRAWING NUMBER
14000	MONITOR	A	A	FIN GOOD	2	10271	78	0	0	12/18/83	A-8647825
15000	PC, SYSTEM UNIT - 128K	A	A	FIN GOOD	2	13158	59	0	0	12/18/83	A-2363475
16000	PC-HD, SYSTEM UNIT - 128K	A	A	FIN GOOD	2	58951	82	0	0	12/18/83	A-6552340
50002	MONITOR CASE	A	B	MFG ASSY	2	2427	62	0	0	12/18/83	B-3638745
50003	PC MOTHERBOARD - REV A	A	B	MFG ASSY	2	4565	179	0	0	12/18/83	B-7354787
50005	SYSTEM UNIT BASE	A	B	MFG ASSY	2	1679	123	0	0	12/18/83	A-6536974
50006	SYSTEM UNIT COVER	A	B	MFG ASSY	2	417	17	0	0	12/18/83	A-6893566
70001	KEYBOARD CASE TOP	A	B	FAB ASSY	2	331	49	0	0	12/18/83	A-4893429
70002	KEYBOARD CASE BOTTOM	A	B	FAB ASSY	2	3415	271	0	0	12/18/83	B-6758769
70009	MONITOR FRONT	A	C	FAB ASSY	2	1327	79	0	0	12/18/83	A-7312652
70010	MONITOR BACK	A	C	FAB ASSY	2	2695	159	0	0	12/18/83	B-7367586
70004	PAINT, PC GREY	B	C	PUR PART	2	5	28	0	0	12/18/83	A-8766486
70005	PC KEYBOARD KEY SET	B	C	PUR PART	2	7	62	0	0	12/18/83	A-2575789
70007	SCREWS, 1/4" X #4	B	C	PUR PART	2	495	4500	0	0	12/18/83	B-6536755
70008	CRT, 10", GREEN	B	B	PUR ASSY	2	21203	257	0	0	12/18/83	A-0086475
70011	MONITOR CORD, 24"	B	C	PUR PART	2	86	391	0	0	12/18/83	B-3645822
70012	KEYBOARD FEET, 1/2"	B	C	PUR PART	2	96	1044	0	0	12/18/83	A-6748148
70013	PC BOARD	B	C	PUR ASSY	2	649	295	0	0	12/18/83	A-5665471
70014	16K RAM CHIP	B	B	PUR PART	2	1385	1259	0	0	12/18/83	B-1496906
70015	8088 PROCESSOR	B	B	PUR PART	2	1381	251	0	0	12/18/83	A-3867462
70016	METAL, 4 X 4, SHEET	B	C	PUR PART	2	2086	316	0	0	12/18/83	A-5482321
70017	FLOPPY DRIVE, 320K, DSDD	B	A	PUR ASSY	2	12925	235	0	0	12/18/83	A-6362718
70018	SPEAKER, 2", 4 OZ	B	C	PUR PART	2	102	169	0	0	12/18/83	A-6746917
70019	POWER SUPPLY, 120 VOLT	B	A	PUR ASSY	2	5940	216	0	0	12/18/83	B-5236833
70020	HARD DISK, 10 MEG	B	A	PUR ASSY	2	43450	79	0	0	12/18/83	A-7563124
140836135-	REAR FILLER	B	C	RAW	2	0	0	0	0	00/00/00	135
LABOR	FINAL ASSY LABOR	B	D	RESOURCE	2	0	0	0	0	00/00/00	
WA3967	PAINT WHITE	B	C	RAW	2	0	0	0	0	00/00/00	
12000	PC - 10 MEG HD - 128K	M	A	FIN GOOD	2	48338	53	0	0	12/18/83	B-5230798
140836135	REAR FILLER	M	C	FIN GOOD	2	0	0	0	0	00/00/00	135
TOTALS						237381	....	0			

FIGURE 4–19 Inventory Control—Obsolescence Report by Part Type Code (Courtesy of Micro-MRP, Inc.).

How the Purchasing Control Module Manages Information

The purchasing control module issues purchase orders, tracks purchased parts, maintains vendor information, and examines cash requirements. This enables a user to know exactly what a part will cost, when the part can be obtained, and what alternatives are available if normal purchasing methods are unable to handle unusual requirements. It provides vendor information, tracks on-time delivery performance, and maintains comparative pricing data.

This module can improve vendor deliveries, track outstanding purchase orders, and aid the expediting function. It can follow up on outstanding orders by part number, vendor, and due date. It also can keep a history of all purchasing transactions for review and audit purposes.

A purchase control module can automatically generate purchase requisitions, assign sequential purchase order numbers, and generate purchase orders automatically. The system also can create cash management reports to indicate the amount of money needed to cover incoming orders, and to show how much has been spent.

A purchase control module provides these features:

- Prints purchase requisitions and purchase orders.
- Tracks and controls each step of the purchasing operation.
- Controls purchases of MRP parts, expense materials, project materials, and other noninventory products.
- Uses exception reports, displays, and reports by part number, vendor, or due date for dispatching and expediting orders.
- Prints a time-phased cash requirements report.
- Maintains a vendor history of all orders placed, and the standard value of materials ordered and received.
- Assigns purchase requisitions and purchase order numbers automatically.
- Maintains separate stocking and purchasing units of measure.
- Separates total purchasing lead time into planning, buying, and stocking lead times.

A typical computer-generated purchase requisition is sorted by order number, part number, or part number by due date. It provides concise information on what needs to be ordered, who is responsible, and when the material is required. The automatically generated purchase requisition is based on MRP approved orders. Only approved vendors are printed on each purchase requisition. Purchase price defaults to vendor's price break, but can be overridden.

A computerized system will then generate a purchase order automatically which contains whatever information is desired.

Figure 4–20 shows a cash requirements report by vendor identification. This report accumulates cash requirements by month based on actual purchase price, or standard cost. This report can be printed by vendor, order number, part number, or part number and due date. The planner, buyer, and vendor can be identified for an improved tracking and reporting purchasing function.

Figure 4–21 shows a purchase order detail report by vendor identification. This report tracks open purchase orders and identifies discrepancies or late deliveries. It can be printed by vendor, order number, part number, or part number by due date.

Figure 4–22 shows a purchase dispatch report by vendor identification. This report prints a list of all open purchase requisitions and purchase orders together, along with their current status, to quickly identify what needs to be done and when. It can be printed by vendor, order number, part number, or part number by due date.

11-05-1985 15:30:50 MICRO-MRP, INC. PAGE NO. 2

MAX * PURCHASING CONTROL * CASH REQUIREMENTS REPORT BY VENDOR IDENTIFIER

PART IDENTIFIER	PART DESCRIPTION	VENID	STA	FPC	ORDER NUMBER	ORIG. QTY	DUE QTY	DUE DATE	PRM DATE	BUY	PLN	ORDER COST
70006	POWER CABLE, 6 FEET	ACME	3	N	700001-01-01	200.00	200.00	11/06/85	11/06/85	CAM	JAN	110.00
70019	POWER SUPPLY, 120 VOLT	ACME	3	N	700002-01-01	100.00	100.00	11/16/85	08/16/85	CAM	JAN	2700.00
NOV	TOTAL											2810.00
GRAND	TOTAL											2810.00

FIGURE 4–20 Cash Requirements Report by Vendor Identification (Courtesy of Micro-MRP, Inc.).

11-05-1985 15:30:08 MICRO-MRP, INC. PAGE NO. 1

MAX * PURCHASING CONTROL * PURCHASE ORDER REPORT BY VENDOR IDENTIFIER

PART IDENTIFIER	PART DESCRIPTION	VENID	STA	FPC	ORDER NUMBER	ORIG. QTY	BAL DUE	ORG DATE	CUR DATE	BUY	PLN	CRIT RAT
70006	POWER CABLE, 6 FEET	ACME	3	N	700001-01-01	200.00	200.00	11/06/85	11/06/85	CAM	JAN	1.000
70019	POWER SUPPLY, 120 VOLT	ACME	3	N	700002-01-01	100.00	100.00	08/16/85	11/16/85	CAM	JAN	1.000

FIGURE 4–21 Purchase Order Report by Vendor Identification (Courtesy of Micro-MRP, Inc.).

11-05-1985 15:31:59 MICRO-MRP, INC. PAGE NO. 1

MAX * PURCHASING CONTROL * PURCHASE DISPATCH REPORT BY VENDOR IDENTIFIER

PART IDENTIFIER	PART DESCRIPTION	VENID	STA	FPC	ORDER NUMBER	BAL DUE QTY	CURR. QTY	DUE DATE	PRM DATE	BUY	PLN	CRIT RAT
70006	POWER CABLE, 6 FEET	ACME	3	N	700001-01-01	200	200	11/06/85	11/06/85	CAM	JAN	1
70019	POWER SUPPLY, 120 VOLT	ACME	3	N	700002-01-01	100	100	11/16/85	08/16/85	CAM	JAN	1

FIGURE 4–22 Purchase Dispatch Report by Vendor Identification (Courtesy of Micro-MRP, Inc.).

THREE ADDITIONAL MODULES IN THE TOTALLY INTEGRATED MRP II SOFTWARE SYSTEM

In addition to the application systems modules discussed previously, which comprise the computer interface between production planning and manufacturing, a totally integrated MRP II software system also includes the following modules as part of its data base.

Capacity Planning. Provides projected loading for each work center. Capacity requirements are generated separately from both planned and released orders. The information can be made available in both summary and detailed formats for ease of planning. Capacity planning can be developed for use in either rough cut format to be used with master scheduling or in detail format for use with material requirements planning.

Sales Forecasting. Required whenever customer orders do not cover the planning horizon. Sales forecasting enables marketing management to participate in the forecasting process at a level in which they can interact easily.

Customer Order Processing. Consists of order entry and inquiry, order promising, order pricing, credit checking, order acknowledgments, and invoice generation.

JUST-IN-TIME MANUFACTURING SYSTEMS COMPARE WITH MANUFACTURING RESOURCE PLANNING

Just-in-time manufacturing systems (JIT) are related to manufacturing resource planning (MRP II) because both have the control of inventories as one of their objectives. The difference is that MRP deals with planning while JIT deals with processes. Thus, MRP II concentrates on reducing inventories based on maintaining minimum lead times, order quantities, and safety stocks while JIT focuses on process improvements.

In an MRP II system, the production control system establishes order quantities for purchased and manufactured parts based on vendor or production lead time, economical order quantity, and reserve stock. JIT, instead, seeks to avoid ordering quantities greater than those actually required.

Another difference between the two systems is work-in-process queues. MRP II tends to accept queues (material on the shop floor waiting for next operation) as part of capacity planning. JIT regards sitting inventory as waste, and aims to correct this deficiency through process improvements that increase throughput.

MRP II and JIT also differ in their approach to safety stock. MRP II maintains and assigns values to safety stock because there is always an element of unpredictability regarding orders on the shop floor. Proponents of JIT believe that reducing inventories in other areas while maintaining safety stocks defeats some of the benefits derived from this methodology.

Although both MRP II and JIT are acceptable management techniques, the main distinction between them is that MRP is broader in scope with a greater overall financial impact, while JIT concentrates entirely on areas within the manufacturing realm.

Benefits of Just-In-Time Systems

An integrated JIT manufacturing system can result in improvements in quality, lower inventories, less material handling, and increased employee productivity. In addition, the JIT system offers some protection against manufacturing down time because it allows for early detection of actual problems and provides the means of anticipating and avoiding process failures.

The JIT manufacturing system is a *total* manufacturing system in which goods are produced and made readily available by a control system—"just in time"—for further fabrication or subassembly. It is a system that eliminates waste through the concept that anything other than minimal equipment, materials, and workers raises costs. The basic goal of JIT manufacturing is to improve productivity and quality. In addition, this system will provide for capital conservation because inventory costs are reduced and some of the machinery and equipment can be eliminated. Investment in warehouse storage for in-process inventory is also reduced or eliminated resulting in the availability of this space for more productive purposes.

Traditional inventory of materials supposedly provides a cushion for unexpected delays due to: unreliable machinery, unplanned down time, unreliable quality from inside or outside suppliers, changing customer demands, and transit failures to and from the production facilities. When large inventories are maintained to protect a company from these problems, the result is frequent reordering of stock because the specific items required are buried in the inventory and are not readily accessible. Even worse, the problems with quality remain hidden until the inventory is used.

How a Typical JIT System Works

The typical JIT manufacturing system uses a lot size of one. To accommodate the use of small lot sizes, it is necessary to reduce setup time and tool change time; cut the distance of material travel within the plant by appropriate rearrangement of the factory facility to ensure optimum efficiency in production lines; assign multifunctional workers who can perform many jobs with proficiency; and shorten lead time from outside suppliers to ensure that outside services or products can be secured within a minimal amount of time. As inventory is reduced, manufacturing problems can be identified and dealt with in a manageable way. With lot sizes of one, the maximum amount of scrap parts in inventory is one. The next step is the elimination of problems as they are discovered. Since the problem cannot be avoided by using inventory, they must be solved when they occur.

Specialized computer programs—custom designed—for specific industries, are available that allow the just-in-time manufacturing inventory control systems to operate at maximum efficiency. For just-in-time to be most effective, data must be gathered on a real-time basis. It is not possible to wait until the end of the accounting period to determine the parts inventory.

For example, some systems are created specifically for repetitive manufacturers' updates of raw material usage and finished parts inventories as they are used and/or produced. This allows a real-time inventory management system.

Some of the basics required by JIT manufacturing are product quality through process reliability and preventive maintenance programs to eliminate unplanned down times. A key ingredient for JIT manufacturing is a statistical analysis of product quality and process performance. A process malfunction can shut down an entire manufacturing operation because inventory is not available. A real-time statistical analysis of product quality and process performance is a necessity. A *statistical process control program* attaches to the machine and/or gauging equipment on the factory floor, allowing production workers to monitor production quality and process performance continuously. A real-time *preventive maintenance system,* when attached to the production machines, can help a manufacturing operation avoid unplanned down time by alerting the maintenance department when service is required or when tools need to be replaced.

How to Implement a JIT System

Once these technological aids to the production control system are in place, the following steps can be followed to implement a JIT system:

1. Evaluate the present manufacturing process.
2. Identify the subassemblies and components that are presently inventoried.
3. Study factory floor layout to develop methods for reducing material travel distances.
4. Meet with suppliers and encourage their involvement, and also determine lead times for JIT components.
5. Work out long-term arrangements with suppliers.
6. Revise paper flow and/or computer software programs to accommodate JIT manufacturing.
7. Start job cross-training for employees.
8. Meet with employees to explain the system and seek their input.
9. Obtain complete commitments from executive management and key personnel regarding the new system.

USING OPTIMIZED PRODUCTION TECHNOLOGY TO ELIMINATE BOTTLENECKS*

One of the more interesting approaches to production planning and control in the factory environment is a methodology referred to as *optimized production technology* (OPT). Basically, the OPT's approach forces production management to identify bottlenecks and

*Based on methodology of Optimized Production Technology (OPT) developed by Dr. Eliyahu M. Goldratt and Creative Output. Permission of Creative Output.

coordinate the flow of parts through a factory, thereby minimizing the impact of any constraints in the manufacturing flow. By identifying bottlenecks in the planning stage, it is possible to determine what can be produced, and when it can be produced. If the results are not acceptable, management can focus attention on increasing production at the bottlenecks, or on smoothing the flow at nonbottlenecks which reduces the amount of work-in-process required, as well as improving response time.

The OPT software can be used in simulation mode to quantify the impact of proposed improvements to the constrained areas. Based on these simulations, management will set policies which determine where the constraints will be. The OPT software can then be used to generate detailed schedules that are consistent with the policies selected by management. Many companies that have used this production technology have been able to reduce work-in-process inventory drastically, and sharply reduce customer order lead times.

How OPT Relates to MRP II

To place OPT in its proper perspective, you should understand its relationship to MRP II. MRP II begins with a master schedule prepared manually. Next, it generates the production schedule, followed by material requirements planning, time-phased production requirements, and capacity planning. Two final steps are procurement of outside vendor parts and shop floor control of parts produced in-house. The activities demanded by these reports may or may not be possible within the constraints that result from policies established by management. OPT, on the other hand, starts with the constraints and automatically produces a master schedule that is consistent with those constraints. An MRP-like module then generates the other reports based on this master schedule. Some OPT users have interfaced OPT with traditional MRP packages, combining OPT's superior scheduling capabilities with MRP's sophisticated data-base and report generation capabilities.

Understanding the OPT Simulation Process

The simulation process begins with data stored in the bill of materials file and process routing file. This data is linked into OPT's product network. A machine file describes all available resources and provides the system with flexibility to reroute parts to different machine tools or work centers as shop floor conditions dictate. The resultant capacity needs, when divided by the number of weeks in the planning horizon, become the average capacity requirements for each machine or work center. When divided by the capacity of the machine or work center, the result is average expected load. The average loads are then arranged in descending order, and the most heavily loaded areas are analyzed to determine whether policies should be changed to smooth out any appreciable imbalance. A new simulation can be run to predict the results of the revised policies. Where management policies allow it, the simulation algorithm will reroute parts to compensate for the bottleneck. If there is no other choice, the OPT's system downgrades the part's scheduled due date. By setting policies which will reschedule work flows, change lot sizes, cut down on work-in-process, reduce part scheduled lead times, shift work orders to different machines or work centers, and prioritize part requirements the plant management achieves optimum throughput.

OPT virtually eliminates the conflict between MRP and finite loading because only bottlenecks require finite loading. To achieve optimum throughput, WIP buffers are established in front of bottlenecks and where the output from a bottleneck joins with some other parts. Maximum utilization of bottlenecks is achieved by running large batch sizes. Low inventory is achieved by running small lots through nonbottleneck operations.

OPT achieves the integration between manufacturing planning and control, time-phases scheduling, capacity planning, and shop floor control. By concentrating on critical bottlenecks in the production cycle through forward planning and feedback of actual shop floor conditions, OPT enables management to constantly manipulate all work orders to optimize production throughput.

How OPT Works

As mentioned previously, OPT requires the development of a product network containing the bill of materials and routing files in the same manner as a typical MRP system. The difference is that the information is combined into a single file, using a network structure. Each operation required to manufacture every part within the family tree must contain setup and run time. In addition, OPT data includes: stock level at each operation, maximum stock limits, minimum batch quantities, auxiliary machines, and scheduled delays. Variable data entered into the network consists of order quantities and due dates. And, for each resource (machine, worker, or combination of machine and worker) substitutes and their relative efficiency, overtime availability, resources required for setup and support machines are defined.

The Function of BUILDNET, SERVE, and SPLIT

Information from the product network and resource description are combined (using a procedure called BUILDNET) to form an engineering network. In an initial analysis, a module called SERVE uses this information and logic similar to that used in traditional MRP systems. SERVE backward schedules to the order due dates specified in the product network. SERVE assumes infinite capacity is available at each machine and user timing information calculated from the setup run time and scheduled delay data included in the product network. SERVE performs the task of generating a report that identifies which machines are bottlenecks. Management decides whether these bottlenecks should be eliminated or tolerated. Any bottlenecks that must be tolerated are reported to the SPLIT procedure.

The SPLIT operation separates the engineering network into two sections. One section includes operations that use the bottleneck resources and all operations that follow the "critical" operations. The other section includes operations that precede the "critical" operations.

Key Point: Buffer "safety" stocks should be placed before each critical operation.

The first section of the engineering network is scheduled using the OPT module consisting of variable batch sizing. Each order is eventually broken down into a number of smaller "transfer" batches that represent the minimum amount of work that can be performed at a particular operation.

Once a transfer batch is completed at an operation it is immediately sent to the next operation and placed in a queue.

Key Point: Parameters are established as to how many like transfer batches can be grouped together to reduce setups.

After critical parts are scheduled, the SERVE module then schedules the remaining parts assuming infinite capacity. Original order due dates are replaced by dates determined by the OPT procedure.

Key Point: SERVE schedules so that material will be available for the first operation in the critical part of the network.

The Nine Basic Rules of OPT

The OPT system is based on adhering to a number of rules which form the basis for its logic. These rules are:

1. Balance flow, not capacity.
2. The level of utilization of a nonbottleneck is determined not by its own potential, but by some other constraint in the system.
3. Utilization and activation of a resource are not synonymous.
4. An hour lost at a bottleneck is an hour that cannot be recaptured later in the production cycle.
5. An hour saved at a nonbottleneck does not result in added production throughput.
6. Bottlenecks govern both throughput and inventory in the system.
7. The transfer batch may not, often should not, be equal to the process batch.
8. The process batch should be variable, not fixed.
9. Schedules should be established by looking at all of the constraints simultaneously. Lead times are the result of a schedule, and therefore, cannot be predetermined.

In summary, conventional MRP systems do not include a finite loading module for developing detailed shop schedules and for shop floor control. Therefore, MRP generates schedules that may not be achievable. By comparison, OPT produces schedules that are consistent with established policies. This allows management to reconsider those policies before actually implementing them if the expected results are poor. Also, OPT is more flexible since it schedules small "transfer" batches, whereas MRP basically schedules full quantity lot sizes. Results have shown that OPT is a viable production planning and control system that can reduce work-in-process inventory, slash production costs, and substantially reduce throughput times.

CASE-IN-POINT

A CAUSE OF EXCESS INVENTORY?

Situation

Two workers are engaged in production of a product and its spare part. The resources required, rates of production, operational steps, and market potentials (demands) are shown in figure 4–23. There is only one Worker V and one Worker W per shift. The plant operates twenty-four hours a day (three shifts), five days a week. The workers cannot substitute for each other and there is no other work except as detailed in the diagram.

1. Schedule the two types of workers for a typical twenty-four hour day (figure 4–24).
2. Estimate inventory accumulation (in units) after one day, after one week, and after one month according to your one-day schedule (figure 4–25).
3. Are there any steps you might take to lower the level of inventory accumulation without losing sales?
4. In light of the above, draw some conclusions about the following ideas:

    Worker "piece incentive" as a reward system

    Worker efficiency as local performance measurement

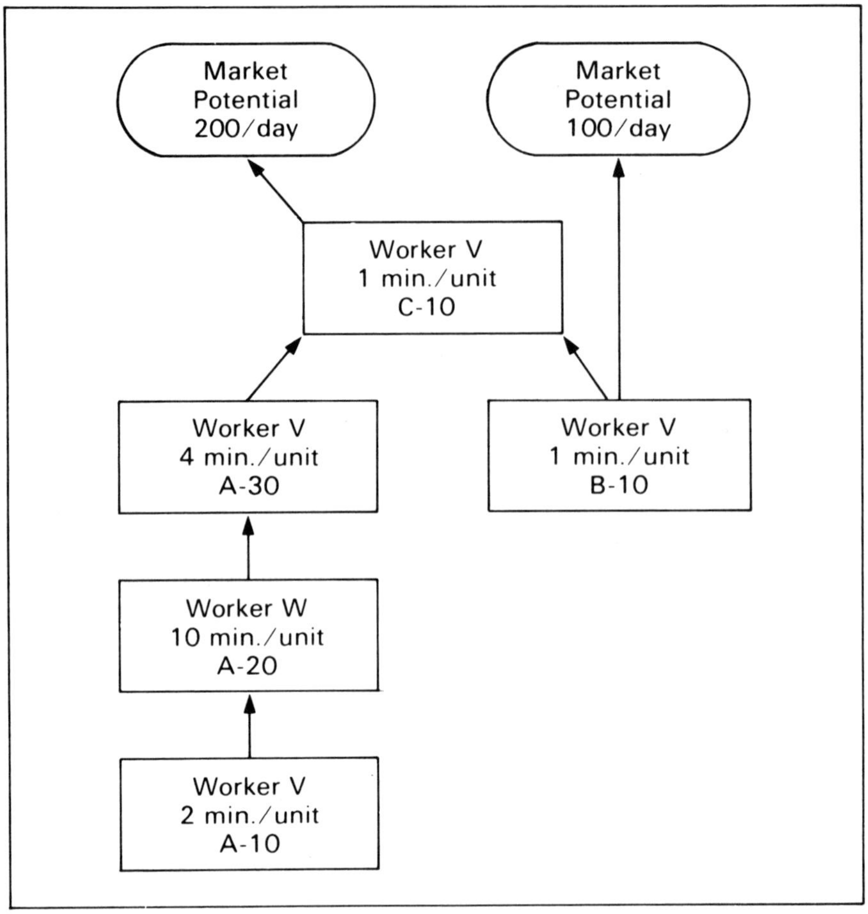

Legend

Resource (e.g. Worker V)
Production Rate (e.g. 4 min./unit)
Operation Number (e.g. A-30)

FIGURE 4–23 A Cause of Excess Inventory—Situation (Courtesy of Creative Output Inc.—OPT).

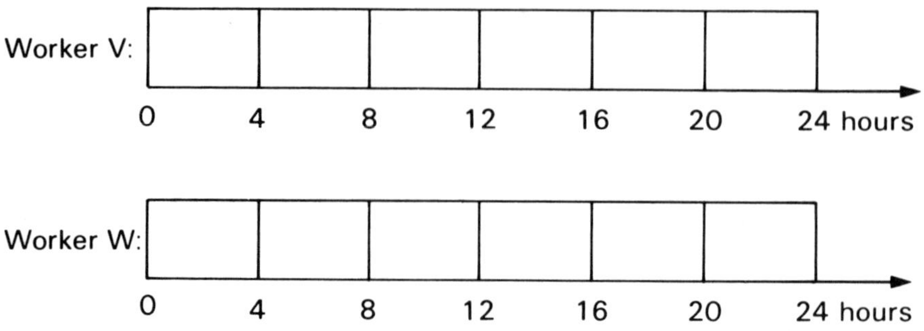

FIGURE 4–24 A Cause of Excess Inventory—Situation (Courtesy of Creative Output Inc.—OPT).

	Day 1	Week 1	Month 1
Operation A-10			
Operation A-20			
Operation A-30			
Operation B-10			
Operation C-10			

FIGURE 4–25 A Cause of Excess Inventory—Situation (Courtesy of Creative Output Inc.—OPT).

Solution

The usual concern in constructing a schedule is to make sure that each worker is scheduled to be kept busy all the time. The OPT approach is different. The first step is to identify the constraints inherent in the problem. A brief examination of our case shows that Worker W is a constraint since he or she cannot produce more than 144 items a day ($24 \times 60/10$) thus limiting the sales of the product to that number. The other constraint is an external one—the market potential for the spare part.

The schedule of Worker V is arranged to satisfy the constraint to produce 144 units/day at operations A-10, A-30, and C-10 and 244 units/day ($144 + 100$) at operation B-10. Any production above these quantities will not yield more sales—just more unneeded inventory. Thus a typical schedule for the two workers will look like figure 4–26.

Notice that Worker V is scheduled not to produce anything for 3.13 hours/day. This is an application of Rule 3 of OPT that states:

Utilization and activation of a resource are not synonymous.

By obeying such a schedule, accumulation of excess inventory is avoided. Trying to activate a worker to his or her full potential will not result in additional sales but only in the accumulation of inventory before A-20 (due to the limited capacity of Worker W) and after B-10 (due to the market restriction and the limitation of type-A parts available for assembly).

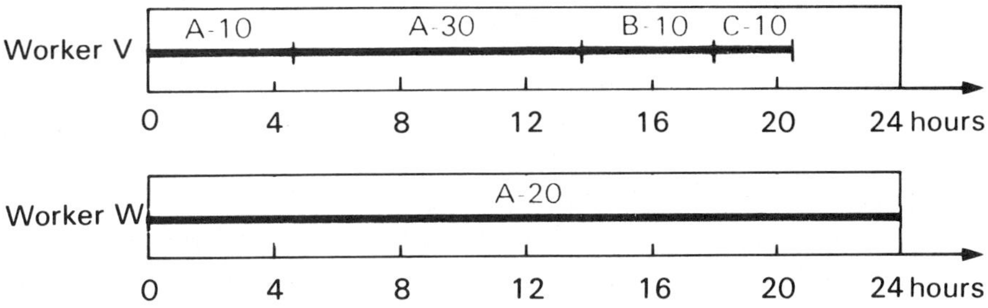

FIGURE 4–26 A Cause of Excess Inventory—Solution (Courtesy of Creative Output Inc.—OPT).

	Day 1	Week 1	Month 1
Operation A-10	47	235	1034
Operation A-20	—	—	—
Operation A-30	—	—	—
Operation B-10	94	470	2068
Operation C-10	—	—	—

FIGURE 4–27 A Cause of Excess Inventory—Solution (Courtesy of Creative Output Inc.—OPT).

If we assume the common practice is followed for activating Worker V's full potential, splitting its spare time in overproducing at A-10 and B-10, the buildup of excess inventory is shown in figure 4–27.

Notice that the traditional performance measurements of a worker, whether piece incentive or efficiency measurements, encourage the schedules, foreman, and worker to use the worker's potential for producing to the maximum. If we want to avoid the buildup of excess inventory, we must replace such performance measurements. A logical replacement is *adherence to the schedule.* Workers whose capacity does not constrain the performance of the plant should produce according to a reliable predetermined schedule, and be encouraged not to overproduce as well as not to underproduce.

Adequate stable levels of inventories should be maintained to ensure that deviations from the schedule will not jeopardize the plant performance. In our example, holding about fifty pieces (60 × 8/10) after A-10 and eighty pieces (50 + 30) after B-10 will give sufficient protection to the plant to overcome deviations of Worker V of up to a full shift of production.

CASE-IN-POINT

HOW SHOULD WE PLAN A JOB?

Situation

An order for 1,000 units was received by your plant. Figure 4–28 specifies resources needed, production rates, and the sequence of operations. The time to set up each resource is ten hours. Notice that Resource V is used for two different operations. Machines U, V, W, and Y are dedicated to this order, along with a crew of four workers per shift. The plant operates twenty-four hours a day (three shifts), five days per week. There is only one machine of each type. Each machine is activated by one worker who is also capable of setting it up.

1. Draw a schedule for the job (figure 4–29).

   Assume that the operating expenses of the plant—for the designed resources and labor power, including overhead—are $7,500 per week and are paid at the end of the week. Similarly, assume that payment for material is made when raw material is actually used at the first operation (A-10). Finally, money is received whenever a finished unit is completed.

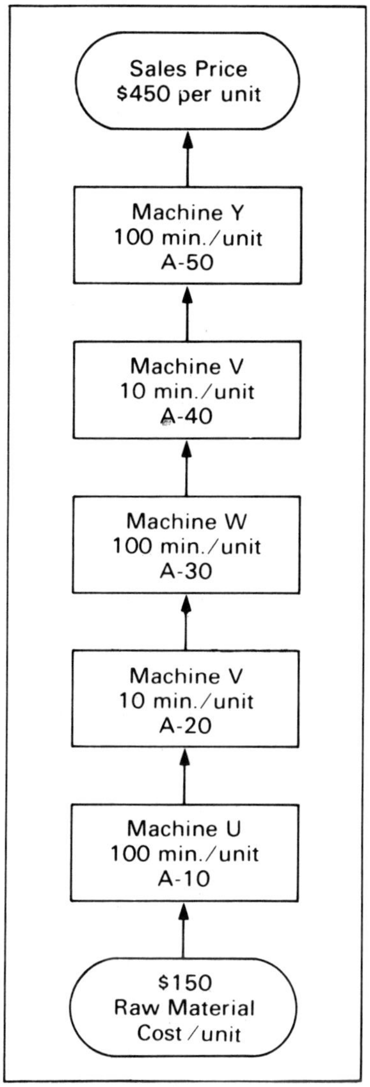

FIGURE 4-28 How Should We Plan a Job—Situation (Courtesy of Creative Output Inc.—OPT).

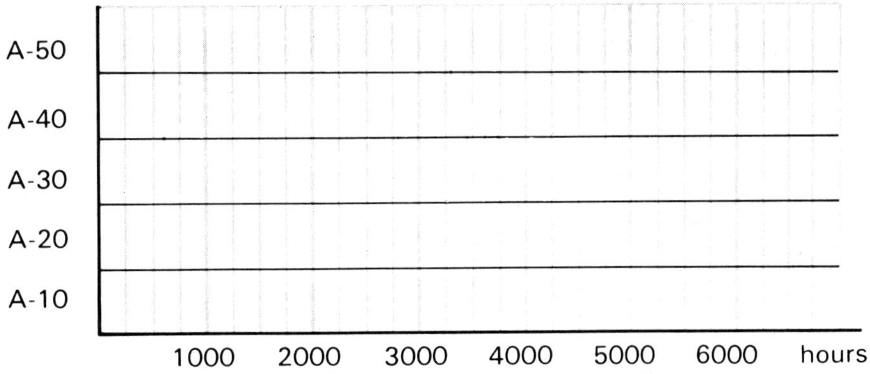

FIGURE 4-29 How Should We Plan a Job—Situation (Courtesy of Creative Output Inc.—OPT).

2. Plot the cash flow results from this schedule (figure 4–30):
3. If the plant has only $200,000 in cash, should it accept this order?

Solution

The conventional approach is to treat each order as a single batch to be done in full at each successive operation. In our case, an order for 1000 would take 5385 hours as shown on the schedule in figure 4–31.

A calculation of the cash flow shows that the money being spent for the first 3708 hours reaches a total cash outflow of $382,500 before the sales revenue begins to reverse the flow. By the end of the period, the job has resulted in a net loss of $37,500 (not taking into account the additional loss due to the interest cost on the money required).

The OPT approach starts by identifying the constraints. Machines U, W, and Y are the bottlenecks that must be fully activated. Machine V is nonbottleneck so we can utilize its spare time very effectively to decrease the batch size with no penalty for the additional setups. At 15 units/batch, Machine V would be utilized fully, but to avoid loss of throughput

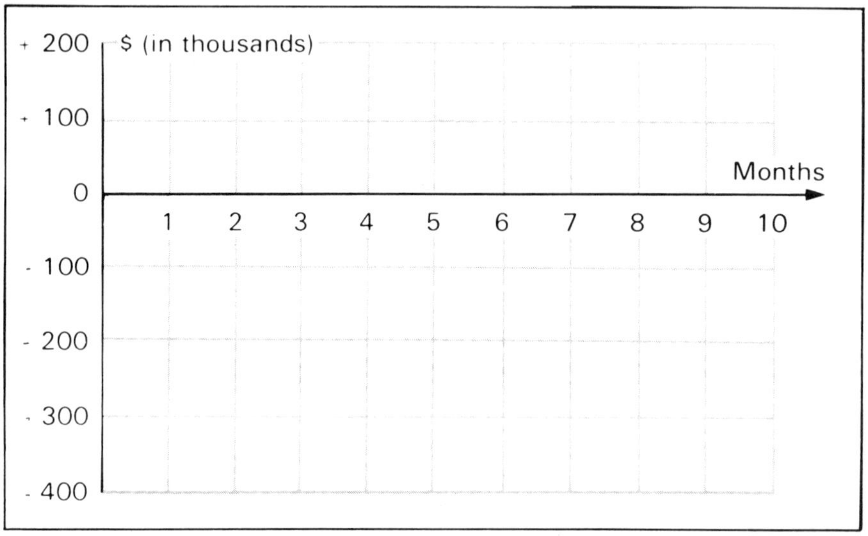

FIGURE 4–30 How Should We Plan a Job—Situation (Courtesy of Creative Output Inc.—OPT).

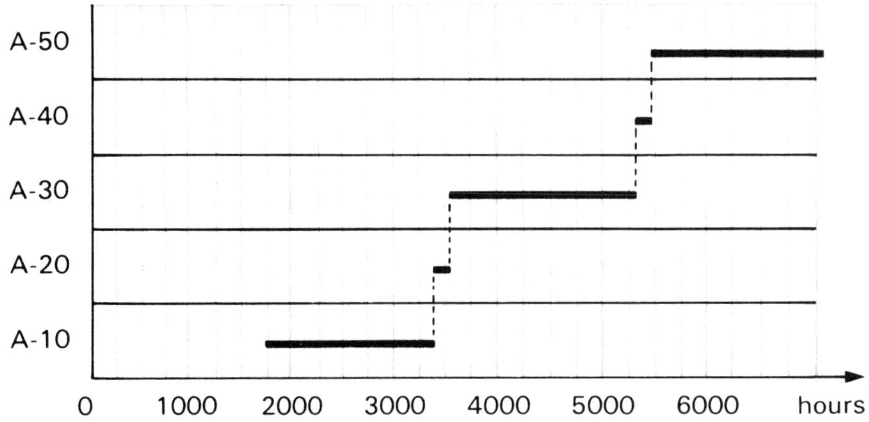

FIGURE 4–31 How Should We Plan a Job—Solution (Courtesy of Creative Output Inc.—OPT).

due to fluctuations in Machines U, V, W, and Y, we should ideally choose a somewhat larger batch size. All major benefits will be obtained even with a batch of thirty. The result is a schedule with one worker on Machine V alternating frequently between operations A-20 and A-40 and with the other three workers dedicated to Machines U, W and Y, all working in parallel. A sample of the schedule is shown in figure 4–32.

The overlapping of operations A-10, A-30, and A-50 is made possible by moving small transfer batches through the system while maintaining large process batches only at the constraining resources. The net effect is a drastic reduction in the length of the project from 5308 to 1807 hours. The $37,500 loss is converted to a net profit of $187,500. The competitive edge of the reduced cycle time should have an important effect on future sales. Also, the initial negative cash flow is reduced from almost $400,000 to only $20,000, as shown in figure 4–33.

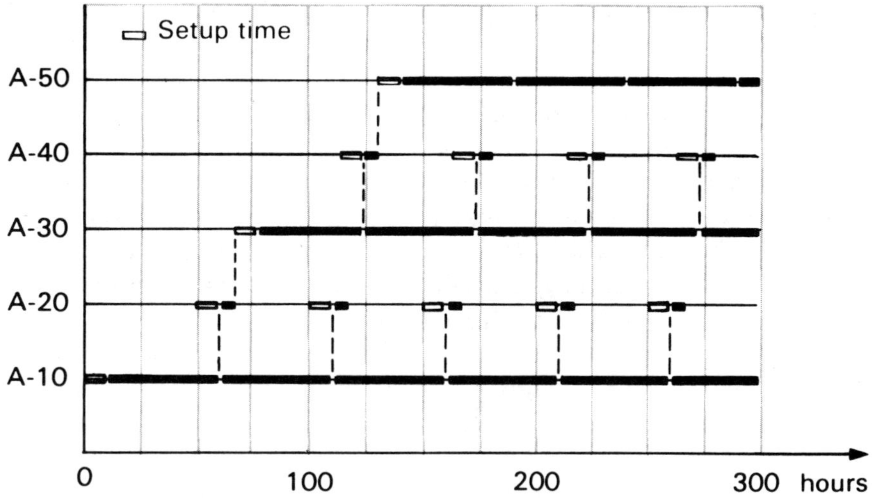

FIGURE 4–32 How Should We Plan a Job—Solution (Courtesy of Creative Output Inc.—OPT).

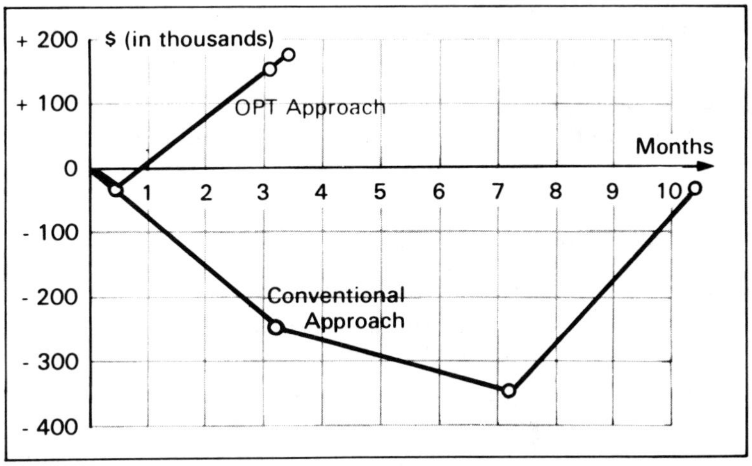

FIGURE 4–33 How Should We Plan a Job—Solution (Courtesy of Creative Output Inc.—OPT).

A very important side effect of reducing the cycle time and inventory is the effect on quality. Consider the effect of a defect in the first operation A-10 which is not noticeable until the last operation A-50. Under the conventional approach, the defect is detected months after it occurred and after the whole A-10 process has been completed and disassembled. What can be done to correct process A-10 after the discovery is too little and too late.

By contrast, the OPT schedule will lead to the discovery of the defect after a few days when only a tiny fraction of the parts at the A-10 operation have been processed. The identification and quick correction of the defect in the A-10 operation can be crucial.

This case illustrates how important it can be to follow OPT Rule 7 which states:

The transfer batch may not and often should not be equal to the process batch.

THE PRIMARY FEATURES OF A MANUFACTURING AUTOMATION PROTOCOL NETWORK

Manufacturing Automation Protocol (MAP) is a seven layer, broadband, tokenbus-based communication specification for the factory environment (figure 4–34). In the following sections we explore what these features mean.

What Makes Up the "Seven Layers" of a MAP Network

The seven layers refer to the International Standards Organization (ISO) Open Systems Interconnection (OSI) model on which MAP is based. The model separates the communications task from the applications task (the top boxes in the two columns of the model). The model then divides the communications task into seven subtasks, or layers.

The system is hierarchical with each layer providing a set of communication-related services to the layer above, or in the case of the top layer, to the application program. Each layer provides these services through its programs and through the services available from the layer below. Layers also communicate with equivalent layers, or peers, in other devices by protocols on how to do a particular task.

The layers in a given device communicate with adjacent layers above and below through hardware and software interfaces.

At present, MAP specifies Institute of Electrical and Electronics Engineers (IEEE) standards for layers 1 and 2, and National Bureau of Standards (NBS) standards for layer 4. Computers and other intelligent equipment are available to support these standards. Standards for the other layers are in various stages of development. Eventually, the MAP specification will contain international standard protocols for all seven layers.

Using a layered approach to design computer networks has the advantage that the layers are relatively independent of each other and can be replaced without affecting the rest of the design.

The layers of the model may be classified into three basic categories: end node specific, routing, and adjacent node specific as shown below.

Layer 7 (Application)	What you do with the data (Interface with application programs which are not part of OSI model)
Layer 6 (Presentation)	End-node specific (Transfers data to and from negotiated standardized formats)
Layer 5 (Session)	Reliable (Synchronized and managed dialogues)
Layer 4 (Transport)	Delivery of the data (Provides transparent, reliable data transfer from end node to end node)

Layer 3 (Network)	Routing (Performs message routing for data transfer between nodes)
Layer 2 (Data link)	Error detection for messages moved between nodes
Layer 1 (Physical standards)	Electrically encoded and physically transfers messages between nodes

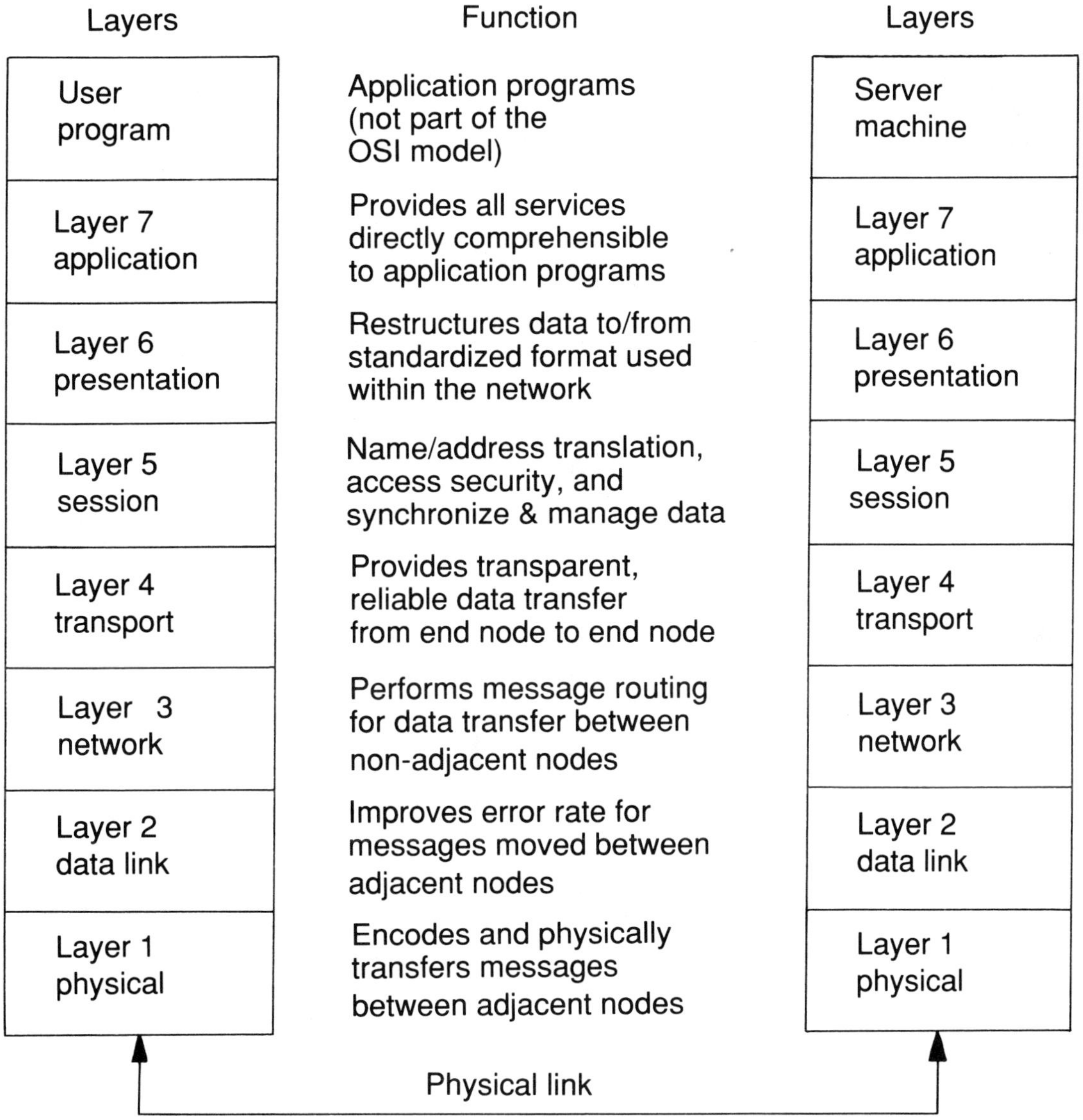

FIGURE 4-34 Seven Layer, Broadband, Tokenbus Based Communications Specification for the Factory Environment (Courtesy of General Motors—MAP).

The application, presentation, session, and transport layers are concerned with application level services, data format standardization, address translation, address security, and the end-to-end processing of messages. Protocols at these levels must be consistent for pairs of communicating and user programs.

The network layer is concerned with how to route messages between the two (or more) communicating applications in the presence of intermediate nodes. Protocols at this layer must be consistent throughout a given network. The definition and implementation of the network layer was crucial to the success of MAP as a nonproprietary, multivendor network standard.

The data link and physical layers are concerned with the physical communication of messages between adjacent nodes. Protocols at these levels must be consistent within the connection of each pair of adjacent, communicating network nodes. The use of MAP standard protocols at these levels allow for the interconnection of nodes from different vendors. The relationship between matched pairs of protocol layers is shown in figure 4–35.

The distribution of functionality is not absolute. As an example, addressing is a multilayer problem which includes locally identified names processed by the session layer, global addresses processed by the data link, and physical layers.

Each layer provides services to the layer above. Two types of information are passed between layers in providing these services: *control* and *data.* The control information is the basis for all the services that are required to process the message. As each layer provides its part of those services, the remaining control information is passed to the next lower layer. This process continues until no control information remains.

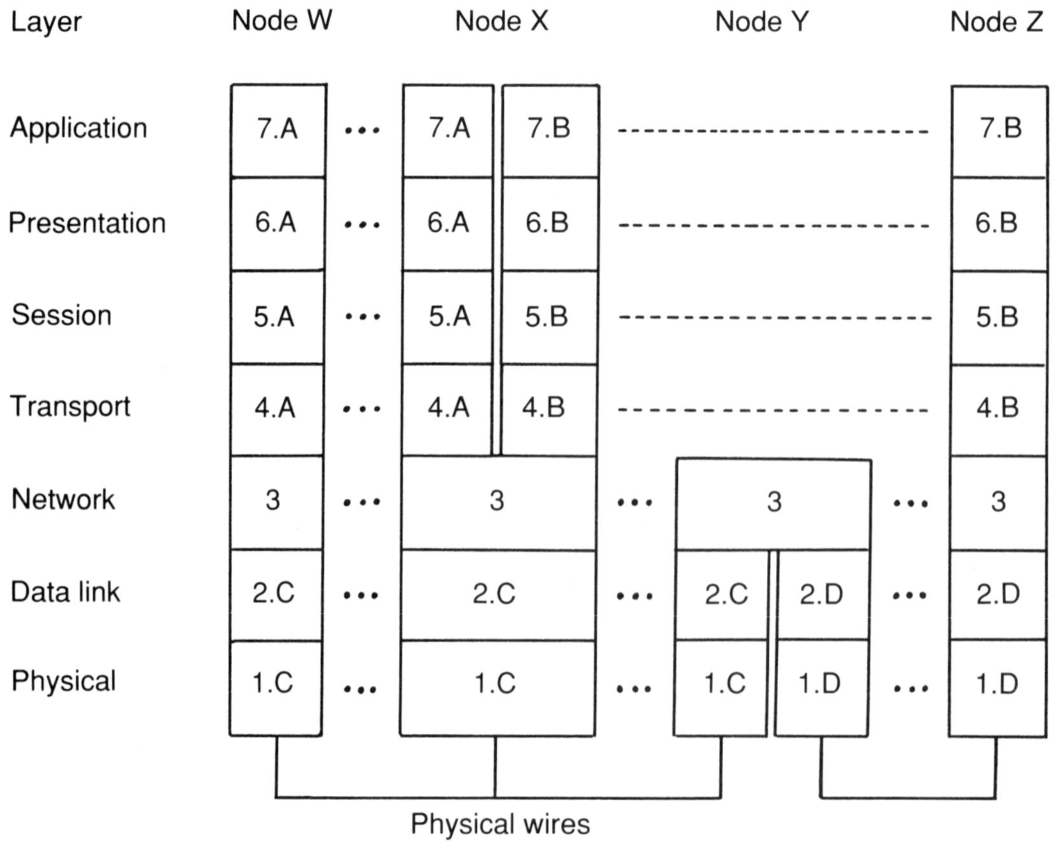

FIGURE 4–35 Matching of Communications Layers (Courtesy of General Motors—MAP).

The data that is passed down to a layer is generally transported transparently. (An exception is the presentation layer whose job is to reformat that data.) Each layer prefaces that data with a control block prior to requesting the services of the next lower layer. This control block is interpreted by the corresponding layer in the receiving node.

As the data is passed to successively lower layers, the size of the data base increases as shown in figure 4–36. As the data is passed to successively higher layers at the receiving node, the control blocks are removed.

The Purpose of Broadband

Broadband is a wide-bandwidth coaxial cable. With it, a number of data signals can be transmitted over a single cable. It eliminates the mass of individual wires or other connectors joining devices in a typical manufacturing environment.

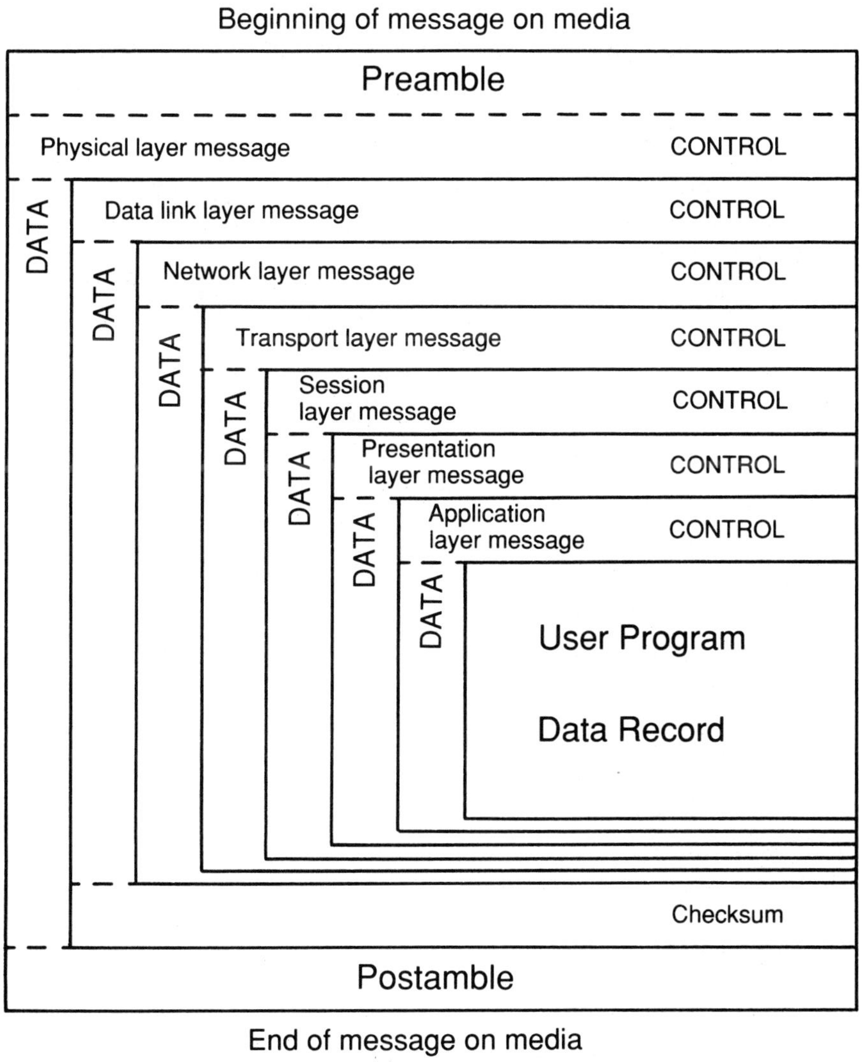

FIGURE 4–36 Nesting of Layer Protocols (Courtesy of General Motors—MAP).

Token-Bus: How MAP Establishes Priorities

Computer networks may be arranged in several configurations. MAP is based on the bus configuration in which devices are all wired together with a common connection, so that all devices receive all messages. The sending device designates in the address portion of the message where it should go. A receiving device examines the address field of the message. If it recognizes its own address, it takes the appropriate action; if not, it ignores the message.

With many messages passing through the network, priorities must be established. MAP uses a token-passing technique to ensure that all devices have an opportunity to send messages. It involves passing a code or token from one device to another. Only a device with the token can transmit messages, and then only for a preset period of time. The maximum time that any network device must wait before it receives the token and can transmit messages is also precisely determined. This is a valuable asset in a network passing many messages.

IMPLEMENTING MAP: A FIVE-STEP PLAN

In the case of General Motors, a leading exponent of this technology, the intent of the MAP specification is to establish a set of standards for a factory *local area network* (LAN) to support communications among computers and other intelligent devices.

The guidelines and specifications in this standard define a local area network for terminals, computing resources, and programmable devices within a plant or complex. The architecture allows for the interconnection of multiple LANs and for connection to wide area networks or digital PBXs for long distance communications.

The five-step implementations plan developed by General Motors is as follows:

Step 1: Centralized Network

- Multivendor connections via a centralized computer node
- Terminal emulation via a centralized computer node
- 2084

Step 2: Local Area Network

- Multivendor connection via a distributed LAN
- Gateway to selected programmable controllers
- 3084

Step 3: Application Services

- Enhance Step 2 LAN with additional application services
- Gateway to wide area network
- 2085

Step 4: Low-Cost Hardware

- Reduce ISO layers 1–4 to hardware
- Proliferate Step 3 on multiple processors
- Add ISO layers 5–6

Step 5: Complete Network Utility

- "Plug compatibility" by majority suppliers

Step 1: Centralized Network (2084). Application programs exist within GM plants on processors such as the IBM 4300, DEC PDP/11, or the HP 1000. Most computers offer serial communications capability, but these three computers cannot communicate with each other as purchased. In Step 1, an IBM Series/1 minicomputer is used to connect existing systems with minimal impact on application software or communications software. The Series/1 is used as a centralized communications node which provides terminal and protocol emulation (figure 4–37).

Without a local area network, Step 1 solves the immediate problem of incompatible computers that cannot communicate. An item worthy of note is the Step 1 connection protocol, ISO Transport, which is required for Step 2.

Step 2: Local Area Network (3084). In Step 2 dissimilar computers from different vendors are attached to a distributed local area network (LAN). Each participating computer vendor implements the same ISO Transport that IBM installed in Step 1. Additionally, all Step 2 computers interface to an external unit that provides ISO layers 1 and 2. The external unit, produced by Concord Data Systems, provides IEEE 802 token bus services over broadband at five megabits per second. The combination of ISO Layers 1, 2, and 4 allows each MAP computer to connect and pass data to any other MAP attached computer (figure 4–38).

Selected programmable controller access to MAP is the second objective of Step 2. Gateways support the existing vendor specific programmable controller nets on one side of

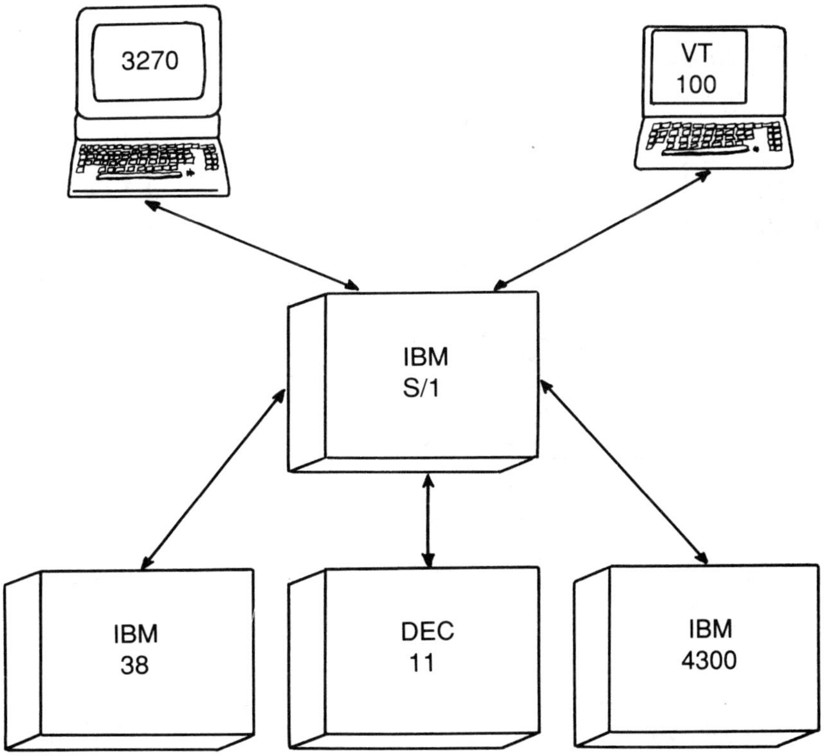

FIGURE 4–37 Centralized Communications (Courtesy of General Motors—MAP).

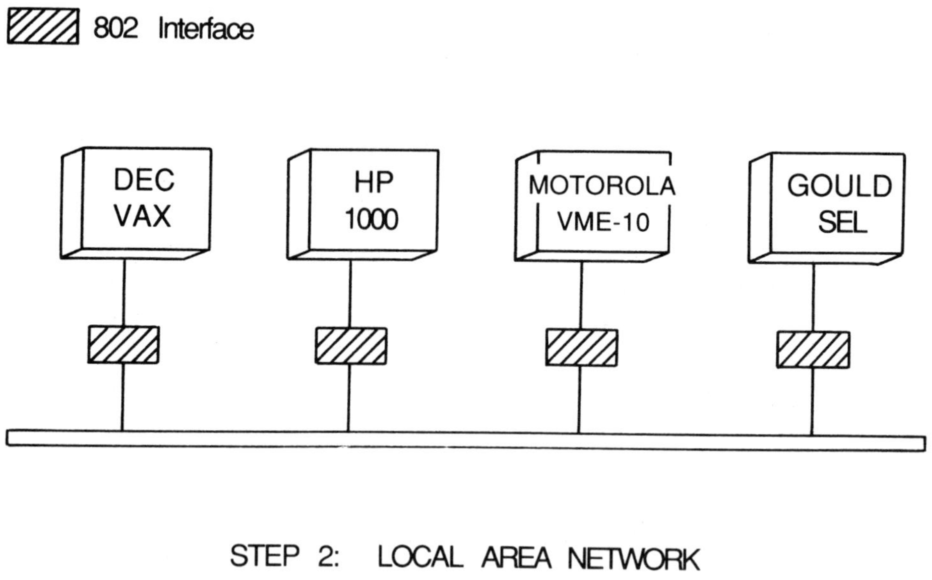

FIGURE 4-38 **Local Area Network** (Courtesy of General Motors—MAP).

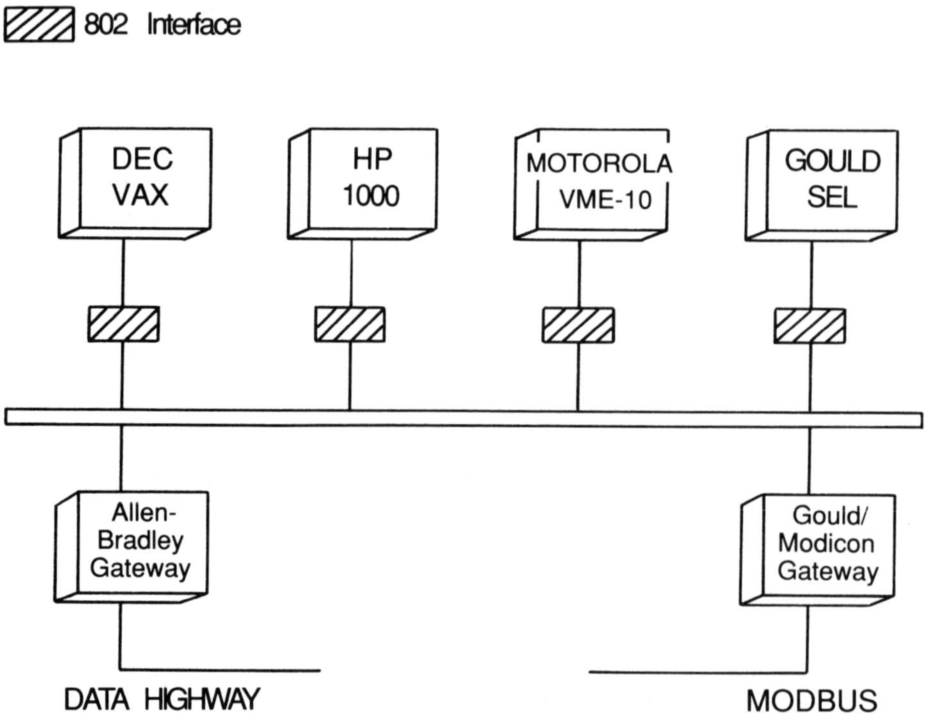

FIGURE 4-39 **LAN With Programmable Controller Gateway** (Courtesy of General Motors—MAP).

the gateway and ISO Layer 1, 2, and 4 on the other side (figure 4–39). The gateway concept allows for graceful migration because a number of installed programmable controllers are already connected by vendor specific network products. The gateway supports the existing systems and provides the additional connection capability to each of the MAP attached computers.

Since the centralized Step 1 processor was a stand-alone system, there is an added benefit of the Step 1 to Step 2 migration. The Series/1 now exists as a node on the MAP network system but still provides connection services to incompatible computers (figure 4–40). For example, the DEC PDP/11 does not support the ISO transport in Step 2, but it will have access to the MAP network through the Step 1 centralized node.

Step 3: Application Services (2085). The Step 3 hardware configuration is similar or identical to Step 2, but significant software enhancements are added. The final goal of a network is application-to-application communication such as remote file access, file transfer, and message exchange. Protocols specified by ISO layer 1 through 4 provide connection and reliable data transfer between computers, but application compatibility is left to layers 5 through 7. Step 3 software is concentrated at layers 6 and 7.

Further MAP work on ISO layers 6 and 7 is based on working drafts from various standards groups. The interim MAP specifications for file transfer, messaging, virtual terminals, directory service, and network management are unique to GM because no international

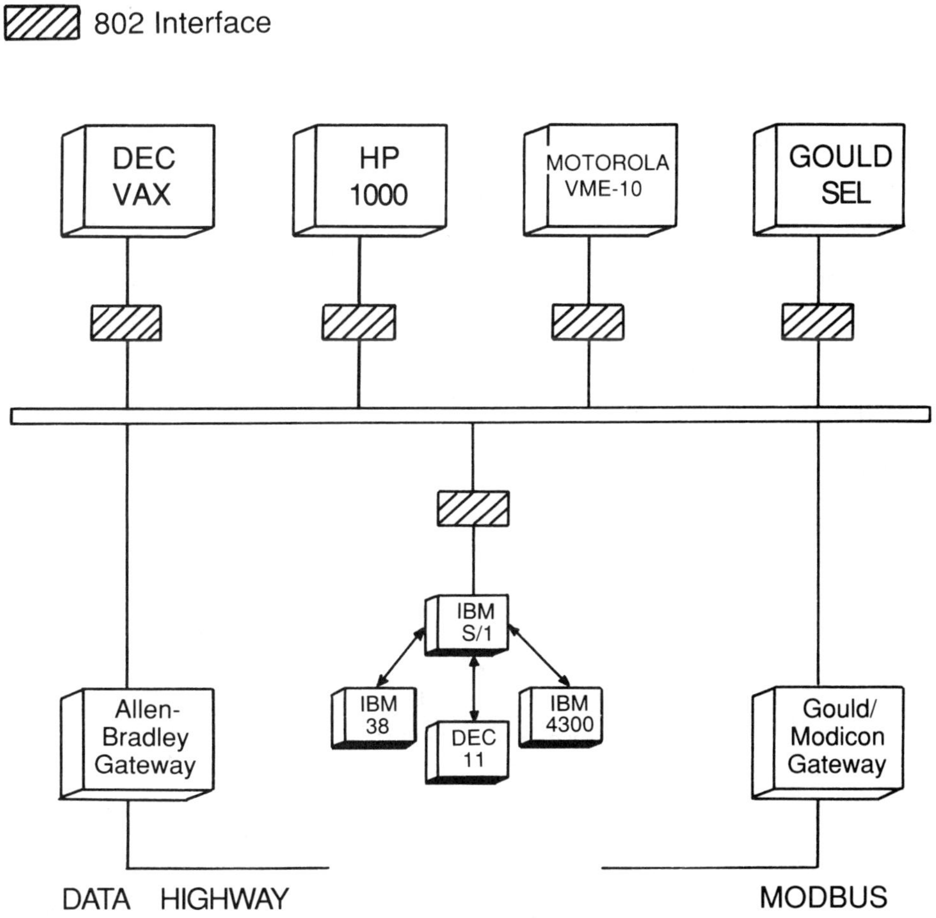

FIGURE 4–40 Integrating LAN (Courtesy of General Motors—MAP).

standards exist. The Step 3 software will be a mix of GM specific and standard protocol based on pilot needs and the approval process of standards groups.

A new MAP gateway is shown in figure 4–41. The Step 3 gateway connects the MAP network to other networks usually at a remote site. The gateway point is the plant PBX which provides access to long-haul telephone-type connections. Joint development has continued between the ICM group and General Motors Information Systems Communications Activity (GMISCA).

Step 4: Low-Cost Hardware. The reason for selecting standard protocols is most apparent in Step 4 where VLSI hardware is to be used. Acceptance of standards generates a high-volume market necessary to encourage VLSI products. Silicon implementation lowers networking costs, and in turn, leads to more installed networks. In this step, MAP networks will proliferate in GM manufacturing plants.

Figure 4–42 shows some computer nodes without Step 2 external hardware performing ISO layers 1 and 2 functions. Board-level hardware for various computer backplanes supporting ISO layers 1 through 4 will exist. A low-cost communications board will allow more processors and programmable devices to participate directly on a MAP network.

Step 5: Complete Network Utility. With an installed utility, computers supporting MAP will "plug-in" and communicate with existing MAP systems. The concept of a totally transparent network utility is the goal of Step 5 (figure 4–43). In this time frame, most computer and control equipment for GM manufacturing should participate on MAP networks.

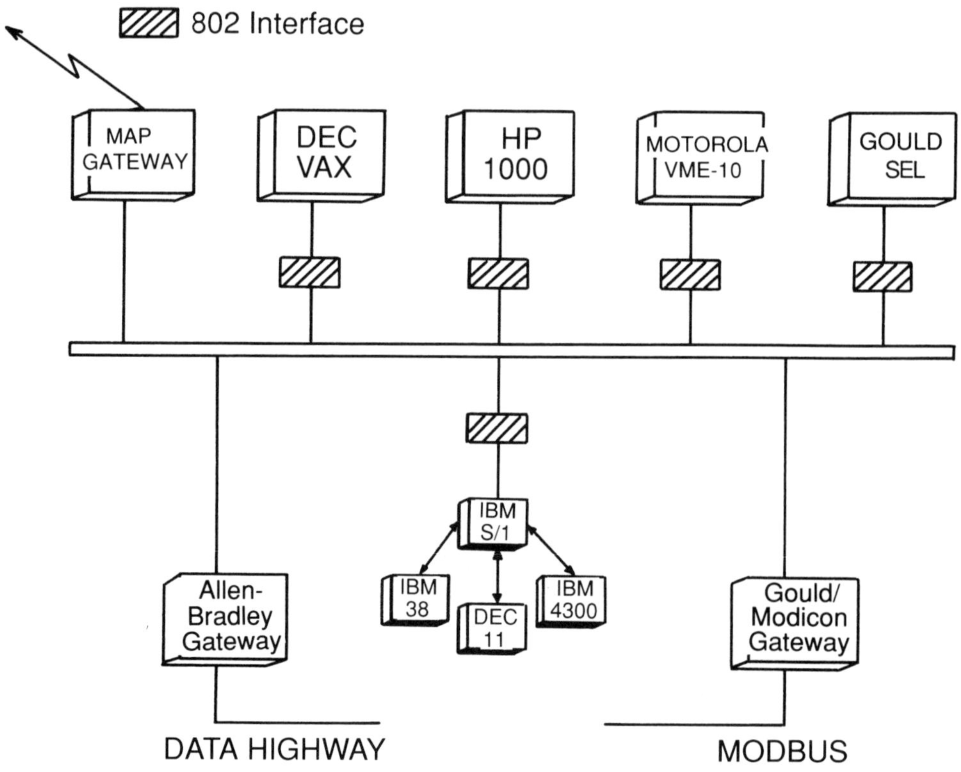

FIGURE 4–41 Application Services (Courtesy of General Motors—MAP).

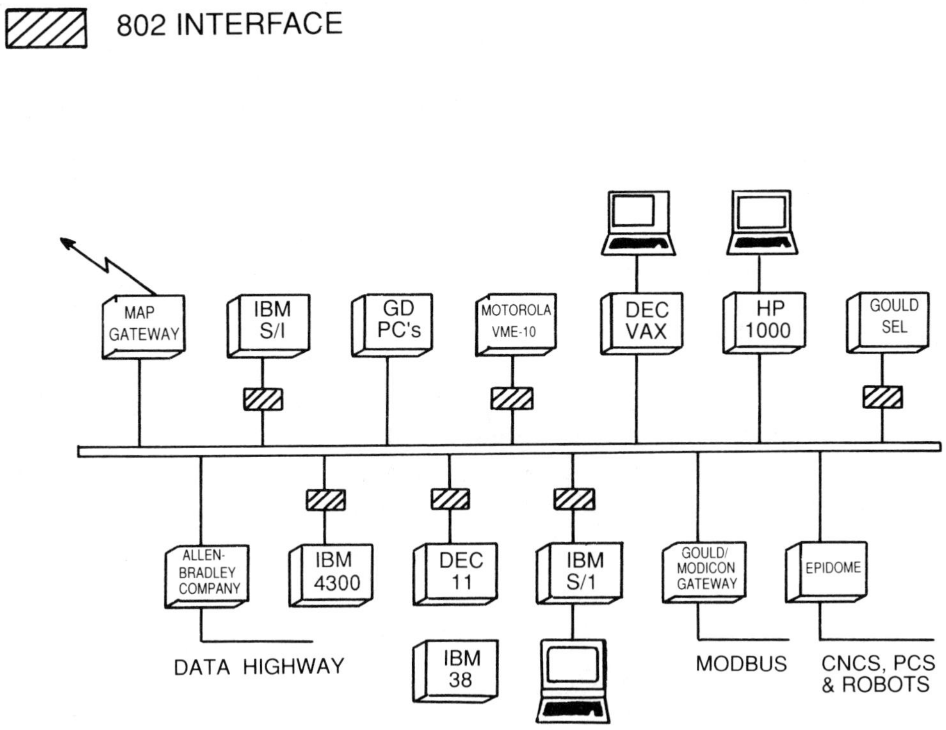

FIGURE 4–42 Low Cost Hardware (Courtesy of General Motors—MAP).

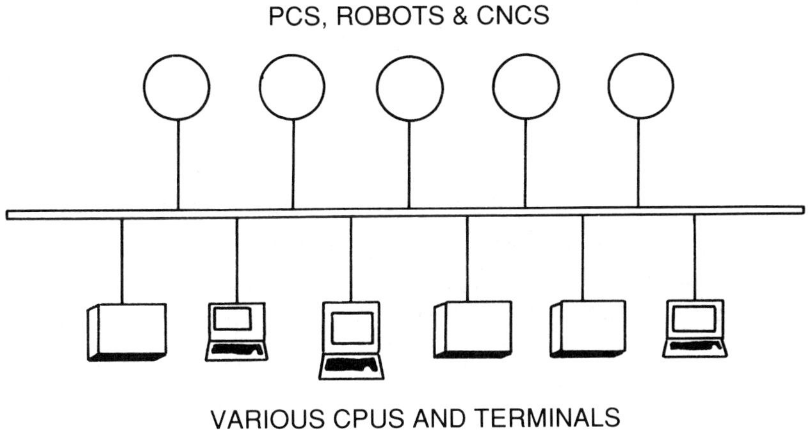

FIGURE 4–43 Network Utility (Courtesy of General Motors—MAP).

Tip for Successful Automation: The Need for a LAN

As the amount of automation increases, the need for a standard industrial local area network also increases. The MAP specification leads to a low-cost off-the-shelf solution. The migration strategy developed by GM provides interim steps and meets short-term critical manufacturing needs. The migration plan reduces vendor risk, thereby making the entire project realistic and attainable.

Network Architecture for ISO/OSI Layers

The network architecture rationale, including logical (peer-to-peer) and physical (broadband, token bus) topologies offered numerous choices to GM project team members. Also evaluated in detail is the use of repeaters, bridges, gateways, and routers for the interconnection of network segments. Protocol selections for the ISO/OSI layers are:

Layer 1	IEEE 802.4 broadband
Layer 2	IEEE 802.2 class 1
Layer 4	ISO transport class 4
Layer 5	ISO session kernel
Layer 6	Null
Layer 7	ISO CASE kernel

Five Elements of Network Management

The basic elements of the network management are:

- Monitoring
- Control
- Configuration
- Problem determination
- Recovery

Why Create a Clear Migration Path?

The MAP project team developed a clear migration path so that a communication system designer would be able to use this specification no matter what level of implementation. It defined the broadband coaxial cable as a physical backbone with MAP gateways for providing MAP capability. Other non-MAP devices are attached to these gateways through attachment links.

Acceptable attachment links include:

- HDLC ABM—for point-to-point attachment
- HDLC NRM—for multidrop attachment
- Vendor proprietary non-MAP networks

Four Minimum Requirements of MAP Communications

The intention of the MAP project is that programmable devices (for example, programmable controllers, robots, CNC machines, weld controllers, and so on) support all seven layers of the ISO/OSI model. Gateways are used until this support is in-place. All application-to-application communications use the manufacturing message format standard and the MAP CASE application layer.

A minimum set of network functions required include:

- Program upload and download
- Storage and retrieval of data
- Status reporting
- Remote diagnostics

CASE-IN-POINT

*KAISER ALUMINUM AND CHEMICAL CORPORATION'S ERIE WORKS MANUFACTURING AUTOMATION PROTOCOL NETWORK**

Situation

Erie Works specializes in the manufacturing of high-quality aluminum forging dies. Forged aluminum is as strong as steel but much lighter in weight, which explains its popularity in the automotive and aerospace industries.

In recent years, Kaiser Aluminum and Erie Works realized that they needed to modernize and move into the front lines of the newest industrial revolution if they were to stay competitive and to survive. This meant investigating ways to speed the design and manufacture of forging dies with an eye to cutting lead times and costs, and ways to improve accuracy and repeatability.

That investigation led to CAD/CAM. Erie Works now has one of the only computer-controlled systems anywhere in the world dedicated exclusively to the design and production of aluminum forging dies.

However, what has evolved since Kaiser and Erie first ventured into the computer age goes beyond mere CAD/CAM. There now exists a sophisticated system of graphics terminals used for designing forging dies; CNC machines that have automated die production; accounting and word processing terminals that handle the business operations of the plant; a complete energy management system; personal computers that access programmable logic controllers (PLCs) on production equipment; and super minicomputers that manage the entire system.

All of the above are connected by a state-of-the-art, high-speed broadband network compatible with MAP and endorsed by General Motors as a communications standard for the automation of manufacturing organizations.

It is also the beginning of true *computer-integrated organizational automation.* Plans are being made that every group and department in Erie Works—from administration to shipping and receiving, from engineering to accounting, from MIS to maintenance—will eventually benefit from this system.

Solution

After considerable investigation, Kaiser and Erie developed a hardware/software benchmark that called for design of a part, mass property calculations, production of the blueprint, and machining of the die that generates the part on a CNC machine.

*Permission granted by Data General.

The majority of the systems evaluated required two days or more to complete the benchmark. Unigraphics® software from McDonnell Douglas Manufacturing Industry Systems Company running on graphics terminals connected to an Eclipse MV/10000™ system from Data General completed the test in a matter of hours. A system of graphics terminals and an Eclipse MV/10000™ system with four megabytes of memory was installed at Erie Works, and was operational almost immediately.

Once the CAD/CAM system was in place, the next step was to establish computerized numerical control of the die sinking equipment to speed die production and to assure greater accuracy. Kaiser and Erie added three Rambaudi vertical CNC machines, which allow die sinkers to machine dies with the help of computer control.

When the Rambaudi equipment was first installed, die sinkers had to translate the design specifications into computer code and then use keyboards on the CNC machines to enter these numerical control commands into the machines' memories. But recently, the CNC machines have been connected directly to the CAD/CAM system via a Masterlink 3000 distributed numerical control (DNC) unit with a Data General Desktop Generation® computer having dual disk drives and 15 megabytes of memory.

This means that the necessary NC commands are generated on a graphics terminal using a computer model of the die created by the forge engineer, and transmitted directly to the CNC equipment, which produces yet another substantial savings of time. The CNC equipment can quickly capture any modifications made on the completed tooling after a test forging has been developed by digitizing the die cavity and saving the data on disk.

In another phase, Erie Works installed a second Eclipse MV/10000™ system with six megabytes of memory to begin the process of offloading the plant's computer operations from the corporate network at the firm's headquarters in Oakland, California.

This fully integrated system handles all accounting, purchasing, work-in-process tracking, payroll, and shop floor control requirements of the plant.

In addition, it provides on-line data entry and inquiry. The system also supports Data General's CEO® software to enhance the productivity of the administrative staff.

With the Erie administrative system up and running, the plant is able to access the corporate network using the Data General's DG/SNA communications package.

Networking with MAP. The element that ties the various devices at the Erie Works together is a MAP-compatible broadband communications network called Token/Net from Concord Data Systems. The primary component of the Token/Net system, which is also fully compatible with IEEE 802.4 LAN standards for token-passing broadband networks, is called the Token/Net Interface Module (TIM).

There are currently 36 TIMs at Erie Works, connecting eighty-five different devices throughout the twenty-six building complex. Plans are for the 150-plus devices to be eventually accessible over the network.

What makes Token/Net so important with regard to MAP is that it was designed to industry standards with the express intent of fostering compatibility among computer products and programmable devices from various vendors. The product also supports high-speed data transfer (up to 19.2 kilobits per second asynchronous) over regular phones, which provides significant cost savings for users.

Kaiser and Erie have found broadband networks to be better suited to manufacturing than traditional modes of communications (particularly twisted-pair wiring). A few devices within close proximity can be hardwired together easily using twisted pair. But in a manufacturing plant, where many different devices exist and are typically spread out over thousands of square feet, the single, coaxial broadband cable provides easy connection and allows greater flexibility.

Using twisted pair in a large manufacturing organization can result in a nest of wires above the ceiling, thus making diagnosis of network problems a monumental task.

It also costs less to add devices to a broadband network than to a twisted-pair network. Kaiser and Erie had a one-time charge to install the broadband cable. It doesn't cost anything extra each time they add a device to the network. If the plant had a twisted-pair network, there would be cabling costs each time a machine was added.

Broadband's flexibility is another advantage. It can handle synchronous or asynchronous devices concurrently; and can connect devices with differing baud rates, data bits, and parity.

However, the addition of the broadband network at Kaiser and Erie also provides another important automation possibility—one, which will in fact help the network pay for itself. The Erie Works chief electrical engineer now monitors and manages the energy use in the factory, using a personal computer in his office which is connected to the broadband network that processes information gathered by remote controllers, to cycle loads and reduce peak load demand. Erie Works saves several thousand dollars each month in utility charges thanks to the energy management system.

One of the most dramatic examples of the broadband's flexibility is portrayed by the use of personal computers in the engineering and maintenance departments. For example, the maintenance foreman can place a "call" through a TIM and communicate through a personal computer to a PLC on a piece of production equipment such as a hydraulic press. This foreman can upload a new program, download a copy of the existing program, or simply monitor the machine's activity using the personal computer. From this same personal computer, the foreman can place a "call" to a process controller on a heat–treat furnace to check a temperature in response to a phone call from a production foreman. The maintenance foreman might also want to monitor the energy management system or the CAD/CAM system, or tie into the administrative system to use CEO. Each call is routed through the broadband and virtual connections are made and broken at the request of the user. It would be very costly to provide point-to-point communications to match these capabilities.

Kaiser and Erie see the existing system as the "beginnings of a fully integrated information system." Other areas that soon could be on the network include computer-aided quality assurance that uses the original engineering models stored in a company data base; checks of completed parts against specifications and designs; on-line cataloging of dies; and records that trace the history of machine repair and maintenance.

There also are plans to implement an MRP package in the near future. Further out, Erie Works sees even tighter integration between the CAD/CAM system and the administrative system when they implement shop floor data collection to handle time card activity and payroll; and, when bids and quotes made by the sales department can be accessed directly by the design engineers who need the information for specifications.

Results

As management has discovered in recent years, truly workable factory automation is based on more than just hardware, software, and data bases. Perhaps the most important element in an automated factory is the ability of computers and software packages from a variety of vendors to coexist, to communicate with one another, and to share information easily and quickly over a network. The network acts as a translator, a sort of "electronic diplomat," ensuring that all the devices understand each other.

However, it is this networking that was the last piece of the industrial automation puzzle to fall into place. Until recently, only a few communications products could really integrate the range of hardware and software available.

Over the last few years, factories have either begun automating because they felt they couldn't afford to wait any longer, or have postponed automating until the question of multivendor communications was answered, or until their vendor-of-choice could offer the hardware and software that fit their particular needs.

Those that waited to automate are typically finding themselves at the back of the pack, increasingly unable to compete in a world of rising costs and decreasing sales.

However, those that automated early, such as GM, have been stung by the resultant "islands of automation"—personal computers on managers' desks that can't talk to the frame in accounting, systems in shipping and receiving that could use information on computers in order processing if they could only get to it.

Industry standards seem to be the solution to many of the problems of factory automation—standard hardware interfaces (such as RS232), standard operating systems (such as UNIX™), and standard communications protocols.

In 1982, GM launched a campaign to standardize telecommunications within manufacturing using MAP. By 1984, a MAP Users Group of the Society of Manufacturing Engineers (SME) was formed in an effort to resolve potential compatibility problems. By 1985, MAP was supported by more than 300 companies, representing virtually all the major electronics vendors: AT&T, IBM, DEC, Hewlett-Packard, Honeywell, and Data General Corporation. When this book went to press, the user group has continued to expand in membership.

MAP's greatest strength is that its specifications have been developed not by any *one* vendor but rather by a *user*—one that uses the equipment from many vendors. GM chose MAP because it conforms to the widely accepted, seven-layered communications model developed by the ISO, which was discussed earlier.

chapter 5

How to Develop an Effective Shop Floor Control System

ADVANTAGES OF EFFECTIVE CONTROL: AN OVERVIEW*

Production planning and control is responsible for developing detailed plans on how and when products will move through the manufacturing cycle, as well as the methods used to control the product during this phase in order to ensure that it will be produced on time and within customer required specifications.

The planning phase of this important function consists of the following elements:

- Sales forecasting
- Master production scheduling
- Material requirements planning
- Capacity requirements planning

The computer-based planning phase should flow smoothly into the manufacturing control cycle. This control mechanism involves the following elements:

- Shop scheduling (fine-tuning)
- Capacity control
- Purchase order scheduling and control of vendor deliveries

Shop floor control is necessary because shop orders released to manufacturing cannot possibly anticipate all of the unforeseen circumstances that may occur in the plant environment. These conditions may involve such things as:

1. Flaws in the scheduling process that did not provide sufficient allowance for emergency orders, changes in product forecasts, seasonality factors, or single-source vendor uncertainties

*Based on material by the Little People's Productivity Center. Used with permission.

2. Inadequate management policies in regard to handling orders flowing through manufacturing
3. Unusual incidence of worker absenteeism
4. Breakdown of a key, special-purpose machine tool, or major disruption in a production cell
5. Inability to obtain vendor materials in consonance with overall production schedule
6. Lack of systematic movement of products through the manufacturing cycle
7. Lack of accurate or up-to-date inventory information, with the inability to locate a specific item in inventory within the required time

This chapter explains how a company can install shop floor control systems that can minimize disruptive influences and ensure that products will flow through manufacturing on schedule. Once these systems are completely operative, it will be possible to achieve the following results:

- Faster throughput of product in the manufacturing cycle
- Reduced direct labor cost
- Improved customer delivery
- More accurate information
- Quicker feedback of data for planning and control purposes
- Tighter in-process quality control

The Need for Integrated Manufacturing Control

Manufacturing control requires computerized systems that involve bills of materials, material requirements, machine loads, and scheduling of shop orders.

This in turn involves the development of controls over sequencing of shop orders, routing of orders, and establishment of order priorities.

Feedback of activities on the shop floor is achieved through various methods of shop order reporting and backlog condition of machine tools, either individually or within work centers. In addition, the movement of products from one machining center to another requires control and feedback on materials handling activities within the factory.

The Right Time for Releasing the Shop Order

It is most important that shop orders be released only when verification has been received in production control that material and tooling are available and that the load factor has been properly calculated to ensure that the shop order can be produced within the allotted scheduled cycle time. When all of these factors have been addressed the shop order then can be released with a reasonable assurance that the customer order commitment to ship the requested quantity by a certain due date will be met.

When all of these conditions have been met, upon release it will then become an open shop order, consisting of the following items:

- Engineering drawing
- Bill of materials
- Process/route sheet

- Shop order
- Material requisition
- Tool requisition
- Labor feedback
- Move ticket

Once the shop order with all accompanying documents has been released it will then be tracked through collection of data as it moves from operation to operation, when an operation is completed (and part inspected), and through various input/output controls.

Six Reasons to Establish Effective Dispatching

As a shop order flows through the manufacturing cycle it is necessary to control its movement through some type of dispatching to ensure that:

1. The shop order is on schedule.
2. Any delays are being recouped through expediting, short cycling, or other methods.
3. Unexpected bottlenecks are resolved so that lost time may be made up.
4. Rejections, or scrap, that impact the order are mitigated (recycled into the system) as soon as possible.
5. Information is provided to production control on any major delays so that a recovery plan can be developed to make up all or some of the lost time, and the sales office is notified so that the customer can be appraised of the affect on delivery due to this unforeseen happening.
6. Assistance is given in developing any changes in scheduling or machine loads as circumstances dictate.

Major Elements of Input/Output Control

Data used to control the flow of a product involves the following factors:

A realistic agreed upon plan	Developed from capacity planning stage
Feedback of data	Obtained from various methods of shop feedback
Tolerance	The limits beyond which cumulative deviation is not allowed to go without corrective action being taken
Corrective Action	Results in increases (or decreases) to the output

Expediting: How It Differs from Dispatching

The distinction between dispatching and expediting is that dispatching is more concerned with providing the necessary materials, tooling and other requirements to shop personnel to meet scheduled requirements. Expediting is involved in tracking emergency jobs, handling problems that occur on the shop floor, and locating lost material, misplaced tooling, or running down any other unusual occurrence. This may involve reconciling differences in order of priorities, shortages, scrap or rework, and other conditions that may require a trade-off between conflicting order demands. Whenever shop order input results in an excessive backlog that cannot be satisfied by productive capacity a reduction in output will result. When this condition occurs, more overtime, addition of another shift, more workers per shift, or eventually the addition of more machine tools is required to handle a continuing situation of inadequate capacity.

TEN BENEFITS OF A COMPUTERIZED SHOP FLOOR CONTROL SYSTEM*

To ensure that all elements of the control phase are being monitored properly, it is essential that an effective computerized system be installed that interfaces with the computer-based planning cycle. This set of controls should enable a user to issue shop orders, track work-in-process, maintain work center information, and analyze shortages and backlogs. Data obtained from this system should result in increased labor efficiency, better machine utilization, less downtime, and more reliable capacity planning and overtime scheduling. In essence, it can result in more complete control of the manufacturing process through greater detailed knowledge of what it will take to finish a job and what alternatives are available when there are deviations from the plan.

More specifically, this system can inform production control when orders are completed at each operation, how much has been scrapped or rejected, when an order needs to be split, and the variance between the planned hours and the actual hours it takes to finish an order. It also can display or print shop order history by part, order, or date, and examine issues, partial completions, and receipts.

A computerized shop floor system can generate routings, manufacture bills of material, and pick lists.

A shop floor control system can provide these features:

1. Maintains data instantly with on-line interactions.
2. Stores, maintains, and prints standard routing data by operation for each manufactured part.
3. Separates total manufacturing lead time into plan, build, and stock times.
4. Tracks standard run times and queue times for batch or unit processes.
5. Prints shop paperwork for picking inventory and order routing.
6. Calculates operation due and start dates using backward scheduling.
7. Prints and displays work center status with start and due date priorities.
8. Monitors work center input and output and reports work center queues to help manage shop lead times.
9. Displays time-phased work center queue and backlog versus capacity to identify capacity constraints.
10. Displays load and queue by work center.

Figures 5–1 and 5–2 show a work center summary report and work center detail report. These reports contain work center information including number of labor hours per day (for capacity planning) and standard queue. It also can contain total work center input, output, load, and queue, to identify potential bottlenecks quickly.

*Permission granted by Micro-MRP, Inc.

11-04-1985 16:59:35 MICRO-MRP, INC. PAGE NO. 1

MAX * SHOP FLOOR CONTROL * WORKCENTER SUMMARY REPORT

WRKID	DESCRIPTION	TYP	STD QUE	HRS MND	TOTAL INPUT	SMOTH INPUT	TOTAL OUTPT	SMOTH OUTPT	TOTAL LOAD	TOTAL QUEUE	ALPHA	DAYS
ASSY	ASSEMBLY	S	2	40	252	50	127	25	1877	125	0.20	46.9
CUT	CUT TO SIZE	S	2	20	0	0	0	0	0	0	0.20	0.0
FOLD	FOLD TO SHAPE	S	2	40	0	0	0	0	0	0	0.20	0.0
INSP	FINAL INSPECTION	S	2	40	52	10	22	4	782	30	0.20	19.5
PICK	ISSUE FROM STOCK	S	1	40	2	0	1	0	1	1	0.20	0.0

FIGURE 5–1 Workcenter Summary Report (Courtesy of Micro-MRP, Inc.).

11-04-1985 17:00:20 MICRO-MRP, INC. PAGE NO. 1

MAX * SHOP FLOOR CONTROL * WORKCENTER DETAIL REPORT

WORKCENTER : ASSY DESCRIPTION : ASSEMBLY TYPE : S STANDARD QUEUE : 2

TOTAL INPUT : 252 SMTHD. INPUT : 50 TOTAL LOAD : 1877 TOTAL QUEUE : 125

TOTAL OUTPUT : 127 SMTHD. OUTPUT : 25 HOURS MANNED : 40 ALPHA : .2

ORDER#	PART IDENTIFIER	PART DESCRIPTION	SEQ#	BAL. DUE	START	CUR DUE	HOLD	QUE	RUN HRS	SETUP HRS	LT REM
500001	13000	PC KEYBOARD	0020	200.00	10/28/85	10/30/85	N	N	0.0000	2.0000	2
500003	13000	PC KEYBOARD	0020	175.00	11/07/85	11/11/85	N	Y	125.0000	2.0000	2
TOTAL									125.0000	4.0000	
DAYS OF QUEUE									3.2250		

CHAR

FIGURE 5–2 Workcenter Detail Report (Courtesy of Micro-MRP, Inc.).

11-04-1985 17:02:09 MICRO-MRP, INC. PAGE NO. 1

MAX * SHOP FLOOR CONTROL * PART ROUNTING BY PART IDENTIFIER

PART IDENTIFIER : 13000 DESCRIPTION 1 : PC KEYBOARD DESCRIPTION 2 : TYPE : A

SEQ#	OPER DESCRIPTION	WRK ID	RUN HRS	SETUP HRS	REV DATE	OPR TYP	STD TYP	OPR ID	QTY PER CYCLE	TOOL REFERENCE
0010	PICK & STAGE	PICK	0.0000	1.0000	12/31/83	U	E		1.00	
0020	FINAL ASSEMBLY	ASSY	5.0000	2.0000	12/31/83	U	E		1.00	JIG 53467
0030	INSPECTION	INSP	2.0000	1.5000	12/31/83	U	E		1.00	JIG 77777
TOTAL			7.0000	4.5000						

FIGURE 5–3 Part Routing Report by Part Identifier (Courtesy of Micro-MRP, Inc.).

11-04-1985 17:05:04 MICRO-MRP, INC. PAGE NO. 1

MAX * SHOP FLOOR CONTROL * ORDER ROUTING DETAIL REPORT BY ORDER NUMBER

ORDER NUMBER : 500003 PART ID : 13000 DESCRIPTION 1 : PC KEYBOARD COMMODITY : FIN GOOD

SEQ#	OPERATION DESCRIPTION	WRK ID	RUN HRS	SETUP HRS	QUE QTY	LOAD QTY	SCRAP QTY	OPR TYP	STD TYP	QTY PER CYCLE	TOOL REFERENCE
0010	PICK & STAGE	PICK	0.0000	150.0000	150.00	150.00	0.00	U	E	1.00	
0020	FINAL ASSEMBLY	ASSY	875.0000	350.0000	25.00	175.00	0.00	U	E	1.00	JIG 53467
0030	INSPECTION	INSP	380.0000	285.0000	15.00	190.00	0.00	U	E	1.00	JIG 77777
TOTAL			1255.0000	785.0000							

FIGURE 5–4 Order Routing Detail Report by Order Number (Courtesy of Micro-MRP, Inc.).

Figure 5–3 shows a part routing report by part identifier. This report details each part's planned routing, and can be printed by part number, part type code, planner, commodity code, or by ABC class code.

Figure 5–4 shows an order routing detail report by order number. This report enables a user to compare actual shop routings against the planned part routing to quickly identify discrepancies on the shop floor. This report can be sorted by completed orders by date, open orders by part, or open orders by date.

FIVE OPTIONS FOR COLLECTING SHOP FLOOR DATA

To maintain accurate, up-to-date data-base files, a method for gathering, validating, and transmitting production data from the shop floor has to be implemented. Timely and accurate feedback is needed for management control.

Shop floor data collection should be complete and stored in a usable format and it must be validated near the source of data.

Types of data collection include:

- *Manual*—move and status tickets record information from the shop floor and send the information to production control. It is error prone, costly, and lacks timeliness.
- *Paperless processing*—through use of bar codes and/or computer terminals.
- *On-line data collection systems*—shop floor data is keyed directly to a host computer using terminals on the shop floor, by automatic, on-line, identification systems, reading magnetic cards, or wanding or moving-beam laser scanners for bar codes.
- *Microcomputer*—for several data collection applications.
- *Software packages*—on-line, real-time processing capabilities.

Key Point: You cannot overemphasize the fact that success of a control system depends largely on the input data for production control use in monitoring results.

In addition to basic control elements, shop floor data feedback is also needed to compare standards against actual direct labor performance, and to seek possible areas for improvement, such as:

- Machine down time
- Setup time
- Yields
- Manufacturing standards
- Scrap control

WHAT INFLUENCES SHOP CAPACITY CONTROL LOAD FLOW*

A manufacturing facility is limited by its ability to handle the amount of work received compared with its machining resources to produce the products required. This can be expressed in the following ways:

- *New orders* result in a backlog of *open unreleased orders.*
- Production control plans the *input rate* that the factory can handle based on its capacity and customer or produce to stock requirements.
- As the shop orders flow through the plant they become *work-in-process,* or *load.*
- Manufacturing lead time is based on the plant *capacity* to handle these orders which then results in the amount of orders completing the manufacturing cycle, or plant *output.*

SIX FACTORS AFFECTING MANUFACTURING LEAD TIME

Orders scheduled for production involve several factors:

- *Process or cycle time*—including setup or make-ready time
- *Size of lot*—as a multiplier of process time
- *Capacity*—convert process time to elapsed time
- *Move time*—transporting product between process steps
- *Queue time*—backlog of work awaiting next operation
- *Allowances*—prepare order, schedule order, issue material or parts, stock parts in temporary hold areas

FOUR METHODS FOR ESTABLISHING PRIORITY

The four basic methods of establishing job priorities are as follows:

1. Operational due dating
2. Critical ratio
3. Slack time remaining
4. Slack time per operation

*Based on material from the Little People's Productivity Center. Used with permission.

1. Operational Due Dating

Calculate due dates for each operation

Example:

5 operation (OP) job
Job due date + OP50 due date
OP40 due date = OP50 due date – OP50 run and wait
OP30 due date = OP40 due date – OP40 run and wait, etc.
Start date = OP10 due date – OP10 run and wait

2. Critical Ratio

Example:

Customer order – need in 30 days = Demand time
Std. lead time data – takes 40 days to make = Supply time
Critical Ratio = D/S = 30/40 = 0.75

This item must be processed through all operations in 3/4 the normal time. It should take priority over all items that have a higher critical ratio.

3. Slack Time Remaining

Example:

Lead Time Remaining – Standard process days remaining
(Due Date – Now) (Setup and run time converted to days) = Total slack time remaining

4. Slack Time per Operation

Example:

$$\frac{\text{Total slack time remaining}}{\text{number of operations}} = \text{Slack time remaining per operation}$$

SIX MAJOR NEEDS SATISFIED BY TACTICAL RATIO AND MONITORING SYSTEM*

The tactical ratio and monitoring system (TRAMS) was designed to control work-in-process lots throughout the manufacturing process. It is completely compatible with many MRP and master scheduling programs. It handles most shop floor problems, including meeting delivery date promises, reschedules, additions, cancellations, customer emergencies, short lead times, machine loads, and priority lots.

TRAMS has many unique features, such as variable queue based on machine load and mix, automatically reduced queue for "must ship" orders, group technology identification of parts, and infinite loading of machine centers.

The TRAMS system was designed to satisfy six major requirements:

1. Meeting customer delivery date promises
2. Priority identification of work-in-process lots
3. Short lead-time capability

*"TRAMS—Tactical Ratio and Monitoring System," by Harry G. Smart. Reprinted courtesy of the Society of Manufacturing Engineers, Copyright © 1984 from Autofact 6 Conference Proceedings.

4. Automatic rescheduling and prioritizing
5. Group technology decision capability
6. Infinite loading of customer orders

The methodology used to satisfy these criteria necessitated considerable shop management involvement to develop "real world" solutions. These requirements are discussed in the following sections.

1. How TRAMS Helps Keep Customer Delivery Date Promises

The TRAM system continuously recalculates estimated completion dates to keep customer delivery date information current and accurate. The system determines the *estimated completion date* of a given work-in-process lot taking into consideration the processing time (setup and production time) and waiting time (queue time) of a given lot for each department and machine type through which it must be processed.

Estimated completion dates are then determined by summing up the queue, setup, and production times in decimal days for each operation and adding them to the current day's date. A date is calculated each time an operation is completed.

Setup and Production Time. Each machine code has its own setup and production times for each part number. To determine the minimum decimal days required to perform an operation without queue time, setup time and production times are added on each machine in each machine code. For example: Machine code 124 – 2-shift operation, eight hours per shift.

$$\text{Minimum decimal days for this operation} = \frac{\text{Setup and production time}}{16}$$

Queue Time. Queue time for each machine code is calculated on a rolling 15-day average in accordance with the following formula:

$$\text{Current queue time decimal days for each operation} = \frac{\text{Current actual department load in standard hours}}{\text{Average standard hours produced per day}}$$

where

Current actual department load (standard hours) is, on a snapshot basis, the total hours of work on the floor ahead of each individual machine code.

Average standard hours produced per day is calculated by dividing the total standard hours produced during the preceding 21 calendar days by 15 days minus the number of holidays in the same period as follows:

$$\text{Average standard hours produced per day} = \frac{\text{Total standard hours produced in preceding 21 calendar days}}{\text{15 days} - \text{number of holidays included}}$$

This formula includes the affect of current average overtime hours being worked, if any.

Estimated Completion Dates. Estimated completion dates are determined by summing up the queue, setup, and production times in decimal days for each operation and adding them to the current day's date. A date is calculated each time an operation is completed.

2. Why the Tactical Ratio Is Essential in Priority Identification

The next step is to determine which job to run next. This is done by calculating a number, called *tactical ratio* which ranks each available job to run in a given department in a sequence which considers its present status in relation to its scheduled due date (customers requested due date as opposed to production's estimated completion date).

The tactical ratio of any job is the number of days between the current day's date and its scheduled due date, divided by the number of days between the current day's date and the estimated completion date as determined previously:

$$\text{Tactical ratio} = \frac{\text{Scheduled due date} - \text{current day's date}}{\text{Estimated completion date} - \text{current day's date}}$$

A lot's tactical ratio is recalculated each time it moves from one department to another and is printed on the receiving supervisor's schedule.

> ***Key Point:*** The production lot with the lowest tactical ratio number has the highest priority. A tactical ratio of 1.0 or greater indicates a job is on schedule; less than 1.0 indicates a job is behind schedule; and a negative tactical ratio indicates the scheduled due date has passed.

3. Procedures for Handling Short Lead-Time Orders

The system provides for customer emergency needs when it is necessary to deliver an order in less than the normal lead time. This is accompanied by recognizing that the only portion of the cycle time that be reduced realistically is the queue time. To determine the percentage by which present queue times must be reduced to meet a given completion date, the sum of minimum decimal days by operation is subtracted from the number of days between the current day's date and the "must ship" date to provide the days available for queuing. By dividing by the sum of the normal operation queue days, it is possible to determine a percentage of normal operational queue days that can be permitted.

$$\begin{array}{c}\text{Percent of normal}\\ \text{operational queue days}\end{array} = \frac{\text{Minimum decimal days by operation (“Must Ship” Date} - \text{Current Day's Date)}}{\text{Normal operational queue days}}$$

The percent of normal operational queue days is then applied to each sequence operation to determine the decimal queue days allowed for each operation.

$$\begin{array}{c}\text{Decimal queue days}\\ \text{allowed by operation}\end{array} = \text{Percent of normal operational queue days} \times \text{normal queue by operation}$$

4. How TRAMS Accomplishes Automatic Rescheduling and Prioritizing

Each day the tactical ratio is calculated for every job in process. Based on the previous day's activities, all jobs are placed in their proper relative position to each other.

> ***Key Point:*** Additions, deletions, pull-ups, push-outs, short lead-time orders, and individual machine loads are all considered in the rescheduling process.

On the basis that the system has captured the current work-in-process load (all jobs that have been released to the initial operation) by machine code, the following steps apply:

1. By summing up the setup and production hour load ahead of each machine code, the current queue time is calculated.
2. All remaining operations in their respective machine codes are summed up for every job by setup, production, and queue times. These times are then added to the current day's date, thus arriving at the estimated completion date.

3. The tactical ratio is calculated for every job.
4. Each job is prioritized in a given machine code in accordance with its tactical ratio.
5. The computer re-addresses itself to the short lead-time jobs and calculates the percent of normal queue time allowed for each of them.
6. The computer calculates the decimal queue days allowed for the machine code in which it currently resides.
7. The computer adds up all setup and production hours of the jobs on the machine code in ascending order of tactical ratio until the cumulative decimal days are equal to or greater than the decimal queue days allowed for the short lead-time jobs.
8. The short lead-time job is inserted in the schedule ahead of the point calculated previously.
9. An operation completion date and hour is printed on the foreman's control sheet.

Key Point: This completion date and hour will remain unchanged until the job is completed at that operation.

10. When an operation is completed and the job is moved, a new date and hour is calculated. This automatically reallocates the remaining allowable queue time in the event of a slippage due to an operation not being run on time.
11. When a departmental due date on a short lead time is missed, an exception report is generated and the computer prints an exception message notifying production management.

5. How TRAMS Utilizes Group Technology Decision Capability

This system utilizes some of the features of group technology discussed in Chapter 2. In addition to TRAMS furnishing production supervision with the tactical ratio and short lead-time ranking on a hard-copy print-out, this report also contains the group technology code for each part number. This allows the floor supervisor to group setups of similar parts, thus reducing cycle time and cost.

The group technology code used in the TRAM system is a twenty-two digit alphanumeric code which identifies eleven characteristics of every part number. By scanning the machine load report, a supervisor can determine like part numbers for grouping setups.

6. Using TRAMS to Predict Overload Conditions

Most, if not all, companies must build into their system the capability of absorbing an overload condition and predicting the need for overtime or additional manpower in the near future. This approach is referred to as *infinite loading.*

In most instances, this is due to the nature of incoming orders where lead times vary significantly. There also may be many different product types sharing the same machine code, so that changes in the mix of product types may cause an ebb and flow effect on specific machine loads. By using the tactical ratio technique of shop floor control to meet customer delivery demands, TRAMS identifies overload conditions far enough in advance to plan overtime as needed. In the event that incoming orders exceed capacity for an extended length of time, the tactical ratio technique has the effect of making all orders equally past due, thus giving the sales department the capability of determining the amount of additional lead time to quote on new orders.

Tip for Successful Automation:
TRAMS Data Base and Input Requirements

The TRAMS data base consists of the following:

1. Production and setup standards
2. Group technology coded part numbers
3. Routings for each part
4. Scrap allowance tables
5. Order backlog and master schedule

TRAMS reports and screen displays are as follows:

1. Order lead-time screen (figure 5–5)
2. Daily machine load report (figure 5–6)
3. Work-in-process status report (figure 5–7)
4. Load hour analysis report (figure 5–8)
5. Capacity requirements summary report (figure 5–9)
6. Routing numbers (figure 5–10)

```
*****************************************************************************************
*                                                                                       *
*          TRAMS7           TACTICAL RATIO AND MONITORING SYSTEM            02/21/86    *
*                                      ORDER LEAD TIME                       6:41AM     *
*       LOT#: 17732-91     FAC: 02 RTG #  20831 STATUS:          FLG      PG   1 OF  2  *
*       PART  #: HL570-6MC                NET-MFG QTY:     99,000  CURR QTY:   103,000  *
*       R/M:    410 PROD:  110 PC DATE: 03/10/00 DUE DATE: 03/14/86  ET-CMP DTE: MS/03/17  *
*       REL:  09/27/85  C/R: M/S C003  OT: 1   GO:      DO:  15 SD:  0375B410     C: 5934*  *
*       REMARKS:                                                                        *
*       COM RMK:  SYS                                                                   *
*       SEQ DPT M/C OPER  OPER DESC          SET-UP     RUN    QUEUE     TOTAL  DAYS DY-CMP  *
*       001  23 110  198  MACHINE              8.00  124.63   103.57    236.20   1.1  16.3  *
*       002  14 680  746  DECREASE - S-700                                        .0  15.2  *
*       003  14 690  318  TUMBLE/VIBRATORY                     16.00     16.00   1.0  15.2  *
*       004  91 952  830  INSPECT, IN PROC                      8.00      8.00    .5  14.2  *
*       005  91 951  842  LABORATORY, Q.C.                      8.00      8.00    .5  13.7  *
*       006  13 617  605  AGE                                  80.00     80.00   5.0  13.2  *
*       007  91 951  842  LABORATORY, Q.C.                      8.00      8.00    .5   8.2  *
*       008  91 953  894  ZYGLO, FINAL                         16.00     16.00   1.0   7.7  *
*       009  14 639  634  ANODIZE TYPE II                      16.00     16.00   1.0   6.7  *
*       010  14 142  273  SORT                                  8.00      8.00    .5   5.7  *
*       011  91 952  827  INSPECT, FINAL                       16.00     16.00   1.0   5.2  *
*       012  14 810  727  CETYL ALCOHOL LU                     32.00     32.00   2.0   4.2  *
*       013  91 951  839  LABORATORY, Q.C.                     16.00     16.00   1.0   2.2  *
*       PF:NEXT=1 PREV=2 MENU=3 KRUP-R=5 KRUP-G=6 KRCC=7 TERM=9 KRPM=10 KRML=11          *
*                                                                                       *
*****************************************************************************************
```

FIGURE 5–5 Order Lead Time Screen (reprinted courtesy of the Society of Manufacturing Engineers, copyright © 1984, from Autofact 6 Conference Proceedings).

```
TRAMS1                                VSI CORPORATION                                    PAGE      84
                               TACTICAL RATIO MONITORING SYSTEM                          DATE  03/24/86
                                  DAILY MACHINE LOAD REPORT                              TIME   3:11AM

 FAC  MACH GRP   DESCRIPTION                     AVG T.A. HRS
 01                                                  3.38

CUST     C/R    LOT NUMBER     PART NUMBER/              RM-DEPT/OPER/    FLAG/ GO/ DA  ED  DH SEO  QUANTITY/  SET-UP  RUN   ACCUM  DYS/
                 ASSET/SCH     GROUP TECHNOLOGY CODE        M/C PROD FH  EXC   OT   SPEC DATA      CURR QTY   SOURCE              DAYS

5172   M/S C222 11478-01   VS336409D40                  070/ 0282                   1   1     011   10,153      3.20  2.24         1.6
                           HCPS6230SBVG334-43X-MVS91-2   253 0792              C

3648      1.556 13895-04   1275M34G01                   004/ 0279         R     10  10     006      366      3.00  1.41         2.9
                           WBNG3267TDQG125-----NGE01-0   240 0850                                  7,835
--------------------------------------------------------------------------------------------------------------------------------------
```

FIGURE 5-6 **Daily Machine Load Report** (reprinted courtesy of the Society of Manufacturing Engineers, copyright © 1984, from Autofact 6 Conference Proceedings).

```
TRAMS2                                  VSI CORPORATION                                  PAGE      28
                               TACTICAL RATIO AND MONITORING SYSTEM                      DATE  02/03/86
                                    WORK-IN-PROCESS STATUS                               TIME    2:33AM
                                            KEY AA

 FACILITY  01
 PART NUMBER      LOT NUMBER GO   C/R DUE DTE PC DTE NET QTY CUR QTY SEQ DPT  M/C OPER FLG  SPEC DATA  DO OT FH EXC ECD

HLT311D15-4        70607-01      2.489 053086          1575     1785 001  01  446 0165                  11       S

HLT311DL5-5        81333-01       .784 030386          1410     1655 004  14  645 0800                   3            03/14/86F

HLT311DL6-13       75994-01       .843 031786 0414     3059     3373 001  01  446 0165      C002.189     8       A    2 HF FEB
```

FIGURE 5-7 **Work-in-Process Status Report** (reprinted courtesy of the Society of Manufacturing Engineers, copyright © 1984, from Autofact 6 Conference Proceedings).

TRAMS3

VSI CORPORATION
TACTICAL RATIO AND MONITORING SYSTEM
LOAD HOUR ANALYSIS REPORT

PAGE 104
DATE 08/16/85
TIME 9:03AM

FAC	DEPT	M/C	DESCRIPTION	AVG HRS/DAY
02	50	184	BAKEWELL HAND FEED TAPPER	.76
	50	203	=2B.T. VERT.W/ROTARY TABLE (3)	6.49
	50	199	NO 2 BROTHER HI-TAP	
	50	187	VERT SNOW TAP-HEAD	
	50	198	NO 3 BROTHER HI-TAP	7.60
	50	202	NO.2 BROTHERS TAP AUTO CRIMP	6.34
	50	189	ROCKWELL TAPPER	.51
			TOTAL	21.70

	AVAILABLE		UNAVAILABLE					
	IN DEPT	PREV OPER	ON HOLD	BAL OPER	TOTAL	AVAIL HRS	+/- HRS	CUMULATIVE
1ST WK	209.86	30.17		114.51	354.54	108.50	-246.04	-246.04
2ND WK				26.89	26.89	108.50	81.61	-164.43
3RD WK				33.21	33.21	108.50	75.29	-89.14
4TH WK				105.91	108.50	2.59	-88.55	
5TH WK		30.76		48.70	79.46	108.50	29.04	-57.51
6-10 WKS				309.82	309.82	542.50	232.68	175.17
11-15 WKS			52.62	384.01	436.63	542.50	105.87	281.04
16-20 WKS			275.10	164.82	439.92	542.50	102.58	383.62
21-26 WKS			108.67		108.67	651.00	542.33	925.95
BALANCE			261.75		261.75		-261.75	664.20
NO REL DATE			120.74		120.74		-120.74	543.46
TOTAL	209.86	60.93	618.88	1,187.87	2,277.54			

FIGURE 5-8 Load Hour Analysis Report (reprinted courtesy of the Society of Manufacturing Engineers, copyright © 1984, from Autofact 6 Conference Proceedings).

TRAMS4 VSI CORPORATION PAGE 6

TACTICAL RATIO AND MONITORING SYSTEM DATE 08/16/85

FACILITY 02 MFG. MGMT. REPORT CAPACITY REQUIREMENTS SUMMARY TIME 9:03 AM

DPT	M/C	M/C DESCRIPTION	ACTUAL TA HRS	TA HRS PER DAY	LOTS	D Y N A M I C Q U E U E PIECES	DAYS	HRS	DUE OR PAST DUE HRS AVAL	HRS HLD	HRS
50	177	INCLINE SNOW TAP AND S (14)	87.02	87.02	11	342462	2.1	214	191		387
	201	INCLINE BROTHERS NUT TAP=1 (3)	15.13	15.13	3	6954	.1	13	13		13
GROUP TOTAL			102.15	102.15	15	373416	2.4	252	229		400
50	183	FOSDICK HAND FEED TAPPER	.00	2.00M	1	360	3.1	6	6		
	184	BAKEWELL HAND FEED TAPPER	.76	.76	1	3320	1.1	25	25	11	9
	187	VERT SNOW TAP-HEAD	.00	.00						73	
	189	ROCKWELL TAPPER	.51	.51	1	990			8	65	
	198	NO 3 BROTHER HI-TAP	7.60	7.60							
	199	NO 2 BROTHER HI-TAP	.00	.00							
	202	NO.2 BROTHERS TAP AUTO CRIMP	6.34	6.34	2	500	.1	2	6	5	
	203	=2B.T. VERT.W/ROTARY TABLE (3)	6.49	6.49	7	33811	8.3	180	170		57

FIGURE 5-9 Capacity Requirements Summary Report (reprinted courtesy of the Society of Manufacturing Engineers, copyright © 1984, from Autofact 6 Conference Proceedings).

TRAMS5 VSI CORPORATION TIME 210

TACTICAL RATIO AND MONITORING SYSTEM DATE 03/31/86

FACILITY 43 ROUTING NUMBERS TIME 1:34AM

PART NUMBER	LOT NUMBER	SEQ	DEPT	M/C	OPER	OT	PART NUMBER	LOT NUMBER	SEQ	DEPT	M/C	OPER	CT
ROUTING NUMBER 96264													
NAS1221-3E13	87186-J1	008	14	685	778								
*** TOTAL LOTS				1									
ROUTING NUMBER 96271													
BAC830NF4-2	89944-J1	004	13	617	630		BAC230NF4-8	95776-J1	001	01	208	042	
*** TOTAL LOTS				2									
ROUTING NUMBER 96273													
NAS6604-1	86751-J1	003	02	221	312		NAS6604-1C	94709-J1	001	01	208	042	
NAS6604-4	94704-J1	001	01	208	042		NAS6604-5	93212-J1	003	02	272	312	
NAS6604-6	94703-J1	001	01	208	042		NAS6604-6	86040-01	004	13	617	630	
*** TOTAL LOTS				6									

FIGURE 5-10 Routing Numbers Report (reprinted courtesy of the Society of Manufacturing Engineers, copyright © 1984, from Autofact 6 Conference Proceedings).

HOW JUST-IN-TIME TECHNIQUES APPLY TO PRODUCTION*

A just-in-time production system attempts to manufacture and assemble discrete products in the same way a process industry is run. It accomplishes this by substantially reducing the time element from receipt of raw material at one end of the plant to shipped finished product to a customer, at the other end of the plant. Thus, the major requirement of a successful JIT program in nonprocess plants (plants making discrete parts) is to take the interruptions out of the production process. Although it is virtually impossible to eliminate all interruptions, ideally, JIT means configuring a production system so that the time required for parts to flow from supplier to receiving and then through the plant to shipping is only as long as the sum of the time of the value-adding steps in the manufacturing process. Conversely, what must be avoided is, for lack of a better way to state it, "just-in-case" production methods. JIC means that regardless of what goes wrong, *there is a backup* to keep production flowing. No matter what might happen, despite every imaginable interruption, the plant will continue to deliver goods.

For instance, JIC allows the creation of huge inventories before assembly lines in the event the company's component production departments, or outside suppliers fail to meet their schedule commitments. Or, JIC provides cushions in the schedule for machine breakdowns, worker absenteeism, and other contingencies. As a result, JIC builds into the production process numerous anticipated interruptions whereas JIT disallows these happenings by introducing various preventative manufacturing methods and techniques (figure 5–11).

To accomplish this, it is imperative to remember that the *driving force* in a manufacturing facility is production—not cost or quality. The two primary time-consuming interruptions are machine setup and lack of machine tool balance. Other secondary interruptions that need to be overcome are:

- Quality checking
- Machine breakdowns
- Waiting for parts delivery

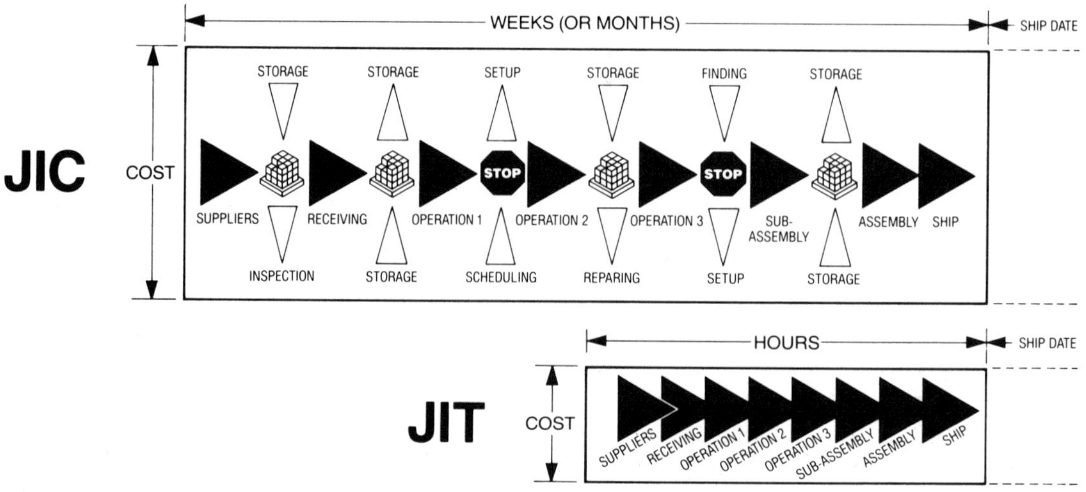

FIGURE 5–11 Comparison of Production Cycle—JIC versus JIT (Courtesy of Ingersoll Engineers, Inc.).

*Permission granted by William Arrott Associates, Inc. for Ingersoll Engineers.

- Waiting for tools or fixtures
- Waiting for an inspector
- Grinding tools
- Poor plant layout
- Shortcomings in the scheduling system

The contradiction between these two opposing approaches may be expressed in this manner: Although JIT seeks to eliminate or reduce production interruptions, JIC actually gets support for them (figure 5–12).

As discussed in Chapter 4, OPT attacks bottlenecks that are one of the primary impediments to JIT production processes. Chapters 6 and 8 discuss numerous advances in robotic technology and contain methods for eliminating or sharply reducing machine setups, the other primary interruption factor.

JIT seeks to accomplish a reduction in costs related to inventory, such as:

- Cost of money
- Space
- Taxes
- Obsolescence
- Material handlers
- Fork trucks
- Expeditors
- Scrap
- Storekeepers
- Paperwork

To install JIT in the production environment, the entire production flow has to be streamlined, interruptions virtually eliminated, and throughput reduced to as close to actual

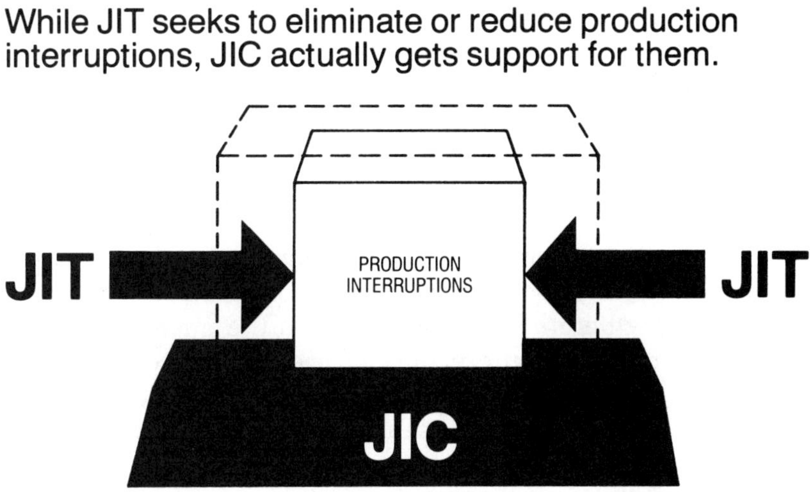

FIGURE 5-12 Production Interruptions—JIC versus JIT (Courtesy of Ingersoll Engineers, Inc.).

machining hours as possible. In conjunction with these changes, a company can now obtain and put in use, a JIT software package.

AUTOMATIC IDENTIFICATION: THE COMPUTER'S SENSES*

Computerization has enabled companies to capture, manipulate, and generate data at every level of manufacturing and distribution. This is accomplished through hierarchal arrangements of mainframes, minicomputers, microcomputers, microprocessors, and programmable controllers.

Automatic identification (AI) acts as the senses of the computer. AI enables the computer to track, monitor, and report happenings involving progress in production and distribution of parts, products, and packages. AI can be developed in a real-time mode, with completely automated feedback capability to the computer.

How AI Works

Automatic identification is the technology of recognizing the identity of an item or package by means of an electronic device and, usually, a machine-readable code, and making use of the information to initiate activity or to compile data. Three important concepts associated with this technology are:

- *Code*—method of representing information
- *Symbol*—machine-readable representation of the code
- *Symbology*—methodology for arranging bars and spaces to represent a code system

Tip for Successful Automation:

Complete industries are pooling their members' expertise to select a symbology (or symbologies) that will accommodate all the codes being issued throughout that industry.

With code, symbol, and symbology in place the next step is concerned with scanning the part, product, or package. A source intense light, produced by a laser, light-emitting diode or an incandescent bulb located in the scanner is aimed at the bar code symbol. The dark bars absorb the light and the light spaces reflect it back to the scanner. The pattern of reflected light is converted to an electrical signal, decoded by a microprocessor in the scanner and transmitted as information to the computer. Defined another way, the major components of a bar code scanner are an illumination source, a photodetector device, and a microcomputer.

A focused light beam is moved across the symbol, the reflected light is received by the photodetector, which, in turn, generates a voltage (analog signal) that is proportional to the amount of light reflected by the symbol. The signal from the photodetector is "conditioned" by the preprocessor circuitry before being presented to the microcomputer for analysis. The microcomputer software program then executes a series of subprograms that will decode the data, make decisions, and transmit data or commands to another device.

Decoding of the scanned symbol usually occurs in several steps. The raw analog signal is preprocessed by the electronics hardware before real decoding begins. The software first decides whether the data it has received is a valid bar code. This is usually determined

*Permission granted by Automatic Identification Manufacturers.

by checking that a proper quiet zone (clear area on either side of the symbol) exists, that a number of bars and spaces is consistent with a selected symbology, and that valid start/stop characters are present.

Along with this process, the analog signal has been converted to a digital signal (zeros and ones) for the final decoding. At this point, some codes need no further processing while others need a final algorithm to be decoded fully.

Five Types of Bar Code Scanners

There are five basic types of bar code scanners:

1. Stationary
2. Portable
3. Fixed-mount
4. Hand-held laser
5. Moving-beam

1. Stationary scanners. Hand-held light pen/wands (contact) or noncontact scanners that read bar codes as the object passes by the scanning device.

2. Portable scanners (portable data collection terminals). Enable workers to engage in such activities as taking inventory in warehouses without a single pencil mark. Portable scanners read bar code. They are data collection devices that accept both contact and non-contact scanners. There are two basic types of light sources for portable scanners:

- *Visible*—Incandescent and LED
- *Infrared*—Helium laser and laser diode

3. Fixed-mount scanners. Incandescent, LED, or laser base light sources are used to sense marks or codes that consist of retroreflective material or bar code. The position of the retroreflective material on the object or the arrangement of material into a digital code, enables a photosensor or bank of photosensors to identify the object. Fixed-mount scanners also can use an infrared light source to read a bar code printed with fluorescent ink. Both hand-held and fixed-mount scanners can have fixed or moving beams.

CASE-IN-POINT

FIXED-MOUNT SCANNER
PRINT/APPLY SYSTEM FOR UNIQUE IDENTIFICATION ON PRODUCT PACKING

Situation

A major manufacturer of electronic office equipment needed to increase the productivity of its intermittently fed packaging line, reduce human errors in identification and labeling steps, and manage inventory data on stream with its mainframe system.

Solution

Robot packaging and automatic labeling of unique identification and item numbers on each boxed unit, aided with laser scanning devices, were used to pack, identify, and record completed units on one conveyor line (figure 5–13).

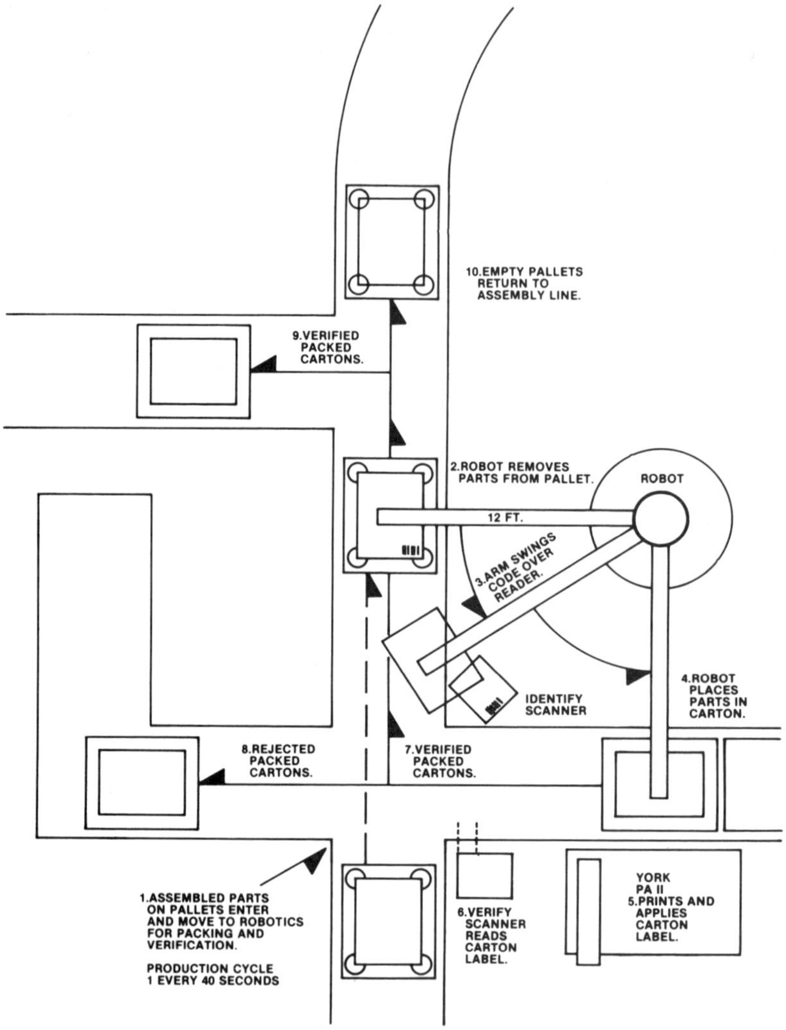

FIGURE 5–13 Print/Apply System for Unique Identification on Product Packing (Courtesy of Automatic Identification Manufacturers).

The system accepts palletized component units from multiple assembly lines on an intermittent work-flow basis. The system is designed to transport palletized-unpacked and box-packed components to the packaging and distribution areas respectively.

To speed the packing process, the installation of the robot arm replaced human packing. The robot opens an approaching empty carton, lifting off the top. Swinging 90 degrees to the conveyor presenting the palletized electronic components, the arm lifts a unit from its build pallet and carries it toward the opened container. On the underside of each component is a label that was previously placed on the unit housing by the vendor. This label has an imprinted bar code identification and serial number. During this return sweep of the arm, a stationary laser scanner reads this label.

Electronically, after label scan, the Print/Apply labeling system is triggered. Then a $1^1/_2'' \times 4''$ label is imprinted in a three-line format (a medium density Code 39 bar code and two lines of human-readable identification). As the component unit is placed in the box along with special packaging, the label is applied to the outside surface of the carton.

The box lid is then placed on the packed carton and the completed unit passes through a verification scanner. This scanner assures uniform packaging and scans the bar code segment of the label on the box. Record is made of the boxed unit and data is passed on to the company's mainframe computer as a record of inventory and daily output.

Features. The labeling process and identification of the packed units are completely automatic. The identifying scanner triggers the label print and apply system. The labeling head prints a scannable label and applies it to the box during the packing process.

Results

Over 25,000 units have been packed to date with a failure rate of .001 percent per day. The speed of the packing and labeling process was brought down to 45 seconds per unit packed.

Prior to the installation of the automatic labeling system, labels were operator-keyed into a computer. With the automatic print and application of the labels, the chance for human misread and miskey was eliminated, along with the need for a human operator.

4. Hand-held laser scanners. In a "gun" configuration scanners are portable and able to scan symbols at greater distances (typically at 3" to 12" from bar code depending on scanner style and density of the bar code) as compared to a wand scanner that requires actual contact with the surface or proximity.

Versatility is the hallmark of hand-held light pens (light wands) and laser (guns). These code readers can be linked directly to a control computer, or connected to desktop or portable terminals to provide automatic data collection almost anywhere throughout the plant.

A worker can use these devices to enter work-in-progress data without using a terminal keyboard or punched cards.

CASE-IN-POINT

HAND-HELD LASER SCANNER
BAR CODING ENHANCES PRODUCTION OF PRINTED CIRCUIT BOARDS

Situation

At one of their plants, a leading manufacturer of minicomputers builds an average of 1000 printed circuit assemblies per day. Because this production volume includes sixty-five models, frequently with more than one revision level, obtaining accurate and timely data on engineering change order (ECO) status, as well as work-in-process inventory and throughput and yield rates, presented a difficult problem.

Solution

The company developed an on-line bar code system to track PCBs throughout the production cycle. To begin the process, the operator enters the quantity of circuit cards to be built, their type, whether labels are desired and other information, such as serial number, manufacturing plant code and revision level. The computer then transmits the label printing data to the label printer, and directs it to print the required labels. An operator applies them to the circuit cards and scans them on-line to the computer. The computer then recognizes and logs these labels, and starts the production control tracking record for the individual circuit cards. A "time clock" starts when the boards are first scanned into memory.

Following the scanning of the labels, the boards are processed in a vacuum oven which pulls the label down tight against the board and cures the smear-resistant bar code printing ink. After vacuum baking, the boards go through a numerical control machine stuffing process for certain components. After this process the boards are scanned and sent through a manual stuffing process for the balance of the components. The boards are then inspected at an on-line inspection area.

Items that pass inspection are scanned and routed into the next scheduled assembly process; those that fail are scanned and returned to rework areas.

The computer program contains routing information specific to each model and ECO level of PCB, and alerts the operator at each scanning station, via the terminal, as to the next scheduled process for the board. Scanning a board's label on completion of a process indicates to the computer that the PCB has been routed to its next scheduled assembly or test area, unless the operator overrides the message by scanning a bar code "menu" applied to the CRT, which contains alternate destination codes such as "rework/repair." The computer knows the exact location of each board at all times. If any defects are found in the board or its components, that information is transmitted to the CPU through the keyboard. This begins a history file for the board, which is used by quality assurance and field service during the life of the PCB.

Completed circuit boards are checked out on automatic testing equipment, and if satisfactory, are routed to final inspection. On completion of all assembly, test and inspection processes, the boards are scanned out of production and into stores inventory and an ECO status sheet generated. The boards are scanned again when shipped out of stores to other manufacturing facilities, thus providing finished goods inventory control.

Results

Throughout this system, whenever an engineering change order is required, it can be determined immediately what boards are in production at each revision level and at what stage of manufacture. Because each circuit card is tracked individually, the computer can keep tabs on the actual implementation of required changes.

The system also provides management with up-to-the-minute information on inventory and throughput and helps identify bottlenecks, as well as providing information on employee performance, repair and rework turnaround time. Before the system was implemented, WIP inventory at workstations was typically sufficient for two days' work. This has been reduced to approximately four hours at many work stations.

5. A Moving Beam Scanner. A moving optical beam that searches for symbols by sweeping across the object in view. A single movement of the beam, which normally uses a laser as light source, constitutes a scan through its angular path. The fast-sweeping movement of the beam permits multiple scans to be made on a passing symbol.

CASE-IN-POINT

MOVING BEAM SCANNER
REAL-TIME TRACKING AND CONTROL IN THE MANUFACTURING PLANT

Situation

A large domestic manufacturer of TV sets, faced with price erosion from foreign competition, decided to investigate the potential savings from automating its manufacturing plant.

A critical component in the final assembly process is the enclosure—the TV case. An insufficient quantity or an incorrect model number TV case can result in a "slow down" or even a halt in final assembly. A survey of the current operation disclosed a very labor-intensive procedure for tracking and controlling the inventory of TV cases coming out of production. Other problems encountered included errors in production counts due to manual data entry, misplaced product in the warehouse, and inaccurate on-hand inventory counts.

Solution

TV cases with the same model number are removed from injection moulding machines and placed in large cardboard containers (3′ × 4′ × 5′). A bar code label is applied to the side of the container to identify TV case part number. Containers are then transported by conveyor past a fixed position moving beam laser scanner. The code is read and transmitted to an inventory computer which updates a production count. Containers next travel past a second moving beam laser scanner interfaced to a sortation controller. The sortation controller tracks and diverts containers into vacant accumulation lanes. Forklift truck drivers using on-board radio frequency terminals receive instructions from the inventory computer through a base station link. The inventory computer communicating through the RF terminal dispatches the drivers to either move the material to final assembly or assigns the warehouse section in which the containers are to be placed.

Features. Both on-demand and automatic reports are available at the system console. Production counts by part number and shift are reported. Production schedulers input final assembly requirements by part number and an exception report is printed to point out insufficient quantities of on-hand inventory. Labor efficiency reports and forklift truck driver audit trails also are available from the system.

Results

Effective integration of field proven technology has provided this implementer with "tighter and better" operating controls. Tangible savings from the elimination of labor enable the implementer to:

- Count the product.
- Sort the product.
- Assign and maintain warehouse space locations.
- Perform daily physical inventory.

Manual record keeping of on-hand inventory has resulted in a savings of over a quarter of a million dollars per year. Furthermore, intangible savings are anticipated from enhanced accuracy of on-hand inventory and production counts and enhanced accuracy of real-time production schedules.

Tip for Successful Automation: Types of Bar Codes In Use Today

It should be noted that there are several bar codes presently in use. They are:

- UPC—Universal Product Code—used in retail (point-of-sale) applications.
- Code 39—fast becoming the de facto standard symbology for industrial usage. It is the military (DOD) standard, government standard (GSA), automotive standard (AIAG), health industry manufacturers standard (HIBC). It is alphanumeric, variable length, and can be printed by most standard bar code printers.
- Other codes in use: Codabar, I 2 of 5, 2 of 5, Code 128, EAN, and others.

Five Additional Forms of Automatic Identification Technologies

Technologies other than bar code scanning for automated identification can be used in separate systems or in conjunction with methods employed with bar code scanners. These other types of automatic identification techniques are as follows:

1. *Mark sense* is an optical technique in which readers are set up to see marks in a pattern. The most common application is the use of strips of retroreflective material on a tote

box. Position of these strips identifies the box's number as read by banks of photosensors along the conveyor line.

2. *Optical character recognition (OCR)* relies on a sensor to differentiate between light-reflecting background and printed data that does not reflect light. Output from the sensor is fed to a set of recognition logics, then forwarded to the computer. An advantage of OCR printing is that it is human-readable. In manufacturing, work orders and routing slips can be printed in OCR and scanned at workstations for tracking parts and assemblies.

3. *Surface acoustical wave (SAW) systems (A subset of radio frequency technology)* employ a reader consisting of a low-power radar transmitter and microcomputer, a bowl-shaped antenna and the small SAW tags. SAW tags are factory-programmed by laser trimming of an aluminum film pattern of interdigitated transducers (IDTs) deposited on the SAW microchip. The IDTs resemble combs with interlaced teeth and are left in or out by trimming to form binary codes.

In operation, the reader transmits radar pulses via the antenna. These pulses are intercepted by the tag's antenna and converted into ultrasonic acoustical waves. The waves pass along the uniquely encoded "teeth" of the tag's IDTs which modify their amplitude according to tag code content. The waves are then converted back into electromagnetic signals for code transmission to the reader.

4. *Machine vision* is thought of less as AI technique and more as integral to robotic productivity. The trade-off is: How much do you spend on a material handling system that precisely orients each part for the robot versus investing in a machine vision system that alerts the robot to the orientation of each part coming down the conveyor?

The eye of a vision system is generally a solid state control known as a charge couple device (CCD) that scans a line of points into computer memory. An image is made up from a two-dimensional array of pixels—a pixel being the smallest element of an image or picture.

Aside from its application as the robot's eyes, machine vision also is useful for identifying and rejecting defective parts.

5. *Voice recognition* is a technology complementary to automatic identification since, as "keyless" data entry, it performs many of the same functions. Voice recognition is being used as part of an automatic identification technology mix in some installations. Where people have to be involved in the initial data input and the original information cannot be read by scanners, voice can be used as a starting point for a system that eventually moves on to some scanning technique.

SPEECH TECHNOLOGY: USER-FRIENDLY COMMUNICATION*

Direct speech communication with computers is perhaps the most user-friendly form of interface between operator and machine. It has arrived through an evolutionary process that has included punched cards and punched paper tape, CRT/keyboard terminal, bar code, OCR readers, graphic displays, light pens, and "mice."

The four speech recognition and voice response technologies now available are:

- Automatic speech recognition (ASR)
- Electronic speech synthesis (ESS)

*Reprinted by permission of Intel Corporation.

- Voice store and forward (VSF)
- Discrete and continuous speech recognition (DCSR)

Guidelines for Implementing Speech Technology

When considering implementation of speech technology in the factory environment, these application selection criteria are a useful guide:

1. Will it benefit the operator directly?
2. Will the change be cost effective?
3. Will the operators be a controlled group that can be given proper training?
4. Will the work environment include acoustic ambiance within an acceptable decibel count? Throughput requirements within the range of normal speaking rates? Minimal operator stress? And no conflict with other necessary verbal communications?

Using the Speech Transaction Process to "Talk" to Computers

Speech communications is based on words and phrases. Thus, input requires additional processing between the recognizer and the application software to bridge the communication gap. This *speech transaction process* is the key to providing a user-friendly, human-directed communications environment for systems operators. Thus, a properly designed application will use vocabulary and syntax that are learned easily and are readily associated with the operator's work task. Also, the design must edit data prior to transmission to the host computer to minimize errors from entering the system. In addition, the *response-time loop* between the operator and the system must be kept under an acceptable increment of time throughout a transaction to avoid perceptible delays that can mitigate operator support.

In summary, the five speech transaction process functions are as follows:

1. Prompting and feedback to provide the operator with a communication context
2. Buffering spoken messages for verification prior to transmission to the application computer
3. Verbally editing previously spoken information (word or phrase)
4. Syntax processing, indicating the expected repertoire of words or phrases valid at any point in the transaction
5. Conversion between the spoken input and the compressed data formats required by the application computer

Figure 5–14 shows the dynamics of the transaction process. Figure 5–15 is an operator's view of a typical transaction.

As illustrated in figure 5–16, speech recognition and voice response technology has many application uses in integrated computer-aided manufacturing (ICAM) at the factory level, cell/center level, station level and process level. Another way to describe its use is to relate speech technology to the *basic functional factory subsystem* where it has application—production control and scheduling, inspection and audit, process control and scheduling, and material handling. Figure 5–17 contains the functional requirements for a speech-based information system for quality inspection.

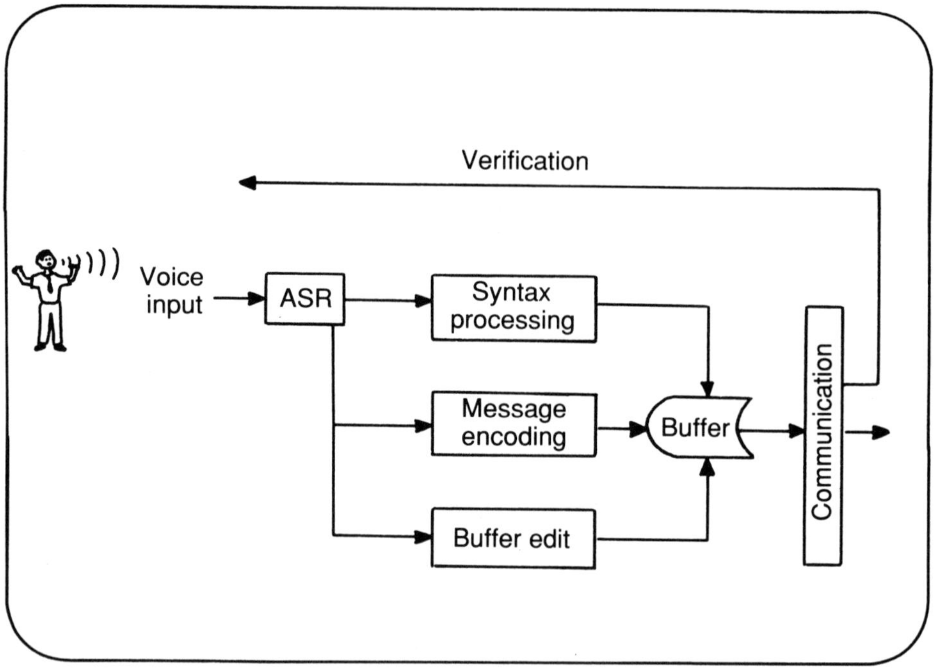

FIGURE 5–14 General Speech Transaction Concept (Courtesy of Intel Corporation).

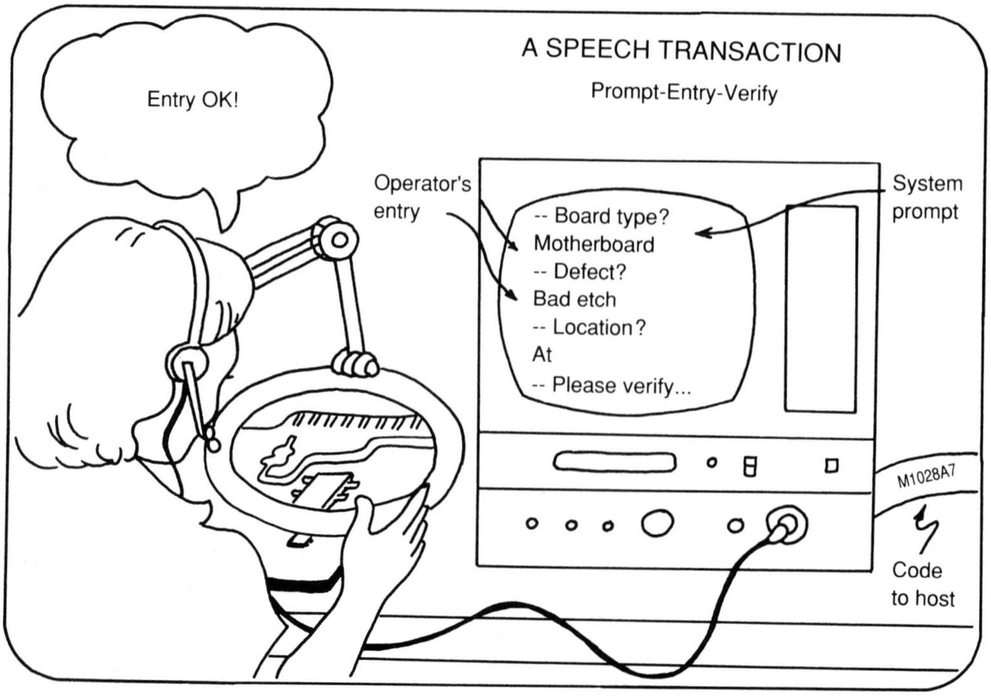

FIGURE 5–15 Operator's View of a Typical Transaction (Courtesy of Intel Corporation).

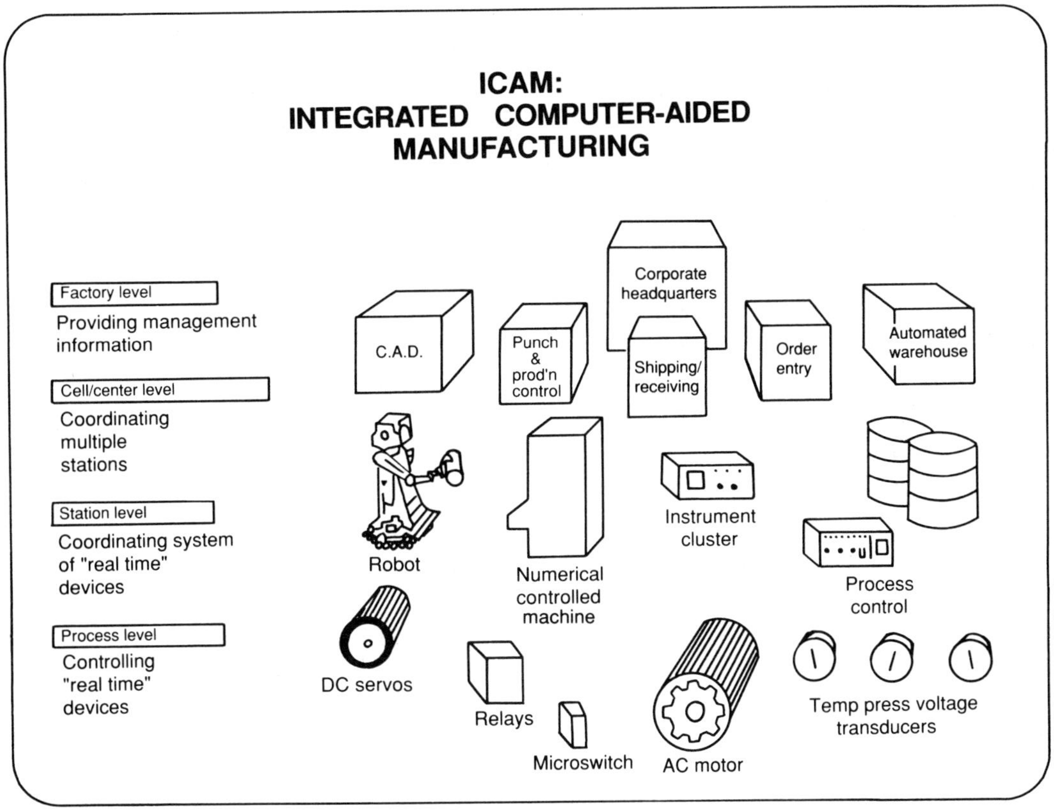

FIGURE 5-16 Speech Products Recognition—Applicational Uses in Integrated Computer-Aided Manufacturing (Courtesy of Intel Corporation).

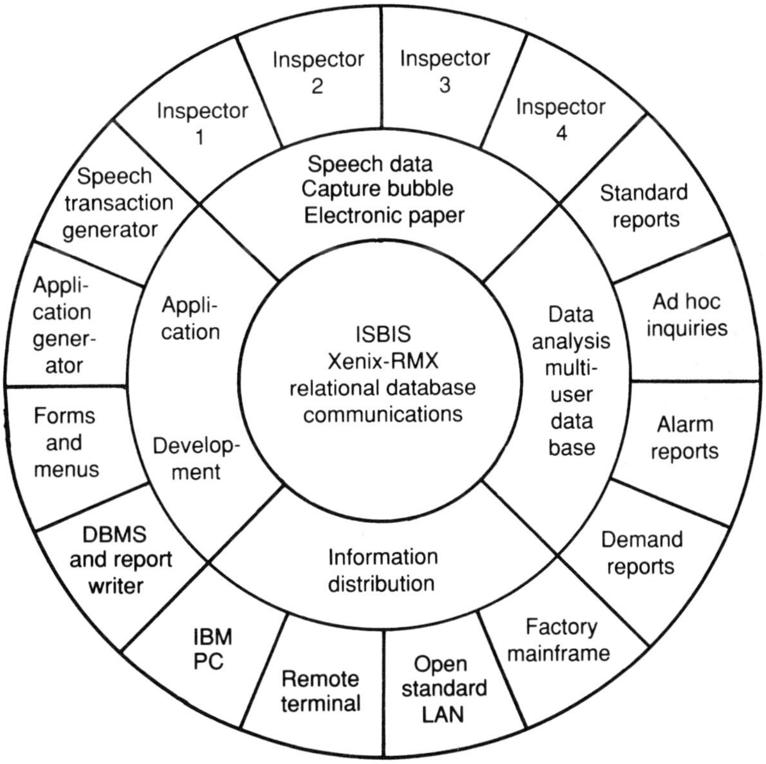

FIGURE 5-17 Functional Requirements for a Speech-Based Information System for Quality Inspection (Courtesy of Intel Corporation).

Although physical conditions in many factories are not necessarily conducive to speech technology, it has made inroads in a number of functions, in particular, tests and inspections. This is because the nature of inspection often requires freedom of eye and hand movement when seeking to capture production-related input data. Inspection needs are usually satisfied by inputting a limited vocabulary from a repertoire of defective types or categories in a structured fashion.

CASE-IN-POINT

APPLYING SPEECH TECHNOLOGY IN AN AUTOMOTIVE ASSEMBLY PLANT

Situation

A major automotive company was of the opinion that inspection worker productivity could be improved in its midwestern automotive assembly plant by using speech recognition technology. This advanced technology would allow an inspector to have freedom of hand movement while inspecting a part, and simultaneously report all pertinent information through verbal communications.

Solution

The company decided to implement a program that would solicit worker cooperation, therefore engaging in a one-year pilot test. During the final formal acceptance test period, neither the speech vendor nor headquarters personnel sponsoring the test were permitted on site. By so doing, there was no way for assembly plant management to detect which method was used in the final work output. At the end of this period, the UAW inspectors on both shifts voluntarily chose the speech technology over the previous method. Most importantly, it was proven to be an effective tool for increasing worker productivity even in a noisy, difficult environment.

Results

As a result of the application of speech technology, the company was able to reduce its per shift work force from three to two people, and thereby achieve annualized savings of $100,000.

This successful installation indicates that system design engineers should consider evaluating and prototyping speech I/O when installing factory automation that includes a human interaction element.

Why Speech Synthesis Expedites Process Inspection

In a typical company, sales orders are received, a manufacturing order is computer generated, information is translated into a sequence of process steps and control actions, and a data base is created and maintained (figure 5–18).

As the product proceeds through the manufacturing process at various points it is checked and adjustments are made, if required, by process inspectors. These inspection workstations are usually in a hostile dirt, heat, and temperature environment of the factory floor. The functional requirement for these human-directed process adjustments and process inspection nodes are shown in figure 5–19. Since speed is critical when inspectors detect a problem, speech synthesis enables the process inspector to input data and obtain verification messages almost instantaneously.

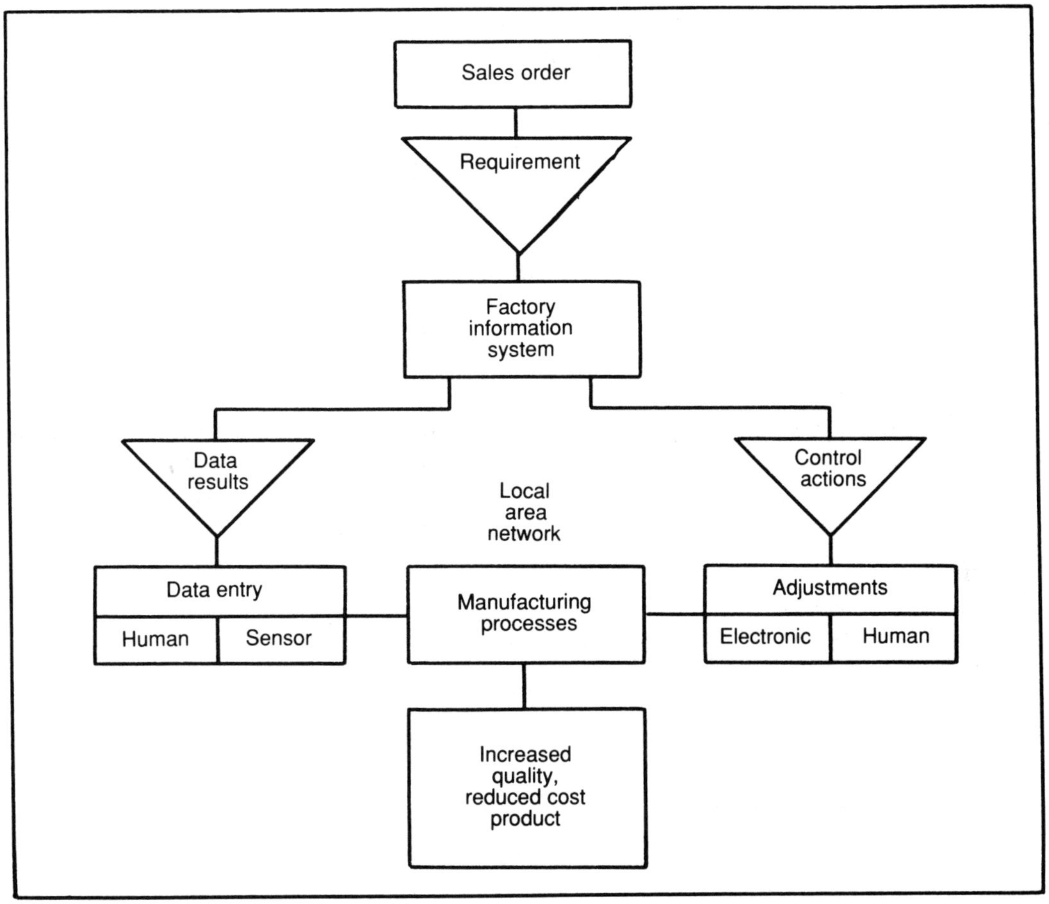

FIGURE 5-18 General Model of the Data Flow within an Electronically Controlled Factory (Courtesy of Intel Corporation).

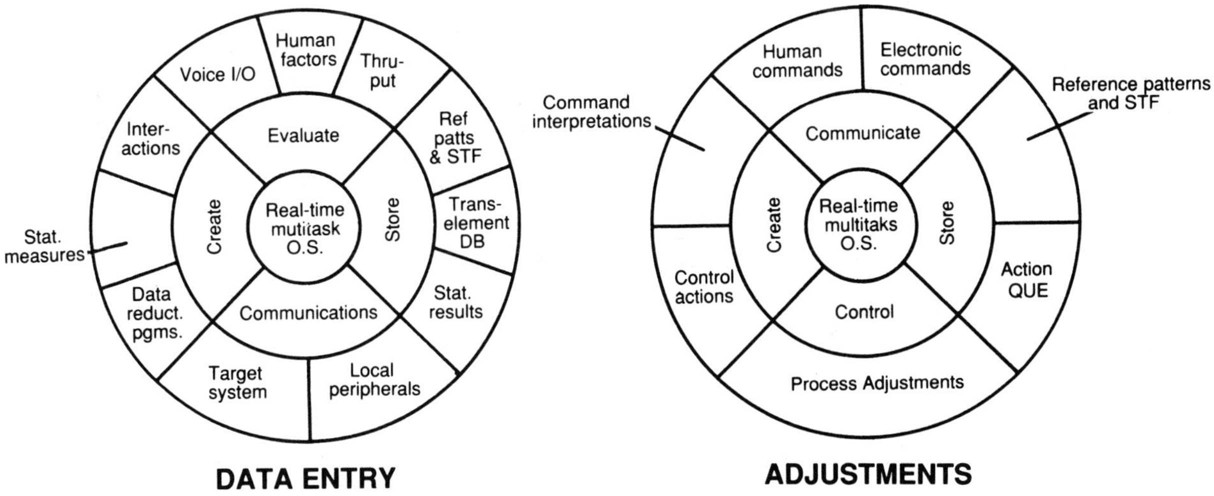

FIGURE 5-19 Generic Functional Requirements of the Human Directed Factory Floor Control Nodes (Courtesy of Intel Corporation).

Tips for Successful Automation: Requirements of an Inspection Workstation

Most factory inspection operations require positive identification of the inspector. Thus, speaker-dependent recognition technology is ideally suited to this application area.

From this, you can deduce that a speech I/O engineering workstation must provide (figure 5–20):

1. A speaker-dependent recognizer
2. A text-to-speech synthesizer
3. A speech transaction for generating and maintaining the application-specific operator interaction
4. A speech transaction that incorporates other I/O in a complementary fashion and executes under the control of a proven real-time multitasking operating system
5. Statistical measures to assist in the selection of natural word and phrase response to operator prompts that are not acoustically confusing
6. Operator training
7. Control of key discrete word parameters
8. A multitasking operating system with a procedural programming language and utility set integrated with speech recognition and synthesis devices as well as a mass storage file system
9. A unifying set of open system standards that enables implementation of the end result in a package that will meet the requirements of the intended factory environment

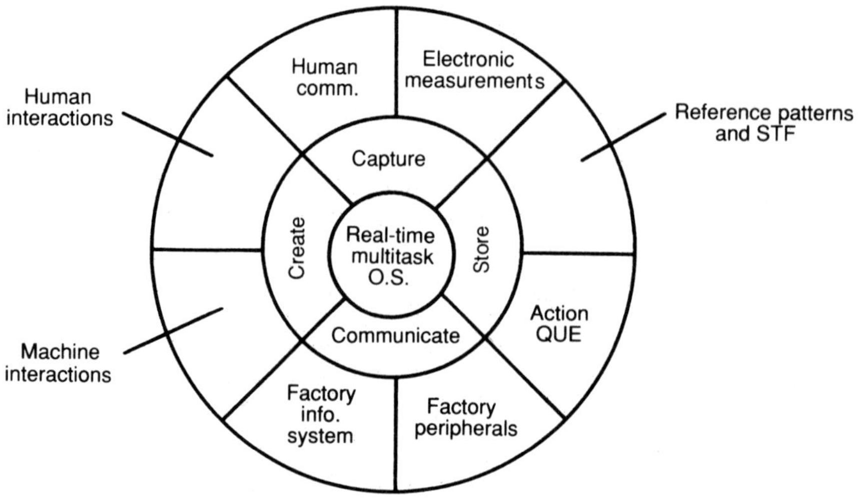

FIGURE 5–20 General Requirements for a Speech I/O Evaluation and Prototyping Workstation (Courtesy of Intel Corporation).

TYPES OF RADIO FREQUENCY IDENTIFICATION SYSTEMS*

Radio frequency (RF) is only one element of a family of identification techniques covered in this chapter. These different techniques span many technologies and each has its strengths and weaknesses. Proper selection of one, or sometimes a combination, will allow the designer to solve most ID problems (figure 5–21).

Basically there are two types of radio frequency units available.

- *Read only system*—information contained in the tag is programmed at the time of manufacture and normally cannot be changed. Even in the units that allow for field programming, the data is not readily changeable.
- *Read/write unit*—designed so that, in normal operation, the information contained in the tag can be changed. A user can add, modify, or delete information easily and quickly.

Primary Differences between Read Only and Read/Write Systems

The basic parts of a radio frequency system are the reader unit with its antenna, the tag, and the host system or other form of controller (figure 5–22). The tag is detected and the read operation takes place. The code is communicated to the host or other controller through a serial communications link. At this point, whatever is done is a function of the system design. The information can be recorded, or trigger some response based on the information, or do both. The operation of a read/write system is the same as the read only system with the exception that information contained in the tag can be read, written to, or a combination of both. The major differences between the read only and read/write systems are in the tags.

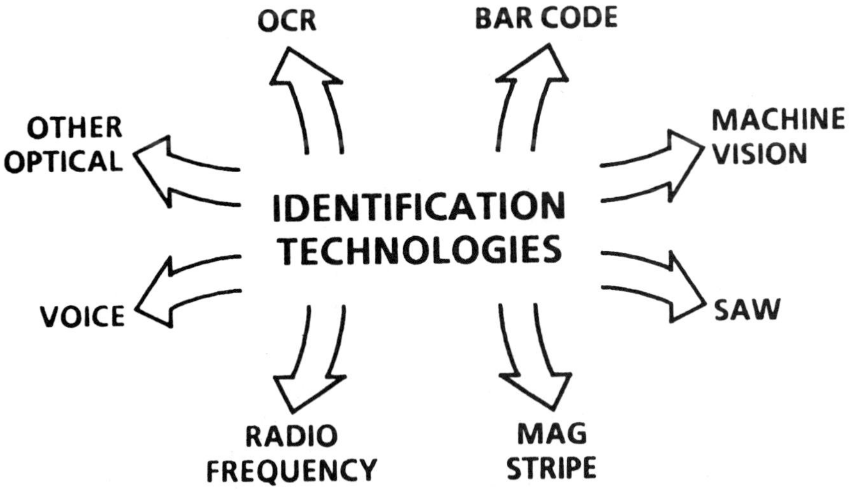

FIGURE 5–21 Identification Technologies (Courtesy of Rochester Products Division of General Motors Corporation).

*Permission granted by Rochester Products Division of General Motors.

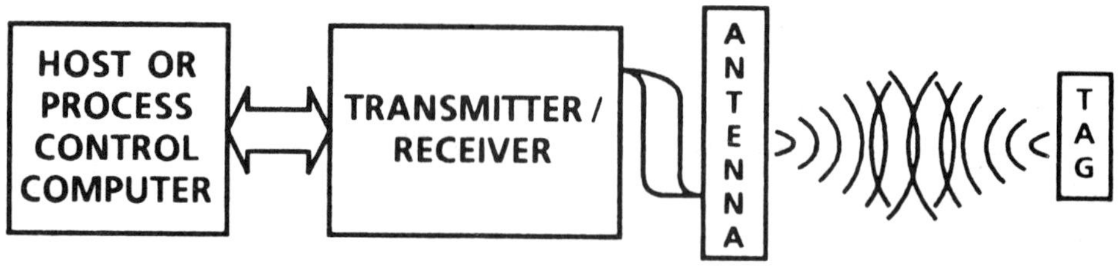

FIGURE 5–22 Basic Parts of a Radio Frequency System (Courtesy of Rochester Products Division of General Motors Corporation).

Read Only Tag. The read only tag normally has a receive antenna, special function circuits, and a transmit antenna (figure 5–23). Some have no power source. The power used to operate the tag is derived from the field energy. The tag enters the field, absorbs enough energy to operate, and sends its code. Once the tag passes out of the field, there is no source of energy available to it. The amount of information that is transmitted to the reader is limited by the design. This type of system can be obtained in a variety of packages, different operating frequencies and temperature, and different amounts of information carried. The units also are capable of a wide variety of distances and data rates.

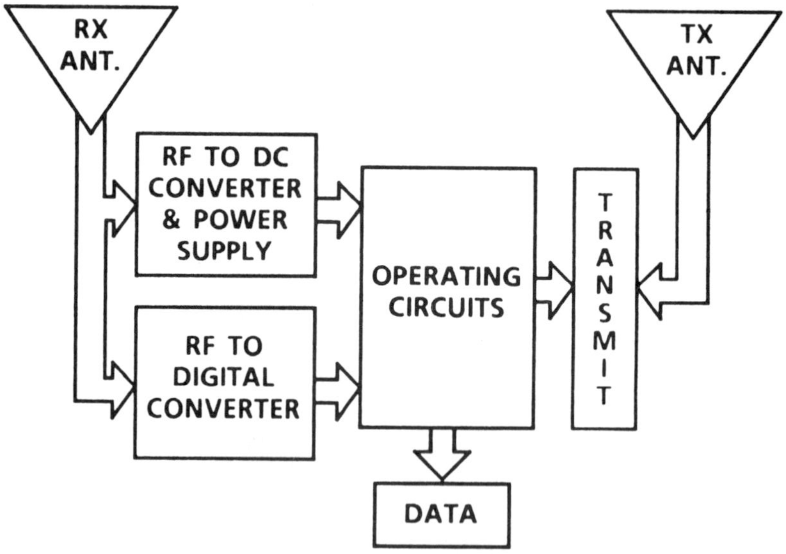

FIGURE 5–23 Read Only Tag (Courtesy of Rochester Products Division of General Motors Corporation).

Read/Write Tag. The read/write tag has two major differences. First, it is more sophisticated because it needs to receive and act upon commands from the reader. It also is more complex because it transmits information in addition to receiving and storing it. Another major difference is that it may have an energy source. The technique for storing the data may be random access memory. This type of memory requires small amounts of power to retain the information when the tag is not in the field. Some units use E2PROM's to store the information. These memory devices do not require power to retain information but are slower in operation, which tends to limit the rate of communication.

Reasons for Using an RF Identification System

The reasons for using an RF identification system are as follows:

- It offers data capabilities from a few characters to many thousands of characters, with the ability to add or modify this information, as desired.
- It is capable of "blind" operations.
- The system can work even when the tag is not visible, as long as the blocking media allows the passage of radio waves.
- It has advantages in distance of operation.
- There are systems available that operate in separation of inches, and there are some available that operate in separations of many feet.
- Another possible advantage is high-speed operation. There are applications of RF identification where the tags pass by the antenna in excess of one hundred miles per hour.

These capabilities might be the reason that RF identification is selected for an application. Possibly, the two most important reasons are the wide range of operating environments that these systems not only endure, but can operate in, and the system flexibility that is possible.

Radio frequency identification systems can be used in harsh environments. Some of the other types of identification techniques might not survive much less operate in these environments.

The main disadvantages of this type of system is that it is more complex and higher in cost than other systems.

RF systems have applications in virtually every aspect of manufacturing, inspection testing, material handling, order processing, storing, and shipping.

How RF Technology Results in Flexibility of Information Systems

RF identification systems offer considerable flexibility in information systems, since they can be

- Centralized,
- Decentralized, or
- Combination of both

Centralized System (figure 5-24)

In a centralized system, all of the information is stored in the central computer. Every time a decision needs to be made, the controller needs to communicate with the central system. This can be on a small scale, where the central system is the computer controlling a process, such as a conveyor. It also can be on a large scale where the central computer is the main plant computer. The important point to recognize, is that there needs to be communication each time within the central system. If the central system is down or there is a communications problem, the process can be halted.

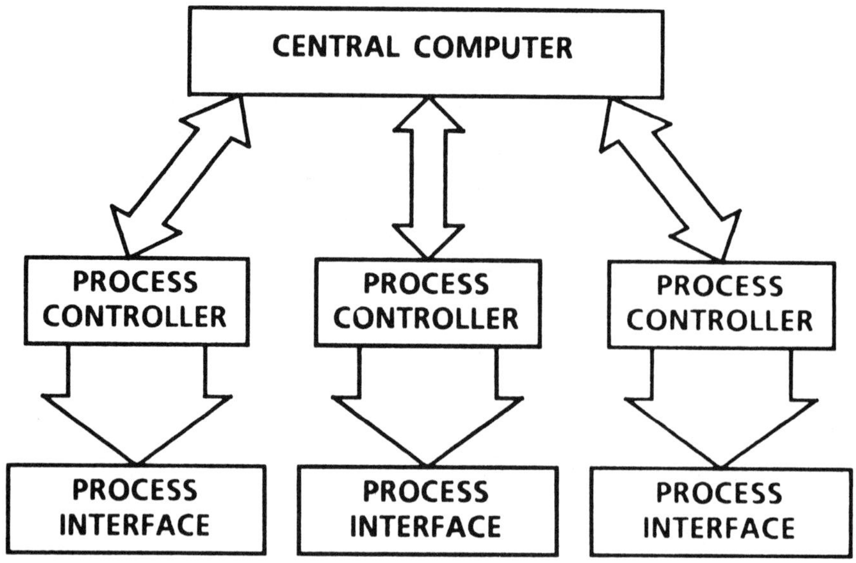

FIGURE 5-24 Centralized System (Courtesy of Rochester Products Division of General Motors Corporation).

Decentralized System (figure 5-25)

In a decentralized system, there does not have to be communication with the central system. In some instances, the central system is used to initiate the whole operation, but after that, it is not required. At that point, the central system can be down with no effect on the process. What makes this possible is the communications links between the lower nodes. This allows each to communicate with each other without using the central system. These communications links can be traditional communications channels between the computers.

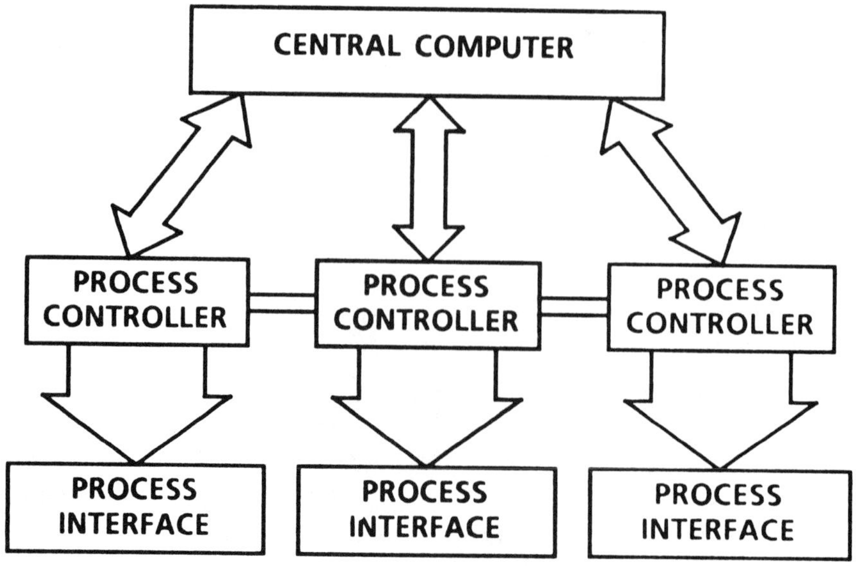

FIGURE 5-25 Decentralized System (Courtesy of Rochester Products Division of General Motors Corporation).

Combination of Both Systems (figure 5–26)

The other approach is the use of RF read/write tags. The function is the same. Information can be stored in a tag, and when the tag arrives at the other area, so does the information. This method is extremely flexible because there is no hard wiring required and every possible node can be connected in every conceivable way. Now the bottom of the information flow structure is tied together creating a decentralized capability from a centralized system.

As an example, figure 5–27 shows how this approach would work on a flexible machining cell.

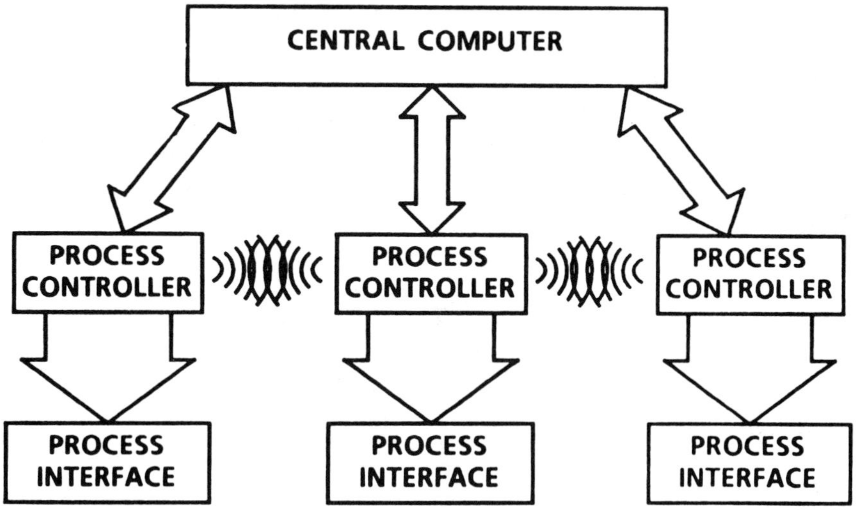

FIGURE 5–26 Combination of Both Systems (Courtesy of Rochester Products Division of General Motors Corporation).

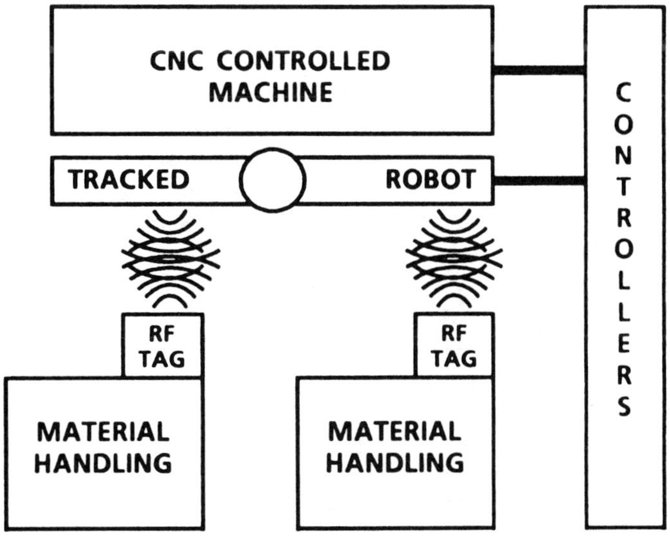

FIGURE 5–27 Combination of Both Systems on a Flexible Machining Cell (Courtesy of Rochester Products Division of General Motors Corporation).

The parts arrive at the cell with the part number encoded in a tag. The central system is notified of its arrival and the parts begin to be machined if the CNC machine has the correct program. If it does not, it is requested and sent by the central system to the appropriate machine. With a read/write system, it even could be possible to have the tool program in the tag as well as the part number. It also could contain the matrix information so the parts could be loaded and unloaded by automatic means. This allows the same tray to be used for both tested and untested parts, even keeping track of the actual test results. All without the necessity of contact with the central system. Besides the advantage of free-flow of data as the part progresses through the shop, it also can provide an on-board audit trail, and real-time QC as it continues from process to process.

In summary, in a read only system, the application can be anywhere where it is necessary to read information as a part passes by in a noncontact method. It works well with a centralized type of system since it has the benefit of operating at greater distances than other technologies, and also the ability to function in "blind" or harsh environments.

In contrast, a read/write system can enable a centralized system to become a more flexible decentralized type of system, if necessary. This is because the system is capable of sending any information desired; and being able to read it, add to it, delete from it, as well as make decisions based on the information. As a result, it provides a user with almost unlimited application capability.

CASE-IN-POINT

*RADIO FREQUENCY DATA COLLECTION SYSTEM USING READ/WRITE TAGS ROCHESTER PRODUCTS DIVISION OF GENERAL MOTORS CORPORATION**

Situation

One of the company's product lines consists of eleven major types. Although all eleven ultimately perform the same basic function, each of the types are mechanically, electrically, and operationally different. Within each type, there are mechanical and electrical variations. Specific to each of these variations is the final calibration. This calibration is unique for not only each customer, but for each application.

As part of the production process, every unit must be individually calibrated. The calibration equipment or "stands" are fundamentally identical but the fixturing and automatic adjustment tooling makes it specific to one of the eleven major types. Each of the stands is computer controlled. The computer makes all of the measurements and controls and all of the adjustment tools. Each of these computers is linked to the mainframe computer by a serial communications channel. The mainframe serves as the storehouse for all the various calibration stand operating programs and all of the calibration parameters (or model specifications). (In reality, the communications concentration computer is capable of, and performs, these functions.) Once the calibration stand computer has the appropriate operating program and model specification, it is self-sufficient and does not require the communications channel. While not necessary for operation, it is used to communicate stand status, production counts, and actual values upwards for production counts and statistical trends.

The original operation of the calibration area was as follows: The product was produced to the calibration stage and was transported to the calibration area by a towline cart system. Each cart contained an XYZ matrix of parts of the same part number. These carts were received at the calibration area, and manually routed to the appropriate bank of calibration stands based on the part number tag attached to the outside of the cart.

*Permission granted by Rochester Products, Division of General Motors.

There was a load operator assigned to each bank of calibration stands. The bank size was determined by the average cycle time. The load operator placed the unit to be calibrated on the stand and pressed two palm buttons to start the sequence. The stand automatically clamped the part and made the necessary measurements and adjustments to calibrate it. At the completion of the cycle, the part would be marked, unclamped, and the load operator would be summoned. The operator would unload the part and put it into a different cart. These parts were then sent to the shipping department.

If the unit could not be calibrated properly the stand would "alarm out" and unclamp. The operator would attach a rework tag to the part, mark the rework code that the computer had determined to be the probable cause of the problem, and put the part into a cart destined for the rework area. The suggested rework would be performed and these parts then went back to the beginning of the calibration process.

When there was a model change, the operator would key the new model number into the stand, and the computer would retrieve the proper operating program and/or the proper model specification (calibration values).

Solution

After considerable study, the company arrived at two types of solutions, both of which were implemented. The selection of the solution type was based on several factors:

- Rework of existing tools and equipment versus new tools and equipment (new customers, added volumes, product design changes)
- Production volumes required and volume versus product life increments
- Overall product life
- Geographic location of plant and existing plant facilities

Solution A. In lower volume, reworked equipment applications, the calibration system is as follows:

The parts are delivered to the calibration equipment in molded trays in an X-Y matrix. These trays are conveyed on a low cost, reconfigurable, flex-link conveyor. The trays provide enough accurate positioning for robotic handling. A radio frequency read/write tag is attached to each tray. The tag contains the part number of the parts in the tray as well as the palletizing information necessary for robotic handling. The tray of parts is delivered to the bank of calibration stands. There is a load/unload robot to service each bank of stands. Bank size is designed around robot reach and average calibration stand cycle time. The robot end effector is double tooled so that it can unload and reload the calibration stand without repositioning or returning to the parts tray. All parts—not calibrated, calibrated, and those that failed calibration—are handled in the same tray. The information regarding each unit in the tray is added to the radio frequency read/write tag. Any failure code information is also stored in the tag.

After all of the parts have been processed, the tray is released and the tray in queue is processed. The released tray is conveyed to the shipping area. If there are any units that failed calibration, these are automatically removed with the rework information in the rework area before they leave the calibration area en route to shipping.

If different programs or model specifications are necessary, the calibration stand computer requests this from the system via the communications channel.

Results

In achieving its targeted results, the company was able to gain the following operational advantages:

- Elimination of all load and unload labor
- Reduced load and unload time

- Elimination of paper (cart tags and rework tags)
- Elimination of keyboard input errors
- Reduction in material handling containers and costs
- Automatic auditing and integral audit trail
- Reduced floor space requirements for same equipment count
- Higher overall machine utilization

Certain things were unaffected, or became slightly different due to these systems changes.

The same labor is required of setup personnel for equipment malfunctions. Computer communications is the same or better. This was necessary to change robot programs for different part numbers, in addition to original communications requirements.

There were some offsetting disadvantages, which are as follows:

- Increased system complexity
- Added equipment investment dollars

However, the bottom line was positive, in terms of overall cost savings, and improved quality. The system reduced damage due to less materials handling, and improved information flow reduced errors in quality indices.

Additional equipment and higher maintenance costs were more than offset by increased machine utilization and elimination/reduction of nonvalue added labor costs.

Solution B. In high-volume new equipment applications, the calibration system is as follows: The parts to be calibrated are delivered in one of two ways. They are delivered individually via a low-cost feed conveyor from an adjacent assembly conveyor, or they can be delivered in quantity via traditional fork truck matrixed pallet methods. In both methods, they are presented to the load station of a power and free, cross-shuttle calibration conveyor. The units are loaded manually into nested conveyor pallets. Each pallet has a radio frequency read/write tag attached to it. All of the stations on the conveyor are computer controlled and have the same communications capability described previously. The load station enters the proper information into the tag; model number, virgin or reworked, and basic production information. The pallet is transported to the first available cross-shuttle station and queued. There are a number of identical calibration stations. The stations have a work position and queue position to minimize the "load" time. The pallet enters the work position, and all of the calibration adjustments are made with the part while in the pallet. Model numbers are checked before operations begin to ensure proper program and model specifications are resident in the calibration station computer. This check is performed because production is normally done in a batch mode. This provides for the capability to run a mixed batch, if desired. It also allows for a model changeover "on the fly" because the station will call for the proper program and/or model specifications if required.

Upon completion, the pallet is shuttled out to the opposite side of the conveyor. The RF tag now contains all of the data necessary for subsequent processing. It also has the actual calibration results. This allows automatic audits to occur on-line. There is a rework spur on the conveyor. Pallets that contain parts that failed calibration are routed onto this spur automatically. The tag is interrogated and the rework operations are performed. The pallet is routed back into the mainflow and starts the process again.

The audit stations perform a rotational audit on passed parts and can do the comparison of audit values to the actual calibration values.

The completed parts are unloaded automatically from the calibration conveyor. The system is capable of matricizing the parts as they are unloading or loading a feed conveyor that delivers them to the shipping area.

Results

The results of this type system also were positive, and included the following advantages:

- Elimination of all load and unload labor (except for the one conveyor load station operator)
- Extremely low station load time
- Elimination of paperwork
- Reduction in material handling containers and costs
- Fully automatic auditing and integral audit trail
- Improved floor space utilization
- Extremely high machine utilization

The primary benefits from this solution were the same as the other type discussed previously. However, in addition, some communications advantages accrued from use of RF tags.

To operate the system and perform the same functions *without* the tags, not only would the pallet need mechanical flags, but the communications structure would have to be configured. Each pallet in the system also would need an individual, machine-readable identification. Although very cumbersome, this configuration could work, and allow for operation independent of the communication concentrator computer.

By using the read/write tags, the communication capability is performed by the RF tag. The dramatic reduction in wiring, hardware, and software reduced implementation costs considerably and increased system reliability.

CASE-IN-POINT

RADIO FREQUENCY DATA COLLECTION SYSTEM
USING READ/WRITE TAGS
ROCHESTER PRODUCTS DIVISION OF GENERAL MOTORS CORPORATION

Situation

The manufacture of an end item consists of the production of component parts, assembly, and individual calibration of each part. All of the part numbers in the product family are basically the same, with minor mechanical variations and different calibrations being the main distinctions.

The product was new, and therefore, required new equipment. All of the various part numbers needed to be assembled and calibrated on the same conveyor.

Since it was a high-volume product it would be necessary to make the power and free conveyor in a cost effective system.

Solution

The company designed and built a power and free conveyor that is a combination of cross-shuttle and series stations. The stations are controlled by either a programmable controller or a process control computer, determined by the station function. The pallets on the conveyor are capable of handling all of the various configurations. All of the parts for the product are fed to the stations automatically, except one, which is loaded manually by the only operator on the line.

Each of the conveyor pallets has a radio frequency read/write tag attached to it, and the stations that need to communicate to and from the tag are equipped with the appropriate hardware.

In this communications hierarchy the stations receive their control programs and necessary model specifications (calibration values) through the serial communications channel. Once this data is loaded, additional communications are no longer required. The programmable controller programs have all of the necessary codes to handle any of the models. The appropriate portions of code are executed based on the information contained in the read/write tag on each pallet. The process control computers contain their programs and all of the current model specifications allowing them to switch to the proper one based locally on tag information.

Results

The equipment is capable of producing any of the models. This includes the assembly of the proper components, calibration, and automatic unload. Failed parts have no salvage value so they are unloaded by a reject station. This conveyor, which includes thirty-four assembly or calibration stations (most stations perform multiple assembly tasks or calibration adjustments) is operated by one load operator and one setup/parts loader operator.

The read/write tag is used as a station-to-station communications channel. Information processed to and from the tag handles all of the pallet routing, assembly variations, calibration variations, actual part values and audit/verification requirements. The same benefits described in the first case apply to this case as well.

Automatic Identification: A Key Building Block

In summary, it is evident that AI technology is a key building block in the development of systems that involve the movement of material through manufacturing and related functional areas. AI applications bridge the gap between the physical movement of the material and what the computer knows about the material. Moreover, AI enables the computer to issue commands for actions to be taken regarding the material, such as machine, test, transfer, divert, weigh, package, store, ship, and report everything that's happening throughout the plant facility.

Taken in this context, the question is not whether AI is needed, but rather what method should be used, and how it should be used.

HOW TO AVOID CONFUSION ABOUT RADIO FREQUENCY DATA COLLECTION

It is important to distinguish between RF terminals, which link to a host computer via a UHF RF link, with the RF identification systems, which use readers, tags, and a short range RF link to identify and store data regarding a part or group of parts. The reason for understanding clearly the distinction between these different approaches is that both systems have been called RF data collection systems.

As data communications and radio technology become interfaced, you can begin to understand the practical use of RF terminals for real-time data collection and inventory control. An interactive mobile computer can speed up gathering of data on the shop floor.

The starting point for determining what information system to use is where your cost centers are. In this regard, how far are workers from data collection units?

Next, you should determine what the objective is for installing this system. Is it to accomplish any or all of the following?

- Reduce scrap
- Avoid stock-outs
- Control WIP inventory

- Improve labor productivity
- Capture shop floor data for a number of reasons, such as monitor schedules, fine-tune machine loads, track orders for feedback to customer service, or obtain more accurate and timely cost accounting data

Once this determination is made, the selection process should consider whether it will be both mobile and an interactive communications system.

Finally, this evaluation should include an investment payback criteria based on the level of sophistication of the system. Will the system be any of the following?

- Manual
- Batch
- Real-time
- RF/real time

As an example, an advanced approach may consist of a mobile real-time processing system that is used in conjunction with radio frequency terminals mounted on light trucks and hand-held terminals. The portability and flexibility of such equipment transmitting over a wireless link has numerous advantages for shop floor control and uploading of data to the host computer.

All the key data, identification, location, quantity and quality, can be entered on RF terminals and transmitted directly rather than batched or written for later updating of the data base.

Mobile terminals mounted on lift trucks can put operators on-line with a host computer. This method can enable a company to update information on stock in a real-time mode, and to establish standards for pick/put away materials handling. It also can reduce deadheading and increase inventory accuracy because the lift-truck operator is required to verify the correct inventory location by wanding or scanning a location bar code. Also, it can speed up handling of "hot orders" that have to be processed expeditiously in the production cycle.

Real-time systems and RF mobile terminals can communicate with a host computer through a computer interface unit/radio frequency unit (CIU/RFU) that can handle thirty to sixty wireless terminals, depending on the computer operating system and number of transactions.

In these RF data collection systems, the CIU manages the flow of information to and from the host, thus relieving the host of traffic management tasks. These systems also are designed to be compatible with standard interface formats of host computers.

chapter 6

Trends in Robotic Technology

ROBOTIC TECHNOLOGY COMES OF AGE IN AMERICAN INDUSTRY: AN OVERVIEW

The use of robots in American industry is beginning to take shape with more and more companies becoming interested in automating the work place. Within this technological thrust each company is seeking customized solutions to satisfy its own particular requirements. Although the robot itself has almost become a generic piece of equipment, each application is unique because of the tooling that accompanies the basic robot. Typically, the robot costs less than 30 percent of the complete system's cost, with the remainder in tooling, manipulators, conveyors, vision systems, computers, safety guards, and other components.

For example, each type of gripper has adaptive characteristics for certain applications. There are exchange grippers that allow for changes between internal gripping and external gripping; grippers that have sensors based on limit switches; proximity switches for sensing by measuring current or pressure in a pneumatic cylinder or hydraulic pressure; very small gripper fingers used to pick up integrated circuits; and large grippers to grasp big objects for the manufacture of major end items.

Just as there are many types of grippers and methods of gripping parts, so too are there many different ways of mating. Some companies mate using a tape written element, with balls that lock in place; others use a straight pin with a rotary cam that locks in place; while every company can transfer electrical connections, and transfer pneumatic connections.

Robotic technology varies from more basic applications, such as welding, paint spraying, and pick and place, to more complex tasks like automating the handling of nuclear fuel elements. Robots are heavily involved in inspection systems equipped with measuring devices, which in the automotive industry can literally measure the "as built" condition of the vehicle body and accommodate the variations when it comes to subsequent electrical components. IBM, for instance, uses IRT 90s, which are huge robots, to handle plasma display screens for the advanced plasma display tube that they have developed.

Most paint spraying robots are specially designed because there are no electrical motors on the robot arm since they are surrounded by the paint containers. As a result, paint spray robots employ hydraulic servos. At present, there are a few companies seeking to get electro-paint spraying robots certified.

Robots in general are becoming more versatile and have greater intelligence through fully utilizing the microcomputer's capability and employing various kinds of sensor techniques. Not only are vision sensors becoming more prevalent, but development work is proceeding with "smart" tactile sensors located on the end of the gripper that will be able to distinguish what part it is grabbing.

Also, the computer operating the robot and updating servo information at 40 milliseconds, is being improved steadily to eventually speed up this cycle to 100 milliseconds. This will mean development of real-time force feedback, which would result in more frequent updates of information.

Currently, very few robots are reprogrammable, multiple-act systems. Other than the automotive industry, where a spot welding robot reprograms spot welding, very few robots ever get put into a manufacturing cell, operate for the life of the product, and are then sold or dismantled. Instead, the robot is remanufactured into another cell doing something different. To this extent, you must regard most robots as single-purpose units. Basically, they are a form of part automation.

Understanding the Dynamics of Robot Technology

To illustrate the dynamics of robot technology, consider the process of component insertion for printed circuit boards. Thousands upon thousands of printed circuit boards are assembled each year. Various engineering studies have been made concerning the problem of how to get the leads so that they will all fit into the holes properly. Over the years, development has revolved around remote center compliance devices. With this approach, when the lead is inserted into the hole, it hits the gripper, and thus automatically centers itself with the hole. However, this technique is being replaced by processes that do not require insertion of parts through holes in printed circuit boards. Instead, companies are developing surface-mounted devices in which there are no holes on the board, but only a pad—with the object placed on the surface mount pad and held there. Without holes on the circuit board, this surface-mounted technology enables the part to be held temporarily with an adhesive. When the board is run through the flow solder machine, the solder melts, reflows, and adheres to the circuit board. What this advancement means, is that companies that manufacture equipment that drills holes in printed circuit boards—that had nine heads on it, moved at a thousand inches per second, and could drill fifty thousand holes per hour—may no longer have a market for their product.

What Is a Robot?*

According to the Robotics Industries Association definition, a robot is a reprogrammable multifunctional manipulator designed to move materials, parts, tools, or specialized devices, through variable programmed motions for the performance of a variety of tasks.

The typical elements of any automated production process include the following:

1. The workpiece
2. Fixturing
3. Tools
4. Tool driving machinery
5. Work infeed system
6. Finished work removal
7. Control system

As applied to a robot application, these elements may be combined in several ways. For example, in a welding, sealing and bonding, or paint application, the tools and related machinery are carried by the robot as end-of-arm equipment while the workpiece fixturing is incorporated into a conveyor system which also does work infeed and removal. This is illustrated in the following GMF robotics applications (figures 6-1 through 6-5):

*Permission granted by GMF Robotics

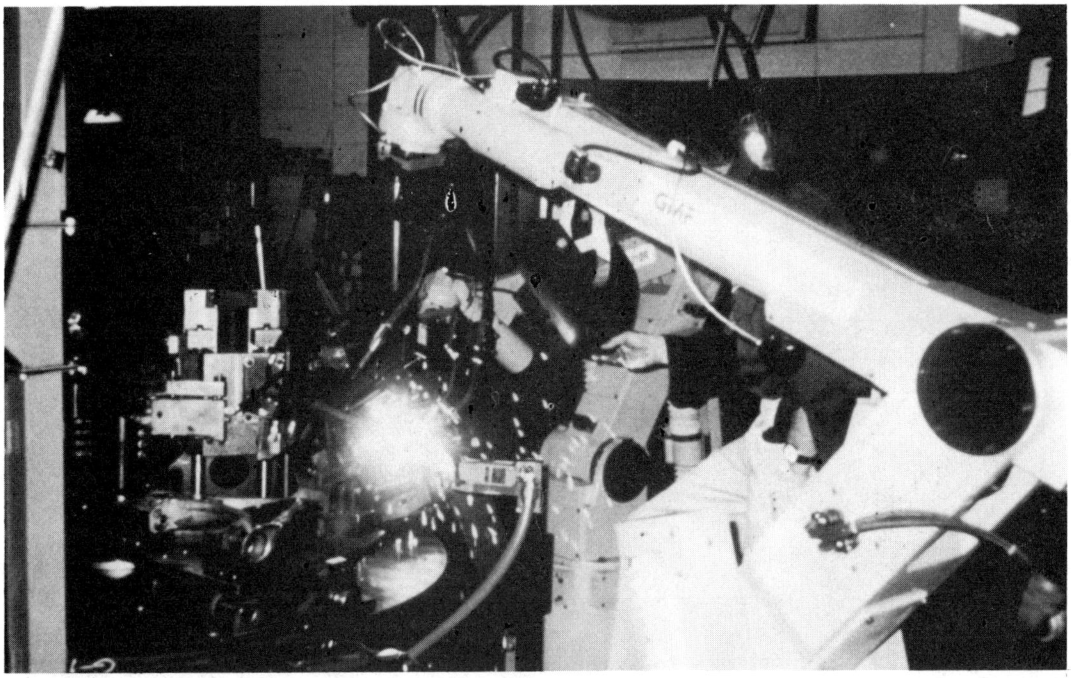

FIGURE 6–1 Arc Welding Robot (Courtesy of GMF Robotics Corporation).

FIGURE 6–2 Arcweld Vision System (Courtesy of GMF Robotics Corporation).

FIGURE 6–3 **Sealing and Bonding Robot** (Courtesy of GMF Robotics Corporation).

FIGURE 6–4 **Numerically Controlled Paint Robot** (Courtesy of GMF Robotics Corporation).

FIGURE 6–5 Electric Paint Robot (Courtesy of GMF Robotics Corporation).

FIGURE 6–6 Robot Performing Pick and Place Operation (Courtesy of GMF Robotics Corporation).

FIGURE 6–7 Robot Performing Assembly Operation (Courtesy of American CIMFLEX Corporation).

In a part handling and processing application the robot usually acts as the fixturing device and often takes care of infeed and output functions. This type application is illustrated in figure 6-6, which shows a GMF M-100 robot performing a pick and place operation of valve bodies. While in an assembly operation the robot arm is positioned to execute its function on an assembly in a fixed position. This is illustrated in figure 6-7, which shows an American CIMFLEX robot assembling a Ford window panel.

Regardless of which application is involved, the important point to consider is that the cost rises exponentially with the degree of intelligence required by the robot. Thus, an investment in a more sophisticated peripheral system can drastically reduce the complexity and cost of a robot.

The Differences between Hard Automation and Robots

Within a broader robot definition, one may include automated handling devices that lack servo control but are mechanically reprogrammable by means of adjusting stops, limit switches, and drum logic controllers.

Hard automation most often takes the form of the part handling type of system. The following comparisons can be made:

Robot	Hard Automation
Lower production rate	High production rate
Flexible programmed cycle	Fixed cycle
Re-usable	Dedicated to one job
Fixed production rate	Flexible production rate
Decision-making ability	Limited decision making

The hard automation device usually is designed to do several things at once, much like comparing a turret lathe to a screw machine. Like the screw machine, however, it is not able to jump around in its programmed sequence. A robot is similar to a machine tool in that retooling and reprogramming is all that is required to convert to a widely different process. Hard automation is more amenable to varying amounts of manual assistance allowing variable production rates at variable labor-power costs. A robot has the highest degree of flexibility for interfacing with fluctuating demand or information from sensing devices to reject parts in process, rework parts in process, or other decision-making functions.

Six Key Characteristics of Industrial Robots

Work Envelope. Industrial robots have arms that move in several directions to perform work. The robot's work envelope is the set of points described by the maximum reach of the robot end effector (or working tool) in all directions.

End Effector (Hand). An end effector is the device attached to the robot arm tool-mounting plate in order to interface with its machine tool or other work environment. End effectors include grippers, tools, welding equipment, or any type of device or fixture attached to the robot arm to accomplish the actual work.

Payload. A robot's payload is its ability to carry, both continuously and satisfactorily, a specified maximum weight, usually defined in pounds or kilograms.

Velocity. A robot's velocity is the maximum speed at which the tip of its arm is capable of moving at full arm extension, usually expressed in inches or millimeters per second.

Accuracy. The accuracy of a robot is its ability to position its end effector, when commanded, at a specified point, without having attained that position previously.

Repeatability. A robot's repeatability is its ability to return consistently and accurately to a predefined and previously attained position.

Basic Components of All Industrial Robots

In basic terms, an industrial robot is a mechanical arm that is bolted to the floor, ceiling, wall, or a machine. It is equipped with an end effector to perform some sort of work, and programmed to repetitive work in a controlled and ordered environment. Except in unusual cases, an individual robot cannot move about in its plant, or see or feel parts upon which it is working. Industrial robots vary widely in size, shape, structure, and capabilities.

All industrial robots encompass three basic components (figure 6-8):

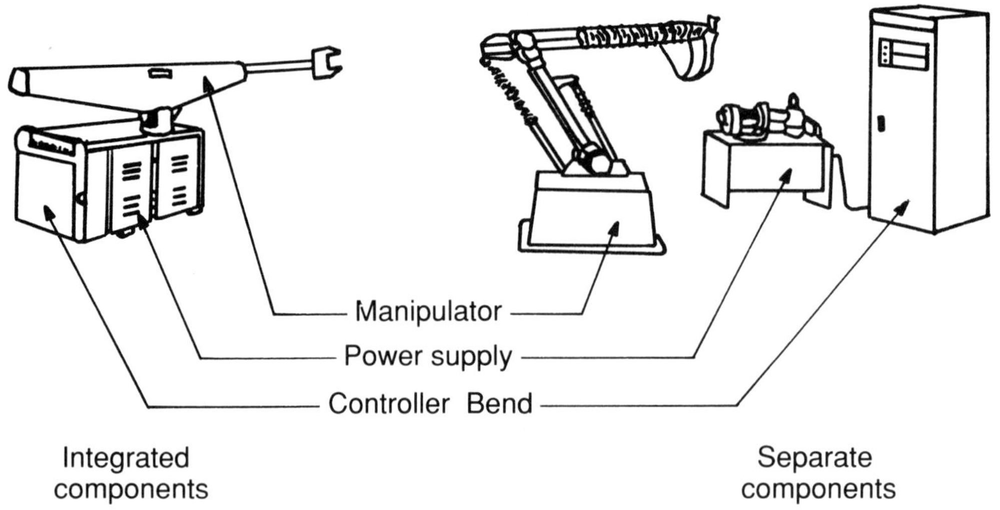

FIGURE 6–8 Basic Components of an Industrial Robot (Courtesy of GMF Robotics Corporation).

- Manipulator
- Power supply
- Controller

Manipulator. The term manipulator refers to all of the robot motion axes that move the end effector in various directions. It consists of these components: actuators, drives, bearings, and feedback devices. The mechanical design of the robot affects its work envelope and motion characteristics.

The work envelope of most robots are defined by the coordinate system of the three major axes of motion which provide: a vertical lift stroke, an in-and-out reach stroke, and a rotational or traverse motion at the robot's base.

In addition to the major axes of motion usually found at the "wrist" of the robot, these axes are most commonly referred to as pitch or bend, yaw and roll, or swivel.

Additional axes also can be provided by mounting the robot on a floor track, mounting the robot overhead, or in end-of-arm tooling for extra reach and manipulation.

Tips for Successful Automation: Understanding How Robots Are Classified

Robots are classified according to work envelope structure defined by the three major axes:

1. Cylindrical coordinate
2. Spherical coordinate or polar
3. Jointed arm or revolute
4. Rectangular coordinate or cartesian

1. *Cylindrical Coordinate (figure 6-9).* The motion of the three major axes in this assembly form a pie-shaped section of a cylinder. This configuration consists of a horizontal arm affixed to a vertical axes which is mounted on a base. The horizontal arm strokes in and out and up and down on the vertical axis. In addition, it can swing clockwise and counterclockwise about the base of the vertical axis. Arm structure may be rigid or flexible.

2. *Spherical Coordinate or Polar (figure 6-10).* The work envelope of a spherical coordinate or polar robot resembles a portion of a sphere. This configuration consists of an arm which strokes in and out and pivots to achieve its vertical stroke. In addition, the arm swings clockwise or counterclockwise about the base.

3. *Jointed Arm or Revolute (figure 6-11).* This robotic design features an arm having two rotary joints called a "shoulder" and an "elbow." The arm is mounted on a rotating base. In this configuration, the robot can fold its arm closely to its base, minimizing floor space requirements.

4. *Rectangular Coordinate or Cartesian (figure 6-12).* The work envelope structure in this configuration is rectangular. The robot consists of a horizontal arm assembled to a vertical left axis, which is mounted on a linear traverse base.

Various arm applications may be used with this type of robot, including:

- Multiple arm
- Rigid arm
- Flexible arms
- Arms/base mounts
  (a) Pedestal
  (b) Overhead gantry

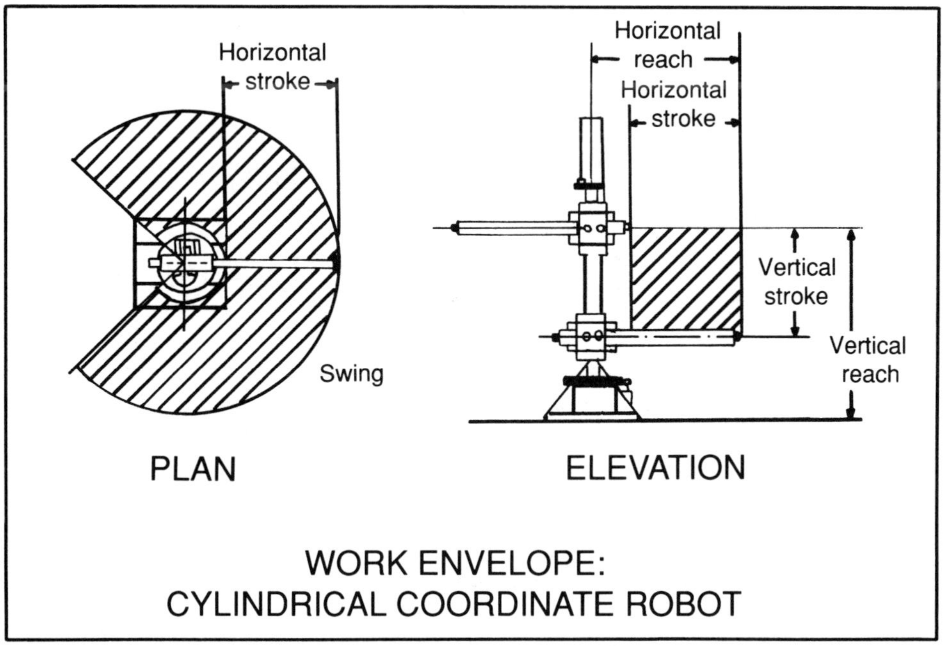

FIGURE 6-9 Work Envelope: Cylindrical Coordinate Robot (Courtesy of GMF Robotics Corporation).

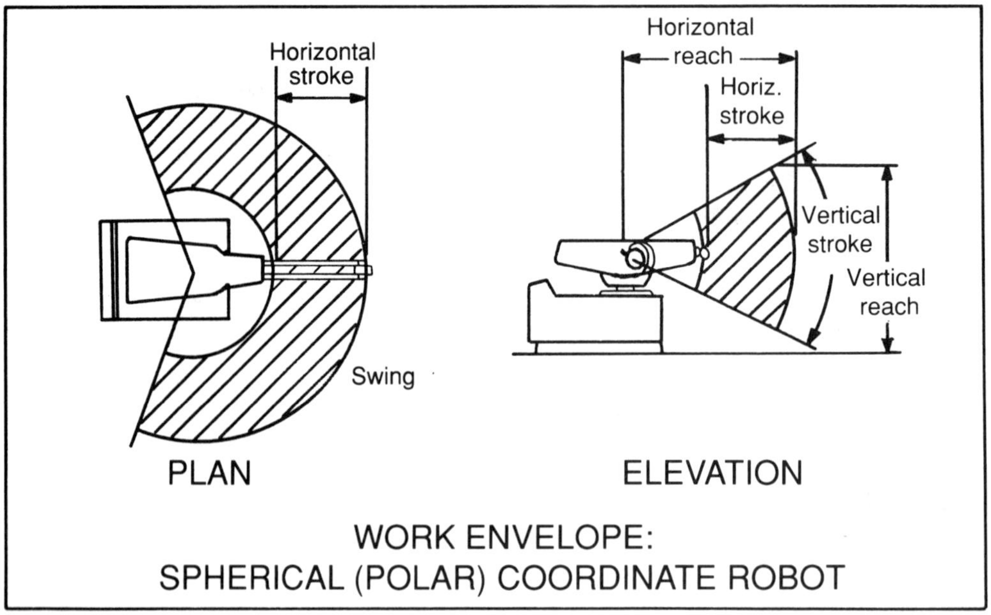

FIGURE 6–10 Work Envelope: Spherical (Polar) Coordinate Robot (Courtesy of GMF Robotics Corporation).

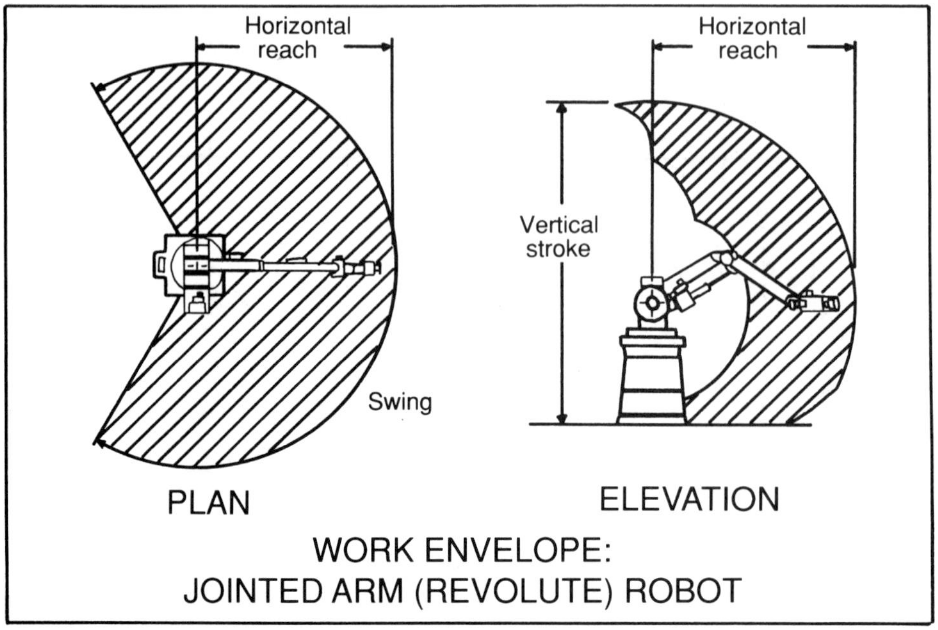

FIGURE 6–11 Work Envelope: Joint Arm (Revolute) Robot (Courtesy of GMF Robotics Corporation).

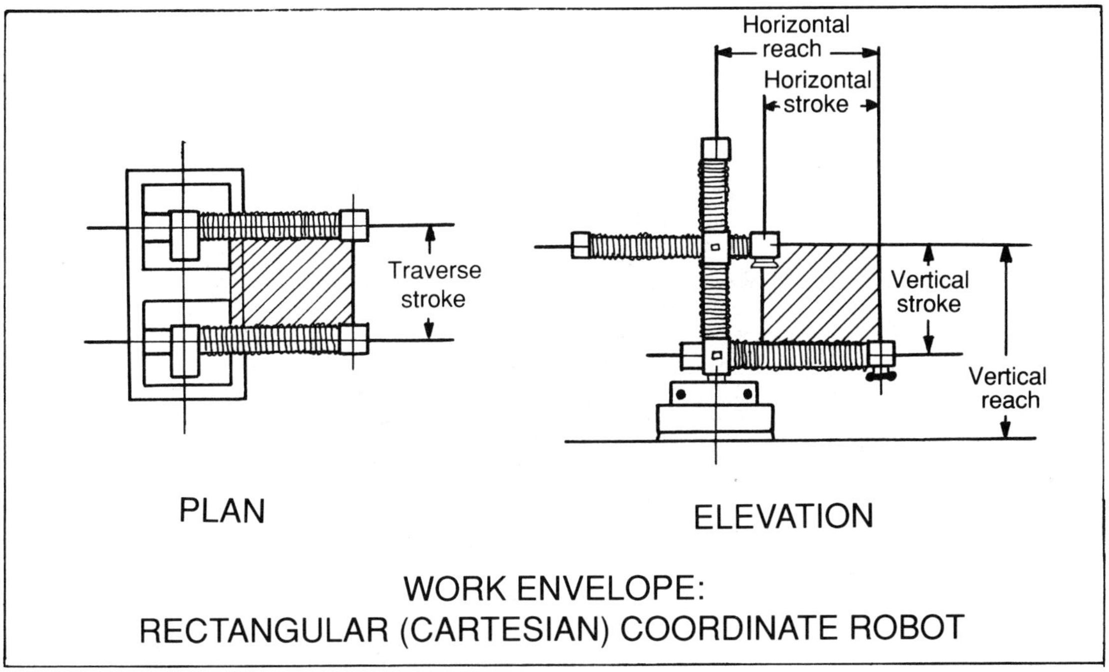

FIGURE 6-12 Work Envelope: Rectangular (Cartesian) Coordinate Robot (Courtesy of GMF Robotics Corporation).

Power Supply. The power supply provides and regulates the energy used by the robot. It is related directly to the payload rating of a robot:

- *Pneumatic*—light payloads, fast cycle times
- *Hydraulic*—very heavy payloads
- *Electric*—quiet, energy efficient, repetitive, does not leak, does not need special start-up maintenance

Controller. The control system starts, coordinates, and ends all robotic motion. It accepts all input and provides the output necessary to robotic functioning.

The three types of control systems are:

1. *Non-Servo Point-to-Point (figure 6-13).* This consists of a system of flow control valves and mechanical stops actuated by pneumatic or hydraulic cylinders or motors connected to a directional control valve. The end points of motion for each axes are sensed by limit switches.

   Sequencing control is performed by a stepping drum-type programmer or a programmable controller. Programming is usually done mechanically by setting up the desired sequences of axes motions and adjusting the mechanical end stops to their desired position on each axes.
2. *Servo-Controlled Point-to-Point (figure 6-14).* Servo-controlled point-to-point controls can perform more sophisticated part handling tasks. The servo-mechanisms allow the robot to stop at an infinite number of points along each axes of motion. Feedback devices that constantly monitor the position and velocity of each axes are:

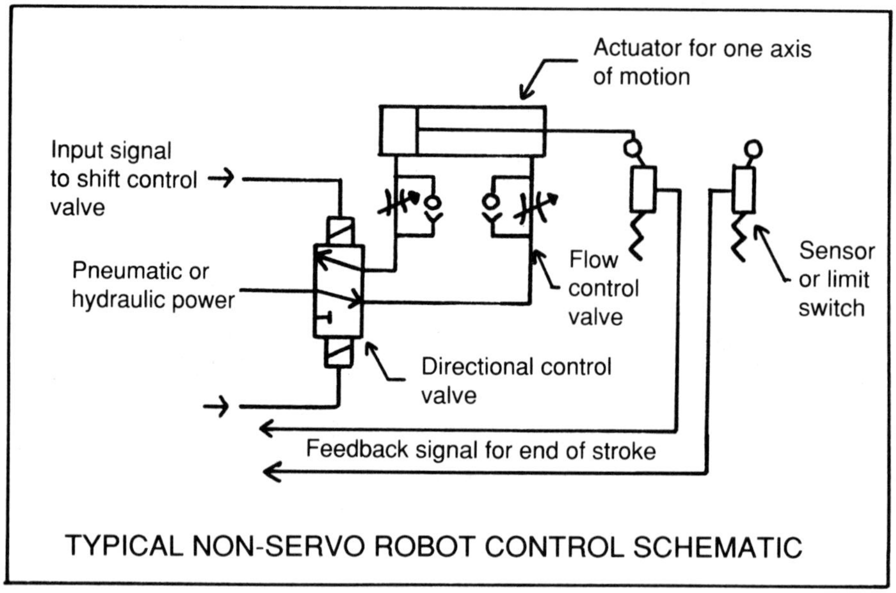

FIGURE 6–13 Non-Servo Robot Control Schematic (Courtesy of GMF Robotics Corporation).

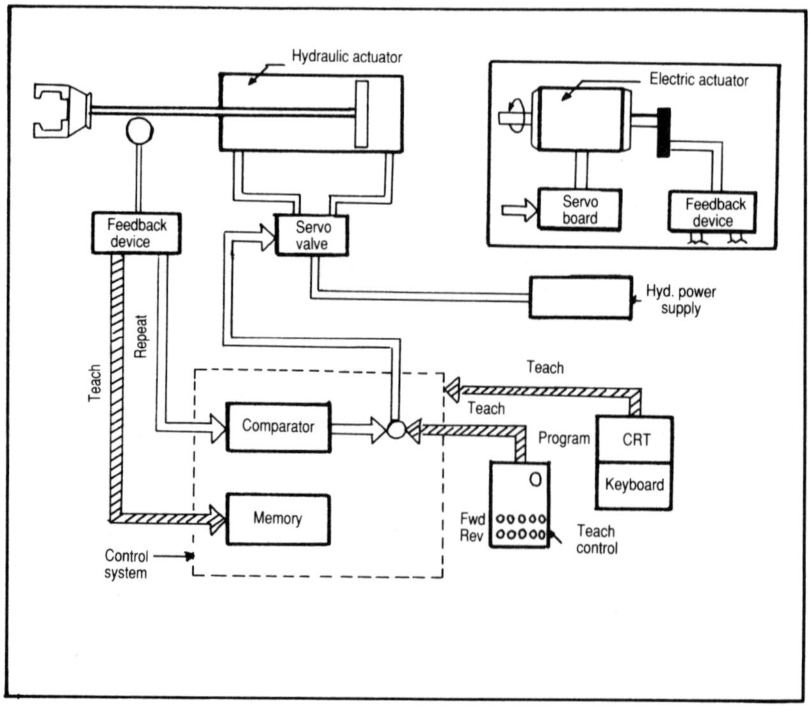

FIGURE 6–14 Servo Controlled Point-to-Point Robot Schematic (Courtesy of GMF Robotics Corporation).

(a) Encoders
(b) Resolvers
(c) Potentiometers
(d) Tachometers

The servo-controlled point-to-point robot is programmed by storing discrete points in space in the robot's memory, which has many more points than nonservo control systems. The programmer depresses an axis button on a teach pendant, which allows the axis to move. The programmer then walks the robot to the next point and releases the button. A record button is depressed, thus recording that point in the robot's memory.

3. *Servo-Controlled Continuous Path.* Servo-controlled continuous path control systems are similar to point-to-point control systems in their feedback devices.

   They differ from servo-controlled point-to-point systems in the following ways:

   (a) Recording programs
   (b) Storing data

   Continuous path control systems store data on a time basis wherein the programmer actually grasps the end of the robot's arm and physically leads it through a pattern of motions. This pattern is then recorded on either a cassette or a floppy disk.

Devices that Enable Robots to Perform Their Work

The device that will grip or secure the part or product is either a vacuum or gripper end effector.

Parts with large, relatively smooth surface areas usually fall in the category of vacuum handling applications. Smaller parts that do not have smooth areas usually are handled with one or more mechanical grippers with fingers designed and contoured to fit the part configuration.

Vacuum Devices

Vacuum cups come in virtually all sizes and shapes. Best results can be obtained from cups that are molded over a metal supporting insert for strength, and have a reinforced lip to resist cutting or tearing.

Gripper Devices

When selecting the proper end-of-arm tooling device, the part to be handled must be evaluated according to the criteria of weight, overall size, length, balance points, and class of the part. The part weight and size should be allowed to determine how many and what size heads to use.

Since the gripper will be loading or unloading the parts directly to and from the fixture, the fingers will necessarily be entering the fixture's tooling area and must not interfere with existing gauges, cutters, and other portions of the fixture.

Specific Types of Grippers

There are many specific types of grippers, including:

- Compressor casing
- Surge arrester

- Tube
- Seal ring
- Ring gear
- Radiator core
- Water compressor casting
- Water jacket shell
- "Standard type"

The following applications are representative of how some of these types of grippers may be used:

Compressor Casing Gripper. This application requires a compressor casing to be inverted and assembled onto a shaft. The clearance between a bearing in the casing and the shaft is ".0005" (figure 6-15).

Surge Arrester Gripper. This application requires the loading and unloading of a family of thirty-six different sized surge arresters to and from a crimping machine for the attachment of a cap (figure 6-16).

Tube Gripper. In this application, the robot system loads and unloads parts during a slot piercing operation. The parts consist of a variety of tubes (figure 6-17).

Seal Ring Gripper. This application requires the loading/unloading of axle seal rings into and out of two CNC lathes. The gripper design must accommodate thirty-three different styles of a part (figure 6-18).

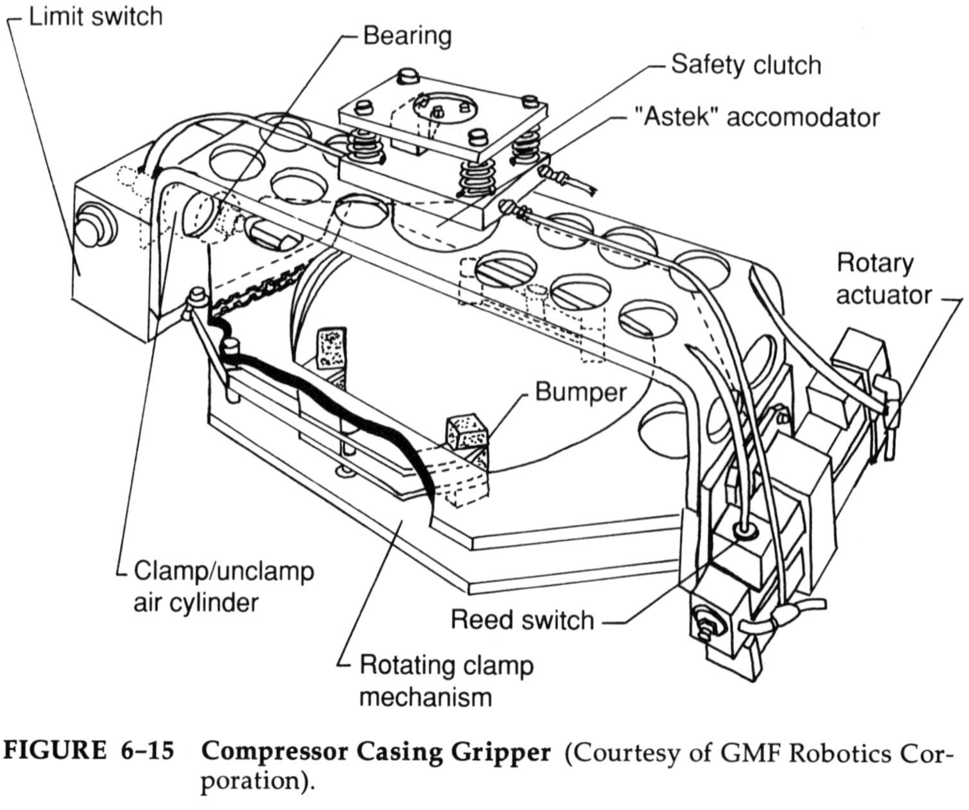

FIGURE 6–15 Compressor Casing Gripper (Courtesy of GMF Robotics Corporation).

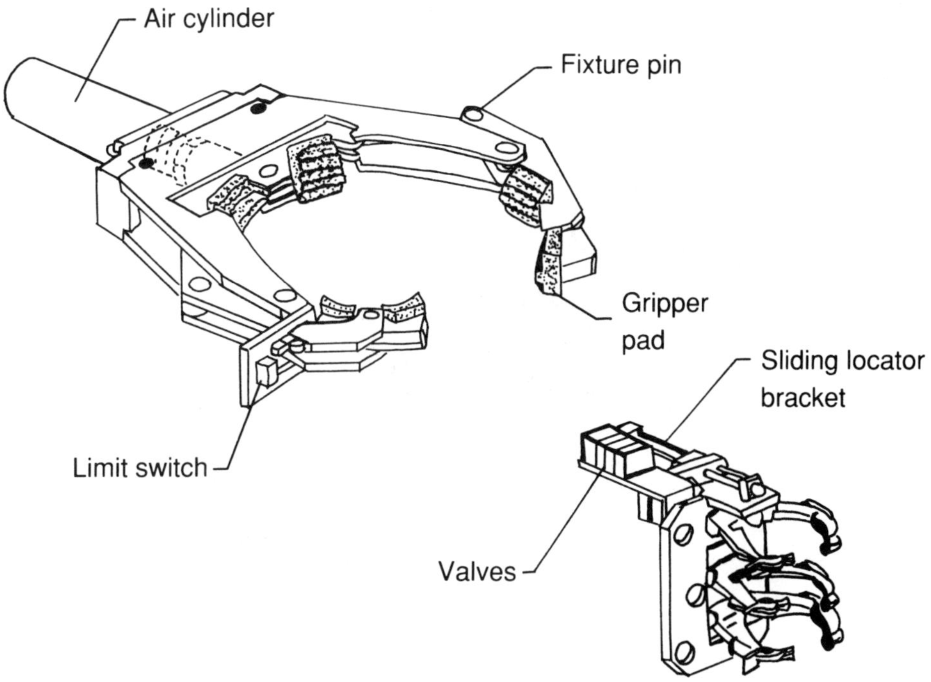

FIGURE 6–16 Surge Arrester Gripper (Courtesy of GMF Robotics Corporation).

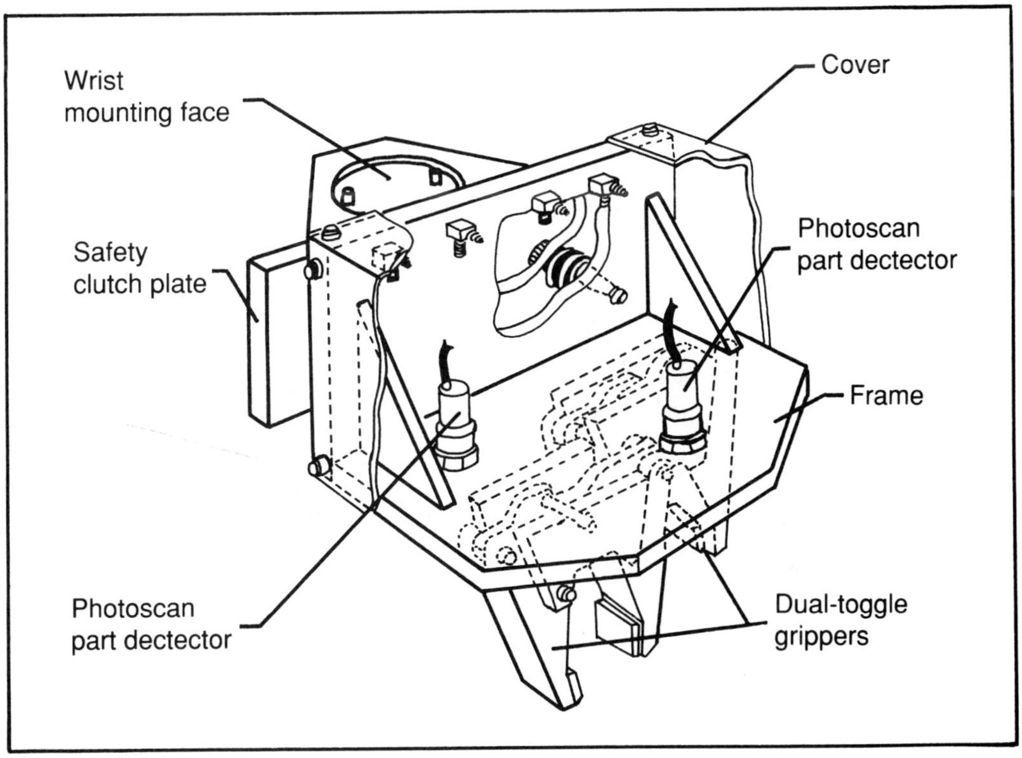

FIGURE 6–17 Tube Gripper (Courtesy of GMF Robotics Corporation).

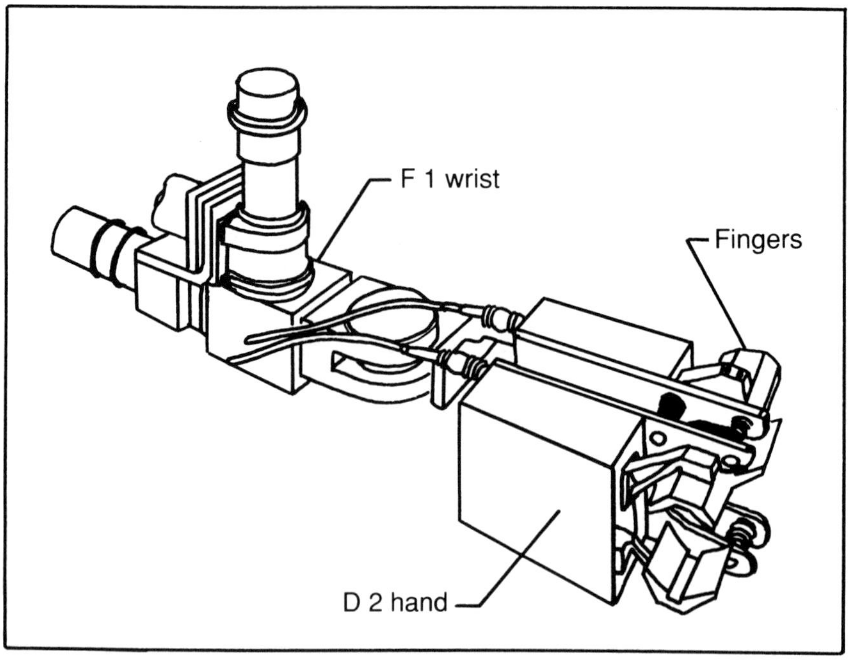

FIGURE 6–18 **Seal Ring Gripper** (Courtesy of GMF Robotics Corporation).

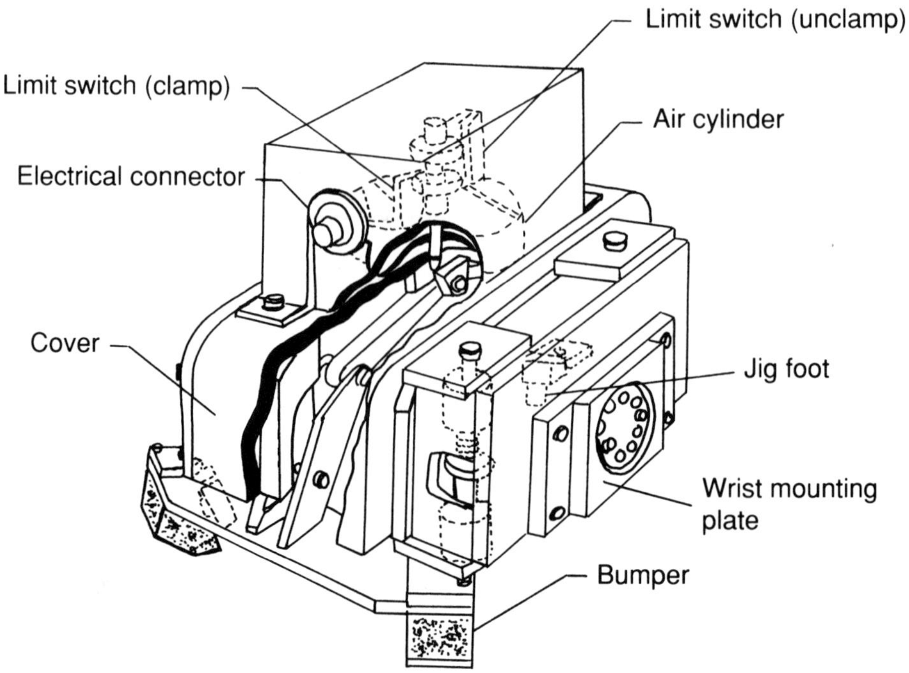

FIGURE 6–19 **Ring Gear Gripper** (Courtesy of GMF Robotics Corporation).

Ring Gear Gripper. This application involves the loading of a variety of ring gears from two different conveyor lines into three dial machines. Different styles of the parts must be routed to a particular machine. After processing, the parts are unloaded and placed onto three separate gravity feed conveyors for delivery to manual unloading stations (figure 6-19).

Water Jacket Shell Gripper. In this application, water jacket shells must be loaded and unloaded from a lathe chuck and transferred via a buffer storage table to a second machine cell where they are loaded and unloaded from a hole punch machine and a stamping machine, and finally placed onto a portable storage cart (figure 6-20).

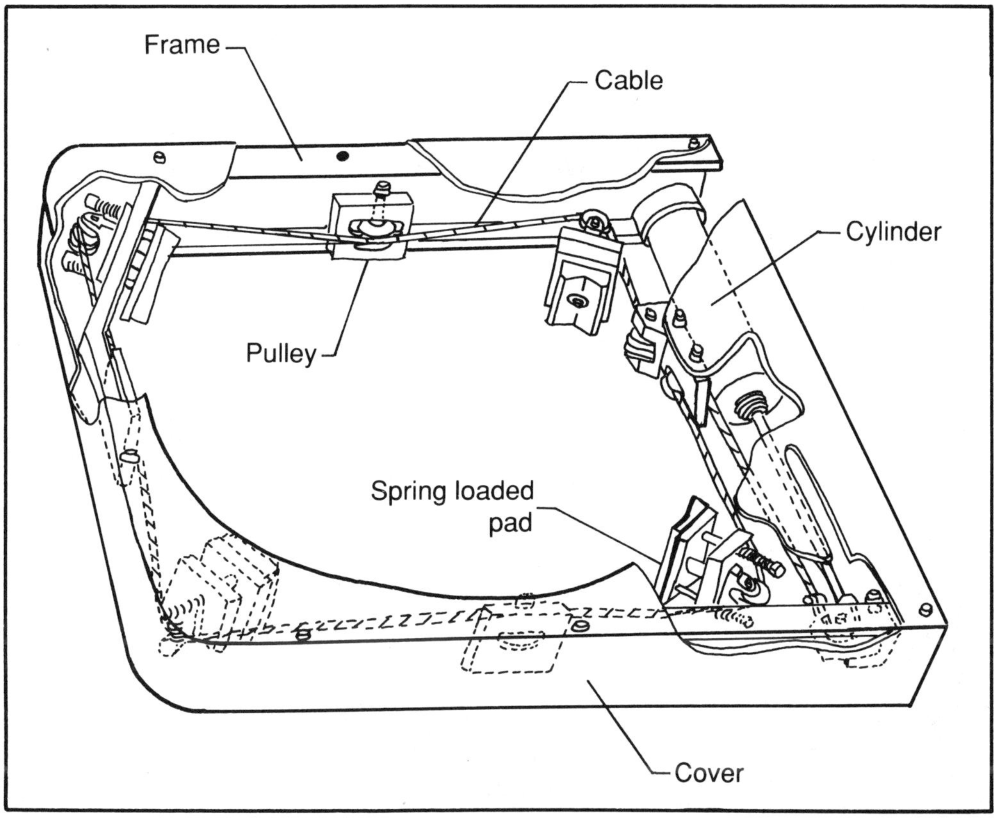

FIGURE 6–20 Water Jacket Shell Gripper (Courtesy of GMF Robotics Corporation).

"Standard Type" Grippers

In addition to specific types of grippers designed for particular robotic applications, one may regard the following "standard type" grippers as generic.

- Grippers that have fingers on them and they either grip on the outside of the part or grip on the inside of the part
- Grippers in which the fingers move in an angular pattern because they are hinged
- Grippers that move in a parallel fashion because they are hung on four bar lengths, or forced to move parallel due to the cam follower bearings sliding in slots, or they use a lead screw that has a left- and right-hand thread on it that causes the fingers to move parallel

- Quick exchange grippers
- Cluster head grippers

Tool-Changing Robot Hands: Advantages of the Foreman Hand*

Industrial robots are becoming increasingly evident in assembly operations. Typical industrial assembly processes involve performing multiple tasks on one product in various states of production. The introduction of robotics commonly demands one task per robot since industrial robotics end effectors are usually custom designed to perform a specific task in the assembly process. Often, a change in product design or model will warrant removal or replacement (or both) of the end effector to be compatible with current running configurations. Sensor capabilities designed into each end effector must also be considered for monitoring cell functions. The principal advantage of using robotics in assembly over other traditional methods is versatility. However, in most instances, the capabilities of a robot in the work cell are limited to functions performable by the current end effector in use.

Over 70 percent of products manufactured in the United States are produced in batches. Although the robot can easily run a variety of motion programs, the end effector is almost always dedicated and must be retooled to meet current assembly line requirements. One of the leading companies in this area is Monforte Robotics, which has developed the Foreman Hand, a tool-changing end-effector system that is capable of performing multiple, diversified tasks intelligently, in one robot work cell, using only one robot. Both the Foreman Series of end effectors, which are lightweight, multifunctional, intelligent end-of-arm toolings (figure 6-21), and the Workman Series, which is designed for heavy-duty industrial robotic applications (figure 6-22) eliminate this restriction by not only utilizing a large set of

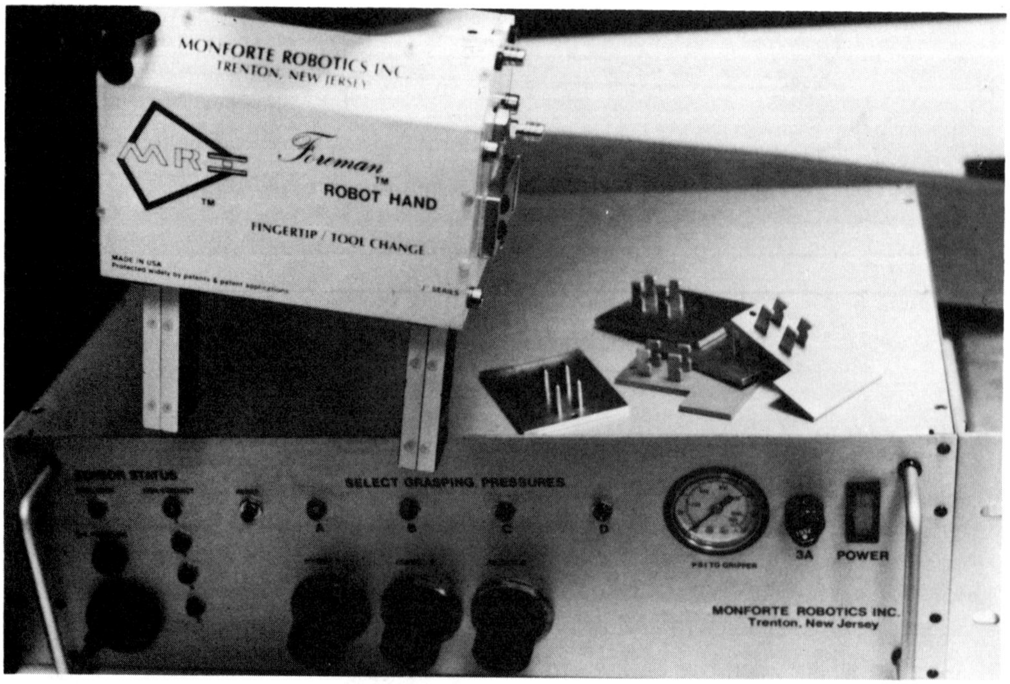

FIGURE 6–21 Foreman™ Automatic Fingertip/Tool Change End-Effector (Courtesy of Monforte Robotics, Inc.).

*Permission granted by Monforte Robotics, Inc.

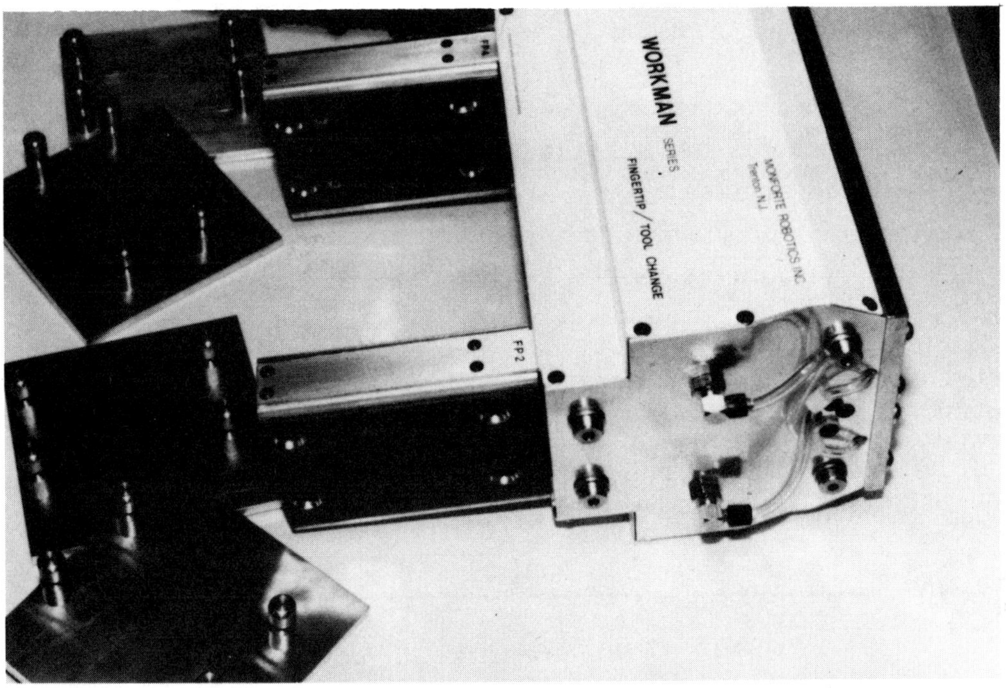

FIGURE 6-22 Workman™ Heavy Duty Automatic Fingertip/Tool Change End-Effector (Courtesy of Monforte Robotics, Inc.).

uniquely styled adapter plates designed to fit onto a universal hand, but by providing the capability to change the different adapter plates by themselves according to the changing needs of the work environment.

The advantages of this approach include:

1. Eliminating the cost and floor space of multiple work cells
2. Fewer line shutdowns for retooling high-cost custom-designed end effectors
3. Generally enhanced production

For a multifunctional, intelligent end effector to be practical and effective, design considerations for the end effector must include:

- Durability in construction, but light in weight to be compatible with the largest selection of robots
- Retention of as much accuracy as possible from the host robot by having the fingertip-to-mounting clearance as small as possible
- Ability to exert enough gripping force to be applicable with higher payload robots
- Ability to grasp a variety of parts or tools using desired gripping forces as needed
- Ability to acquire workpieces or tools in a variety of modes of pickup
- Ability to acquire workpieces or tools larger than maximum finger opening
- Ability to retool itself under automatic control
- Ability to work with parts or tools in a work cell in a manner similar to custom-designed end effectors

- Ability to engage or disengage custom toolings or work tools in less than half a second
- Cost effective retooling method
- Interface method compatible with the largest selection of robots
- User-friendly as possible
- Selection of sensors to monitor gripper duties
- Selection of features to perform complex assembly

The Monforte Robotics Foreman Hand satisfies all these design guidelines with a series of twelve end effectors, with common internal mechanisms, sensors, and other features, capable of handling a wide range of assembly operations. The Monforte Robotics heavy-duty Workman series end effectors are capable of performing a variety of tasks including palletizing, mechanical assemblies, material handling, machine loading and unloading; and are adaptable to most power tools, such as welders, drillers, water knives, and others.

CASE-IN-POINT

MANUFACTURER OF HAND-HELD CALCULATORS

Situation

A company that manufactured hand-held calculators desired to automate the assembly process to reduce labor costs and increase productivity throughput.

The hand-held calculator unit consists of a face plate, onto which a plastic window is glued. A printed circuit board containing electronic components and keyboard is located and screwed into place. A battery clip is then sprung into place on the PC board. Finally a back plate is snapped in, and the complete assembly is sent to a test area.

Solution

A simulated operation was made of two robotic-assembly approaches in making an evaluation of this task: one using task-dedicated robots and the other using multifunction robots equipped with tool-changing robot hands.

In the task-dedicated line eight robots would be needed: the first picks up the face plate and puts it onto a conveyor, the second squirts epoxy, the third picks up the window and places it on the epoxy bead, the fourth picks up and places the PC board, the fifth carries an automatically fed screwdriver and fastens the board down, and the sixth picks battery clips from a chute and inserts them, the seventh picks up and pops in a back plate, and the last robot uses another specialized gripper to pick up the entire calculator and carry it to testing. (See figure 6-23.)

By comparison, with the cluster of tool-changing robots, each robot has access to eight tools and is equipped with various sensors to signal its controller that the tools are in place and parts are indeed picked up and deposited. The three or four seconds required for each tool change is integrated into the cycle to assemble each calculator. This configuration only requires three such robots. (See figure 6-24.)

Results

The conventional assembly line would require eight robots, at $47,000 each, or a total investment of $376,000. In addition, every three robots would require the services of one system engineer at an annual cost of $25,000 each, or $75,000 in engineering talent. Also, one robot technician would be required for every five robots, thus, the line would need two technicians

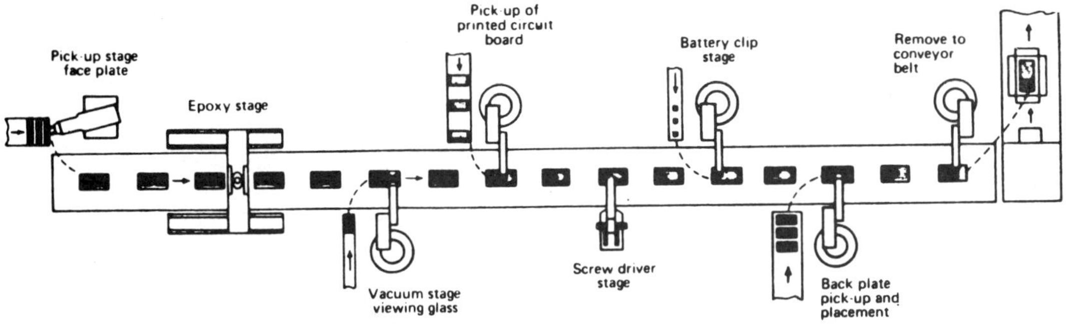

FIGURE 6–23 Assembly of Small Calculator Using Eight Robots (Courtesy of Monforte Robotics, Inc.).

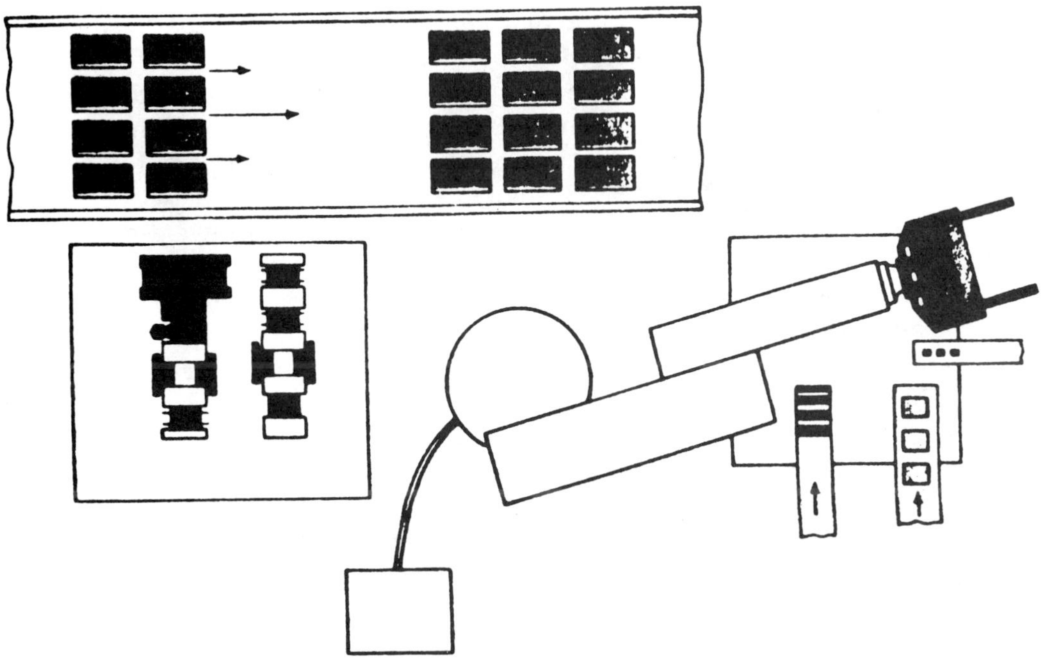

FIGURE 6–24 Alternate Assembly Method Using Only Three Robots (Courtesy of Monforte Robotics, Inc.).

at $18,000 each, or $36,000. Eight grippers at $1500 each would amount to $12,000. The grand total is $499,000 in equipment–acquisition costs and first-year salaries.

In contrast, the three tool-changing robot line would require only a $141,000 expenditure for the $47,000 robots, one engineer at $25,000, one technician at $18,000, and three $14,000 grippers at $42,000. The grand total of this configuration is $226,000, or less than half of the conventional robotic assembly system.

The task-dedicated robot line could produce 240 calculators per hour, while the tool-changer robot system would produce 308 assemblies each hour. Part of that disparity is based on the estimated average downtime. If one robot breakdown or a part feeder gets stuck on the dedicated-tooling line, the entire flow stops until a worker stands in for the machine that is down. Whereas, the tool-changer line would continue to produce at two-thirds efficiency.

*Innovations to Watch: Specific Application versus Sensor Integration Current Trends in Robotic Technology**

Although application of robotic technology appears unlimited, two distinct trends are evident. Companies are either installing specialized robots that perform a specific application, or they are moving toward sensor integration. In some instances a combination of both approaches is being employed.

Specialization offers the advantage of accelerating a company's movement into advanced technology through enhanced robot capability. For example, in a minimum period of time, a company can install a specialized arc-welding system.

These systems have gained acceptance in industrial plants because they have demonstrated an ability to increase productivity in harsh work environments.

As an example, American Cimflex Corporation has developed the Merlin™ arc welding robot with integrated seam tracking for arc-welding applications. The robot's six axes give it the flexibility and circular motion ability of an expert arc welder. The versatile three-axis wrist enables Merlin to reach into the tightest corners and still maintain the correct welding orientation between the arc and the weld joint.

Problems can arise in an arc-welding environment due to the harsh conditions created by the application. Seam tracking, an important development in arc welding, can be adversely affected by fumes, smoke, and spatter created by the arc-welding process. For this reason, American Robot has developed an arc-welding system, integrated with a proprietary seam tracker developed by Cyclomatic Industries, that is sturdy enough to operate in severe thermal, mechanical, and electrical conditions.

The system utilizes a joystick with a teach pendant, which is used to teach the weld path, and thus eliminate tedious point-to-point programming. The seam tracking system then corrects for any weld joint deviations encountered later in the welding process. Customer tests have shown Merlin's repeatability to be ± 0.001 inches and straight line deviation at welding speeds to be better than ± 0.010 inches.

Also, the power of the Merlin Universal Controller allows a user to program multiple weld schedules via the teach pendant to change settings for different weld joints and out-of-position welding. By doing this, one can create a complete flexible welding cell that operates in the harshest conditions, and withstands the general abuse usually encountered in a welding shop.

CASE-IN-POINT

MAJOR SUPPLIER OF DISK-HANDLING EQUIPMENT

Situation

A major supplier of disk-handling products was concerned that scrap rates can skyrocket in disk-handling applications due to the large amount of human contamination and error involved when the operation is performed by people. After a detailed study of this important activity, the company decided to prevent this condition from occurring by integrating four intelligent robots into the manufacturing process.

*Permission granted by American Cimflex

Solution

In the application, two intelligent robots are mounted on twenty-five-foot driven tracks and integrated to the controller. These robots perform precision disk-handling tasks at multiple disk-testing stations in a clean room environment. A conveyor feeds disks to the proper robot for testing and assembling disks on to spindle assemblies. While the robots are doing this, the controllers interface to the factory programmable controllers monitoring the position of the parts in the manufacturing process, and exchange information with the factory floor data collection system. In addition, the two other robots perform disk assembly functions, which require high precision and repeatability.

Results

It has been estimated that the robots reduced the scrap rate by 50 percent. This resulted in a significant cost savings, and therefore paid for themselves within a year. All this was accomplished without a reduction in the work force.

By incorporating the robots into the manufacturing process, the company has been able to solve many of the problems associated with manufacturing disk drives by virtually eliminating scrap generated by human error.

HOW THE THREE-DIMENSIONAL MACHINE VISION ARC WELDING PROCESS COUPLES ROBOTICS AND VISION SYSTEMS

Another technology involves three-dimensional machine vision systems designed specifically to fulfill the demand to automate the arc welding process. These systems are comprised of an advanced vision sensor, a vision processor, an individual robot, and specialized welding equipment.

In operation, the vision sensor, mounted on the robot arm, scans the seam to be welded and generates a three-dimensional image of the seam within the vision processor memory. Based on the seam's geometry and a preprogrammed weld path, the vision processor generates a weld sequence suited to the viewed seam. This new sequence includes adjustments in velocity, voltage, wire feed rate, and/or weave amplitude required to adapt the weld process to the new requirements of the current workpiece.

Advantages of This Technology

The advantages of this system include:

- Compensation for ambiguities in the position of the workpiece, up to ± 1 inch
- Compensation for variations in part fit-up
- Suitability for multipass welding processes
- Calculation of seam width and cross-sectional area to adaptively control weld parameters
- Built-in diagnostics capability to minimize system down time and maintenance

Examples of Vision Systems Applications

Arc welding is only one type of system employing vision sensors. The following examples are indicative of the variety of applications that vision technology embraces.

Spray Painting. This application is very difficult because the robot has to be able to resort to touch sensing, by touching a couple of locations on the product, and then relaying the data to the computer which then calculates whether the arm is positioned properly or not. As shown in figure 6-5, GMF Robotics has developed the P-150 electronic paint robot to solve this problem.

Inspecting Almonds. This unique application involves inspecting almonds with a vision system. The purpose is to inspect the shelled almonds for blemishes or chips, and grade them into levels of quality. The fewest chips become premier grade. The almonds are punched down to a fibratory feeder conveyor, then a snapshot is taken using a strobe light, and the almond analyzed for grading.

Cleaning the Undercarriage of Rail Cars. This application involves using a sensor to locate dirt and grime in the undercarriage of rail cars, and then automatically removing the matter with solvent robot directed and activated.

Inspection of Sutures. This application was developed to ensure that sutures were surgically safe. The need for this system is obvious: If a needle is inserted into someone and a chip of steel falls off and enters the bloodstream and lodges in the brain, the result can be catastrophic. Hence, the need to inspect needles automatically to minimize or eliminate human error, is an operation where thousands of needles have to be inspected each day.

CASE-IN-POINT

*VISION SENSING APPLICATION FISHER BODY DIVISION OF GENERAL MOTORS**

Situation

The Fisher Body Division of General Motors was manufacturing door frame reinforcements for midsize cars manually. Management decided that the operation should be automated to increase production rates even though the planned change would slow line speed.

Solution

The line was retrofitted to incorporate ten robots, eight positioning fixtures, and special sensors. Although line output was cut from 290 parts per hour to 275 parts per hour to accommodate a limitation of robot speed, the slower line speed, however, was more than offset by extended uptime.

The Fisher Body press line is tended by GMF cylindrical coordinate M-1A robots. (Refer to figure 6-9.) It operates on a two or three shift basis, and has the capability for unmanned operation.

Blanks are presented to the first robot in the 140-foot long line by an indexing conveyor equipped with twelve pallets that hold approximately one hundred pieces each.

The first robot, equipped with vacuum grippers, senses the top of the stock with a limit switch while a fiber optic sensor detects whether there are any remaining blanks on the pallet. If there are none, another pallet indexes the position.

As the first robot turns to load the forming press, it passes the blank through a sensing device mounted near the die opening. This sensor verifies that there is only one blank. Upon confirmation, the first robot proceeds to load the blank into a draw die which makes the basic form for the part.

Robot 2 unloads the part and immediately turns to load a gap press for restrike. It then removes the part and places it on the first staging fixture so that the third robot can load and unload the next press. Robots 4 through 9 repeat this procedure, transferring the part through trim and pierce presses.

*Permission granted by GMF Robotics

The last staging fixture collects four finished parts before a signal is sent to the tenth robot for palletizing. It also incorporates a fiber optic sensor to check the quality of the first part placed on the stand, eliminating the need for visual part inspection.

With mechanical grippers, the last M-1A robot grasps the stack of four parts and palletizes them in shipping containers. Standard GMF palletizing software gives Fisher Body several pattern alternatives. Using vacuum grippers, this robot also separates layers of parts with cardboard sheets. When one basket is full, a flashing red light alerts the operator, and the robot begins palletizing in the empty alternate container.

The staging fixtures are integral elements in the automated press line. Equipped with 90- or 180-degree rotational axes, they position workpieces so that each robot will load parts into the dies in the correct attitude. Without this intermediate orientation, each robot would have to have additional axes.

To ensure maximum uptime, the production line is monitored continuously by a supervisory PLC. This unit also serves as the diagnostics system and identifies and locates faults in presses, fixtures, and robots.

Another innovation for keeping productivity high is a unique method for changing dies. Production runs last about two weeks. Eight of the robots are mounted on a track. Before beginning a new run, robots can be moved aside by removing four bolts and two locating pins. One operator can then easily push the robot aside to make room for die changes.

Results

As a result of replacing a manual operation with a completely integrated robot production line, Fisher Body was able to obtain a payback within three years on the robots and ancillary equipment. Not only was the payback period within acceptable financial investment return, but the company can fit the line with new tooling, reprogram it, and use it on another job.

FEATURES TO CONSIDER WHEN INSTALLING A ROBOTIC SEALANT DISPENSING SYSTEM*

Another area of interest of many industrial concerns is robotic sealant dispensing systems. These systems have been able to demonstrate tangible gains in improved quality, increased production flexibility, lower manufacturing costs, and better working environments.

In any robotic dispensing system, there are a large number of system characteristics that have to be analyzed, the proper integration of which yields an optimized dispensing system. Major characteristics include considerations such as type of robotic manipulator, cell control system, process material and equipment, vision guidance, and system integration.

Robot Manipulators: The Need for Six Axes of Motion

In the majority of applications for adhesive and sealant dispensing, the robot manipulator has six axes of motion. This is because in most dispensing jobs it is not only sufficient to maintain an accurate seam path—which a five axes system can do, but also maintain the orientation of the bead or spray with respect to the seam. There are, however, a few simple applications that can be accomplished with as little as three axes of motion, such as those available on small assembly robots.

Typically, it has been found that in automotive, appliance, and aerospace manufacturing plants, articulated type robots satisfy a large need followed by gantry-type robots. Variations of these, such as articulated robot on a track or mounted overhead and dual bridge

*Permission granted by American Cimflex

gantries are tailored to meet specific customer requirements which are driven by issues such as performance, pricing, and flexibility.

For example, in the case of applying adhesive to firmly mounted surface devices to a printed circuit card, a six-axes articulated type robot is required. It can be floor or ceiling mounted, depending on issues such as floor space availability, access, work envelope, and so on. In the case of applying adhesive to the headers for bonding the roof to the car body, the preferred choice has been the gantry robot because it can access all parts of the roof using one robot.

Two Common Cell Control System Options

Depending on the size and complexity of the robotic dispensing system, two general types of cell control are typically implemented.

For small size or simple systems, the cell control function is implemented using the robot controller. Here, the input/output requirements to sensors, actuators, and peripheral devices are small and are limited to less than sixty-four points.

For multirobot systems, or for single-robot systems with complex control requirements, the cell controller is typically a programmable controller or other factory computer. The cell controller not only coordinates and synchronizes robot and dispensing equipment, but also controls transport devices, fixturing, safety devices, operator interface, and communication to other computer devices on the plant floor.

An example of a simple system is an adhesive application where a variety of printed circuit boards (PCB) required adhesive to bond surface mounted devices to the board. The control architecture consists of the robot controller being the cell controller and communication to a bar code reader through a communications link to obtain the type of PCB fixtured; and to dispensing equipment, sensors on fixtures, operator panel, and safety lockout devices through optically isolated discrete I/O.

For a larger adhesive system consisting of two gantry-type robots, transport devices, tactile sensors, dispensing equipment, interfacing to the assembly line, and so on, the central cell coordinator is a programmable controller that communicates with various devices using discrete I/O. In addition, the programmable controllers select the program the robots execute, depending on the model type, and schedule one robot to perform all of the operations in case the other robot is taken off-line for preventative maintenance, replacement of supply material, or cleaning the adhesive gun.

Process Material

Integrating a robot into the dispensing process often requires a detailed evaluation of the material to be dispensed robotically. When it comes to evaluating the adaptability of the material to be dispensed by a robot, the primary concerns are the metering, mixing, and dispensing. Single-part materials that have medium viscosities—that is, viscosity like typical greases, are relatively easy to meter and dispense accurately. These tend to be the applications that are robotized first and include single-part urethanes, adhesives, and hot melts.

In the case of two-part adhesives that have large variations in viscosity, for example, where the epoxy has the consistency of peanut butter, and the catalyst is like water, the problem of mixing, metering, and dispensing accurately can get quite involved and requires extensive experimentation on the part of the system supplier to attain the correct set of equipment to provide a quality bead.

Another example can involve a two-part adhesive system, which is being dispensed manually, but has to be changed to a one-part heat curing adhesive system before it can be robotized. The key problems are accurate ratioing and mixing. A static mixer and a dynamic mixer cannot be used because the size and vibration it produces requires it to be mounted away from the robot. This distant mounting causes the adhesive to cure before it can be delivered to the part.

Dispensing Equipment

In the past, most dispensing process equipment has been designed for manual or automatic operation. Integrating the dispensing equipment with a robot adds a new dimension of control and capability that is presently being addressed by process equipment suppliers and robot vendors. Issues include flow control, monitoring, and packaging.

Most previous manual and automatic systems were set up with one flow rate that could be manually adjusted during start-up. Then, whenever major process parameters changed, such as wear on the metering pump or change to input pressure, minor adjustments made occasionally satisfies the consistent flow control requirements. Currently, with robots being used in the system, it is required that the robot control system adjust the dispensing rate according to its speed, which varies significantly depending on whether it is a simple straight path or a complex 3D seam with orientation changes.

As this book went to press, very few robot systems offered the capability of referencing the instantaneous or programmed tool speed to flow rate that is a requirement to meet the cycle time and quality demands of typical automotive applications.

For example, systems that have been implemented show clearly that for straight line paths, robots deposit a correct bead; but for complex seams, the bead size increases in proportion to the decrease in speed as robots maneuver through complex paths. In these cases, the systems have been accepted as they produced better quality than manual systems.

An important consideration for dispensing is the capability to monitor accurately the flow of material being dispensed mainly for quality control and material savings reasons. Some problems that are encountered include air bubbles that result in a gap in bead path, increase or decrease in the flow either due to wear on gear metering pumps, input pressure, leaks between metering pump and dispensing gun, or slow down or speed up of the metering pump due to electrical malfunction.

Some monitoring systems that have been installed in earlier systems have had to be disconnected as they tended to be overly sensitive in some cases and missed other problems, thus causing unnecessary production slowdown.

Similar to the trends the welding equipment industry went through in the early 1980s to satisfy robotic welding needs, dispensing equipment has followed in its path. Early dispensing systems tended to be bulky with very little controls capabilities other than some simple on/off features. Instead, the various robot and dispensing equipment suppliers are coordinating their effort to develop dispensing equipment with more controls and flexibility and compact packaging.

How Vision Guidance Reduces Problems with Robotic Dispensing Systems

In any manufacturing process, you must deal with part tolerances, tolerance stack-up, and part fixturing issues. For manual dispensing systems, human operators adapt to a vast array of inaccuracies and inconsistencies. However, when trying to robotize, blind robots expect parts to be consistent and to be fixtured accurately. Generally, this is not possible because of the high cost associated with major process changes to produce consistent parts or to provide accurate fixturing. This results in three types of problems that robots have to deal with, namely:

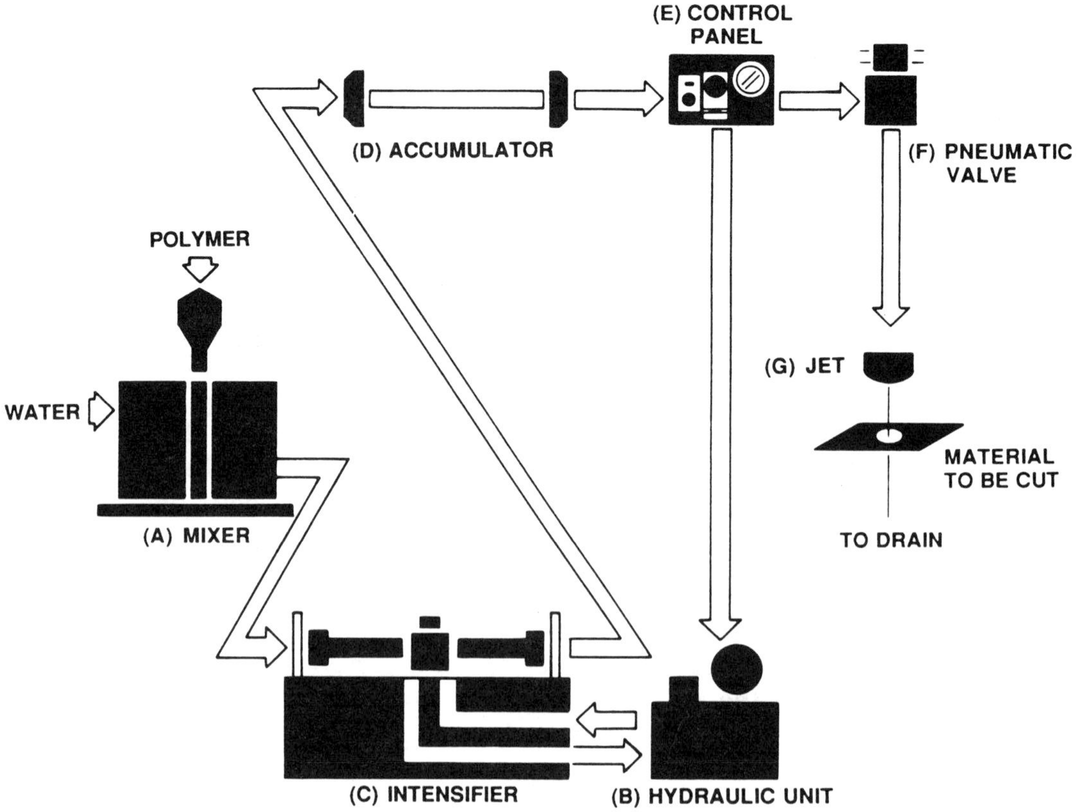

FIGURE 6–25 Waterjet Equipment (Courtesy of Ingersoll-Rand Inc.).

mechanical relief valve. An emergency high pressure safety dump valve is used to relieve system pressure when an "emergency stop" button is pushed (figure 6-26).

With a .007-inch orifice, the power required is roughly 10HP. Waterjet is designed to be integrated and/or retrofitted into most current machine manipulators. A portable intensifier has been developed for transporting with other cutting system components for on-site field work.

Water is pumped from the intensifier through a high grade sapphire nozzle at more than twice the speed of sound. The powerful combination of concentrated force upon a very small area—diameters ranging from .020 to .003 inches—is what produces the superior cutting activity.

Nozzle orifice and pressure are determined by the type and thickness of materials being cut. Softer ones are easier and faster to work than hard metals. The size and speed of the cut is affected directly by the diameter of the waterjet nozzle, as well as the pressure applied to the cutting surface. Speeds are also related to the nozzle's angle, orientation, and its actual distance from the material.

Why a Waterjet Cuts Smaller Than Traditional Systems

Because of the waterjet's small diameter, the cut made by waterjet cutting creates a much smaller kerf than traditional cutting. For example, a typical kerf is nominally in the .005- to .011-inch range, depending on the nozzle diameter.

In addition, the cut is omnidirectional, and needs no starting hole or tooling changes between operations. No matter how difficult or how contoured the cut, waterjet makes it possible.

Water pressure is far different from the pressure exerted in ordinary cutting operations, so the waterjet process results in far less waste and deformation on material. With waterjet cutting, there is little or no fraying. No heat affected zones are created, and there is no clogging.

In addition, since the cutting is produced by an easily manipulated, steady stream of water, it maintains high accuracy. Its precise omnidirectional cutting capability allows it to complete intricate patterns in only minutes. Unlike a die cutting operation, waterjet eliminates the need for costly, time-consuming retooling when switching from one material to another.

A major automotive company's hardware trim plant offers a good example of waterjet's superior cutting capabilities at work. The company places high emphasis on finished component appearance, as well as economy and automation. For use with six ASI five-axes robots that trim thermoformed wood fiber substrate interior door panels, the company selected waterjet over conventional cutting systems, such as steel rule cutters, which produced rougher edges than waterjet.

FIGURE 6-26 Waterjet Cutting System—Ultra High Pressure (Courtesy of Ingersoll-Rand Inc.).

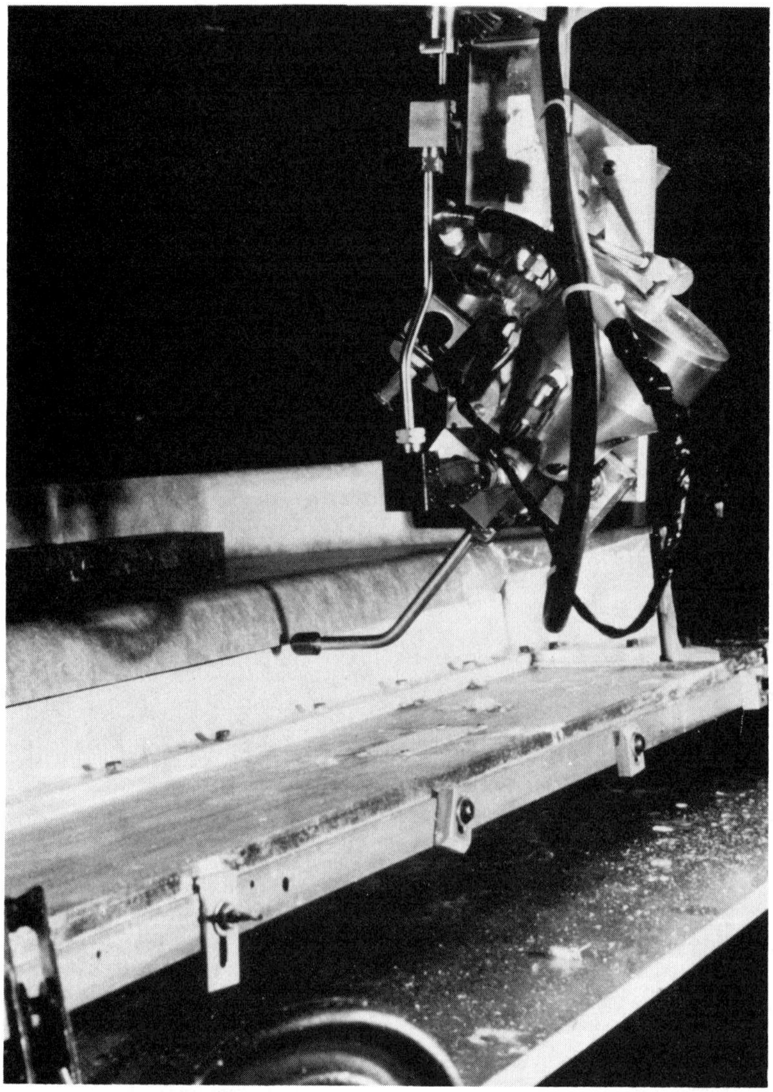

FIGURE 6-27 Waterjet Nozzle Interface with Robot (Courtesy of Ingersoll-Rand Inc.).

In addition, waterjet was more attractive to the company because the nozzles easily interfaced with robots, are omnidirectional, have no side loads, and do not require retooling for separate cutting tasks (figure 6-27). Each robot is "armed" with a waterjet cutting system incorporating a 0.008-inch nozzle that cuts into the wood material at a minimum of 800 inches per minute. The robots perform all needed periphery cutting functions on the door, including the hole insertions for handles, locks, and lamp openings (figure 6-28). The system's 45,000 psi pressure requirements are supplied by six intensifiers—one for each gantry station holding two robots.

Time-Saving Benefits of Waterjet Systems

Because it saves time in several important ways, waterjet cutting technology has increased productivity in the manufacturing of a great number of products employing diverse

FIGURE 6–28 Waterjet Robotics Cutting Operations (Courtesy of Ingersoll-Rand Inc.).

materials, including titanium, steel, aluminum, concrete, Kevlar®, Nomex®, ABS graphite composites, polycarbonate and rubber, and ranging from presintered ceramics to various bakery products.

In many instances, waterjet cutting is much faster than conventional methods, so overall production time is reduced. The time devoted to the cutting operation also is lessened because downtime for tool change is eliminated. With waterjet cutting, there are no blades, knives, or saws to sharpen, replace, or clean.

General equipment maintenance is easier and less expensive with waterjet cutting than many other methods. In addition, because it cuts so sharply, waterjet cutting creates no dust or other material residue. With waterjet cutting there is no need to clean up dust in the environment. This is especially important when working with hazardous materials, which would otherwise necessitate costly and even dangerous dust cleanup.

How Waterjet Cutting Reduces Environmental Concerns

Waterjet cutting technology offers a means of solving a number of critical environmental problems that affect safety and productivity.

First, waterjet cutting virtually eliminates dust and leaves no wasted or spilled material and few scraps.

Waterjet equipment creates less noise than most cutting systems, and because it is relatively compact, it can be sited remotely to reduce noise emissions when necessary.

A waterjet can be started or stopped instantly, eliminating the potentially dangerous time interval while a conventional cutting instrument coasts down to a stop.

Because the nozzle is positioned immediately above the material to be cut, it can be guarded by a fence, electronic beam, magnetic field, or other device. Depending on the material being cut, this feature can be vital.

A waterjet cutting installation at an automobile company's plant where brake-shoe linings are fabricated, illustrates several safety features of this technology. Six adjustable waterjet nozzles slice conveyor-carried asbestos strips to specified lengths. Not only does the system eliminate the major problem of airborne asbestos dust, but in addition, the jets are enclosed within a plexiglass shield, and if the shield is removed, the system automatically shuts down. The systems, which cut 1½-inch linings, also minimize kerf (to 0.010 inch) while increasing cutting efficiency an estimated 30 to 50 percent, at an estimated $25,000 annualized savings.

How Hydrobrasive Systems Work

With the addition of an abrasive system, waterjet can be used much like a laser. Its abrasive effect is the same as would be produced with a 40-hp grinding wheel, $1/16$-inch wide, 8 inches in diameter, running at 25,000 rpm—which is neither safe nor possible.

The resultant abrasive erosion is strong enough to cut, trim, slit, deburr, and drill metals, ceramics, and composites without altering their physical characteristics in any way. Metal alloys, such as titanium, which experience heat affected zones and/or a poor surface finish when cut by handsaws, plasma arc, or lasers, show no deformation or distortion when using abrasive waterjet cutting. With abrasives, waterjet can cut steel over three inches thick, as well as concrete up to 12 inches thick. It machines many types of metal, composites, and nonmetals in aerospace and other metal-shaping industries.

Various kinds, sizes, and amounts of abrasives, such as garnet and silica sand, are used depending on the finish desired, cutting rate requirements, and the economics of the job.

The abrasives are added to the waterjet nozzle after the intensifier pump pushes water through the orifice at great pressure. This makes a waterjet nozzle powerful enough to cut through $1/8$-inch titanium at 18 inches per minute, or 10-inch reinforced concrete at speeds of more than 1 inch per minute. One-inch-thick glass plate can be cut at 20 inches per minute without cracking. Heavy marble and quartzite slabs, ranging in thickness from 2 to 5 inches, can be cut at rates to 16 inches per minute.

Hydrobrasive nozzles can be retrofitted easily to CNC robots, X-Y table positioners, track-mounted routers, mills, lathes, large drill operations, concrete cutting and scarifying equipment and profile torch cutters.

An example of how abrasive waterjet cutting can result in significant time and cost savings is demonstrated by an application at a large aerospace firm: The company purchased a waterjet cutting system equipped with a hydrobrasive nozzle to cut titanium and other hard metals into bomber components. It has two different cutting stations: a manual pin-routing setup for cutting curved and compound contour components, and a CNC X-Y positioning table for cutting flat sheets of titanium and other high technology materials.

Compared with the previous cutting method, the company estimates production cost savings of up to 50 percent. Formerly, manufacturing titanium sheetmetal components involved inking the surface, scribing, hand-sawing, and filing to the trim line. The waterjet system completely eliminates inking and scribing and uses various nozzle sizes, depending on the material being cut, its thickness, the type and amount of abrasive being used and the finish desired. The company achieves a cutting rate of 12 inches per minute for 0.063-inch titanium using about 1.5 pounds per minute of #60 grit-red garnet abrasive.

Specialized Applications of Waterjet Cutting

Some special features of waterjet cutting that have had considerable impact on various applications are:

- *Robotics*—provides higher-speed cutting, easily interfaces with robots, is omnidirectional, has a lightweight nozzle with no side loads, and requires no tooling changes for separate kinds of cutting activity.
- *Aerospace*—eliminates dust, provides a much better edge on many composite materials, and performs at high speed. In addition, because of its edge quality and no delamination with hydrobrasives on bonded metals and composite, it is particularly attractive for a variety of aerospace applications.
- *Automotive*—the combination of speed, edge quality, and soft automation makes waterjet most suited for numerous applications like dashboards, headliners, and door panels.
- *Building Materials*—waterjet cutting systems offer fiberglass insulation processing operations reduced waste, no dust, and increased productivity by cutting with water. Using high-speed jets of water, waterjet also cuts virtually any ceiling board material cleanly with minimal loss. It jets through corrugated boards with different densities without crushing, shattering, or compressing the product.
- *Food Processing*—a major benefit is that no clean-up is necessary on animal fat, cake frosting, or pie filling from the cutting blade. Waterjet reduces waste and increases productivity by cutting with water. It jets through foods with different densities without crushing, shattering, or compressing the product.
- *Electronic/electrical*—the combination of speed, edge quality, precise omnidirectional cutting capability, and clean operation makes waterjet cutting systems excellent for numerous circuit board production applications, such as rigid and flexible circuit boards, component housings, and boards with surface-mounted components. Waterjet eliminates prerouting and tabs, dissipates static, and optimizes board nesting.

In addition to these industrial applications, other possible uses of waterjet cutting technology range from bone surgery—where waterjet cutting eliminates heat damage friction—to mining.

CASE-IN-POINT

WATERJET CUTTING TECHNOLOGY
*CHEVROLET ADRIAN, MICHIGAN PLANT**

Situation

Chevrolet CK-body pickup truck sales had shot up dramatically, and as a result the company needed to increase component production to meet this accelerated demand. One of these areas involved increasing the output of plastic fuel tank shields.

Solution

After analyzing various alternative approaches, the company decided to install a flexible automation system for manufacturing plastic fuel tank shields. A unique system was developed, integrating two state-of-the-art technologies, programmable robots, and waterjet cutting.

The line designed by Chevrolet consists of four GMF Robotics Inc. Model S-360 robots, two of which are equipped with routers to cut holes and slots in the part, and two of which are equipped with waterjets to trim the periphery of the shields.

*Permission granted by GMF Robotics

Initially, all four robots were equipped with waterjet cutters, but the edges of the holes and slots produced with the jets in the first cutting station were too rough. Switching to routers solved the problem. A smooth edge along the part perimeter was not as critical, hence the waterjet continued to perform the trimming operation (figure 6-25).

At Station 1 in the line, 195-inch thick polyethylene thermo-formed sheets are secured by clamp rails to provide a sturdy fixture for routing twelve holes and slots in the part. Specified hole tolerances are maintained because of the ±.020-inch repeatability of the S-360 robot.

At the next station, two S-360s tooled with waterjet cutters trim the periphery of the tank shields and divide the workpiece into two finished parts by directing a stream of water through a .008-inch diameter sapphire nozzle at 55,000 psi. Maximum cutting rate, determined by the diameter of the nozzle orifice, is ten inches per second.

The robots maintain accuracy in keeping the waterjet nozzles at least $^1/_2$-inch from the cutting surface. A jet catcher collects the water after it has penetrated the polyethylene. This device allows the jet to dissipate, reducing noise levels and safety hazards.

At the cycle's completion, the transfer rails the index forward to drop the two finished shields, with the remaining offal, onto a belt conveyor. An operator dusts the shields and stacks them in shipping baskets. The offal lands on a scrap pallet and is eventually recycled.

The four robots in each line work in such close proximity to one another that their work envelopes overlap. Collisions are avoided by special software interlocks and a supervisory PLC that continuously compares positional data from each robot controller. Should a robot violate its work area parameters, all robots in the system stop, and an appropriate "pull-out" program keeps the robots from colliding while returning to their "home" positions.

Results

In the project's cost savings justification, the robotic cutting system was compared with traditional hard automation consisting of a line of stamping presses coupled with either automated transfer devices or manual handling. Such systems would have required regular die maintenance, considerable floor space for presses and transfer equipment, and frequent die changes.

Compared with the cost of presses, dies, and automation, the initial investment for robots was calculated to be lower. Utilizing routers and waterjet cutters eliminated the need to frequently resharpen and replace trim and pieces dies. And, most importantly, laborpower requirements were reduced from five people for a manual press line to one person for the robot line.

Another factor in favor of robots was their flexibility to accommodate part changes and changeovers. The controller's memory contains four different cutting programs. This storage capacity allows the company to change programs instantaneously from left-to-right-hand versions of the sixteen- or twenty-gallon tank shield, whenever production schedules dictate. If an entire product change occurs, the robots and cutting equipment can be reprogrammed easily to make the new part.

In summary, the robot line was able to produce parts in six months from the beginning of project engineering to the start of production. A major reason for the robotized line's success was that Chevrolet sent employees to robot school, thermoforming school, and waterjet cutting classes before the equipment ever entered the plant. This led to a great degree of anticipation—and efficiency—among the employees. As a result of this technological advance the Adrian plant was able to not only reduce initial labor costs, but by installing a second twin line, created the capability to triple production and become the prime supplier of fuel tank shields for all of GM's North American operations.

WHY MATERIAL HANDLING AND LOAD/UNLOAD TECHNOLOGIES ARE THE BASIS FOR ROBOTIC APPLICATIONS

In the final analysis, material handling and load/unload technologies, in one form or another, are the basis for virtually all robotic applications. The primary function of any robot is to move an object, be it a tool, inspection device, or workpiece, from one point in space to another.

The definition of material handling is: applications in industries other than electronics involving only movement or handling of material, exclusive of assembly.

Material handling may be further defined as follows:

- Indirect interface between automation and equipment
- Involving transfer, containerization or packaging of pieces/parts
- Processing of the parts through the cell does not physically change individual shape or size
- Usually the process is effectuated with hard automation equipment due to high volumes and rapid throughputs

The definition of load/unload is: applications in industries other than electronics where the robot is tending, or loading and unloading a machine or process.

Load/unload may be further defined as follows:

- Direct interface between automation and machine
- Involving tending equipment or machines where the part changes shape or size through the effect of the process
- Sometimes facilitates operations within a large flexible automation system or manufacturing unit

What the Process Is

Material handling and load/unload processes may be broken down into these five tasks:

- Locating the workpiece to be moved
- Grasping the workpiece
- Moving the workpiece through a prescribed path
- Orienting the workpiece
- Depositing the workpiece in a prescribed location

What to Consider When Designing a System

Material handling and load/unload, though comprised of a series of simple tasks, involves some subtle considerations, such as:

- Locating the workpiece (it may not be presented with a known location and orientation)

- Moving the workpiece
  - (a) Weight of workpiece
  - (b) Static moment
  - (c) Inertia
  - (d) Desired path
  - (e) Velocity
  - (f) Acceleration and deceleration
- Geometry of the workpiece
  - (a) Allowable grip point
  - (b) Method of gripping
- Initial and terminal workpiece position
  - (a) Clearances
  - (b) Workpiece orientation

Types of Parts Commonly Processed by Robots

Applications of both single robot cells and multiple robot systems will be required to handle piece parts in the following processes:

- Loading/unloading of special machines
- Broaching equipment
- Grinding equipment
- Robots as integral parts or machining centers
- Metalforming (transfer or progressive press applications)
- Loading/unloading of test, inspection, and gauge equipment
- Turning equipment
- Die casting
- Investment casting
- Injection molding
- Forging
- EDM/ECM/ECD machines
- Material handling to and from conveyance equipment
- Box handling, palletizing, product pack-out to boxes

CASE-IN-POINT

MATERIAL HANDLING AND LOAD/UNLOAD TECHNOLOGIES
*PRATT & WHITNEY**

Situation

The Pratt & Whitney installation, although a load/unload job, was complicated by the workpiece configuration, number of machining stations required, and the variety of compressor blades to be processed.

*Permission granted by GMF Robotics

Compressor blades have a complex geometry with virtually no flat surfaces for fixturing and robot handling. To overcome this problem, blades are encapsulated in a soft metal matrix prior to machining. This matrix provides the flat surfaces necessary for handling and fixturing, and also dampens vibration during machining. The metal matrix is removed following machining to permit subsequent processing.

The size of the broaching machine presented another problem. It involved six machining stations. To ensure uninterrupted operation, it was necessary to design four queue stations between the broaching stations to give the robot sufficient time to reorient parts before transfer. In operation, a robot unloads a part from Station 1 to Queue 1, then loads Station 2 and so forth until the compressor blade is completely machined.

The queue stations are mounted on the broaching machine but designed to pivot out and away to accommodate tool changes. For the same reason, the robots are mounted on mobile dollies.

The need to handle sixteen different size compressor blades further complicate system design. After being encapsulated in soft metal, workpieces ranged from a $1\frac{1}{2}''$ cube up to a $9'' \times 3'' \times 2''$ rectangle and weighed from 8 ounces to 14 pounds.

Solution

After a thorough analysis of various alternative methods, Pratt & Whitney specified a GMF S-360R robot for the broach loading application. A number of smaller GMF robots had the workpiece capacity and reach to handle the job. The S-360R, however, is a standard unit with a 3-axes wrist—a requirement for this application. It also was a rugged proven performer, which would minimize any start-up problems.

A special pneumatic parallel-type gripper with interchangeable fingers for grasping the workpiece was used. The fingers compress the encapsulating soft metal matrix to ensure parts are seated securely when placed in machining fixtures. This design proved effective for all size compressor blades handled, and contributed to the ease of part changeover.

Because of the number of different parts involved, programming the S-360R robot was reduced to a simple 3-axes positioning job by setting up a simple X, Y, Z hand coordinate system relative to the angular face of the fixture, and using the robot's remote teach pendant. The machine operator simply presses one of three buttons on the pendant X, Y, or Z to manipulate the robot to the desired position. This allows the operator to establish program coordinates while working in only 3-axes instead of the normal 6-axes.

Software logic was developed to accommodate all sixteen compressor blade configurations. These programs are stored in the robot controller's magnetic bubble memory and are automatically called up whenever a part change is scheduled. This facilitates programming of the new part or touching up the position data of an existing part program.

Results

The major advantage of applying robots to a broaching operation is increased productivity. Although an operator can load/unload parts faster than a robot so throughput is not necessarily increased since broaching is normally a long run operation, it will operate continuously without frequent tool change as long as there is an adequate supply of parts. This continuous operation offsets load/unload time.

This robot broaching operation serves as a model to future industry applications because it involves more complex parts, part variety, machine size, fixturing problems, and workpiece changeover requirements than most other broaching applications.

UNDERSTANDING THE CAPABILITIES AND LIMITATIONS OF ROBOT SOFTWARE*

Advances in robotic technology have been supported by corresponding developments in robot language and software. These language and software developments can be categorized based on a hierarchy of complexity, as follows:

- Standard teaching robots are at the beginning of the hierarchy followed by the graphical teach programming. First generation manipulator languages are extensions of the BASIC language that support robot motion and control.
- The second generation manipulator languages are more powerful languages based on advanced, structured computer science languages such as Pascal.
- Lastly, research is progressing on task or object-oriented languages in which the language commands correspond to the commands one would give a person carrying out a similar task.

Teaching Robots: A Choice between Two Systems

The vast majority of robots in factories around the world today are programmed by one of two methods of operator sequencing, lead through programming, or teach box programming. A robot program is a path in space composed of multiple points. At each point in the program, an operator indicates whether the robot is to stop or pass through the point. In addition, the operator has the ability to instruct the robot to wait for a signal at any point or to generate a signal at such a point. In lead through programming methodology, the operator physically moves the robot through the path in space. While the robot is moving, a computer controller samples feedback signals from the actuators at each of the robot's joints to record the robot program. In teach box programming the operator positions the robot using buttons or toggle switches on a teach pendant, and indicates when the robot has reached a point that is to be entered into the program.

Both methods of operator sequencing have been augmented by vendor-supplied software, sometimes called blackbox software, to ease the burden of teaching locations and to help in the editing of programs. However, such capabilities must be considered rudimentary at best. Programs taught by the lead through method are almost impossible to edit because of the large number of points closely spaced in time, which make up the program. Thus, it is frequently the case that one must break the program into multiple segments such that each segment is taught as a separate sequence of points. The segments are then concatenated to form a complete program. Editing then requires that only individual sequences be retaught to replace the faulty program segments.

Operators using teach box methodology frequently have access to a terminal or keyboard arrangement which facilitates the use of blackbox software. This software is organized for nonprogrammers and accepts parameters in a simple format. Blackbox software includes a collection of routines commonly used in robot programming. Typical routines may ease palletizing operations or facilitate the use of binary sensors in tasks such as finding the top of a stack. Other routines provide for simple error recovery and permit the definition of relocatable program segments. The last category is different from the relocatable subroutines found in standard computer programming languages because a relocatable program in a robot is a series of motions that can be replayed at any point in space. The importance of being able to

*Based on "An Analysis of Robot Software and Plans for Its Enhancements," Technical Report #85-001 by Dr. Roger N. Nagel, MSE, June, 1985. Used with permission.

teach relocatable program segments can be easily seen in a repetitive operation such as palletizing.

Most factory robots are programmed today with augmented teach box software. The programming capability is supplemented by simple operating systems that allow storage and retrieval programs and performance of rudimentary editing operations. This methodology is used successfully to program a wide variety of applications such as machine loading, spot welding, and painting (as discussed previously). However, since each vendor has unique human interface and blackbox software routines, there are currently no standards for the representation of robot programs developed in this manner. Thus, it is difficult to create a generalized robot programming capability, even though generic robot programming elements can be identified clearly. The lack of standards goes beyond program representation. Most robots have an RS232 communications port on their controller; but there are no standards for the information that might flow across that interface. The quest for standards is a recurrent research theme at each level of the hierarchy.

The Graphics Teach Method—A Technique from CAD/CAM

Many vendors of CAD equipment have begun to realize that their graphical method of programming NC machine tools has some application to the programming of a robot. As a result, they have developed and introduced specialized graphic robot programming systems to the commercial market. These commercial offerings are being supplemented by a large number of research efforts at universities around the world. Such a graphics system begins with a library of three-dimensional robot models which includes the geometry of each robot, along with a model of the constraints on the robot's motion. The models also contain approximations of the control algorithm for each robot. At a typical session, the user selects a robot from the library and models a work cell by using an additional library of other devices, such as conveyors and machine tools, typically found on the manufacturing floor.

The user sits at a graphics console and uses a combination of alphanumeric input and a joystick (or light pen) to indicate placement of the parts and components which make up the work cell. The user also uses the graphics console as the method of programming the robot with a generic teach pendant. In some systems users can employ a simulated method of lead through programming to program the robot in combination with the teach pendant. On other emerging systems, actual language commands for particular robots can be interspersed with the graphic instructions. The key benefit of such a system is that the user can see a graphic animation of the working robot cell. Because of the lack of standard robot interfaces, these systems do not yet produce robot programs unless special arrangements have been made between the vendor of the graphics system and the vendor of a particular robot.

The graphic animation, which displays hidden lines, is used primarily to select an appropriate robot or to lay out a manufacturing cell. The animation also facilitates the planning of robot programs, and can be used to a limited extent for visual collision detection. The systems are quite powerful and are efficient at evaluating alternative robots that might be used for a particular task. Users also have been able to avoid purchasing a robot which would not be applicable for a specific task. If several robots can perform the task, graphic simulations may also be used to compare them and choose the one that best meets the task requirements. The graphic simulation method of doing cell layout effectively facilitates checking for the appropriate location of all the components of the cell and evaluating alternative layouts for the optimization of material flow or time.

Graphics systems can be used for visual collision detection, but must be done with care since collisions can occur and not be observed by the user. Research is ongoing to develop heuristic methods of detecting collisions. For example, researchers at Lehigh

University have surrounded the hand of the robot with a sphere and checked for interference between the sphere surrounding the hand and objects in the molded system. When interference is detected, the actual computations are made to determine whether the hand or any object it may be holding is, in fact, colliding with something else in the cell.

The example above is a good heuristic, but it is not complete since you can easily have collisions between the cell and parts of the robot that do not involve the hand or its surrounding sphere. For example, robots with elbows could bang their elbow into parts of the cell and this would not be detected. This process of putting a sphere around the elbow and other places likely to be involved in collisions can be continued, and work is progressing along these lines. However, researchers are now looking into methods of detecting all interferences to more accurately detect collisions in the robot's world.

Planning the robot program to discover timing problems also is possible on a graphics system. But if the timing from the graphic cell is close to the timing tolerances required in the actual cell, then further investigation is needed because the models of robot controllers are not very accurate. This is further complicated by the constantly changing capability in robot controllers as they are made more sophisticated.

Functions of First Generation Programming Languages

The first generation of formal languages for robot programming can be characterized by the VAL language introduced by Unimation, a subsidiary of Westinghouse. They can be thought of as early robot programming systems created by extending BASIC to provide for robot motion, lock–step coordination, elementary sensor usage, and a minimal operating system. They have commands in them that allow users to move the robot and permit the user to specify motion along a straight line. Some have the ability to follow circles or arbitrary curves. Using lock–step coordination, they have the capability of generating, or waiting for, a binary signal at predetermined points in the robot's program. This is used to coordinate the motions of the robot with other devices in the environment. Consider, for example, the need for a robot unloading a press to be in sync with the press as it cycles.

Elementary sensor usage implies the use of binary or on/off sensors that are typically connected to the robot controller through digital input/output (DI/DO), bit-oriented logic. The powerful REACT command was introduced in VAL to provide for adaptive behavior with the use of binary sensors. This allowed you to program the robot to react when a sensor goes high or low by executing a special subroutine. You have the capability in such a system to force the robot's motion to be stopped immediately when the sensor trips, or after the robot completes a physical motion to a destination. The special subroutine can then either continue the interrupted motion sequence or initiate a new motion sequence, thus providing programmable adaptive behavior for the robot. This capability is limited in that you can only deal with a binary sensor and not with semantic information such as might come from a more advanced sensor, such as a vision sensor.

The operating systems supplied with first generation programming languages were designed to be self-contained and to be usable by operators without computer science training. Typically, only simple operating system capabilities are provided, such as program editing, program storage and retrieval on disk, and uploading/downloading a program to another computer. The operating systems are very limited when compared with those used by computer scientists; they are rudimentary even when compared with those used on microcomputers today.

First generation languages have serious limitations that restrict their robotic applications: They are unable to perform complex computation, and they are difficult to interface with vision and other complex sensors. The languages provide only limited communication with other computers and controllers, and they are not extensible in a way that allows commands and build capabilities to be added into the language. The human interface was designed to allow for flexibility in programming while not requiring expertise of the programmer. Thus, it serves two opposite masters and is compromised in both domains. This class of language can be thought of as allowing the sophisticated user to define his or her own blackbox software. The languages remove the rigidity of augmented teach box software but provide very little new capability. Nevertheless, these languages are quite popular and do an effective job in the factory. VAL, for example, has been mimicked by many vendors in the United States, Europe, and Japan. Unimation has recently introduced a second generation language called VAL II.

What's Here and What's Coming in Second Generation Programming Languages

The next step in the hierarchy of robot programming languages is the group known as *second generation programming languages.* These languages were developed by computer scientists to overcome the first generation languages limited extensibility, communication, and computational power. Thus, they have the computational complexity of a modern structured computer language. Some clearly show their origins. For example, the robot language Rail looks very much like its parent, Pascal. Other languages embody new concepts, such as the structured interpreted robot language AML. Some of the most recent languages, like VAL II are built upon earlier robot control languages. While these languages have the expected robotic extensions for motion, sensor communication, and control, they are enhanced by improved operating systems with more powerful editors and file-handling capabilities. Unfortunately, many second generation languages are supplied by their vendors as "closed" systems; this needlessly restricts their capabilities for computer communication and coordination. Some of them handle geometry, but in a fashion similar to the APT language developed for NC machine tools. (See the detailed description of APT in Chapter 8.)

The main advantage of second generation software lies in its extensibility. The languages are extensible in the same way as current computer science languages; they can be used to create an application package so tailor-made that the user thinks the software was designed especially for him or her. Through the use of subroutines and functions, new commands can be added to the language that give capabilities—transparent to the user—that didn't exist before. After a systems programmer is finished tailoring a second generation software package, the user will never know all the awkward complexities and syntax demanded by the raw language. This compensates for the fact that second generation robot programmers must be well trained in computer science to master the full power of the language. Advanced applications involving the use of complex communication and coordination, such as the integration of complex sensors and the communication with other computers, further demand the power of structured computer science languages and the services of highly skilled programmers.

The most serious of a number of deficits of second generation languages is the difficulty of programming with them. However, these weaknesses offer a challenge to researchers to design software features that meet the needs of the computer-integrated workplace. With the exception of APT-like robot languages, second generation languages cannot yet deal with part or cell geometry as is necessary in the work environment. There is still no way to safely debug a robot program. In contrast to the computer programmer who may discover a program bug from printed output, a robot programmer may discover the bug only when the robot breaks something in the cell, damages itself, or injures its human operator. Thus, you cannot determine whether something can go wrong until the incident occurs. Robotic applications should be able to access the wealth of data that resides in a CAD data base, but current versions of second generation languages can handle neither the communications, the different data structures, nor the interpretation of the geometrical data. Similarly, the languages currently do not facilitate the integration of robots into flexible manufacturing systems which use computer control to coordinate combinations of NC tools, robots, inspection stations, and material-handling systems. Finally, the vast experience compiled by the fields of human factors engineering and ergonomics regarding how people interact with automated and computerized equipment has yet to be applied to the user interface of second generation manipulator languages. When this occurs significant improvement can be expected in the ease of use of robotic software.

APT, the part programming language used for NC machine tool control, is frequently cited as a model for the integration of geometry into a robot programming language. Ideally, the same control language could be used for both NC machines and robots. If robots are thought of as just another kind of machine tool, then it makes sense to try to create a robot programming language by extending APT, which already handles geometry, with some robot motion and sensor integration commands. Unfortunately, the geometry used by APT is infinite geometry, which bears no resemblance to the part geometry captured in a modern CAD data base. Another problem with the APT approach concerns the dynamic manner in which modern robots respond to their environment; they make real-time decisions based on sensory input. Machine tools, on the other hand, can be regarded as static in that their programs are repetitive and fixed. Various attempts have been made to extend APT into a robot language but without lasting success. In fact, the successful addition of desirable robotic language features, such as adaptive response to sensory data and to NC machine tool programs indicates a trend in the opposite direction.

New and more advanced robot software is expected to be developed soon that will overcome present limitations. The following features are regarded as desirable in a second generation programming language:

1. It must be a full structured computer science language with appropriate extensions for robot motion and control.
2. The communication with, and control of, complex sensors must be facilitated.
3. The language must be extensible to facilitate the writing of tailored application software with advanced user interfaces and to be able to adapt to rapidly evolving future needs.
4. A full-featured, open, operating system will be necessary which facilitates computer-to-computer communications.
5. Three-dimensional geometry manipulation and reasoning must be developed that can access and interpret the part geometry data captured in CAD data bases.
6. Graphic simulation of the robot and its work cell will be required for the testing of alternative cell layouts and programs, and for the safe debugging of robot programs.
7. Graphic simulators will need an understanding of physics (including gravity, mass friction, and acceleration) to realistically model the robot's world.

No robot software product is available yet that incorporates all of these attributes, although researchers are actively engaged in the development of some of these features.

Innovations to Watch: Task/Object-Level Languages

Robot software is now positioned for advancement to the next level of hierarchical complexity, that of task-level or object-level languages. Teach box, blackbox, graphical, first generation, and second generation software all have dealt with people translating the needs of their application to the capabilities of the robot. Statements in those languages correspond to specific low-level movements of the manipulator. This is in striking contrast to the way in which a person is directed to accomplish a task. The operator should be able to direct the robot in the same way a person should be directed to perform a series of tasks, which the software would then determine how to carry out in detail. This task-level approach can be compared with computer languages. In contrast to FORTRAN and Pascal, which require step-by-step specification, simulation languages and graphics languages are structured to perform more closely to the task level. The statements in those languages correspond to high-level tasks that the program decomposes into executable actions. Manufacturers would like statements of that kind in robot software which would correspond to tasks the robot is to accomplish.

With such a language, eventually it may be possible to tell the robot something like, "Mate the part with the hole in it with the part with the peg in it, so that the peg and the hole are aligned, and the corners of the surfaces are aligned." A less hypothetical example would be for the robot to understand and carry out the instructions on a manufacturing process sheet. This is the set of instructions, or something similar to it, that a robot of the future would be given. This task-level approach is sometimes referred to as object-level because what happens to the part may be specified rather than what task is to be accomplished. Although such software is not now available, research facilities around the world are working hard on its development, and some research languages, such as AL include object-level features.

There are two fundamentally different theoretical approaches to the development of task-level programming. One relies on the artificial intelligence (AI) community to develop expert systems that will perform tasks according to formal rules acquired from human experts. This approach uses established AI techniques, but faces the anticipated difficulty of finding human experts in the fast moving field of robotic assembly. The other approach may be termed task-decomposition; the task is analyzed from the top down and is decomposed hierarchically into subtasks that the system knows how to perform. One advantage of the task-decomposition method is that algorithms may be applied recursively to a task until the resulting subtasks are at the level of manipulator motion that can be accomplished with present software. The National Bureau of Standards (NBS) is pursuing this second track. Regardless of the chosen method, task-level languages will become a reality as serious research issues are resolved.

Issues That Confront Robotics Researchers

The rapid drive to improve robot control software and to implement a full task-level language has resulted in the identification of several research areas critical to the fulfillment of this effort. This activity involves developing solutions to such things as: How can robot-independent software be implemented on robots of differing geometries? How can 3D geometry knowledge be used by a program for path planning, object grip decisions, and collision avoidance and detection? How can complex sensory data be interpreted and translated into meaningful feedback? How can data bases be used to create a world model for the robot software system? How can an advanced robot deal with uncertainty, particularly the uncertainty created by more intelligent adaptive behavior? How can modeling and simulation software, the human interface, and error handling be improved? How can robot software be fully interfaced to a FMS for the optimization of cooperation and scheduling? Will one language perform all of the above tasks, or is it reasonable to assume that families of languages will evolve to meet needs as different as the clean environment of precision electronic assembly and the dirty environment of arc welding?

Benefits a Robot-Independent Language Could Produce

Currently, each robot make and model has its own—usually proprietary—programming software, and programs created for one robot cannot run on another make or model. This situation prevents the establishment of a strong off-the-shelf application software market, because an application package would have to be written, marketed, and stocked for each robot. A robot-independent programming language would permit the programming of a task for a generic robot so that practically any robot could be used to perform the task. The computer industry has long benefited from the standardization of programming languages such as FORTRAN-77 and COBOL. More recently, computer independent programming languages, such as Pascal, C, and Ada, are being used because they are more or less portable between different types of computers, thus eliminating the expense of reprogramming for each kind of computer. Defacto standards even exist in the BASIC language. The robot industry has not developed robot-independent software because of the difficulty of deciding at what point the language instructions are to be translated from robot-independent instructions to the movement instructions required for execution by a particular robot. This incompatibility is rooted in the geometry of robots; a program written today for a robot with an elbow couldn't possibly work on a gantry robot because their physical configurations are so fundamentally different.

It has been suggested that a generic language be developed that can run on any robot. This would be an interim step because, although the language commands would be the same, a program written for one robot would still not run on a robot of a different configuration. The programmer would still have to account for the differing geometries and work spaces of each robot. IBM advocates the use of its AML for the development of such a generic robot programming language.

Advocates of one commonly advanced solution to this issue believe that the machine tool industry has paved the way toward robot-independent programming languages with their APT language. APT produces cutter location data (CLDATA), which is machine tool independent. CLDATA is put through a post-processor to make it specific to the machine tool on which it will actually be used. (Refer to Chapter 8 for a more detailed discussion of CLDATA.) This strategy works with machine tools because there is no difference in the geometries of the machine tools which would be used to perform a given operation. In contrast to robot environments, machine tool environments are constrained to be highly deterministic with respect to points in space, the zero–zero line, and intruding obstacles. An even greater problem with the APT approach is that no geometrical reasoning is done in the translation from APT to CLDATA to machine tool specific instructions. Complex and subtle geometrical reasoning is demanded by any robot-independent programming scheme because so much more than the location of the tool must be taken into account: the location of the rest of the arm, the location of fixtures and permanent parts of the work cell, and the temporary locations of parts in process. The solution to making the robot independent/dependent interface at the time when the conversion is made from world coordinates to the joint coordinates will not work because so much geometric reasoning must be done that is specific, not only to a particular robot, but to the particular work cell where the program will be used.

The most promising approach to robot-independent programming yet advanced involves the creation of tables that capture the geometry of robot and the work space. The task program would then be run through a task compiler which would reference the table for the respective geometries of the target robot and its immediate environment. Although this is very different from post processing, it seems like the most productive direction to date.

A related issue is the methodology of interfacing complex sensors to the real-time interpretation and decision-making capability that is necessary in a moderately intelligent robot. There is no way that a simple post processor can handle this task of large scale interfacing, real-time interpreting, and decision making. NBS is probably on the right track with their hierarchical communication and control system in which more than one computer controller is used. Microprocessors may control each robot joint, the robot controller may coordinate the joint processors, another computer may process vision or tactile data, and yet another may coordinate the work cell's robots and machine tools. The task of operating the robot is then distributed hierarchically so that the lower the position in the hierarchy, the more specific the information is about the particular tool and arm being used.

Understanding the Three-Dimensional Geometry Issue in Robotics

There is more to the 3D geometry issue than robot-independent software. The geometrical data necessary to describe a part currently resides in CAD data bases. For years the robotic community has asked the CAD/CAM community for access to geometrical data needed on the shop floor, implying that if such data were available, great progress could be made. Here again, the cross-disciplinary, pioneering nature of robotics first advanced a simplistic solution to a profound problem. Deeper inspection of the issue reveals that many CAD systems are only 2D or 2 1/2D systems that cannot capture semantic geometry in their systems. Their description of a part is captured by the human–machine pair; the CAD system displays three views and an isometric, and the human interprets the necessary geometry. Since so much of the geometric interpretation is in the human's mind, another computer program cannot access the CAD data base and construct a geometrical description of the part. Although older CAD systems clearly do not meet the needs of task-level robot software, state-of-the-art CAD systems now can capture the 3D shape of a part. (Refer to Chapter 2 for more details on this subject.)

Modern CAD systems use several different methodologies to describe completely the shape of a part so that a solid model of it can be displayed and manipulated on the CAD screen. One technique of solid modeling is called constructive solid geometry (CSG), another is the boundary value (BV) approach, still others use some combination of the two. The problem is that there is no standard in which like the IGES standard, allows the reliable communication and interpretation of 3D CAD data. Work is now underway to revise the IGES standard to allow the communication of more sophisticated geometric descriptions than is presently the case. A standard 3D data format is an absolute necessity to making real progress in task- or object-level robot software. Until the standard is written and accepted, progress will be awkward and slow.

If you assume that the 3D standard data format problem has been resolved, and that a robot program can access and interpret the 3D data from the CAD data base, you must then think of how to use this new knowledge. Unfortunately, 3D data is only part of the information the robot must have to perform geometrical reasoning. The program must have access to information about the part's materials, its weight and density, its flexibility or rigidity, the forces it can withstand, its sensitivity, and its stable states. An example makes this clear: Compare the way a robot would have to handle an aluminum egg that may be a billet with the way it would have to handle a chicken egg that may be breakfast. The shape data for each may be exactly the same, but the robot program would have to treat them very differently. Roboticists have now begun to realize that they have yet to define either the exact nature of the data needed from part data bases, or how to get that data from the data base into the robot software system.

Three-dimensional geometrical reasoning is also necessary to resolve issues of collision avoidance and detection, problems whose current solutions are too expensive and time-consuming to be practical. There is work in progress in very narrow parts of the larger problem of how to plan robot motions: path planning, grip planning, and collision avoidance. Path planning research, as discussed previously, is concentrating on the path of the hand and elbow. Grip planning concerns the automated decision of where to pick up a part and with what force in the fingers. This is another difficult problem even when the 3D solid model of the part is captured and the geometry of the gripper is known.

The concept of collision avoidance has been discussed along with collision detection, but the two are quite different; collision avoidance is much more difficult. How can the kind of "watchdog" sensors be put into a robot that can decide that a collision is probable, but has not yet happened? Detecting the collision is not our goal. Consider why a driver education class would not lump collision detection with collision avoidance. Robot software should decide when a collision is imminent and should avoid it. The open domain approach has not been successful in developing task-level software with today's technology, but progress is now being made by various researchers who have taken the approach of confining and defining the problem.

WHY SENSORY DATA IS BECOMING A MORE IMPORTANT ELEMENT IN ROBOTICS

Rapid progress is being made in sensor technology: vision and tactile sensory systems are becoming more powerful, with new capabilities. (Refer to Chapter 9 for a more detailed discussion.) This progress is impressive by itself, but the next problem is how to use that data. What kind of interface and standard will allow communication of this complex sensory data to the robot system? What is the best way to create a control system that depends on this kind of data in a flexible manner?

Complex sensory data is used in many levels of the robot system. It requires complex data analysis and requires the program to decide what parameters, like speed and direction, to change and by how much. Advanced task- or object-level software would be expected to integrate data from multiple complex sensors, but the sensors may not agree with one another because of noise in the data. How are such conflicts to be resolved? It may be even more difficult to resolve conflicts between a wide-angle or far-away sensor whose purpose is to generate a broad perspective and a telephoto or close-in sensor that is used to examine detail but can't see the whole object. Developing a requirements definition for a task system that can integrate and understand this complex sensory data is a real challenge.

Integrating Data Bases and Robotics

One of the least explored issues critical to the implementation of task-level software is the integration of data bases with factory floor robotics. For example, how can a robot system gain access to a CAD system so as to meaningfully determine what a vision sensor is looking at? A data base accessible to the robot must have a world model that contains, at a minimum, the current locations of all the things that the gripper can presently manipulate. When an object is moved, the data base must be updated, and the actions and the possibilities that result from those actions must be interpreted. If the task requires the use of a part that is under another part, the top part must be moved to a new location before the bottom part may be used. The software must be smart enough to decide where the part is that it needs and how to get it, and the data base must keep track of the new location of every part.

To use the data from data bases in interpreting complex sensory data, a world model must be created for the robot system that knows where the things are, the effects of its actions, and physical laws. AI techniques will undoubtedly be used to create this world model, which will have to include physics concepts such as gravity, mass, and momentum.

Why Robots of the Future Will Have to Deal with Uncertainty

Robot systems of the future will have to deal with an order of magnitude more adaptive to human behavior than do robots in today's factories and research institutions. The task domains of present robotic systems are highly restricted so that all of the possible decision parameters necessary for adaptive human behavior may be anticipated by the programmer and entered into a relatively deterministic program.

As an example of current deterministic systems, consider the development of sensors for robot arc-welding systems. The parts to be arc welded do not come from exact machine processes, so there is some variance as to the location and shape of the gaps. An arc-welding sensor shows the robot the path to take, and may even check the width of the gap, commanding the robot to slow down or speed up. This arc-welding space problem is so highly constrained that the variables have been identified and defined completely, which is why the requirements for the sensor can be specified and the data received from it can be used.

In contrast to the deterministic arc-welding example, consider what a person must do in the simple act of walking from one part of a building to another. A person uses his or her eyes to detect whether doors are open or closed and whether roller skates are on the stairs; his or her brain automatically reacts appropriately. How is a machine programmed to perform tasks of this nature where the space problem is loosely defined? When adaptive robot programs are written today, the programmer must state explicitly how the robot will test for each of the anomalies which may occur, and what to do for each anomaly if it is detected.

In task-level programming the robot will have to know that if parts don't fit properly on the first try, it should jiggle or massage them until they do fit. How does a person insert a peg into a hole when the fit is tight? How is a robot told to "squeak" the peg into that hole? Uncertainty, when a person is not acting to anticipate anomalies, is an order of magnitude more difficult than today's programming of responses to sensors that is now called adaptive behavior.

Why Researchers Need Better Simulation Techniques

Current simulation techniques must be greatly expanded to be able to simulate the behavior of task-level systems. Researchers need to develop methods of simulating sensor behavior, particularly for complex vision and taction sensors; they need to simulate physics. These are the same priorities that have been identified for the task-level robot system; all of the data and decision-making capability is needed in the task-level robot system that is in the simulation system. Eventually, the task-level system may become so intelligent that it routinely models its environment and its behavior to anticipate and avoid collisions, deal with uncertainty, and generally reason about its task. When this happens, the simulation system may cease its independent existence and become an integral and necessary part of the robot system.

How Far Have We Come toward Human Interface with Robots?

The human interface to a robot system is an important issue because autonomous robots are so far in the future. While automatic programming techniques are being developed, robot system designers must take full advantage of the best capabilities of both humans and machines. To optimize the human–machine partnership, the designers will have to involve ergonomics specialists in the development team.

The move toward automatic programming will involve the development of task-level languages which may be English-like languages, the so-called "natural language" human interface. Unfortunately, the English language has many ambiguities that humans normally resolve by contextual clues; sometimes the ambiguities are not resolved and an argument develops. The main research issue in the development of a natural language interface is the problem of eliminating these ambiguities while retaining the full semantic content of the dialogue. How will the robot feed back to a person what it thinks it has been asked to do in a way that is not dangerous? How will operators get confirmation that the robot is working properly and not running amok? The human-factors people who specialize in the human-machine interface will have a lot to contribute to the future of robotics.

The human-machine interface to an advanced robotic system may involve the use of a teach box, a graphics display with light pen, natural language, or AI. Which is the "right" approach? Many researchers believe that future languages may include all of these approaches, using each mode of control in its most appropriate form. For example, no robot language has been developed to date that does not at least use a teach box. Human factors research has shown that a joystick, such as that on a teach box, is one of the most effective and unambiguous methods for a human operator to control mechanical motion.

Computer users have discovered the ability to use graphics to enhance human-to-machine communication. For example, the spreadsheet metaphor used in VisiCalc and Lotus 1-2-3 enables the user to see what "looks like" a spreadsheet on the screen and therefore grasp the content of the information almost intuitively. The numbers can be graphed to present an even more interpretable image. However, there comes a time when the spreadsheet metaphor isn't comprehensive enough to handle the desired task. CAD graphics systems are bound by the same limitations of the graphic metaphor. Although CAD systems have excellent interactive graphics systems, every one of them also has a meta-language for use when the graphics methodology is not appropriate. Like the teach box, a graphics system can facilitate human–machine communication up to a plateau, above which lies a complexity beyond the reach of the graphics metaphor. Beyond that plateau you again depend on language, possibly because the higher level metaphor that comes after graphics has not yet been identified.

It is probable that a combination of input tools, such as teach box, graphics, and language, will be used by people to communicate with semi-expert robotic systems until some ultimate expert system is developed. These systems will depend on the intelligent interaction of humans and machine, such that AI developers will help create the semi-expert system with which humans will interact through the most appropriate input mode: teach box, graphics, or language.

Job description in the robotics field will develop into a hierarchy such as is found in the computer field. At the top of the hierarchy will be the highly trained robot system programmers, fluent in robot programming languages. The system programmers will develop new primitive commands and functions that extend the language in ways that are usable to application specialists, the next step down the hierarchy. Application specialists will develop generic application packages, the robotic equivalents of VisiCalc and WordStar, for arc welding and other general task families. Manufacturing engineers will be responsible for the installation, interfacing, and customization of the applications software so that it can be optimized for use in a particular FMS, performing the kinds of feedback and feedforward communication tasks necessary to the future factory system. The actual day-to-day supervision of the robot system will fall to the robot operator, who may know none of the robot language, but who may interact with the robot primarily through a very high level interface such as a vision system. Although this four-level occupational hierarchy is not unique to the robotics field, its adoption will speed the successful implementation of advanced robotic systems on the factory floor.

The Technological Challenge of Machine-Machine Interface

A final challenge in the development of advanced robotic software is the machine-to-machine interface, a problem that has been largely ignored to date. This is a problem with current versions of second generation languages because flexible manufacturing systems and islands of automation cannot be integrated into higher level entities without the required hardware and software interfaces. Convenient means must be found to interface robot software with hierarchical controllers and complex sensors. Easy ways must be created for interfacing the robot system with the machine tools, inspection stations, and material-handling systems of an FMS. Effective communication interfaces must be developed that allow the robot system to share information with other computers, data bases, AI programs, and simulation systems, as will be necessary for the distributed decision making and reporting that can integrate a task-level robot system with the rest of the factory of the future.

The most challenging problem facing those developing robot software may be to create a requirements definition or manufacturing specification for intelligent robot software. When this is accomplished, it will be possible to specify and standardize the hardware and software interfaces that are necessary to integrate the parts of the robot system into a working whole. Commercial and research institutions will then be able to divide the job into software modules they want to develop, and they will know that their modules will be compatible with the intelligent system as it evolves. This is such a large-scale task that it can be best achieved by a broadbased, cooperative project. When the interface problem is solved, robot systems in an FMS can finally be integrated with shop floor recording systems, MRP systems, information systems, and decision-support systems.

Tips for Successful Automation: Where to Find State-of-the-Art Robot Software

A fine example of current state-of-the-art robot software is found in the modules developed by McDonnell Douglas which allow companies to design, simulate, and program industrial robots off-line. These software modules enable manufacturing engineers to evaluate robotic cells quickly and thoroughly before building them, and create, test, and down-load programs to a robot controller.

The systems result in lower production cell setup costs, faster installations, and smoother operations. Engineers can try several solutions for robotic cells before purchasing any equipment. Because they can program robots off-line to perform exactly as simulated, costly mistakes can be avoided which are an inherent feature in the "teach mode" method of programming robots.

McDonnell Douglas software modules include the following:

1. *Place*—an interactive graphic cell design and simulation system
2. *Build*—a system to create the joint motion equations of robots for the simulation processes in PLACE

Recent Developments in Automated Guided Vehicles

Automated guided vehicles (AGVs) are flexible and relatively easy to integrate with computer-based systems. Some recent advances in AGV use are:

- Mobile work platforms that carry work to various assembly operations
- Fork-equipped vehicles that lift and stack loads to various heights
- Special-purpose designs for applications in specific industries, for example, newsprint handling, "clean room" handling in disk drive, and semiconductor production
- Other designs that integrate on-board robots, manipulators, conveyors, turntables, and other devices to fit the user's particular requirement
- Significant improvement in the cost and reliability of control software including user-friendly routines that minimize the complexity of guidepath or vehicle routing changes

More detailed discussion of AGVs appears later in this chapter.

Conveyors That Offer Flexibility, Integration, and Control

Transporter systems that provide flexibility, integration, and control of a family of modular systems to support kitting and small parts assembly are:

- Chain or belt driven bidirectional transport conveyors
- Microprocessor controlled, single or double air-actuated transfers that can be easily moved or interchanged
- Preassembled modules (i.e., beds, transfers, and workstations delivered with integral air, electrical, and communications lines) to speed installation and simplify maintenance as well as adaptation for new requirements
- Low noise level and "clean room" compatibility
- Direct communication with operator workstations and user-host computers

The integration of these systems with contemporary small parts storage systems has led to significant improvement in throughput control, inventory reduction, and space utilization.

Applying Robotics to Material Handling

Robotics as material handling devices consist of advances in the following areas:

- Machine tool and test station loading and unloading
- Design of sophisticated, quick-change grippers
- Expanded use of machine vision applications
- Robot servicing of carrousel and miniload storage systems
- Marriage of robotics and AGVs

Currently, not too many AGV robotic applications are found in industry. In both modes, a robot aboard an AGV or a vehicle attending stationary robotic work cells will require applications that demand a high degree of accuracy.

AGVs offer accuracies comparable to those of robots and battery-charging systems which are able to equip AGVs with robots and integrate them to work together.

Future developments in such vehicles will include enhancement of their mechanical capabilities, such as more active lift/push and robotic capabilities with arm-like features built on them. Some existing applications, such as in semiconductor manufacturing, have mounted robots onboard their vehicles.

For example, the AGV vehicle using an inertial–guidance system coupled with an optical distance encoder arrives at a docking station, scans a mirror, and locates its position and that of the robotic arm from the reflected image. The arm not only carries out such materials-handling functions as transporting and transferring semiconductor-wafer cassettes and boxes, but also plugs in an umbilical power cord into a power source and recharges the vehicle.

TRENDS IN AUTOMATED STORAGE SYSTEMS

Although the application of just-in-time delivery has lessened demand for large-scale automated storage systems in the manufacturing sector, it also has fostered growth and innovation in other storage system approaches including miniload, microload, tote stackers, and vertical and horizontal carrousels.

These innovative technologies add to a wide range of more basic approaches such as unit load systems for finished goods warehousing and computer control in conventional flow rack picking systems. We will discuss some of the more significant trends in storage applications in the next four sections.

Optimizing Throughput Rates with Unit Load AS/RS

Developments in this area are associated with optimization of throughput rates through simulation and software that maximizes productivity on the basis of operation-specific requirements such as first-on, last-off trailer loading.

Latest Advances in Miniloads, Microloads, and Tote Stackers

Latest advances in these areas are:

- Increased reliability with a very high percentage of uptime
- Lateral as well as end-of-aisle input/output stations for assembly operations
- Lower software costs with increased focus on "velocity loading" and higher transaction rates
- Standardization or "pre-engineered" hardware and software modules targeted at reducing costs and delivery time
- Designs that are somewhat easier to relocate than earlier versions

Why Carrousels Work Well in Manufacturing

Vertical and horizontal carrousels are important elements in integrated systems. Carrousels have the advantage of bringing the "pick location" to the worker. Combining carrousels with transporters, robotic loaders and unloaders, and other devices offers not only faster, more efficient material flow to kitting and production, but also the tighter stock control that permits reduction in inventory. Other features that assure carrousels of a continuing role in the manufacturing environment and distribution include:

- Low maintenance cost
- Modularity and flexibility
- Innovative designs such as the horizontal carrousels whose tiers move independently for increased throughput
- Continued developments in basic controls, interfaces, and standard software

Benefits of Flow Racks and Paperless Picking

Digital display control systems increase the speed and accuracy of split-case picking wherein order pickers can select items from flow rack for conveyor or cart transfer to packing and shipping. Benefits achieved in these systems can show:

- Dramatic pick–error reduction
- Sharp reduction in time per item picked
- Simplification of new picker training and corresponding reduction of training time

All of the storage systems discussed are modular (flexible), can be readily integrated with existing equipment, and offer a high level of control of both inventory and operations.

INNOVATIONS FOR EFFECTIVE, TIMELY CONTROL SYSTEMS

Controls for material handling and material management include refinement of sensors, motor starters, communication links, and other methods for direct machine control, as well as advances in hardware and software. Control involves automated handling systems, programmable logic controllers, and minicomputers. All of these innovations are directed toward linking AHS with real-time information systems in the plant and warehouse environment.

Effective material control depends on the accuracy and timeliness of information exchange at and between the operations level and other elements of the corporate hierarchy. This in turn requires the implementation of a real-time system, such as the one shown in the following example.

Example of Automated Control: A Warehouse System

The basic warehouse control system focuses on tracking materials from receipt to storage, stock location record maintenance, pick generation, and inventory adjustment upon shipment. Communications with supervisory personnel are conducted via printers and CRTs, and with order pickers via hand-held or lift truck-mounted RDTs. Most of these systems have a batch or real-time communications link to the user's host computer.

In operation, the systems accept orders from the host while stock receipts are entered via CRTs at receiving. The systems organize daily activities and allocate work on the basis of demand and available resources. Operators receive tasks via RDT to store, move, or pick materials. They confirm completion of each task before the systems will assign the next activity. Inventory files are updated by item and location upon completion of each transaction. Reports are generated routinely and on demand for supervisory personnel. All transactions are recorded on an inventory transaction log which is periodically transferred to the host.

Additional Uses and Benefits of Control Systems

Other functions of these systems include:

- System or operator selection of storage location
- Shipment planning
- Item reservation
- Batch or single order picking
- Kit picking
- Stock relocation
- Forward pick area replenishment
- Cycle counting
- Backorder and cross-dock handling
- Lot control
- Productivity analysis

System benefits include the ability to interleave material storage, picking, replenishment, and inventory checking on a single lift truck cycle, thereby eliminating deadheading and increasing operator and vehicle productivity. Also, these systems eliminate paperwork and increase inventory accuracy. The modular construction of the systems permits the addition of both software and hardware features to meet expanding requirements with minimum disruption of operations. Finally, these systems permit the user to generate additional reports peculiar to his or her specific requirements.

HOW IDENTIFICATION METHODS APPLY TO MATERIAL HANDLING

A number of identification methods are used extensively throughout manufacturing and related activities. Although several of these ID technologies were covered in Chapter 5, a brief review is in order since they are also employed in the area of material handling.

Bar Code

- *Auto-discrimination:* The ability to decode electronics to automatically recognize, read, and verify multiple symbol formats.
- *Hand-held laser scanners:* Lightweight, noncontact, portable code readers that provide greater operator flexibility.

Radio Frequency (RF) Systems

- *Passive tags:* An antenna positioned along the path of tagged item movement and a reader–decoder. The tags absorb energy as they pass through the antenna's field. This energy is used to power-up the tags which then transmit their unique codes back to the reader.
- *Battery-powered* tags can be written to as well as read from at ranges of up to a few meters. These read/write tags can store up to 8000 characters of information. They can carry instructions to the station(s) at which the item they identify is to be processed and can receive and store data along the way, thereby creating an audit trail that can be readily checked at quality control or prior to release for shipment.

integrated circuits, as well as flexible manufacturing cells where grease and metal chips define the environment. Vehicles have been designed to carry 50-pound loads as well as 100,000-pound loads. There is a great deal of variety in vehicle design, but most fall into one of the categories discussed below.

*Towing Vehicles (figure 7–1)**

These vehicles consist of an AGV with no-load carrying facilities but with a hitch or tow bar that can pull trailers, carts, pallet jacks, and wheeled racks. They are used where there is a large volume of product to be moved or in retrofit applications where the product has historically been moved by trailers. Towing vehicles have been used to move rolling loads up to 50,000 pounds. If the trailer must follow the AGV exactly, then special tracking trailers must be used. These types of AGVS can follow an electronic guidepath which can have selectable multiple destinations. They can interface with other AGVs units, host computers, or other system controls.

Unit Load Transporters (figure 7–2)

These vehicles are designed to carry individual loads on-board the vehicle. Unit load transporters can have an extremely versatile deck design which permits them to be equipped with rollers, belt conveyors, power lifts/lowers, special fixtures, or on-board robot arms. Push/pull power stations can be used to move loads to the nonpowered deck of the vehicle. These AGVs can either be bidirectional or unidirectional and are used in warehousing as well as the factory floor. Loads of up to 12,000 pounds are common and loads of up to 60,000 pounds have been carried by unit load transporters.

FIGURE 7–1 Towing Vehicle (Courtesy of Mannesmann Demag Corp., Mannesmann Demag Material Handling Systems Division).

*Permission granted by Mannesmann Demag.

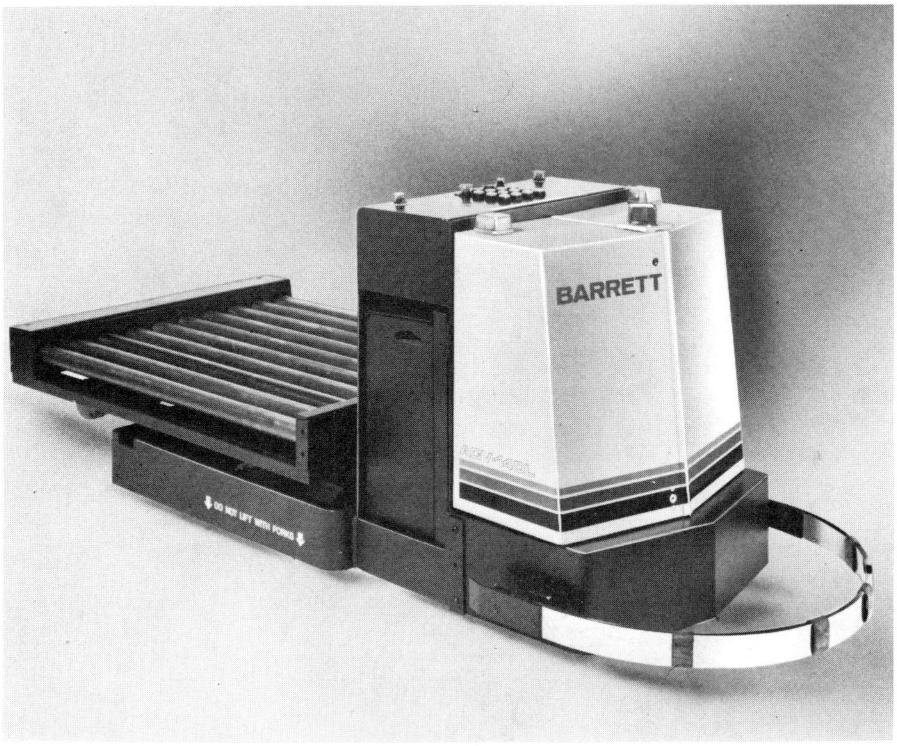

FIGURE 7–2 Unit Load Transporter (Courtesy of Mannesmann Demag Corp., Mannesmann Demag Material Handling Systems Division).

Heavy duty bidirectional load carriers can transport loads almost any distance without operator intercession. They follow an electronic guidepath and may operate in concert with other driverless units as part of a complete materials handling system. These AGVs may be controlled by a built-in microprocessor or a user-friendly PC.

Unidirectional load carriers move loads to predetermined locations on a guidepath. They can load or unload automatically at stands or conveyors. These AGVs can interface with other AGV units or other automated equipment.

Standard Automatic Guided Pallet Trucks (figure 7–3)

These vehicles are designed to service palletized loads to and from floor level positions. The shadow fork version has limited fork travel and is designed to move pallets to and from floor positions exclusively. The fork truck version has fork travel of up to 20 feet and can move pallets both at floor level and on stands or racks. Ride-on standard fork lifts can be retrofitted with AGV controls to give them most of the features of automatic guided pallet trucks.

Assembly-Line Vehicles (figure 7–4)*

This vehicle has a fixture on-board that accepts the frame or initial parts of the product that is to be assembled. The vehicle is routed through the various manufacturing stations

*Permission granted by Interlake Conco-Tellus AGVS..

FIGURE 7–3 Standard Automatic Guided Pallet Truck (Courtesy of Mannesmann Demag Corp., Mannesmann Demag Material Handling Systems Division).

of the factory where parts and assemblies are added to the product. Opportunity charging is often utilized as is bar code scanning so that the vehicle does not have to get out of sequence for battery exchange and that all parts and assemblies are tracked and accounted for.

These AGVs can provide total automatic transfer of materials. Equipped with a shuttle transfer device, they can automatically load products from pick-up stations and then carry them to designated drop-off stations.

AGVs are preferred as a materials handling system on automated assembly lines rather than conventional conveyors, because processing is stationary. Moreover, unlike other handling systems, AGVs have special features. They can skip assembly sections if equipment in a particular section breaks down; accommodate high-product differentiation, as well as short-product life cycles; handle high production volumes; incorporate work group buffers prior to and after workstations; minimize index time and idle time; and interface easily with other systems of automation.

There are three types of AGV-linked assembly lines: synchronous, asynchronous, or a combination of both. The synchronous line, characterized by the absence of queues between workstations, is balanced by the longest in-station assembly-cycle time in the line. However,

FIGURE 7-4 Assembly Line Vehicle (Courtesy of Interlake/Conco-Tellus AGVS).

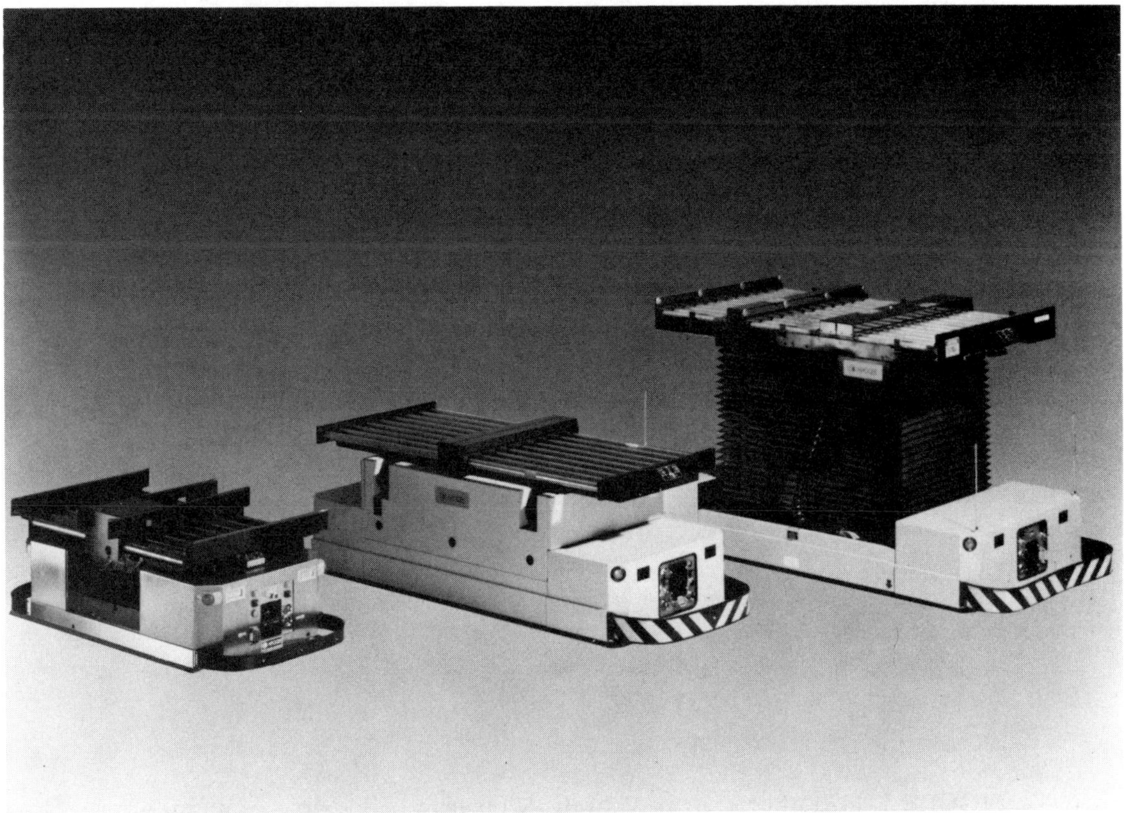

FIGURE 7-5 Three Types of Light Load Transporters (Courtesy of Apogee Robotics). (a) Left—Dual Conveyer Vehicle (b) Center—Dual Conveyor with Lift Powered Conveyors (c) Right—Three Conveyors Handling Six Tote Boxes with Lift

this causes it to be inflexible. Hence, synchronous lines usually represent low-product differentiation.

An asynchronous assembly line, on the other hand, features queues, which act as buffers, before and after workstations and is balanced by the line's average cycle time. There are three workstations/vehicle configurations for an asynchronous assembly system: fixed station build, which delivers a workpiece to a fixed assembly station; mobile work platform, wherein the piece remains on the vehicle throughout the assembly process and is removed only when the entire unit is assembled; and robot build, which consists of a robot on-board or interfacing with an in-station robot.

*Light Load Transporters (figure 7-5)**

The vehicles are designed to carry tote boxes, baskets, small parts, corrugated board boxes, and so on. They may be outfitted with a flat top, shelves, conveyors, custom handlers,

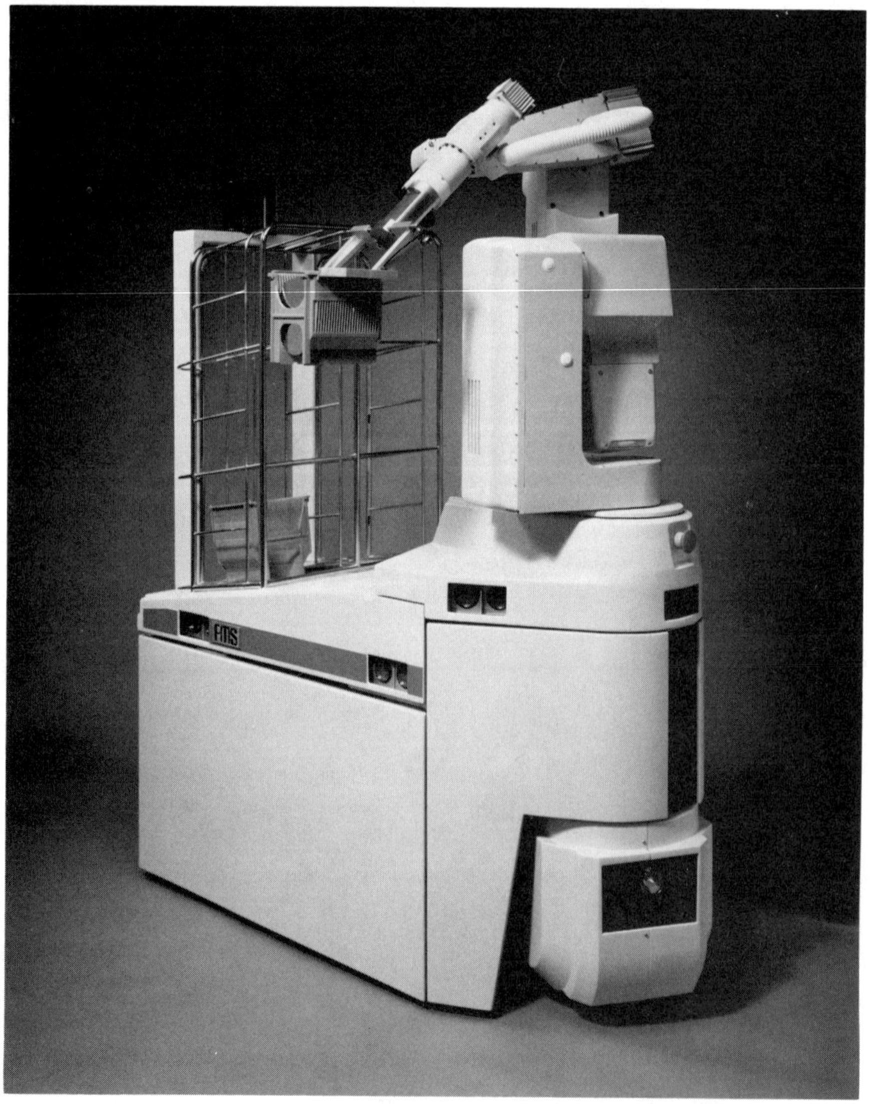

FIGURE 7-6 Robotic Arm Vehicle (Courtesy of Flexible Manufacturing Systems).

*Permission granted by Apogee Robotics.

or with any other unitized container. The light load transporter generally has a small footprint allowing its use in tight spaces and narrow aisles. These vehicles have been used for a wide range of functions from mail rooms to clean rooms in every type of manufacturing and office environment. This particular example consists of three types of light load transporters: left-dual conveyor vehicle, center-dual conveyor with lift powered conveyors, and right-3 conveyors handling six tote boxes with lift. An automatic fork truck would be used to pick up material from the floor.

Robotic Arm Vehicles (figure 7–6)*

These vehicles are light-load vehicles that have a robotic arm attached to the mainframe of the AGV. This vehicle has the ability to be reprogrammed to perform a variety of tasks. A typical application is the handling of silicon wafers or wafer carriers in clean handling rooms. The potential uses for an automatic, reprogrammable arm *with legs* are almost unlimited.

Flexible Machining System Vehicles (figure 7–7)**

These vehicles transport tools, fixtures, and raw materials to the machining cells and finished goods from the machining cells. They require a high degree of positioning capability,

FIGURE 7–7 Flexible Machining System Vehicle (Courtesy of Interlake/Conco-Tellus AGVS).

*Permission granted by Flexible Manufacturing Systems, Inc.
**Permission granted by Interlake/Conco-Tellus AGVS.

FIGURE 7–8 Automatic Fork-Lift Vehicle (Courtesy of Interlake/Conco-Tellus AGVS).

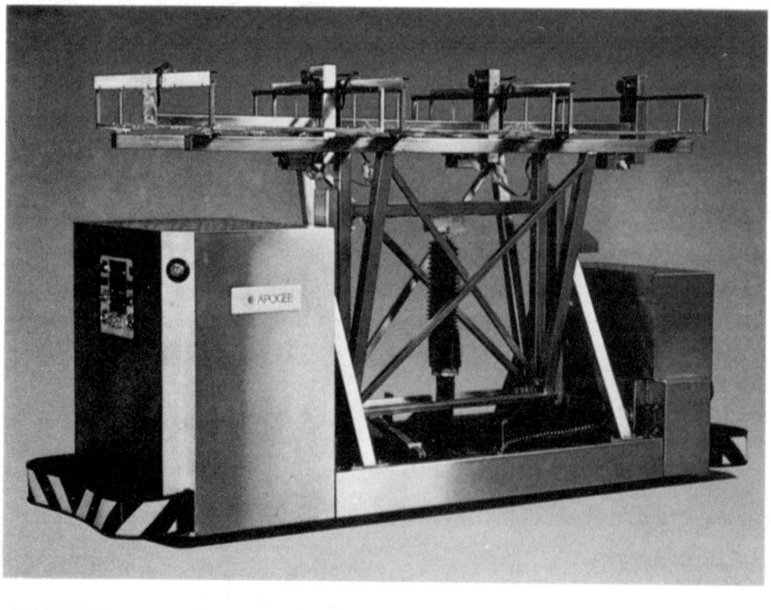

FIGURE 7–9 Custom-Designed Vehicle (Courtesy of Apogee Robotics).

and after docking, are usually pinned to the machining cell. With the application of these vehicles a lights-out factory is achievable.

Automatic Fork-Lift Vehicles (figure 7–8)

These vehicles can automatically pick up and deposit unit loads at both the floor level and on stands or racks. They can be programmed to automatically position its forks at any height within its lifting range. This type of AGV can have automatic guidance control, a choice of blocking controls, and destination selection.

Custom-Designed Vehicles (figure 7–9)

In addition to these basic AGVs, there are numerous custom designs that fit a particular need requiring precision positioning, such as shown in the illustration.

Automatic guided vehicle systems were not meant to be used for nonbatch, high-volume, continuous manufacturing processes like the beer or coal industries. The application of AGVS to batch nonsynchronous industries for horizontal material movement and information generation and gathering seems to be more appropriate. These systems are clean, quiet, and safe. They operate in hazardous environments and can be repaired easily.

An Innovation to Watch: Autonomous AGVs

The future of AGVs indicates autonomous vehicles that can communicate with other vehicles in the system and with the factory controller. They will have vision capabilities for obstacle avoidance and part recognition. These vehicles will have the ability to diagnose their own malfunctions and to order and perhaps replace the assemblies that are defective. They will be able to sense and warn operators of fire, sabotage, and unauthorized intrusion.

AGVs use the same technologies as most machine tools: microprocessors, servo-amps, DC motors with tachs and encoders, stepping motors, programmable logic controllers, hydraulics and relays, and others. What separates AGVS from most other machines is guidance technologies, wheels and their geometries, and the basic system control functions of AGVs: routing, traffic management, load transfer and material handling, and system management.

ADVANTAGES AND DISADVANTAGES OF SEVEN TYPES OF GUIDANCE TECHNOLOGIES

A number of guidance technologies that are available now or in an advanced developmental stage are buried wire, optical, dead reckoning, infrared, inertial, positioning reference, and ultrasonic and laser imaging. A description of each technology along with its advantages and disadvantages follows.

*1. **Buried Wire.*** This technology uses wire that is either buried in the floor or surface mounted. A radio frequency is induced into the wire that can be sensed by receivers aboard the vehicle. Two different methods are used by suppliers to sense this radio frequency: amplitude sensing and phase sensing. When using amplitude sensing, two antennas mounted at the front of the vehicle are used to keep the amplitude of the signals received by both antennas balanced by keeping the wire between the two antennas. This is achieved by processing these signals and evoking the proper response from the drive wheels to keep the antennas and the vehicle centered over the wire as the vehicle moves. When using phase sensing, an antenna mounted at the front of the vehicle senses positive and negative phases. As the vehicle moves over the wire, processing equipment on-board evokes the proper response from the drive wheels to attempt to keep the antenna directly over the wire by seeking a neutral phase.

The major advantage of buried wire is its reliability. It is a tried and proven system. If the vehicles, for some reason, lose the guidepath, the vehicles will come to an immediate halt. Communication between the vehicles and the host computer is done over the wire in the floor. This method involves simple equipment, is resistant to interference, immune to the factory environment, operates in real time, and provides a communications network that can easily involve every piece of equipment in the system. Wire systems also resist oil, grease, and water. They are immune to physical abuse such as high traffic areas involving fork lifts and carts. Buried wire is also immune to dirt and paper covering the wire path. Most AGVs in use today are wire guided.

The major disadvantage to buried wire is that some factories cannot have their floor cut. Typical examples are retrofitting a clean room, plants with metal floors or metal grates, and floors with rebar close to the surface. Other disadvantages are that it is not easy to add or change paths relative to the nonwire techniques, and if there is a need to keep vehicles running if the system fails, then wire systems are not appropriate.

Currently, when comparing total costs of sophisticated systems where advanced communication, routing, traffic management, and system management are involved, buried wire is still the most inexpensive method. However, if the system is simple, especially in single vehicle systems where point-to-point delivery is used and communication with the factory computer is not needed, then optical systems may be more economical. Systems without a guidepath are not yet competitive. As sensor and computing technologies become more economical, this cost comparison may change.

2. Dead Reckoning. This method is generally used in conjunction with buried wire. It uses optical encoders on the drive wheels or on instrument wheels. These encoders measure wheel rotation. If wheel rotation is the same for both drive wheels, then the vehicle is presumed to be going in a straight line. By controlling relative rotation of both wheels, algorithms can be developed to produce different turning radii. This method is used where many spurs are off of the main path or great flexibility in spur position is anticipated. Dead reckoning is generally used only for short distances because wheel slippage can cause positioning errors.

3. Optical. There are a number of optical guidance technologies available. Some use foil tape for a guidepath, others use ultraviolet sensitive tape or chemicals for the guidepath. These guidepaths are easy to install, economical, and can be changed or expanded with little effort.

Vehicles using optical guidance must have more intelligent on-board control systems. They use preprogrammed maps of the system, optical encoders, and tape placed perpendicular to the guidepath, which is read by sensors on the vehicle somewhat like reading bar codes, to determine the location of the vehicle in the system.

The major disadvantage of optical systems is that the path becomes unreadable when covered by dirt, grease, or particles. Pedestrian or wheeled traffic will also obscure the path over time. Another disadvantage is that these systems are not easily integrated into sophisticated manufacturing systems where communication, routing, traffic, and system control is vital.

4. Infrared. Infrared guidance systems are still relatively untried in an industrial environment. One of these systems uses an on-board camera that sights off of LEDs mounted at the end of aisles. The vehicle is guided by centering the image of the LEDs in the pixel array in the camera. The LEDs come in pairs set at a standard distance from one another. Determining the distance between images on the pixel array in the camera allows deduction of the distance between the camera (vehicle) and the infrared sources.

Another system uses infrared technology that places cameras around the periphery of the room or building. Two LEDs are on-board the vehicle—one placed at each end of the vehicle. The cameras track the images of the paired LEDs as the vehicle or vehicles travel about the system. Position of each vehicle is deduced by way of triangulation from two or more known camera locations. Vehicle identification and possibly communication is accomplished by modulation of the LEDs. Vehicle routing, traffic, and system management are accomplished by comparing the vehicle position with a system map located in the system software.

Another infrared guidance system has both the infrared source and receivers on-board the vehicle. Reflectors are placed overhead along aisle lengths and the images of these reflectors are counted and compared with an internal map on-board the vehicle so that positioning can be determined.

The major advantage of infrared is reduced installation costs. Infrared is also insensitive to sunlight and fluorescent lamps which sometimes confuse optical systems. This technology would appear to be useful in clean room applications and in buildings where the floor cannot be cut.

The disadvantage of this technology is that guidance requires line of sight so that disruption of guidance occurs when someone or something gets in the way of the camera.

5. Inertial. There are several successful installations using inertial guidance systems. These AGVs have optical encoders to determine vehicle position when compared with an internal preprogrammed map. An on-board gyroscope senses acceleration due to deviation from a straight line. This data is processed and corrective measures are taken to counteract wheel slippage and other error-producing effects and to keep the vehicle on its preprogrammed path.

The major advantage of this technology is that it allows clean installations and can be reprogrammed easily so as to allow much flexibility in path changes.

Disadvantages are the lack of system control and the high cost of the vehicle.

6. Triangulation. This method of guidance uses triangulation to deduce the position of the vehicle relative to three or more reference sources placed about the installation. This method has been applied to AGVs on an experimental basis only. Sources that have been tried are infrared, ultrasonic, microwave, and light.

The apparent advantage of these systems is lower installation cost. The critical factor in installing these systems is the ability to develop operative software programs.

7. Ultrasonic and Laser Imaging. Both of these technologies use sensor arrays and sophisticated software to analyze reflected signals to build computer images of their surroundings. These computer images are compared with maps that are preprogrammed into the vehicles' on-board systems. These systems look for reference positioning and obstructions. Programs utilize artificial intelligence to manage the huge amount of data that has to be analyzed. There are examples of uses of both laser and ultrasonic imaging, however, at this time application to AGVs does not yet exist.

The major advantage of using these technologies comes from the ability to have completely autonomous vehicles. Their use is particularly appealing in hazardous areas.

The major disadvantage is the per unit cost of the application of these technologies.

TYPES OF WHEEL GEOMETRIES

There are two basic steering geometries: differential and steered wheels.

Differential drives have two driven wheels side by side and the vehicle is steered like a military tank. Differential wheel velocity determines turning radius. The advantage of this drive configuration is that it will allow a vehicle to turn about a point half way between the two drive wheels. This is particularly useful where space is at a premium.

Steered wheel drives use a motor to steer a single drive wheel. This drive type is found in all types of AGVs and has a greater ability to follow guidepaths.

Another type of wheel geometry, not in use as much as those discussed previously, is two drive wheels placed front to back. The advantage of this geometry is that the front and back of the vehicle will follow the guidepath exactly, minimizing vehicle sweep, and when the drives are turned 90 degrees to the normal direction of travel, the vehicle can crab sideways. This saves floor space that is normally reserved for the docking maneuvers.

DECIDING BETWEEN ON-BOARD VEHICLE AND CENTRAL CONTROL SYSTEMS

System controls fall into two basic categories: on-board vehicle and central control. Each of these categories must address basic functions that together define and control AGVs. These functions are routing of vehicles, traffic management, load transfer or materials handling, and systems management.

How On-Board Vehicle Systems Operate

Due to low-cost computing power there are now more systems in-place with various on-board control options. This approach is generally used on relatively simple systems where there is a limited quantity of vehicles and less need for traffic control. These controls are characterized by program panels or toggle switches which correspond to destination points within the system. The vehicle can be dispatched to one or more destinations by an operator activating the appropriate switch. This approach is used by both wire guided and off-wire technologies.

Routing of vehicles is accomplished by placement of coding points before each branching of the guidepath. In optical systems this involves use of tape laid perpendicular to the path. The tape is read like a bar code and identifies that particular branch. This coded position point is matched with data that lies on-board the vehicle in a preprogrammed map. The vehicle, knowing the destination point from operator keyboard entry, decides which half of the branch it must take to get to the correct destination. The vehicle will always make the same decision regardless of location of the other vehicles in the system. This is a negative aspect of on-board control if a vehicle is down in the path ahead, the destination point no longer needs servicing, or priority servicing is needed elsewhere. These negative aspects add inefficiencies to the system, thereby adding costs to the user. With buried wire systems, each path at branch points has a different frequency or multiple frequencies in multiple wires that identify and differentiate one path from the other. The approaching branch is identified with passive code plates set in the floor. The vehicle, knowing what branch it is at and where it wants to go, looks at its on-board map to see what frequency or combination of frequencies it has to follow. When the vehicle finds the correct frequency code, it chooses that path and is on its way to the proper destination. Buried wire and optical on-board control systems suffer from these same inadequacies.

Traffic management with on-board control systems is quite limited because the whole system is not viewed from a single perspective. With optical systems, communication between vehicles must be provided so that when vehicles meet at an intersection, they can decide which has priority and which vehicle may pass first. With wire guided systems an intersection zone is established. When a vehicle enters this zone its presence is sensed and all other paths leading into this intersection are blocked to traffic.

Vehicles for both the wire and optical guide methods are kept from running into each other when multiple vehicles are sharing the same guidepath and are going in the same direction, by forward sensing technologies. These sensing technologies are field disruption, ultrasonics, infrared and contact bumpers. They all detect forward obstacles and prevent damage due to collisions.

Traffic management with on-board control will not work for larger systems. As the number of vehicles in the system increase, the chances of locking up the system increase.

Load transfer for both the wire and optical methods is very nearly the same. As the vehicles get near their destination, the pick-up and delivery station, a code plate, or tape on the floor informs the vehicle of the approaching station and may even identify the type of station. Identification of the station becomes important if there are different sequences the machinery has to perform at different stations. After the vehicle reads the code plate or tape, it will slow down so that positioning accuracy will be improved. The vehicle will then begin to look for a signal that will prompt stopping. This is provided by a magnet of photoeye. After stopping, the vehicle will begin its load/unload routines. It knows to load or unload by reading load sensors on the vehicle and/or on the pick-up and delivery station.

System management, when using an on-board system, it is limited to the on-board dispatch method described earlier and off-board call systems. Off-board call systems consist of a button/switch located at the workstations. When this button is pushed, a stop mechanism is activated (photoeye, stop frequency, and so on). As a vehicle makes its way around the system, and when it passes a workstation that has a call, the vehicle stops. A destination is then entered into the program panel on the vehicle, or the vehicle seeks out destinations previously programmed into its system, or both. When a central computer, and communication with this central computer is added, it introduces a second category of systems controls—central control.

How Central Control Systems Operate

Considerable benefit can be derived by the application of AGVs in eliminating labor-moving materials from point to point. However, the greatest benefit is derived from the integration of AGVs with computerized manufacturing. Not only does this technology free labor from unskilled work, but it can track all inventory and work in process. All pertinent data can be gathered from the factory floor and interfaced to the MRP system. Central control of AGVs can tie together all islands of automation and provide real-time control of the factory to meet changing market demands for the product.

With central control, one source can integrate routing, traffic management, order management, communications with vehicles and stations, communications with the host computer and real-time graphic representation and emulation of the system.

Routing of vehicles is accomplished in a number of different ways. Systems using optical guidance communicate by use of microwave, radio, or infrared transmitters, and receivers that are located at strategic spots throughout the system. Branching is done by the same method utilized with on-board control systems, but destination points can be changed and cancelled by the central controller at the communication points. Wire guided systems can communicate by use of microwave, radio, or infrared, but most often communicate by way of the guide wire, by way of additional wires that are parallel to the guide wire, or by loops of wire in the floor at strategic places called *information points.* These systems may do routing in much the same way as described in the on-board system description of the frequency select method, except the activation of the frequencies is under central control; or they may use the path switch select method. In the path switch select method, only the path to be taken by the vehicle is switched on by the controller so that the vehicle has only one path that can be followed. All other paths are turned off. Location of the vehicle is assumed until communication can be made with that vehicle, then the next path to the proper destination is activated.

Traffic management for AGVs with central control is more flexible and efficient. The central controller contains a map program of the factory floor and it knows the relative position of all the vehicles in the system. It also knows the destinations of all the vehicles and what they have to do when they get there. In this way, from a central location, vehicles are prevented from colliding with one another, vehicles infrequently wait for other vehicles to clear intersections, and vehicles can be optimized dynamically in real time to perform the tasks that must be done. Some suppliers have programs that emulate vehicles performing orders before the controller executes the orders, thereby ensuring success of the mission.

Load transfer is accomplished in the same way as described in on-board control, except with central control it is possible to detect malfunctions at the station if verification of order execution is not forthcoming. In this manner help can be sent to rectify the problem, and vehicles can be rerouted to service other stations.

Systems management can use on-board dispatching, off-board call systems, remote terminals, a central computer, or a combination of all of these methods when using a central controller. On-board dispatching and off-board call systems are the same as on-board control, except the central controller is notified of the orders and synthesizes the orders with its routing and traffic routines. When using remote terminals, an operator can control the entire system by use of a terminal. The operator must be made aware of the location and status of all the vehicles and stations throughout the system. This information is provided to the operator by status charts or graphic representation of the system on his or her computer screen. When using a central computer, the computer assumes the responsibility of the operator using a remote terminal and the system becomes automatic. The central computer may be directing all facets of the manufacturing process including AS/RS, resource management, and all other forms of automatic machinery. The computer may control and record everything from receiving to shipping.

When using a central computer there are many other advantages. By placing fire sensors on-board the vehicles, it is possible to obtain remote and dynamic information on any fire activity in the plant and help can be automatically summoned. Battery levels are monitored automatically and vehicles can be sent for battery change or plug themselves in for recharging. If a vehicle goes down then another can be sent in its place automatically. There are disadvantages, however. The more complex a system becomes, the higher the cost and the lower the reliability.

Tips for Successful Automation: Criteria for Selecting AGVs

The selection of the most applicable AGVs is of utmost importance. These are some of the criteria to use in this selection process:

- *Plant materials flow*—number of pick-up and drop locations, distances between these points, load profiles, and materials flow density.
- *Internal restrictions*—space available, safety regulations, and building parameters.

- *Performance requirements*—rate of load transfer, vehicle travel speeds, expansion capabilities.
- *Financial requirements*—cost of investment, operating expenses including maintenance and servicing, equipment amortization, relocation and/or plant expansion.

WHY SIMULATION ENCOURAGES EFFECTIVE USE OF AGVs

Simulation is a valuable tool for use in providing information on the planned AGVs. A simulation can enable a user to develop the following data:

- Point-to-point routing protocols for all combinations of pick and drop locations
- Determination of guidepath blocking and control points
- Control of algorithms to avoid traffic congestion
- Enroute protocols for avoiding vehicle interferences
- Empty vehicle management algorithms
- Vehicle scheduling algorithms
- Activity profile for each vehicle
- Average and maximum queue sizes for all pick and drop locations
- Component utilization for vehicles, pick and drop facilities, and guidepath segments
- In-station transactions consisting of response and execution time
- Number of pick and drop activities by station

In addition to the generation of control logic directly from the simulation model, advancements also have been made in the simulation output, mainly in the areas of graphic animation, vehicle emulation, and control-logic generation.

SPECIAL AGV APPLICATIONS IN HEAVY PAYLOAD*

As this chapter indicates, there are many types of AGVs, and numerous applications for their use. This segment of the chapter deals with special applications in heavy payload AGVs—10,000 pounds and higher. A special application is any nonstandard AGV system.

The following applications are illustrative of AGVs that fall within this category:

Robot Transporter Automated Guided Vehicle System

This type of application is important because providing mobility to robots will be critical to future advancements in automation technology.

In one instance, the task was to provide a mobile platform for a painting robot. The robot's job was to paint an airplane. It had to be moved around the plane by the AGVs in a fixed, repeated U-shape path, 60′ × 40′ × 60′.

The robot is 25 feet tall with a painting arm 15 feet long. A vehicle had to be

*Based on address given at Fourth International Conference on Automated Guided Vehicle Systems, Chicago, IL, June 25, 1986. Permission of Production Machinery Corporation, Mentor AGVS.

designed not only to carry the robot, which weighs fifteen thousand pounds, but to provide stability to the robot while it is painting.

Power for the robot's movements was to be provided by dual hydrostatic drives located on the transverse centerline. It was decided in the early stages of development that the same hydraulic pump could provide power to the AGV drive motors.

Because of the limited movement overall, A/C electric power for the pumps could be provided by Festoon cable brought in overhead.

The paint sprayed by the robot was a special volatile compound. This required the vehicle and the robot to be explosion proof. All relay circuitry and switches had to be contained in specially designed enclosures.

How the System Works

An aircraft on a trailer is located by a tugger at a precise location in the plant facility. The AGV receives a command to move to position one.

When the move is complete, a second order is given to latch the floor. Special hydraulically actuated levers mounted on the vehicle latch corresponding notched plates specially located in the floor at each address.

Having moved and latched, the vehicle is shut down, and the hydraulic servo valve redirects the power to the robot for painting. Then, when the first position paint cycle is complete, the vehicle is commanded first, to unlatch from the floor and move to Position 2, relatch and signal the robot to paint. Twelve moves are required to complete a cycle.

The primary gain in efficiency that this concept brings to spray painting a large object is that only one robot is required to do the job because of its mobility. If the robot were stationary twelve would be needed. If tracked robots were used, then at least three would be required.

As you can readily ascertain from this application, providing mobility to robots will be a key area for future AGVs developmental activity.

Carrying Pallets of Sheetmetal Blanks, Ford Motor Company

This heavy payload application, developed for the Ford Motor Company, consists of a 16,000-pound sheetmetal blank handling unit (figure 7–10).

The blanks are carried on a steel pallet 60 inches wide by up to 110 inches long, weighing up to 15,000 pounds. The project is used to simulate a full system for the benefit of all the plants within Ford's Metal Stamping Division. Plant engineers and managers from other plants see how blank handling by AGVs works and how they can incorporate this method of blank handling for safety, for maximizing press utilization, and for computer integration of blank handling along the head of the draw presses.

Powered rollers on the vehicle permit automatic loading and unloading of the vehicle from powered stationary conveyors.

Ford required that the AGV be able to pick up pallets from present conveyors that are no higher than twelve inches off the floor, and that the vehicle be able to discharge the load onto conveyors that are forty-eight inches off the floor, or any height in between.

To achieve this, a double ball screw was incorporated into the design for lift. Then with an LED on the lift table, the table could be raised until it "saw" a retroreflective unit on the conveyor to be served, which matched the height required.

The Ford facility has wood block floors throughout. In most installations where there is wood block, a concrete runway is laid to eliminate trying to lay a wire in shifting, heaving wood blocks.

FIGURE 7-10 16,000-pound Sheetmetal Blank Handling Unit (Courtesy of Production Machinery Corporation, Mentor AGVS).

In this application, the wood was removed, the guidepath was cut into the concrete subfloor and asphalt blocks were put down to replace the wood.

Asphalt does not heave the way wood block does when it gets water soaked. The asphalt blocks are usable the moment they are put in place, unlike concrete which has a long cure time.

On-board antennas need gain adjustment to receive the signal four inches away, but the concept proved workable.

How the System Works

As a fully developed system, the vehicles move palletized sheet metal blanks from the blanking line, where coil steel is turned into blanks, to and from storage to a draw press line.

The pallet of blanks, after being unloaded at the press line is transferred under the destacking automation and loaded robotically into the draw press.

The AGVs replaced material handling performed by a number of forklifts, which have the following drawbacks:

- Damage to the product
- Unpredictable productivity of machine operators
- High labor cost
- Inability to computer integrate the material-handling process when it is performed manually
- High forklift repairs

A typical approach for testing new technology has been through costly trial and error. As mentioned previously, simulation has gained support by letting the technology prove itself on the computer screen before it hits the plant floor. Another middle ground involves combining the best of both: a physical product actually going through its paces, coupled with the security of easing into the technology, and carefully testing it over the broad range of possibilities.

This method has proven to be a popular way to use AGVs for special applications in heavy payloads! Obtain one in a limited system, see how the design works in the physical environment, make the necessary changes, then go on to a full up-system.

Blank Handling with AGVs at General Motors, CPC Group

Blank handling with AGVs at General Motors' CPC Group has taken a similar route to the one pursued at Ford.

There are enough differences and similarities to the Ford project to make it worth comparing. As with Ford, a one vehicle pilot system contract was let to demonstrate the practicality of the method of blank handling. This was another physical simulation conducted to prove the concept's viability. All affiliated GM plant groups observed the physical simulation pilot and then evaluated its merits.

The GM load is heavier than at Ford; 20,000 pounds at GM compared with 15,000 pounds at Ford. The load is carried the same way—on steel pallets. It is picked up and delivered the same way, with side transfer powered roller conveyor on the vehicle matching powered off-board sections.

How the System Works

The system works in the same plant location, picking up from the blankers, delivering to the ASRS, taking from the ASRS to the head of the draw press line, where other automation loads single sheets into the presses.

It differs in the following respects. The AGVs have a fixed load/unload height of 28 inches. Off-board P/D stations match the vehicle's height. (At Ford, heights could vary from 12 to 48 inches for load and unload.)

GM required ready access to all high maintenance components for repair and replacement. This gave them a more usable product.

GM also required the vehicle to be steam cleanable, making it necessary to design component mounting compartments able to withstand 180 psi water jets.

Vehicles are interchangeable plant by plant, yet systems are tailored and unique for each operation.

Tool Room Modernization at General Motors

A special application, heavy payload AGVs, has been installed at a GM plant for moving raw and in-process castings.

CNC machining centers are the center piece in the project to modernize the way stamping dies are manufactured.

An AGV system was purchased to move the dies mounted on air bearing pallets from operation to operation. This involved development of a flat decked vehicle, large enough to handle a pallet 98″ × 137″. All dies in-process are mounted on this standard pallet. The combined weight of a typical die, fixture, and pallet is 60,000 pounds. (See figure 7-11.)

FIGURE 7–11 70,000-pound Capacity Deisel Hydrostatic Pallet Truck Transporting Raw and In-Process Castings (Courtesy of Production Machinery Corporation, Mentor AGVS).

How the System Works

The tool room consists of twenty-four workstations which the AGVs must serve. Critical X,Y,Z accuracy requires cones to be embedded in the floor at each load/unload table. The vehicle frame has inverted cones within sturdy legs mounted to air bags. When over the in-floor cones, the air is released from the bags, and the legs drop mechanically to comply with the floor cones. This gives X,Y,Z accuracy $\pm$ one millimeter.

The second step in an unload maneuver comes when an operator plugs in local plant air. The air bearing pallet with the die mounted on top rises from the flush vehicle deck. Locating holes on either end of the pallet engage into locating pins. These pins are on the end of a rack and pinion arm on either end of the load deck.

Now the arm, driven by an electric motor, moves the pallet off the vehicle onto a waiting work table. The air pallet is deflated, and the work on the die can proceed.

When that phase of work is complete, the cycle is reversed.

Steer drive motors are mounted on the longitudinal centerline of the vehicle, one on each end. Three swivel casters each are mounted on six swivel plates, forming the major weight bearing mechanism for the load.

This configuration permits the vehicle to move 90 degrees from its normal bidirectional mode. The crabbing maneuver allows the vehicle to pull up flush to the work tables. Without the ability to meet flush, the air bearings would deflate over the crack between the vehicle and the table. This is one reason why the cones in the floor are necessary to achieve high accuracy for the air bearing transfer maneuver, as well as for the machining centers.

An AGV system used in this way greatly speeds the material handling. The workstations are located in two separate crane bays, such that with the AGVs the workpiece would require riggers to ready the load, a crane to pick up, and if the next workstation was a crane bay away, then the piece would have to be put on a transfer car. After shuttling the load to the next bay, Crane 2 now has to be available, and so on.

The previous methods for handling dies without AGVs are among the slowest material-handling activities in use. In addition, these methods require mill-wrights and riggers, with very specialized skills. Because their task is dangerous, they move cautiously, and should something slip—aside from someone being hurt—expensive dies may be damaged.

AGVs offer considerable savings in productivity and labor in die handling, as well as instituting safety measures that eliminate the possibility of a crane-related die being inadvertently dropped.

Moving a Mixing Bowl Loaded with Rocket Fuel

This application involves moving a 70,000 pound mixing bowl loaded with rocket fuel to mixing and curing operations for an aerospace supplier. The mixing bowl is 10 feet square on its base and 14 feet high with its lid on.

Each manufacturing operation is located in discrete buildings, hundreds of feet apart, and separated by covered concrete runways in heavy earthen bunkers.

Any incident that might possibly occur is planned for, such that only one building or operation would be affected. As a result, the buildings and runways extend over nearly one mile.

The AGVs must transport these mix bowls between the buildings. Because the runways connecting the buildings are unheated, ambient temperature in winter can vary as much as 100 degrees Fahrenheit.

The vehicle is separated into two discrete elements: the cab end, which is much taller, and the carrying deck which is 9 inches when lowered.

In the cab end, a 105 horsepower explosion-proof diesel engine is resident. The diesel powers two hydraulic pumps which in turn power the two drive wheels. By altering the speed of the two drive wheels, steering is effected differentially.

When going from building to building, the AGV travels cab first. When picking up or placing a bowl, the vehicle performs a rotation of 180 degrees and enters the mixing facility, bowl first, like a forklift truck.

The bowl must be placed $\pm$ 1/16 inch on the X,Y axis. Traditional wire guidance cannot be used in many of the mixing rooms because of pits and lead-lined floors. Three cameras are on each vehicle. A vision system is used for guidance to complement the wire guidance where necessary. Vision also is used to read bar codes stenciled on the bowls.

Communications between the vehicle and the host are done through leaky coaxial cable. Leaky coax is regular coax cable with segments of insulation stripped off. In those bare areas, two-way communication can be exchanged with FM. In the mixing rooms, the vehicles are in continuous communications with the host. In the runways, communication is intermittent.

The computer system for controlling vehicle routing consists of dual micro VAX II computers. One is 100 percent hot backup in case of primary computer failure. The redundant backup can take over the system management in less than 10 seconds.

Color graphic CRTs display critical diagnostic and system status information to the main computer room where technicians monitor the operations.

The fuel is made in batches, (campaigns) that have a beginning, middle, and end phase. Each campaign begins with a vehicle and systems check-out process, where critical areas of the vehicles and systems are examined for readiness.

Dozens of critical functions are monitored and judged according to acceptable norms. Any deviation requires corrective action prior to beginning the campaign. This ensures minimal performance failure during the campaign, where system down time could impact batch cure time.

This AGV system is significant in the following ways:

- It is the heaviest AGV ever employed.
- It uses new vision technology in addition to traditional wire guidance.
- Hot computer backup is employed.
- Explosion-proof vehicles are used with diesel hydrostatic drives.

Because of these features, it represents the leading edge of AGV technology and opens new horizons for users and potential users of AGVs.

chapter 8

Examples of Flexible Manufacturing Systems

WHY COMPANIES ARE MOVING TOWARD FLEXIBLE MANUFACTURING SYSTEMS

The accelerated movement toward development and implementation of flexible manufacturing systems is based on the need for companies to fully automate work cells to gain maximum flexibility and increased manufacturing productivity. This need is dictated by the marketplace which demands a wide range of versions of products and speed of reaction in production. However, flexibility in automated systems has its price, and only can be cost justified if the FMS is kept as simple and rigid as possible. For this reason, one of the most important tasks in planning of automated manufacturing systems must be to determine the actual requirements with respect to flexibility from the existing range of parts and to estimate the requirements for the future (based to a large extent on the company's strategic and long-range plans).

The use of highly developed products leads only to partial successes in terms of rationalization and automation if the newest developments in organization and planning are not implemented simultaneously. In modern technologies, for example, the use of process computers can be utilized fully only in combination with highly developed machines. In today's competitive environment, considerable effort is being made to utilize the advantages of NC technology in producing small- and medium-sized batches. The accompanying advantages, such as reduction of down times, can be regarded as a satisfactory solution only for small quantities. To achieve optimum results, other factors such as tools, material flow, and organizational structures in the technical and organizational design of the overall system must be included in these considerations. Installation of FMS is predicated on achieving complete integration of all components in the production process within a defined work area of activity.

AN INTRODUCTION TO FLEXIBLE MANUFACTURING SYSTEM

A flexible manufacturing system is defined as a group of numerically controlled machine tools that are interconnected by a central control system. The various machining cells are interconnected via loading and unloading stations by a transport system such as rail vehicles or inductively controlled transport trollies. Tested and optimized part programs are stored in a central data station for each workpiece of the range, which means that the interchangeable or neutrally supported NC machine tools are capable of, with the aid of the transport system, executing all manufacturing tasks on workpieces of one or more families of parts. Starting at the loading station, there is an automatic workpiece sequence that can be optimized in accordance with various criteria.

Measuring machine, automatic tool breakage monitoring devices, machine-talk monitoring circuits, lifetime monitoring facilities, optimization and strategies of the availability of single-system components changes, and modules for operational data recording, including the system diagnosis, are integrated in the machining and information flow. This makes it possible to continuously match the system to design and machining modifications without having to accept time-consuming conversion work that occurs in the case of transfer lines. A flexible manufacturing system is capable of processing small batches or even single items in any required sequence. These manufacturing systems are flexible, not only with respect to matching with various production tasks, but also with respect to the sequence and batch size of the workpieces passing through the system for machining.

Seven Prerequisites for Flexible Automation and How to Satisfy Them

Up to now, the production structure has been able to exist through development of relatively small incremental methods improvements within definable limits. However, recent advancements in product and production technologies and heightened worldwide competitive pressures have made it necessary for companies to fully automate production facilities as rapidly and completely as possible. Therefore, to become a low-cost producer by introducing FMS into the work place, a company must be able to satisfy these market-driven requirements:

- Produce more of different products.
- Produce more variations of product.
- Anticipate that products will have a shorter life cycle.
- Reduce design time (manufacturing lead time) so that new design introduction is faster, thereby allowing the company to be more responsive to a rapidly changing market.
- Produce in smaller quantities with faster delivery turnaround.
- Continually improve productivity due to constant cost pressures exerted by customer demands.
- Develop plans that will provide long-term stability against the inherent insecurity of the marketplace.

To meet these requirements, installation of a flexible manufacturing system must effect the following changes in the production facility:

- Automate production processes
- Achieve flexibility of machine tools
- Integrate control of work-in-process materials
- Integrate computerized information system

- Integrate preventative maintenance
- Provide job enrichment for members of the FMS team

Why Flexible Automated Production Should Encompass an Entire Area

To be totally effective, flexible automation must encompass the entire material flow and information system of a particular production area. Application of flexibility in segments or acceleration of material flow alone usually leads to an unsatisfactory result because a large organizational effort is required at interfacing points with the conventional structure.

Flexible automation requires very complex control and information systems. Reliable interaction between these systems and the mechanical systems, as well as each other, and the degree of availability is limited by the quality of maintenance programs.

Obsolete labor-intensive production processes must be consolidated. Planning, programming, and control are activities that call for the consolidation of independent areas in the plant.

The qualifications of plant personnel requires substantial upgrading to install and maintain a flexible manufacturing system.

BASIC ELEMENTS OF A MODERN, FLEXIBLE AUTOMATED FACTORY

What a Layout Should Include

The layout of a flexible, automated production area should contain these various components: elevated storage, robot stacker, transportation system, automated material handling, automated machine tool, and computer interface.

The overall structure of a flexible, automated production area should consist of: computer-controlled shelf storage systems interacting with automatic transport systems and flexible production systems, assembly, testing, and shipping activities linked with the storage area by means of transport systems.

Guidelines for an Effective Material Flow System

From both a technical and economic standpoint, a well functioning material flow system should adhere to the following guidelines:

- Transportation pallets should be standardized.
- Transportation pallets should be used from the entrance to the exit of the FMS.
- There should be no stacking of pallets.
- There should be no fixtures for material float.
- There should be a standardized level of material-float system.
- Material-float control should be done by means of computer software.
- Disposition of material should be by the assembly area.

UNDERSTANDING FLEXIBLE PRODUCTION SYSTEMS

Flexible production systems are the central components in flexible, automated production areas. These flexible production systems contain the following features:

- Diverse numerical controlled units
- Systems control by a process computer
- Autonomous material-float control
- Applicability for a workpiece family
- Different lot sizes
- Inconsistent mixture of workpieces
- Automated material float
- No time for presetting

Flexible production systems may consist of production cells or production islands. A production cell is an individual NC machine with automated loading and unloading devices for workpieces, tools, and process aids which may be interfaced with supplemental, nonprocessing machines, such as measuring devices.

A production island is an organizationally independent production unit comprised of conventional, as well as numerically controlled machines. Integration of the control system and material flow is not required.

Operations, including production cells as well as production islands, are considered solutions transitional to future types of production systems. Future improvements in software should optimize information material flow systems. Through this particular feature the benefits of flexible automation will accrue more readily.

Selecting the Best Layout for Flexible Production Systems Connected by Robots*

Essentially, flexible production systems configurations differ in regard to the material flow linkage. Previously, linkage had been accomplished by means of horizontal conveyor systems such as roller or belt conveyors. Today, increasing numbers of robots of various designs are utilized and interlinked with conventional conveyor, transport, and storage systems.

The use of robots offers certain alternatives as shown in figure 8–1.

Configuration Number 1 is preferable if:

- Transport weights are excessive
- The machine arrangement is changed frequently
- Complementary machines are interlinked
- Only *one* loading and unloading direction is required
- Only workpieces are handled
- The available area is not an issue

Configuration Number 2 is preferable if:

- A nonchanging layout can be assumed
- Floor space must be preserved (or kept clear)
- Only *one* loading and unloading direction is required
- Only workpieces are handled

*Permission granted by Planring U.S.A., Inc.

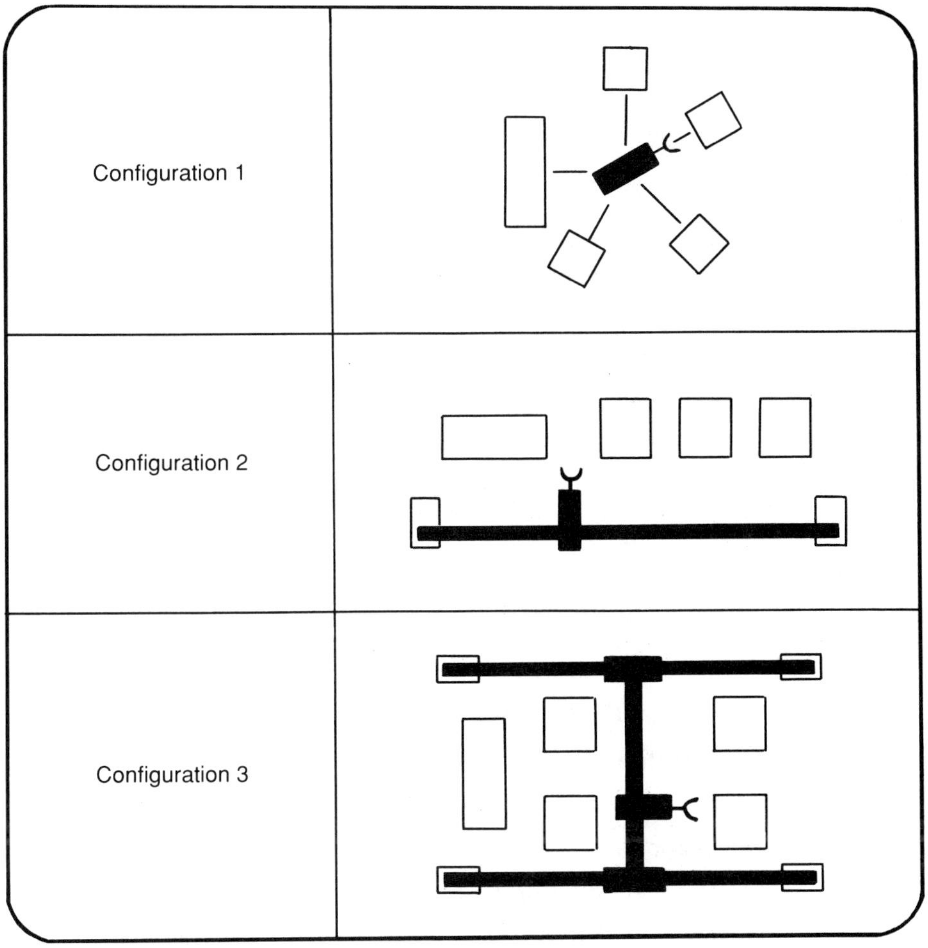

FIGURE 8-1 Basic Robot in FMS Alternatives (Courtesy of Planring, USA).

Configuration Number 3 is preferable if:

- A nonchanging layout can be assumed
- Floor space must be preserved (or kept clear)
- The machine is to be loaded and unloaded from different directions
- Not only workpieces, but also work-holding tooling and cutting tools are to be handled

The Two Types of FMS Control Levels

FMS control levels are divided into *machine controls* and *systems controls.* The functions of these two control levels are as follows:

Machine Controls	NC	SPS
Machine tool	X	
Measuring machine	X	X
Robot	X	
Store		X
Transportation system		X
Preparation station		X
Coolant system		X

Systems Controls

Level 1—Function

Coordination of units

Level 2—NC Data

Administration

Distribution

Dialogue

Modification

Level 3—Organization

Material float planning

Material float controlling

Operating data processing

Preventive maintenance

Diagnosis

Documentation of maintenance

Tool organization

Fixture organization

Quality assurance

Monitoring

Presetting data

Tips for Successful Automation: How to Implement an FMS

Although implementation of FMS should be based on overall plans for the final stage, the actual transition should be done in a step-by-step fashion. Since employees in FMS are involved in a broader range of activities than in labor intensive conventional configurations, their qualifications also must be upgraded by means of continuous training before implementation. This should also follow a step-by-step procedure in consonance with the physical changes that are made.

Traditional, structured line organizations must be adapted to human needs and capabilities, as well as to the technological/economical requirements.

UNDERSTANDING THE BASIC ELEMENTS OF FMS

Flexible manufacturing systems usually focus on the machining of large standard parts in moderate volume. The principle elements of FMS include:

- NC machine tools
- DNC (distributed numerical control)
- Automated material handling
- Supervisory computer control

Although it is imperative that the selection of applicable machine tools, and material handling systems be researched completely and selected properly, these are basically mechanical systems that are dependent on the processing requirements, and therefore perform their task if the plans and support systems are in place. Thus, the computer control system is the linkage needed to transform a group of machines and stand-alone systems into an effective FMS. The functions performed by the computer control system can be categorized into either machine support or planning support.

Items such as NC part program storage, distribution of the part programs to the individual machine tools, traffic control, and tool control are considered requisites to support the machine directly. Functions such as routing, scheduling, and system performance monitoring and reporting are needed to plan the employment of resources. The planning support level is where some differences in approach that may develop between functional area processing (represented by MRP and MRP II) and cellular manufacturing (represented by GT) may be readily seen.

Differences between MRP and GT

MRP is an inventory management tool utilized to obtain the right material at the right time in support of a master business plan. The advantage of using MRP is to reduce inventories by not having finished parts ready that do not have immediate use. Thus MRP is driven by the end-product needs whereby the run quantity is a function of the amount required to satisfy the final end product.

In contrast to MRP, GT is a method of grouping and analyzing discrete manufactured parts for the purpose of improving design and manufacturing productivity, thus lowering operating costs. GT focuses on the reduction of work in process by improving the throughput time to get a part through the shops, thereby reducing the queue time. The rationale of GT is based on the fact that similar parts should be produced together to minimize set up cost. In this context, GT is product independent.

Since MRP focuses on *effective* manufacturing, and GT is directed toward *efficient* manufacturing, there is the inherent conflict between these two methodologies as they impact flexible manufacturing systems. Thus, installation of FMS must arrive at the proper balance between these two approaches to achieve optimum results.

The Three Phases of the Shop Schedule

The interface between MRP and GT is the shop schedule, which involves the following three phases:

1. *Process determination*—the selection of raw materials and the processes needed to transform the material to meet the customer's requirements.

2. *Sequencing*—an approximate process flow, or a sequence in which the processes are to be performed. Usually, there is some flexibility, but it can be arbitrary.
3. *Scheduling*—the procedure of actually routing the part through the different shops in the factory based on the available machining cells and other workstations. Each sequence as it is generated, has to be matched against the constraints imposed by actual conditions in the manufacturing environment.

This should be looked at as three unique, yet dependent problem areas. Their dependence lies within each other. Attention must not be directed to one area without viewing the impact on the remaining ones. Thus, the objective is developing three systems collectively that accurately reflect the decision-making process based on knowledge about the shop and the current status of orders and machines. To achieve this objective, the problem of transferring knowledge must be recognized.

Six Benefits of FMS

The major benefits of a flexible manufacturing system are as follows:

- Increased labor productivity
- Increased quality
- Less scrap, rework
- Increased production flexibility either by product design or production volume
- Reduced operating costs
- Higher machinery utilization

GUIDELINES FOR UPDATING THE FACTORY FROM DNC TO FMS*

DNC is an often confused acronym, as there are two definitions: *direct numerical control* and *distributed numerical control.* The two terms are very different. In a typical direct numerical control system, all of the axis servo-drive mechanisms and auxiliary function magnetic switches of from one to eight or more machine tools are connected to a large mainframe class processing computer. These computers have phenomenal throughput processing speed and resident storage capability, and execute all the machine part programs simultaneously. The inherent concern in this real-time processing/operating system is that any failure, or even a slight error, can cause severe consequences.

A distributed numerical control has all the advantages of a direct numerical control system without any of the inherent risks. In a distributed numerical control system each machine tool has its own dedicated microcomputer or CNC, complete with enough memory storage to handle a whole part program. The microcomputer of CNC is comparable in function to the mainframe computer in the direct system. The microcomputer of CNC communicates with a minicomputer that handles distributed processing between the host mainframe computer and the CNC. This minicomputer or distributed processor handles part program transfers and management status data collection, hence the word "distributed." For the remainder of this discussion the acronym DNC will refer to the more state-of-the-art distributed system as opposed to the antiquated direct system.

*Permission granted by Charles Rosbeck, Fairchild Republic Corporation, and Michael Hitchcock of Manufacturing Technology Division of the ATR Force Wright Aeronautical Laboratories, Wright-Patterson AFB.

DNC now has evolved into the FMS technology. The reason for this evolution is that the best solution to the challenges of achieving manufacturing productivity involves an integrated systems approach which attempts to synergize all of the various technologies. DNC is a critical and large portion of the technologies that reside under the FMS umbrella as DNC addresses the hierarchical integration of all the computer technologies involved in FMS.

Technological developments, especially in the areas of minicomputers and microcomputers, hardware communications, CRT displays, mass storage devices such as floppy disks and EPROM memories, have not only promoted the evolution of FMS but also have allowed its practical application through reductions in cost. These cost reductions have made the hard-wired logic controller, for all practical purposes, obsolete.

The specific concern that DNC seeks to mitigate, is the excessively high cost and length of time required to produce machined parts as well as the time and cost for some processes for assembly manufacturing where NC technology has been applied. Examples of this are positioning devices for spot welding and drilling (Gantry robots); and programmable routing equipment such as sheetmetal and wood routers, plasma, laser, flame, waterjet, and knife-cutting machines. In the case of FMS this also includes programmable controllers and other computer equipment used for the control of guided carts, pallet shuttle systems, material-handling devices such as robots, and in some instances in-process inspection equipment.

Advantages of DNC

DNC can eliminate the following inherent NC problems:

- Tape punching errors at the computer
- Tape reading errors at the control
- Tape transportation and inventory
- Excessive new tape proving time
- Excessive tape turnaround time caused by:
  - (a) NC programming
  - (b) Computer processing
  - (c) Tape punching
  - (d) Transportation and routing
- Reprocessing to execute on a different machine

A premise of any DNC system is that the installation has a CAD/CAM graphics system. The reason is that this allows secure storage and transfer of engineering drawings which are the input and foundation of any DNC/FMS system because drawings are used to develop the part programs of the system. It is very difficult, if not impossible, to manage a system of hardcopy (blueprint) drawings securely.

What a DNC System Should Achieve

A DNC system should achieve the following list of objectives:

- Ability to offload work from one machine tool to another without any reprocessing
- Interface to CAD/CAM systems to allow utilizing that data base and its resources
- Reduced down time and maintenance of production equipment

- Increased reliability and productive capacity
- Reduced vulnerability of production schedules from equipment failure, schedule changes, and engineering design changes
- Improved product quality and control
- Reduction in paperwork flow
- Accurate and timely collection of productivity and capacity utilization data
- Improved part inventory tracking
- Reduced tool-proving effort
- Better procedures for optimizing and analyzing part programs and tooling to increase productivity
- More timely access to progress/schedule tracking data
- More secure, maintainable master data base accessible by all affected organizations
- Accurate historical records
- Flexible and adaptable to future growth and operational climate changes
- Standardization
- Allow for incorporation of operational systems such as computerized process planning, computerized machined part inspection, computerized order writing and scheduled release, and time standards
- Allow for incorporation of other computer equipment besides CNCs and any configuration of computer equipped manufacturing equipment. The limitation is the support of the protocol format by other equipment.
- Fast return on investment (phased implementation)

The Important Role of Computerized Numerical Control in a DNC System

The single most important criteria in the design and functional success of a DNC system is the functionality of the CNC. CNC capability will dictate every design aspect of a DNC system: programming, program portability, data transmission, processing, accuracy and integrity of data, data collection, realized productivity, maintenance, and reliability.

CNC has replaced the hard-wired logic controller due to increased capability and decreased cost. Most of the increased capability is the result of CNC processor's ability to perform real-time calculations such as acceleration and deceleration rates; linear, circular, and helical interpolation; cutter compensations; grid offsets and alignments; and self diagnostics. Previously these functions were accomplished in post-processing or were not available. Besides being extremely reliable, CNCs can be purchased with storage media other than punched tape readers. The most reliable, fastest, and lowest-cost portable storage media for CNCs are floppy disks.

Why "CL Data" Is a Valuable Feature

One of the most valuable features of CNC technology is CL data formatted part programs. CL data formatted part programs are programs that are processed by a CL data converter rather than a conventional post processor. CNCs equipped with this option offer tremendous value to the machining production environment because all CL data formatted part programs are universal—that is, machine independent. Any CL data formatted part program can be executed on any machine tool equipped with a CL data CNC, and is physically capable of producing the part without reprocessing. Having this capability allows work in process to be moved to different machine tools to accommodate new work or a machine tool out of service, without reprogramming or reprocessing.

It is also desirable to have the capability of editing part programs at the CNC during tool proving. This capability allows the NC programmer to not only make minor corrections to the part program quickly without reprocessing, but also to evaluate the performance of those changes before final processing. These changes are not written on the floppy disk but are instead written into an executable buffer that can be dumped to a line printer or communicated to a remote terminal in the NC programming area to be incorporated later into the source part program before being released to production.

Using Pictorial Operators Instructions to Reduce Scrappage

Another very desirable feature for CNCs is pictorial operators instructions. This capability requires the use of a CAD/CAM computer graphics system. This consists of a graphic display of the part created using a CAD system. The CAD drawing consists of a view that shows clamp and bolt locations and contains information about cutting sequences and tool changes. These can be in addition to PRINT statements. This graphic display would appear on the control's CRT and is flagged to the part program such that the pictorial displays can be paged, thereby executing cycle start displays that contain the correct instruction sheet for the sequence to be executed. This capability eliminates operator instruction sheets that indirectly are one of the largest causes of part scrappage.

How a Hierarchy of Computers Comprises a Distributed Network

A DNC system should consist of a hierarchy of computers configured into a dedicated point-to-point network. There are four levels to the hierarchy: the host computing level (HLC), the part programming level, the distributed computing level (DLC), and the CNC level. The HLC handles part program processing, management information processing, report generation, control of all systems diagnostics including automatic IPL of any distributed level computers or CNCs, all peripheral devices such as line printers and terminals used for report generation, status inquiries, maintenance requests, and communications with all DLCs.

All source and processed part programs are stored at the HLC. The HLC is the correct level to control part program release and security. The DLCs handle all communications between the HLC and the CNC level. This is also the level at which it is desirable for order release to be incorporated. A sufficient number of part program copies to support the particular production schedule should be stored at the DLC to be down loaded to the CNCs by request. The DLC also pools the CNC automatically at time intervals for status information.

The CNC level is basically the machine control that can be characterized as a 16-bit microcomputer terminal with a CRT and MDI keyboard. The CNC level handles communications and collection of performance and status data with the DLC. The CNC also performs mechanical-error monitoring which is used to signal maintenance requests. The part programming level involves everything that has to do with developing and maintaining part programs. It also is from this level that part program source is submitted to the host level for processing. Operator instruction graphic displays also would be generated from this level. This level also handles part program proving and incorporates engineering changes. This level is characterized as being the most people-intensive of all of the levels.

HOW SIMULATIONS FACILITATE FMS DESIGN*

When you analyze the function of flexible manufacturing systems, the key aspect is its ability to adapt to change, rather than the amount of automation involved. A fully operative FMS is flexible in terms of the quantities of parts it can produce, the types of products it produces, the sequence that these parts are processed, and its ability to reroute parts back into flow paths.

Flexible manufacturing systems consist of a number of workstations capable of performing different operations. Each of the workstations may vary in terms of combinations of automated, semi-automated, and manual operations. FMS workstations are linked together by an automated material handling system which is computer controlled. A central computer monitors, controls, and schedules the entire operation. When considering the installations of a FMS, the potential configurations are almost unlimited, depending entirely on what you expect to accomplish from this advanced technology.

Using Simulations as Planning Tools

To develop an appropriate and cost-effective FMS solution, you need a planning tool which permits the study of the dynamic behavior of the FMS. Such a tool must be able to analyze the effectiveness of the physical system, the control system, and the ongoing competition for the critical system components. Furthermore, it must provide for an evaluation of the impact of changes in key parameters of the system, should this occur.

This planning mechanism can be achieved through the use of computer simulation. Simulation can enable one to describe each FMS design and predict its performance without actually building the physical system or the control system. It can be used to study the effect of varying throughput rates, queue sizes, processing times, part routings, processing sequences, and inventory practices. It can aid in the identification of the required material-handling equipment and associated fixtures and pallets. It also allows you to evaluate control strategies and operating rules. In addition, today's advanced simulation technology enables you to portray FMS design alternatives in animated screen display fashion and to summarize the results in standard business graphics format.

How Simulation Models Contribute to the Design of FMS

To automate a factory, an integrated approach must be taken, which considers engineering, manufacturing, and information systems. In essence, this means that you must consider the entire manufacturing operation to avoid optimizing parts of the system and then ending up with a suboptimal whole.

The design of an FMS consists of two subsystems: the physical subsystem and the control subsystem. The physical system contains the machines, operators, material-handling equipment, and in-process storage facilities. The control system is the logic that governs the operation of the physical system including the introduction of parts into the system, the assignment of machines and operators to the jobs, and the routing of parts and traffic control. When designing an FMS, selection is from alternate physical configurations and control algorithms. The two are very closely related and must be evaluated in combination.

There are design issues associated with each of the FMS subsystems. Within the physical subsystem, determinations need to be made such as the products to be produced, numbers

*Permission granted by Pritsker & Associates.

and types of machines; numbers and types of material-handling equipment; number, size, and location of in-process storage devices, and personnel requirements. Within the control subsystem, shift schedules, machine and personnel assignments, part routings and part decisions, production schedules, and material-handling decisions are selected.

To evaluate various FMS configurations a yardstick is needed against which to measure increases or decreases in system performance. This yardstick is the same set of critical measures that are commonly used to gauge the actual manufacturing system's performance, including system throughput, equipment utilizations, inventory levels, and the system's ability to meet deadlines. System designers are able to attach a value of one configuration relative to another based on each system's performance.

Simulation models are able to calculate the same measures of system performance for hypothetical system configurations as are used to judge real systems. Thus, FMS designers can try out various FMS configurations without actually building a system. Alternate vendor FMS designs can be evaluated before selection. It is much more cost effective to evaluate alternate configurations with a simulation model than with actual hardware.

What the Simulator Needs to Be Effective

To be an effective design tool, an FMS simulator must have certain key characteristics. First, it must be able to represent the physical FMS accurately. Second, it must be able to describe the control logic that governs part routings through the physical system. Third, and perhaps most importantly, the simulator must be easy to use. A completely data-driven simulator has definite advantages. It can be learned by manufacturing engineers who have little or no simulation experience, often in a matter of days. In addition, models can be built more quickly when the computer programming and code debugging have already been completed.

Conveyors are frequently used in FMS as material-handling devices to transport parts between workstations. In some systems the conveyor also is used as a temporary in-process storage device where parts circulate until they are able to have the next processing step performed. Both closed or open loop conveyors may be present in an FMS. Parts may accumulate at a station along the conveyor to await processing. Alternately, parts may bypass a station if it is not able to process a part that has arrived for processing. A conveyor path may split into several paths, or alternately, several paths may join into one conveyor path. A simulation tool that is appropriate for modeling FMS will be able to represent all of these possible conveyor configurations.

Fixtures also are common in FMS. A fixture is a device used to hold a part or group of parts during machining or transport. Once a part or group of parts is loaded into a fixture, it usually remains in the fixture for several consecutive machining operations and subsequent transports. Following one sequence of operations, parts in fixtures may be turned over or rotated before proceeding to the next sequence of operations. Parts may use several types of fixtures during their processing, but typically they are loaded into only one fixture at a time. Some types of fixtures are dedicated to a particular product type, while others may be used by several types of parts. A simulation must be capable of representing a variety of fixture uses to support the design effort.

The control logic to route parts through an FMS can be very complicated and is almost always dependent on system status. Decisions regarding part processing are frequently made based on the status of the part itself, machines, operators, material-handling devices, or in-process inventories. The control logic for an FMS is usually provided by the FMS hardware vendor. A simulator must be able to represent a variety of control system strategies to be of use in choosing among various FMS designs.

How Simulations Help Justify Buying an FMS

The task of justifying the purchase of a flexible manufacturing system is much more difficult than justifying a more traditional manufacturing system. Since the payback is longer, more variable, and of a different nature than traditional manufacturing systems, traditional ROI analysis cannot be the sole determining factor in an FMS justification. The big payoff is increased flexibility.

One of the main advantages of an FMS is its ability to adapt to unpredictable changes in such things as product life cycles, production requirements, raw material availability, and product designs. Given the difficulty in predicting what an FMS will be required to do, you can understand the reason why FMS is difficult to justify, and why it is considered to be a high risk from a financial point of view.

Given all the unknown factors in the future operations of an FMS, simulation analysis is even more critical in its justification than it has been in justifying traditional manufacturing systems where more factors are known. An FMS needs to have the flexibility to adapt to such changes as unforeseen product mix changes, product design changes, new product introductions, and unknown new technology. It may not be possible to demonstrate an FMS's ability to successfully adapt to any type of change, but with simulation the risk of an FMS can be reduced by quantifying the impact of all anticipated changes.

A simulation analysis should be part of any FMS proposal. The analysis should demonstrate the proposed design's ability to perform as needed, given today's production requirements, and the proposed system's ability to adapt to a variety of possible future changes. Once a vendor has been selected and a particular FMS installed, the same simulation model can be used to identify the best way to take advantage of the system's flexibility to respond to a particular change. When certain changes can be anticipated, the model can identify appropriate responses to such changes before they occur so that you can be ready for them. This advanced knowledge of how to respond to change will enable you to know what production adjustments should be made sooner, and to be able to respond competitively to change.

CASE-IN-POINT ILLUSTRATIONS

Pritsker & Associates, Inc. is a leader in the use of simulation methodology to factory automation technology. The following four cases are illustrative of various types of FMS applications where Pritsker & Associates has applied its simulation methodology successfully. This methodology involves the employment of certain proprietary software technologies, such as SLAM II simulation model, TESS simulation and communications analysis tool, and MAP/1 simulation model.

These examples are indicative of the capabilities that simulation methodology can provide a company engaged in an FMS undertaking, to minimize unforeseen problems during the various physical stages of the project. Experience has shown that excessive cost and time overruns in installing an FMS are due primarily to inadequate planning and lack of simulated test procedures.

In addition to these relatively brief Pritsker and Associate case illustrations, the author also has included a more detailed example of the use of simulation methodology applied by Grumman Data Systems Corporation for the specification and design of a proposed flexible manufacturing system for aircraft machining.

CASE-IN-POINT

CLIENT: ACME-CLEVELAND CORPORATION
FMS DEMONSTRATION SYSTEM

Situation

A flexible manufacturing system was designed by Acme-Cleveland to provide an area where emerging FMS technology could be developed and displayed, and where customer applications could be studied and demonstrated. The layout of this FMS is shown in figure 8–2. It consists of six machines: two horizontal milling machines, one vertical milling machine, one turning machine, one chucking machine, and one lathe. These machines are interconnected by three material handling devices; an automated guided vehicle, a robot, and an overhead gantry.

Parts arrive to the system in various batch sizes and are loaded onto pallets. The pallets are picked up at the buffer storage area, shown in the lower portion of the figure, and delivered to the appropriate machine via the automated guided vehicle. If the machine is busy, buffer storage is available at several locations. Once processing is complete, the pallet of parts is transported to the next station via the appropriate material-handling device. After the parts have completed processing, they are placed in finished part storage and later removed from the system.

Solution

- Illustrate system performance via a simulation model
- Demonstrate the use of simulation in shop floor scheduling
- Enable customers to examine the capabilities of the FMS, taking into account:
  - (a) Schedule changes
  - (b) Machine breakdowns

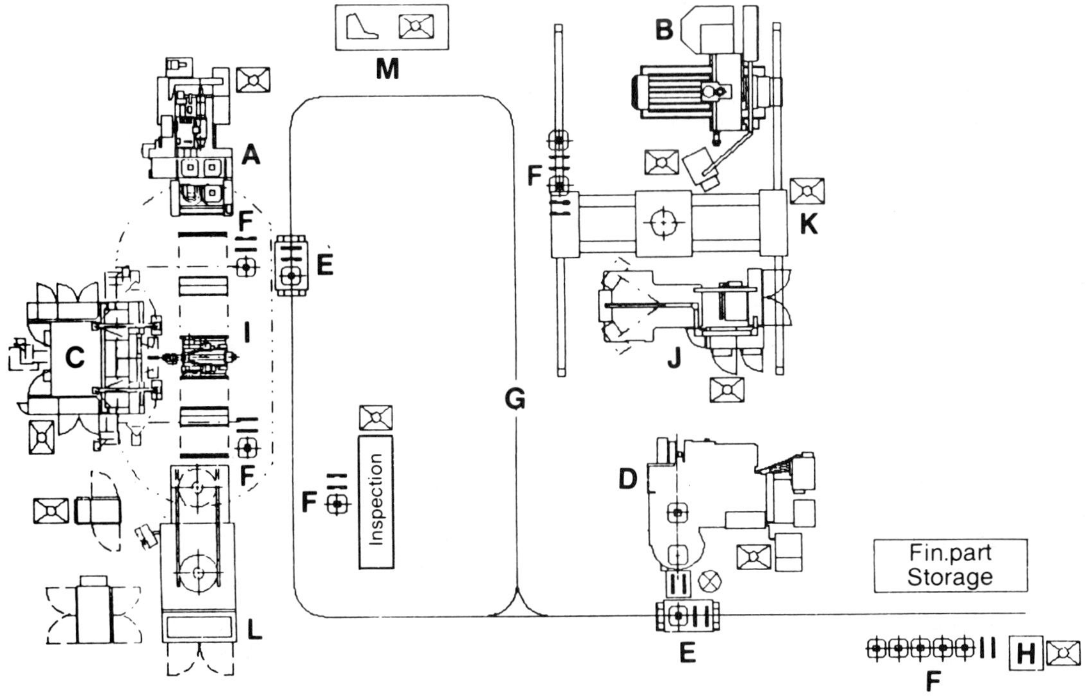

FIGURE 8–2 FMS Demonstration System (Courtesy of Pritsker & Associates).

(c) Machine maintenance
(d) Rework
(e) Shift schedules

Results

A SLAM II simulation model, together with TESS, were used to represent a broad range of scenarios. Customers are able to load the model with their particular product mix using a series of menus. An animation graphically portrays the movement of parts through the system. Accompanying plots describe system performance over time. Customers are able to easily alter the system and study the effects. The result is a better understanding of the benefits to be derived from a flexible manufacturing system.

CASE-IN-POINT

INJECTION BODY FMS DESIGN EVALUATION

Situation

A flexible manufacturing system that performs machining operations on castings is shown in figure 8–3. Castings are loaded initially onto pallets (carrying sixteen parts each) and sent to one of two lathes via conveyor. Upon completion of the turning operation, the parts are transported, again by conveyor, to a wash–load area before being sent to the machining center via a wire-guided vehicle.

The machining center consists of ten identical horizontal milling machines that can perform any one of three operations. The three operations are referred to as 10, 20, and 30. For a particular part type, the machines are set up to be either dedicated—that is, assigned to only one operation, or flexible.

Two types of fixtures (A and B) are used in this system. Fixture A is used for Operation 10 and Fixture B is used for Operations 20 and 30.

Before a set of parts is sent to a machining operation, each part in that set is attached to a fixture and sent through the wash station. When a buffer space becomes available at one of the machines capable of performing the next operation, this set of parts is transported to that machine's input buffer. After the parts have been machined, they return to the wash–load area. The parts are then attached to a new fixture to await the next

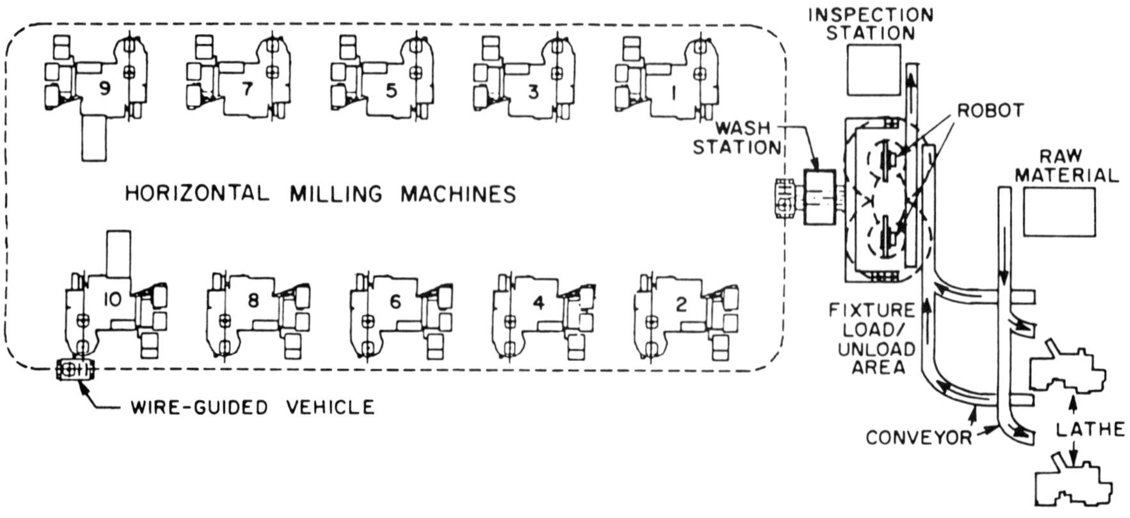

FIGURE 8–3 Injection Body FMS Design Evaluation (Courtesy of Pritsker & Associates).

machining operation. (In the case of Operation 30, the parts are rotated 180 degrees on the same fixture.) After all three machining operations have been completed, the parts are sent to final inspection before departing the system.

Solution

- Evaluate system balance and productivity
- Determine the need for additional equipment
- Determine which under-utilized resources can be eliminated
- Evaluate machine mix (i.e., number of dedicated and flexible machines)

Results

A MAP/1 simulation model was used to evaluate alternate milling machine center configurations. With a configuration consisting of nine dedicated machines (five for Operation 10, two for Operation 20, and two for Operation 30) and one flexible machine, simulation experimentation revealed that several pieces of equipment (one cart, four fixtures, and several sets of tools)—at a total cost of well over $200,000—could be eliminated while still maintaining the desired production goal of 157,000 finished parts per year.

CASE-IN-POINT

PUMP HOUSING FMS DESIGN EVALUATION

Situation

A flexible manufacturing system that performs various machining operations on pump housings is shown in figure 8–4. Housings enter the system on the top conveyor and move to a drilling operation. Once drilling is complete, the housings proceed to the bottom conveyor where they are loaded onto pallets. On this bottom conveyor, the housings undergo a series of operations, including rough milling, rough machining, internal machining, drilling, tapping, turning, and finishing. Some of these operations are bypassed while others are

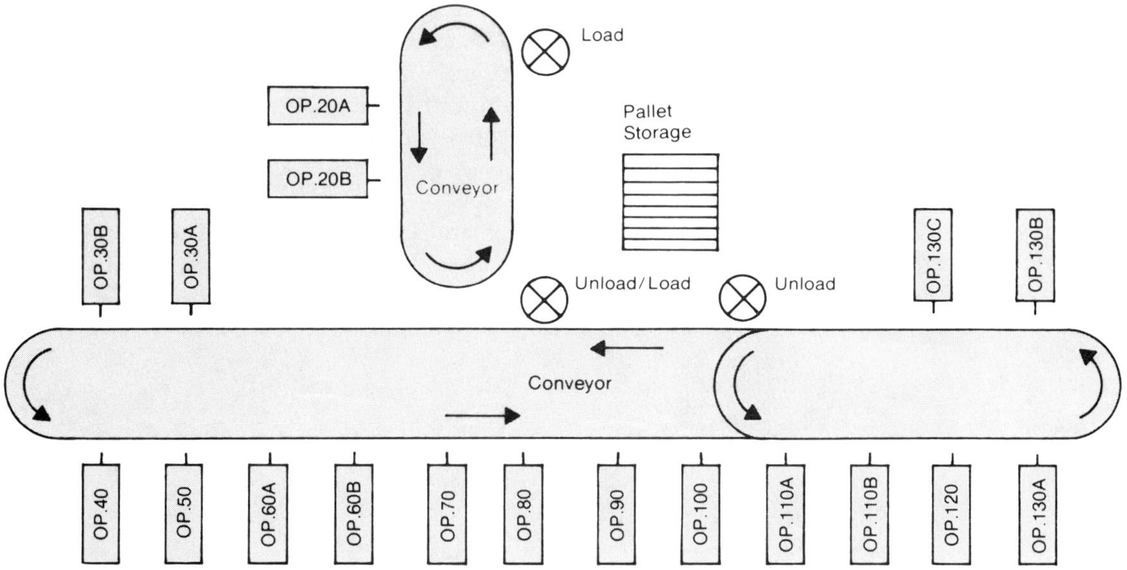

FIGURE 8–4 Pump Housing FMS Design Evaluation (Courtesy of Pritsker & Associates).

revisited, depending on the type of housing being processed. Once a housing has completed all of its required operations, it is removed from the bottom conveyor, taken off its pallet, washed, and placed in storage.

The system is capable of performing operations on ten different pump housings. There also is a limit on one housing per pallet.

Solution

- Determine throughput capacity
- Determine pallet requirements
- Determine buffer requirements
- Evaluate machine mix

Results

A MAP/1 simulation model was used to determine that the production goal of 199,000 pump housings per year could be reached. The machine mix was found to be appropriate and the buffer storage in front of each machine was deemed adequate. The study also revealed that the pallet requirements were not as great as anticipated. The required throughput could be attained by using at least twenty-one fewer pallets of each type. A cost savings of over $800,000 was realized.

CASE-IN-POINT

FMS DESIGN EVALUATION

Situation

A flexible manufacturing system that inserts wire coil windings into insulated stator cores to be used in electric motors was proposed to replace the current labor-intensive operation. The proposal was an attempt to improve the plant's cost effectiveness, competitive edge, and market growth. The flow of components (stator cores and insertion tools) through the system is shown in figure 8–5.

Each week's operations begin with a new set of orders. Stator cores enter into a buffer storage area where they wait for the insert operation to become available. At the insertion station, windings are inserted into the stator cores from the insertion tools. Upon completion at the insert station, stator cores continue to the form operation (coil centering) if all insertions have been done, or the drift operation, if further insertions are required. After drift, the cores travel back to the inserter via a conveyor where the next winding is inserted and the process is repeated. After form, the cores leave the system via a conveyor.

Insertion tools wait at a tooling up elevator for a new winding order to be entered. The tools proceed to the winders via a conveyor and enter an available winder where the winding operation is performed. After winding, the tool travels on a conveyor to the inserter, having the wedge-insert operation performed in the interim. After insertion, the tools proceed to a tooling down elevator and then back to the winders via a conveyor.

Solution

Develop an analysis tool that allows the design engineer to easily test out operational issues such as:

- Number of tools required
- Number of winders required
- Type, number, and sequence of orders to be processed

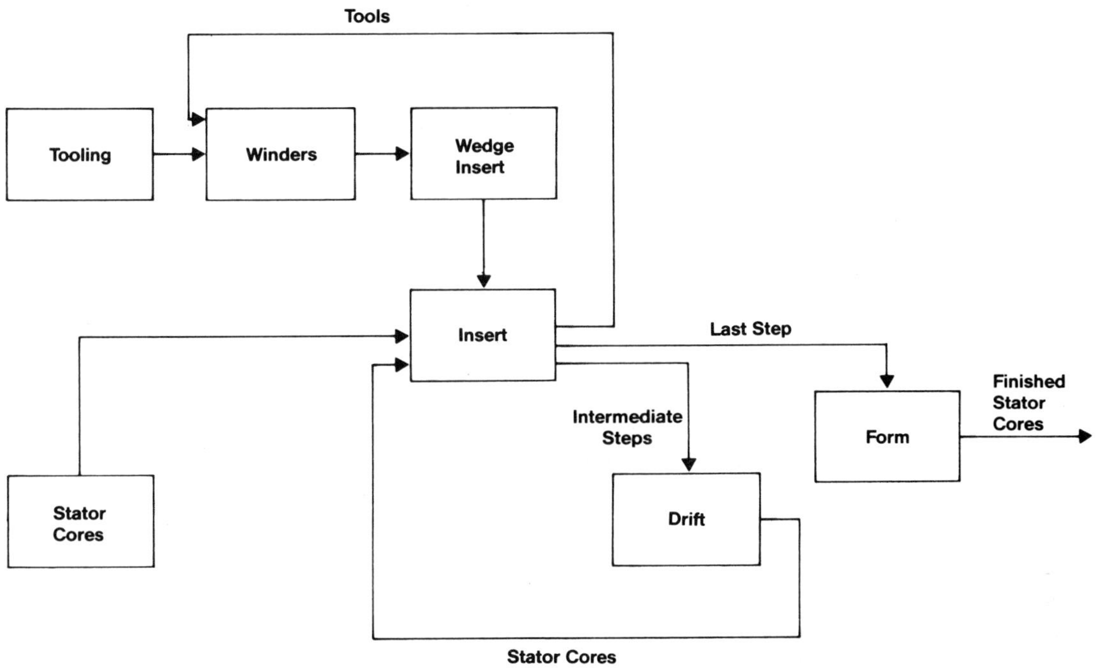

FIGURE 8-5 FMS Design Evaluation (Courtesy of Pritsker & Associates).

Results

A SLAM II simulation model was developed to allow experimentation with key parameters (orders, tools, and winders). It was found that the original estimates of eight tools and a conveyor speed of 20 feet/minute were inadequate to service the system. Although the tools cost approximately $20,000 each, it was shown that additional investment in tools would be beneficial in the long run. Also, the winder area was redesigned to reduce the travel times of the tools which, in turn, increased overall system productivity.

CASE-IN-POINT

FLEXIBLE MANUFACTURING SYSTEMS ANALYSIS THROUGH COMPUTER SIMULATION

By Phillip A. Angelillo and Michael C. Dunn
*Grumman Data Systems Corporation**

Introduction

Computer availability and power and new manufacturing processes are changing the basis of operation of all manufacturing firms. These advances are providing companies with the required technology to significantly improve productivity and quality. The cause of the accelerated change is in large part the availability of new and impressive automation equipment resident in computer-aided manufacturing (CAM), robotics, flexible manufacturing systems (FMS), and other manufacturing-related methods. These expensive and complex technologies, if applied properly, offer the opportunity to increase the productivity of manufacturing labor greatly.

Because of this demanding environment, techniques of operations research are being applied more and more to the solution of automation design and management problems.

*Used with permission of Messrs. Phillip A. Angelillo, Manager, Modeling and Simulations, and Michael C. Dunn, Modeling and Simulations Analyst, The Data System Division of Grumman Corporation.

These techniques supply a rational decision framework as an alternative to the often haphazard and inconsistent judgments common in many design efforts. Instead, the application of operation research techniques can provide the optimum combination of labor power and manufacturing equipment for the expected workload. Although truly optimum systems are rarely achieved, systematic design analysis can drive design efforts to that goal by providing a clear and explicit evaluation and trade-off framework within which key design questions concerning system characteristics and costs can be resolved.

This case provides an overview of Grumman Data Systems' work effort on a simulation-based analysis technique for the specification and design of a proposed flexible manufacturing system for aircraft part machining. The core of the approach is a simulation model written in a general purpose simulation system (GPSS).

A method of analysis using the model is presented in which FMS component size and throughput requirements are determined relative to a global operating objective of maximizing FMS throughput or spindle utilization. The method consists of a systematic incremental evaluation of the components of the FMS for a given set of operating conditions and workload. The analysis produces a series of curves in which the sensitivity of the FMS throughput relative to component capacity is developed as are the demand characteristics and utilization of each component.

The closing sections of the case illustration discusses several FMS design issues that remain to be addressed and suggestions on potential fruitful extensions to the methodology.

Description of Proposed FMS

Airframe machining today is essentially a small quantity batch manufacturing process. Despite the fact that the machining technology utilizes advanced numerically controlled machine tools, the small batch mode of operation and wide variety of part types and shapes produces poor machine utilization. NC machining, introduced in the early 1960s improved productivity considerably over conventional, manual machining, but overall productivity in machining has lagged. Analysis of the process has revealed that the problem lies in the lack of an adequate machining support system—material handling, tool management, chip disposal, and quality control that would allow full exploitation of potential NC technology productivity. The proposed FMS is an approach to imbedding the high productivity NC machines into such a fully integrated and flexible machining support system.

The basic unit of measure of machine tool productivity is *spindle utilization.* This is the ratio, expressed as a percentage, of metal-cutting time to the total time available for machining on the machine tool. In today's airframe machining environment, spindle utilization averages approximately 20 percent for stand-alone NC machining operations. The primary reasons for this relatively low utilization lie in the manual activities performed at the machine tool—the necessity to load or reorient the work piece, change cutters, perform inspections, and remove chips. All of these activities usually require that the machine tool spindle be made inoperative for the time required to perform the activity, thus losing potential metal-cutting time. Herein lies the opportunity for fruitful automation to improve machining productivity.

The underlying rationale of the FMS is to provide automation that (1) removes the workpiece from the machine tool for operations that can be performed off the machine, thus freeing the machine for metal-cutting, (2) minimizes the time of noncutting operations that necessarily must be performed on the machine tool, and (3) in any way possible, keeps the machine in a "chip-making" mode. Figure 8–6 presents such an FMS system. It consists of numerically controlled 5-axes machining centers featuring automatic part load/unloading and a local buffer stock queue, automatic tool changing, and the capability to store up to sixty cutters. All manual work is performed in a manned setup area in which workpieces are mounted onto fixtures, reoriented as necessary during the machine cycle, and removed when completed. An NC robotic deburring machine is provided as well as an automatic coordinate measuring machine for quality control inspections. An ASRS provides work-in-process storage as the workpieces cycle through the system. Also provided is a wash station in which

FIGURE 8–6 FMS Layout (Courtesy of the Data Systems Division of the Grumman Corporation).

parts are degreased prior to entering ASRS. Workpieces are transported through the system on AGVs.

Each workpiece moves through the system in a predetermined sequence of machining and support activities or operations, described in a process sheet for the part. A simplified process sheet is shown in figure 8–7. The process sheet is developed for each part by experienced methods engineers and reflects the best approach to machining the part. The sheet shows the sequence of automatic and manual operations that the part must undergo.

The Design Problem

Given the physical equipment and operations described, a basic design problem arises in determining the economic capacity or size requirements of each component of the FMS required to maximize throughput. How many setup area personnel, wash facility, debur facility, metric measurement stations, and AGVs are required to support the NC machines working at full spindle utilization? Similarly, what should the ASRS capacity be? How should cutters be managed and transported? And what is the best way to schedule and operate the system? These and other questions surface immediately as the design process proceeds.

For a one- or two-product workload, typical of high-volume "hard automation" systems, many of these questions would be addressed in a straightforward way using line-balancing analysis techniques. The FMS, however, representing low volume, soft, programmable automation presents a more challenging problem because of the wide variation in part routing, batch size, and machining requirements.

KEY OPERATION	OPERATION	ACTIVITY TIME (MIN.)			ACTIVITY
		FMS CUT	CUTTER CHANGE	MANUAL	
040	040			10	Load Fixture
045	045	4.5	1		Mill
050	050			10	Invert
055	055	4.5	1		Mill
	060	3.0	1		Drill
	065	1.5	1		Ream
200	200			30	Q.C/Reload
210	209	2.0	1		Mill
	210	12.8	1		Mill
	220	30.6	1		Mill
	230	11.0	1		Mill
300	300			30	Invert/Clamp
310	310	30.6	1		Mill
400	400			50	Q.C./Reload
410	410	20.0	1		Mill
	420	42.4	1		Mill
500	500			10	Reload
505	505			12	Debug
510	510			30	Debug
515	515	26.5	1		
	520	18.0	1		
	530	14.0	1		
540	540			20	Remove
TOTAL		221.4	14	438.4	

FIGURE 8-7 Typical Operations Sheet—Aluminum Fittings (Courtesy of the Data Systems Division of the Grumman Corporation).

The proposed FMS, for example, will ultimately have to process many different part numbers, of three different materials (aluminum, steel, titanium), varying in size from large bulkheads to small fittings, in batches varying from one to twenty pieces.

Solution

The approach in use at Grumman Data Systems to address the problem utilizes a GPSS simulation model of the FMS through which various combinations of workload can be run and for which capacity requirements of each component of the FMS can be determined. For each combination of workload, the component capacity requirements are determined by finding the capacity necessary to support maximum spindle utilization of the machining centers—the primary components of the FMS. The model will be exercised over a spectrum of part mixes and batch sizes to reflect typical combinations of expected workload expected in the next ten years.

A simplified flow chart of a GPSS model is shown in figure 8–8. The dashed boxes represent input data defining part mix characteristics and detailed part routing for each part of the mix as contained on the operations sheet. The solid boxes represent the physical components of the FMS in which the simulated work must spend a discrete amount of time. The time duration depends on the planned activity time plus time spent awaiting the availability of the desired component.

A part enters the ASRS and awaits the availability of the required FMS component (setup area, machining center, metric measurement station, debur station) specified on the part operations sheet. The first operation for all parts requires a move to the manual setup area. For this activity the part moves on a dedicated conveyor to the setup area where it spends a period of time equal to the setup time contained on the operation sheet. At the completion of the first operation, the part moves back to the ASRS and its operation sheet is indexed to assign the part to its next operation, usually a machining step. The part remains in

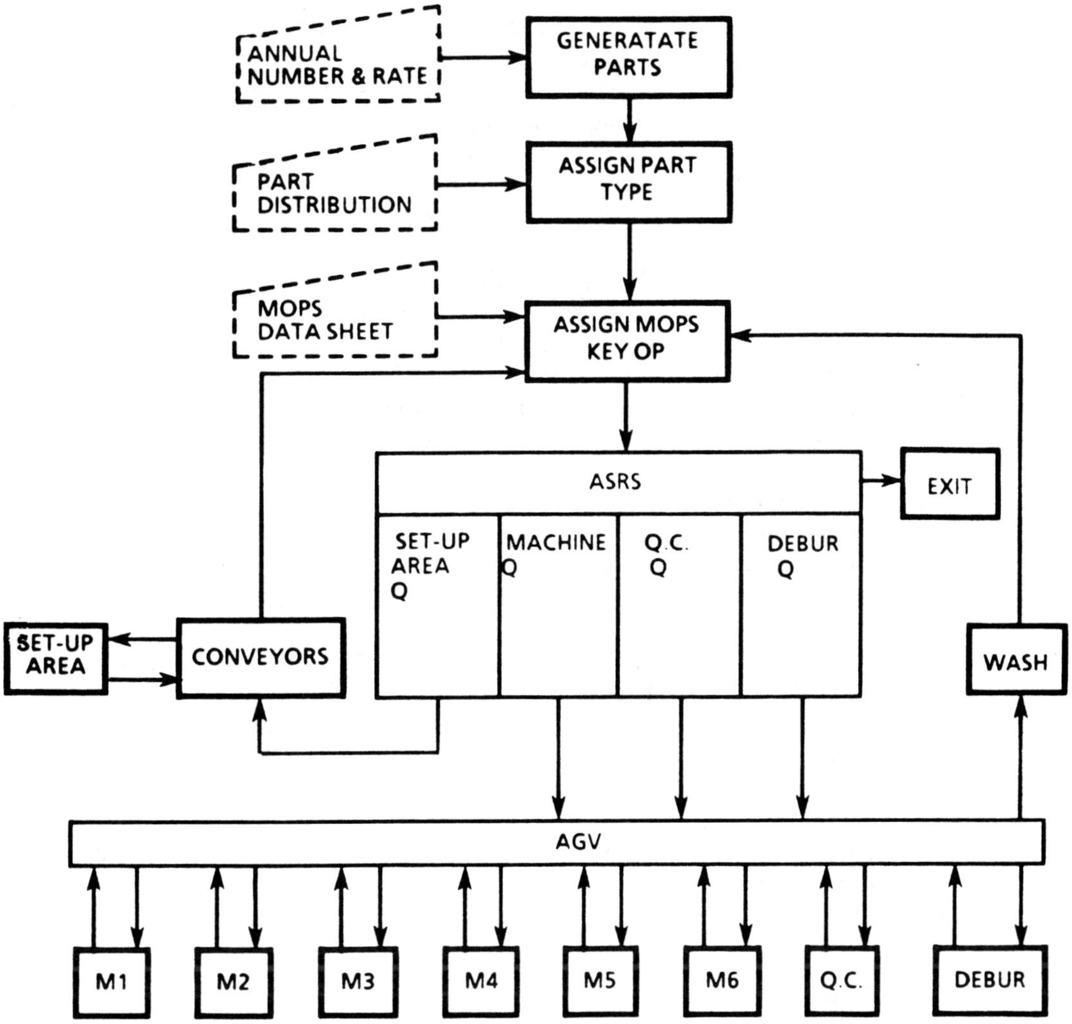

FIGURE 8–8 FMS Model Flow Chart (Courtesy of the Data Systems Division of the Grumman Corporation).

the ASRS until a machining center is available at which time the part obtains an AGV and is moved to the open machining center. It stays at the machining center for the duration of the cut time for that operation. After the machining operation, the part again obtains an AGV, moves through a wash activity, and returns to the ASRS while the operation sheet is indexed to the next required operation. The part cycles this way through all operations called for on the operation sheet, using as necessary, AGVs, machining centers, the deburring machine, the coordinate measuring machine, and setup area. At the completion of all operations the part leaves the system ("exit").

It is pointed out that each FMS component is represented in the simulation as a resource with a specific, limited capacity. Excessive demand on the resource results in the formation of queues and consequent delays in flow. In the FMS, as defined, all waiting occurs in the ASRS with no provision for local storage—a design feature that will be changed in later configurations.

Design Analysis

The design analysis approach determines the required capacity of each FMS component necessary to support maximum spindle utilization of the primary FMS components—the machining centers. A simulation scenario is first established that loads the system with a representative workload at a rate great enough to provide cutting requirements producing 100 percent utilization of the machining center spindles. A scenario for a six-machine system is shown in figure 8–9. The design analysis procedure is shown schematically in figure 8–10. Component capacity is determined by systematically varying the size of each component, in turn assuming all undefined components have unlimited capacity. For a six-machining center case, for example, setup area requirements can be investigated by first making all remaining components of the FMS—that is, the wash station, metric measurement station, deburring station, AGVs, and ASRS be of unlimited capacity and therefore no constraint to flow. The sensitivity of spindle utilization of the machining centers to setup area capacity (expressed as the number of workers in the setup area) is then explored systematically from a situation of low capacity, in which the setup area acts as a blockage in the system, causing less than full utilization of the machining centers, to higher and higher capacities until a value is reached at which spindle utilization is maximized. The results, shown in figure 8–11, indicate that a setup area of less than four workers would cause the setup area to act as a bottleneck; alternatively, no more than six workers are required to remove all setup area blockage, thus allowing maximum spindle utilization. Also shown in figure 8–14 is the utilization of the setup area. As expected, setup area utilization falls off as capacity increases and spindle utilization (and part throughput) level off.

The decision as to exactly what level of labor power to select is based essentially on economic considerations. A clear trade-off of spindle utilization versus labor power requirements is presented in figure 8–11 to assist in the determination. For the production environment of this analysis, the cost of idle spindle time (causing reduced throughput of longer work-in-process time) far exceeds the marginal labor cost at the setup area. This might not be the case, however, in all production environments in all industries. For the situation in which the product is of low value, or more broadly, where the ratio of product value to labor cost is lower, the cost comparison might indicate lower labor-power staffing with lower, but economically justified, spindle utilization and throughput.

At this point in the analysis, it can be stated that for the six-machine case and a given

- **6 MACHINES**
- **110 HOURS/WEEK**
- **50 WEEKS RUN TIME**
- **180 PIECES/WEEK**
- **5 PARTS**

PART NO.	MAT'L	% OF POP
A	ALUM	21.3
B	ALUM	29.9
C	ALUM	34.2
D	STEEL	11.6
E	TITANIUM	3.0
		100.0

FIGURE 8–9 Simulation Scenario (Courtesy of the Data Systems Division of the Grumman Corporation).

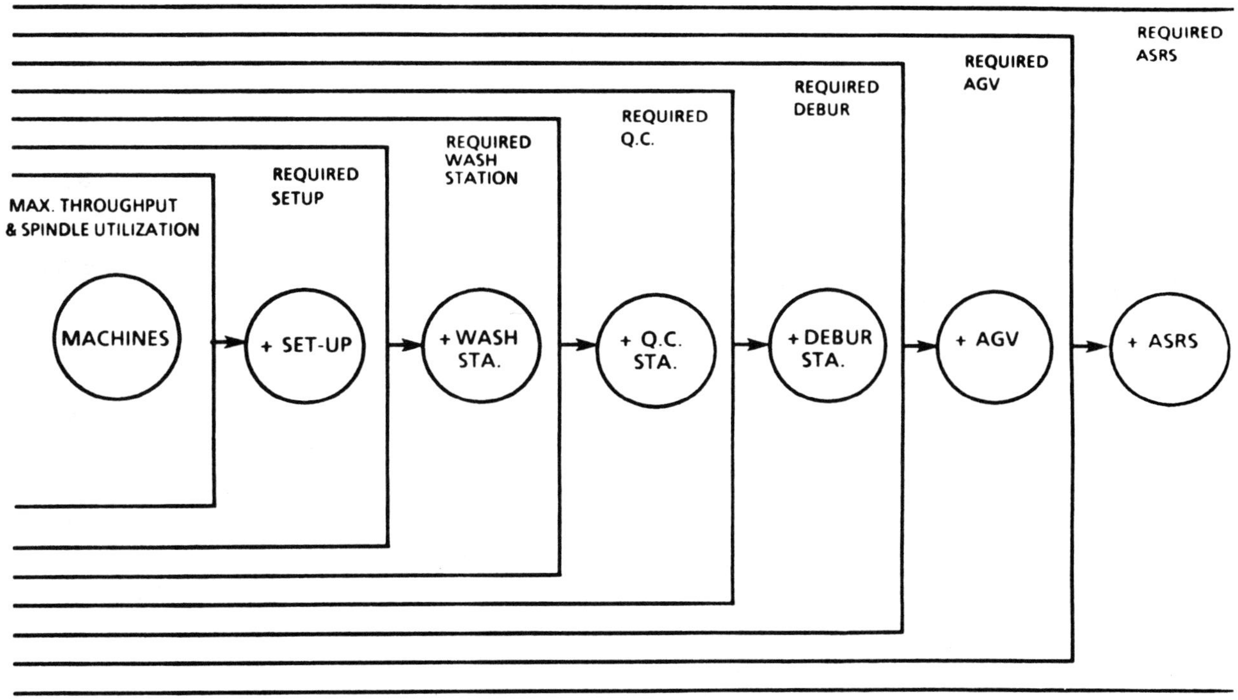

FIGURE 8–10 FMS Design Analysis Procedure (Courtesy of the Data Systems Division of the Grumman Corporation).

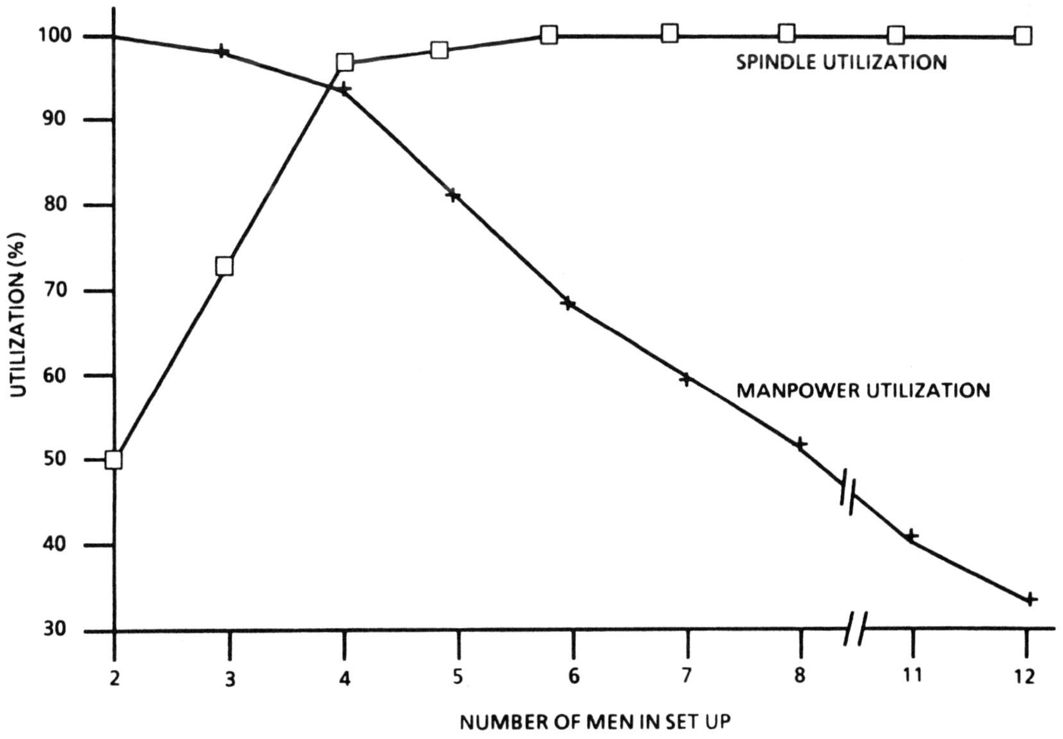

FIGURE 8–11 Setup Area Sensitivity (Courtesy of the Data Systems Division of the Grumman Corporation).

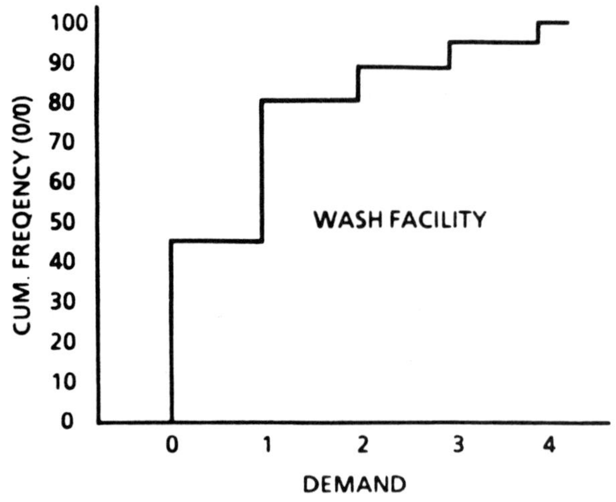

FIGURE 8–12 Facility Demands—Wash Facility (Courtesy of the Data Systems Division of the Grumman Corporation).

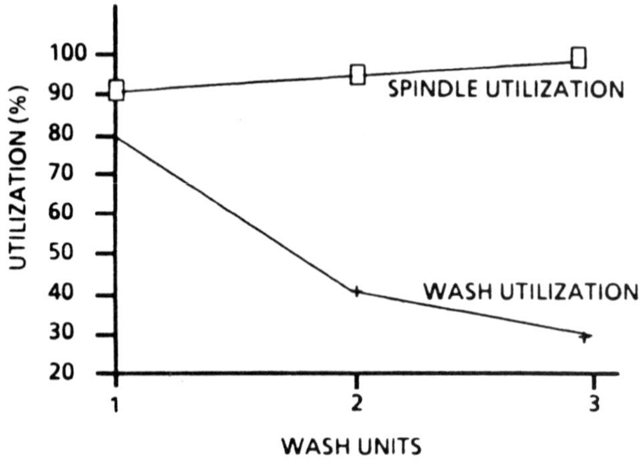

FIGURE 8–13 Utilization Sensitivity—Spindle Utilization to Wash Facility (Courtesy of the Data Systems Division of the Grumman Corporation).

workload mixer, a six-person setup area is required to achieve full spindle utilization, assuming all other components of the system are of unlimited capacity. The process then can be repeated for the next component of the FMS—the wash station.

The task of initial sizing for the next component is eased somewhat by the availability of data produced by the simulation of the proceeding component such as that shown in figure 8–12. The figure presents cumulative frequency diagrams of the demand on the wash station resulting from the six-machine, six-person setup area case, with all other FMS components unconstrained. It can be seen that 45 percent of the simulated time, there is no demand on the wash facility and that demand never exceeded four. Figure 8–13 shows the sensitivity of spindle utilization to the size of the wash facility (expressed as number of units). The wash station is not a serious constraint in the system since spindle utilization is high even with only one wash unit. Two wash units are suggested for maximum utilization.

The process is repeated for each FMS component in turn until each component has been sized. Figures 8–14 and 8–15 show demand and utilization plots for the metric measurement, debur, and AGV components. Figures 8–16 and 8–17 show demand and utilization of the ASRS.

The ASRS is of special interest. The demand plot shows a wide range of demand and the first design reaction is to select a size somewhere in the mid to high-end of the range. The utilization plot, however, shows that spindle utilization is extremely sensitive of ASRS size.

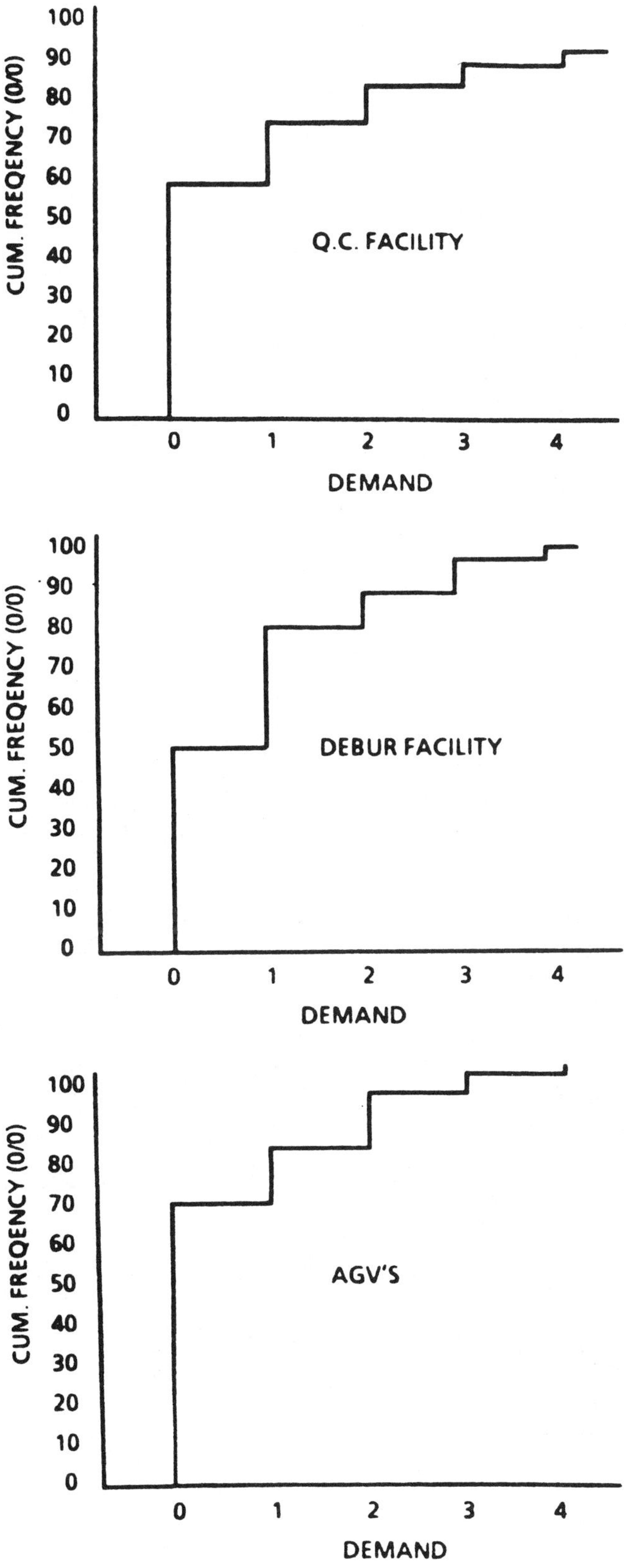

FIGURE 8-14 **Facility Demands—Metric Measurement, Deburr, and AGV Components** (Courtesy of the Data Systems Division of the Grumman Corporation).

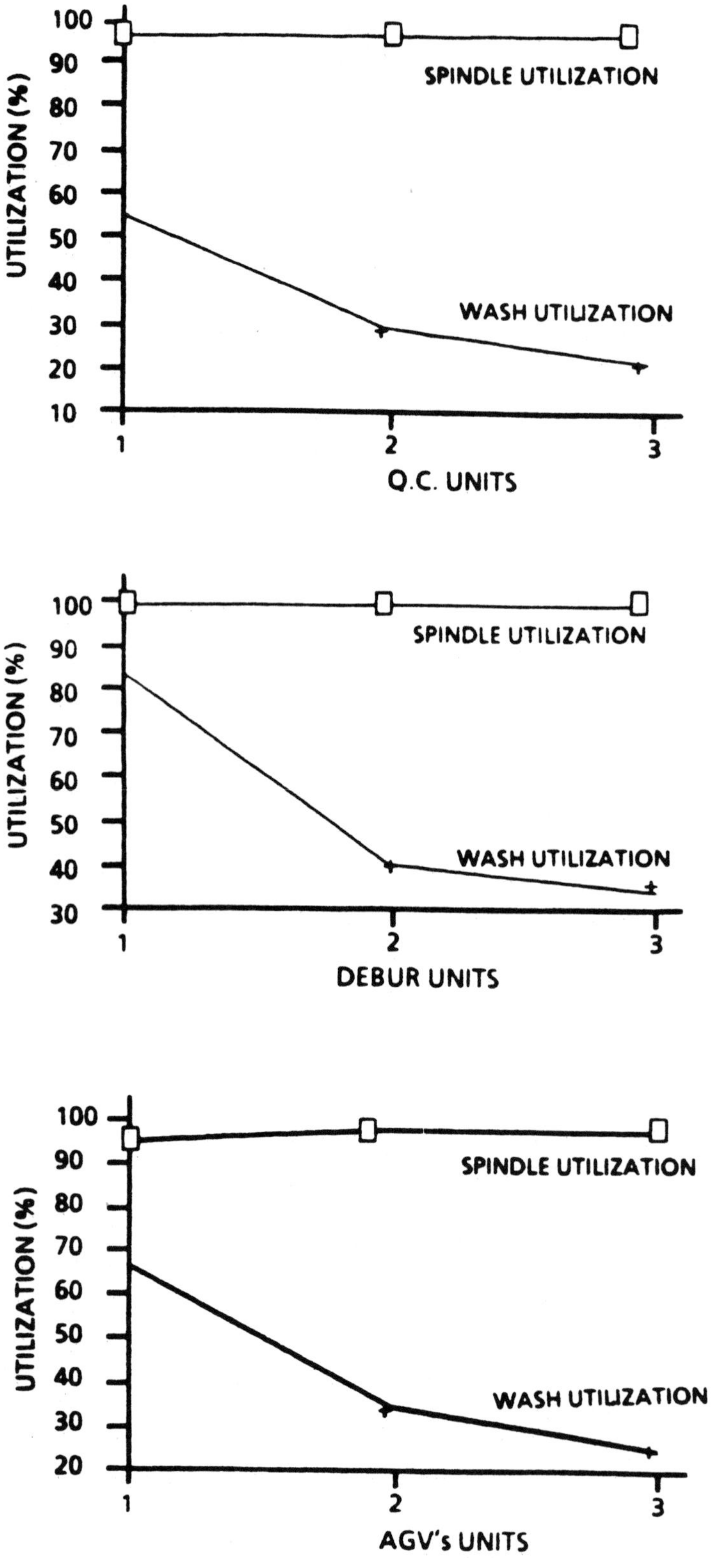

FIGURE 8–15 **Utilization Sensitivity—Metric Measurement, Deburr, and AGV Components** (Courtesy of the Data Systems Division of the Grumman Corporation).

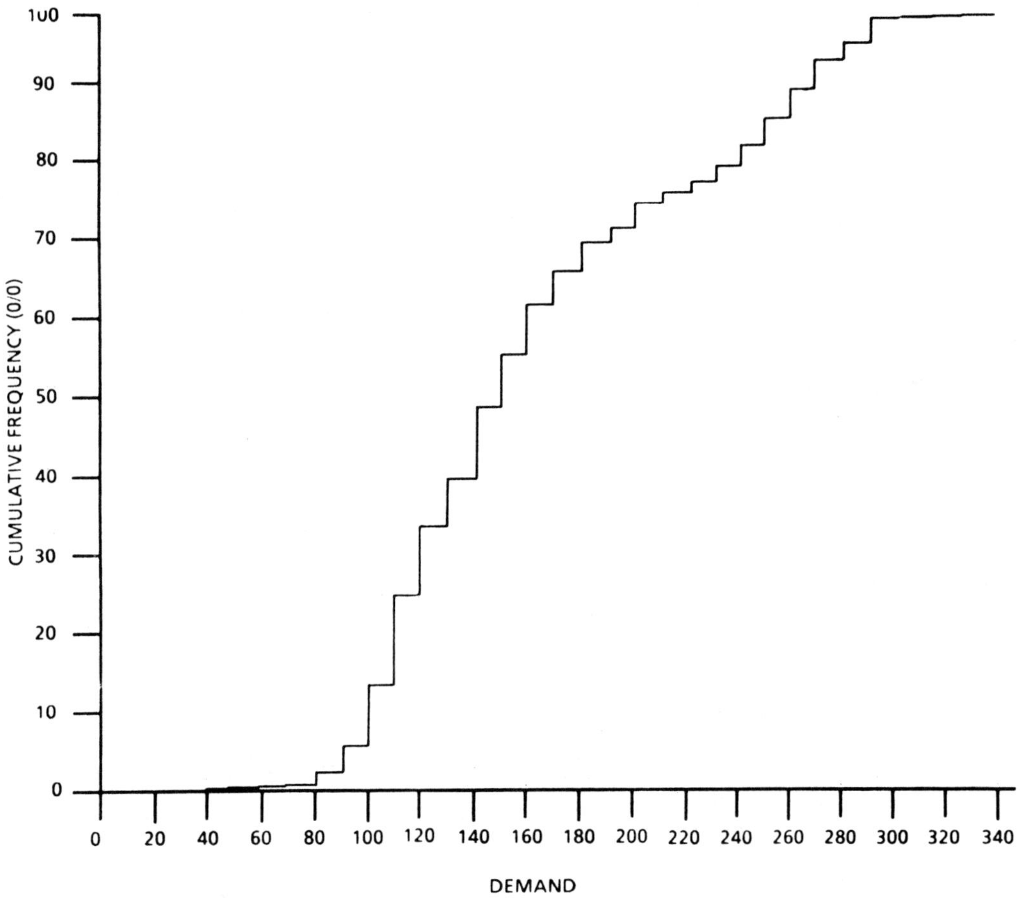

FIGURE 8–16 ASRS Demand (Courtesy of the Data Systems Division of the Grumman Corporation).

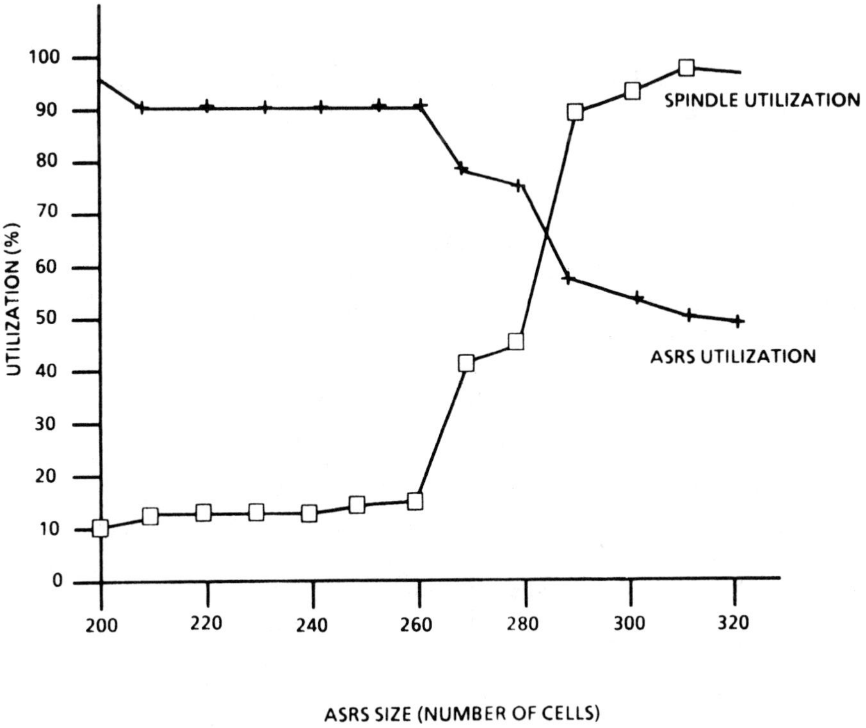

FIGURE 8–17 ASRS Sensitivity (Courtesy of the Data Systems Division of the Grumman Corporation).

- **SCENARIO**
  - **6 MACHINES**
  - **110 HOURS/WEEK**
  - **50 WEEKS RUN TIME**
  - **180 PIECES/WK**
  - **5 PARTS**

RESOURCES REQUIRED	UTILIZATION (%)
• 6 MACHINE TOOLS	96.4
• 6 MAN SET-UP	67.1
• 2 WASH UNITS	42.1
• 1 Q.C. MEASURING UNIT	55.1
• 1 DEBUR UNIT	80.0
• 2 AGVS	31.5
• 320 CELL ASRS	49.4

FIGURE 8–18 **Simulation Results Summary** (Courtesy of the Data Systems Division of the Grumman Corporation).

Spindle utilization drops drastically if the ASRS is sized at less than 300. At very low levels of ASRS size (approximately 120 cells) the system experiences gridlock and all motion ceases. These ASRS results are useful not only to size the ASRS but also serve to warn of potential operating problems with the ASRS as the system is configured, particularly in the design feature limiting all queuing to the ASRS. Alternatives to this approach are currently being evaluated.

The FMS configuration resulting from the simulation for this scenario is shown in figure 8–18. Machining center spindle utilization reaches 96.4 percent while the utilization of all other components at the sizes show lower values. In effect, high utilization of these lower cost resources is foregone to achieve the primary aim of maximum spindle utilization.

Remaining Tasks

Several tasks remain in the FMS analysis. These consist of three general types: the finalization of the specification of FMS components, the analysis of alternative cutter management approaches, and the evaluation of ways to group and schedule the FMS workload.

As pointed out earlier, full specification of the FMS requires that a variety of part mixes be considered that reflect workload requirements over a reasonable future time period. Accordingly, several packages of workloads containing different part mixes are being assembled for running through the model. The characteristics of most interest within the part mixes are variations in the ratio of machining to manual activity times and the number of moves from machining center to support activity areas. Both of these create greater or less demand on FMS components than the part mix described in the preceding sections.

Alternative cutter management systems will be investigated as part of the design finalization effort. Two questions are of primary interest: the required size and staffing of a centralized cutter management facility, and the best means of transporting cutters from the central cutter management area to the machining centers. Two candidate transport systems are being considered: dedicated overhead conveyor and the use of the AGV system extended as necessary to transport cutters. To investigate these questions, the simulation will be expanded to include the necessary system components, and merged with the existing model to capture the dynamics of cutter storage and transport activity.

The final area of interest applies the simulation to the problems of FMS operations management, in particular in the investigation of effective operating procedures and scheduling approaches in view of potential changes in part mix and total workload. Typical areas of interest include the best way to batch work (if necessary) and if appropriate, methods of grouping work using the principles of group technology; how best to address sudden changes in the operating environment (machine failure in particular); and determining an aggregate FMS loading schedule for the medium term that meets schedule requirements and maximizes spindle utilization.

CASE-IN-POINT: ILLUSTRATIONS OF SUCCESSFULLY IMPLEMENTED FLEXIBLE MANUFACTURING SYSTEMS

The following material consists of detailed descriptions of four successfully implemented flexible manufacturing systems. These cases were selected because of their universality of character, which can therefore provide the reader with specific approaches and proven methodology, as an aid in satisfying your particular needs and circumstances. Although Cases 2 and 3 deal with the production of similar parts, the scale and dimension of the two applications are quite different. Thus, these cases provide the reader with a comparison between a single flexible production line (Case 2), and a flexible manufacturing system consisting of forty-three flexible cells (Case 3).

Case 1: Flexible Machining Cell: LTV Aerospace and Defense Company, Vought Aero Products Division, Dallas, Texas

This advanced FMS consists of a number of machining centers, coordinate measuring machines, wash station, automated wire guided vehicles (AWGVs), and complete computer control. In addition to covering all aspects of flexible manufacturing, it also contains specific vendors that provide hardware and software support, and considerable discussion on the use of automated guided vehicles in the cell. It also indicates how the work effort involved in implementing this cell can be the beginning of a series of planned systems leading to total factory automation of a production facility.

Case 2: FMS Installation for Complete Production of Conveyor Shafts for Combine Harvesters: Planring, USA, Farmington Hills, Michigan

This illustration covers all of the steps involved in designing a flexible production line for the complete production of conveyor spirals and drive shafts for combine harvesters. It describes in detail the machining tasks, required machines and automation, individual operations, and methods involved in manufacturing the parts.

Case 3: FMS Installation for a Major Manufacturer of Combines: Planring, USA, Farmington Hills, Michigan

This case is excellent for comparison with the previous case in terms of scope and dimension. Whereas the previous case dealt with one production line, this application was designed to produce drive shafts and augers in forty-three flexible cells each equipped with robot gantry loaders almost fully automated. It describes in detail shaft end production, shaft assembly and welding, and the operation of the flexible manufacturing system in three modes: fully automatic, semi-automatic, and manual.

Case 4: Advanced Robotic Application: Flexible Assembly System, Fairchild Republic Corporation, Farmingdale, New York

This case illustration consists of an ICAM 9401 Gantry Robotic Cell used as a basic element or "building block" for clusters of cells. Whereas the previous cases dealt with the manufacture of discrete parts, this case involves the application of a flexible assembly system. It contains a series of photographs that provide the reader with a visual presentation of each progressive step in the automated manufacture of a subassembly in the robotic cell.

CASE-IN-POINT

*FLEXIBLE MACHINING CELL LTV AEROSPACE AND DEFENSE COMPANY, VOUGHT AERO PRODUCTS DIVISION, DALLAS, TEXAS**

Abstract

The flexible machining cell (FMC) comprises machining centers (fixtures, tooling, and cutters), coordinate measuring machines (CMMs), a wash station, automated wire-guided vehicles (AWGVs), complete computer control, and the capability of machining more than 500 different parts with an economic order quantity of one. The cell is the first in a series of planned systems for total factory automation. The AGV system transports B-1B parts on machining pallets and has been configured for future integration of an automatic cutter delivery system to transport tool magazines within this cell.

Introduction

The FMS is defined as an intelligent group of NC machine tools (machining centers) that can randomly process a group of parts having different process sequences and process cycles. Using integrated material handling and distributed computer control to balance resource utilization and availability dynamically, the system can adapt automatically to changes in part production mixes and levels of output while concurrently ensuring and recording resultant quantity and quality of the system's output through the use of in-line measurement equipment.

This case-in-point presents an overview of the FMC and its capabilities. The type of AGV system used and its application are described in detail in relation to the ultimate objective of the FMC: "Maximize the total number of quality parts obtained within the least acceptable financial cost and time."

LTV Aerospace and Defense Company Vought Aero Products Division (VAPD) developed the FMC as a unique and bold productivity improvement effort in conjunction with the United States Department of Defense, the United States Air Force (USAF), and Rockwell International (as general contractor) for the B-1B technology modernization project. The project entails machining more than 2,000 parts in aft and aft-intermediate fuselage sections.

Flexible Machining Cell Objectives

The goal of the FMC is total serial production with the least possible cost of production. To this end, five critical areas were targeted.

Maximize Throughput. This is the primary focus of all cell activities and is achieved by minimizing wasted and idle time, rejections, costs, and human intervention, and by eliminating spindle down time caused by resource unavailability.

Reduce Labor. This essential objective removed operators from the machine site, eliminated dependence on highly skilled machinists, and provided automatic material handling to support any unmanned operations.

Improve Machine Utilization. Optimal machine utilization dictates negation of machine setup time, substitution of automated features for human intervention, and use of quick-transfer devices to maintain machines in the cutting cycle.

Improve Operational Control. Operational control is afforded by the automation of work orders, scheduling of tool requirements, and a reduced dependence of human intervention.

Reduce Inventory. This objective is achieved through reduced (serial) lot sizes, improved turnover, and as-needed provision of resources for production.

*Acknowledgment for use of FMC material granted by Mr. James A. Leitch, Project Manager Facilities Systems, and by LTV Aerospace and Defense Company, Vought Aero Products Division.

As-Is Machining Activity

Conventional (batch) machining can be described in terms of three activities: cut material, machine parts, and process and inspect—all in a batch mode.

Figure 8–19 is a USAF IDEF [I for integrated computer-aided manufacturing (ICAM); DEF for definition] diagram showing functional and resource relationships in conventional machining. Note that material receipt, work order scheduling, blanks fabrication, machining, and processing and inspection are handled on a batch basis. This obviously precludes economic production of small-quantity orders and increases the overall rejection rate. In addition, the figure shows that work orders and production–fabrication planning papers are handwritten—a costly and error-prone activity. Finally, inspection and processing are performed manually, with resultant cost impact and inferior level of control.

The inevitable result of as-is machining is unacceptable throughput time: Up to 95 percent is "waiting," with an additional 3.5 percent in setup, leaving only 1.5 percent in actual run.

Automated Machining Activity

Figure 8–20 is an IDEF diagram of the FMC automated machining activity. The functions are streamlined to fulfill the fabrication needs of the defined part population.

The system functions without human intervention except for the part loading/unloading operation and support areas such as fixture building and cutter loading and delivery. Various components of the FMC are discussed in detail in the following sections of this case; the following paragraphs describe the automated operation scenario.

An average part requires three passes through the FMC. The first involves milling one side of the part blank for tooling holes. The second pass is scheduled by the computer system after availability of all resources, such as NC part programs, fixtures, cutters, and machining center, has been verified, and this pass produces the first features of the part. The third pass finished the actual part production.

Part production begins when a load/unload operator loads blanks to be machined onto a pallet according to instructions displayed on a computer terminal. The parts, which

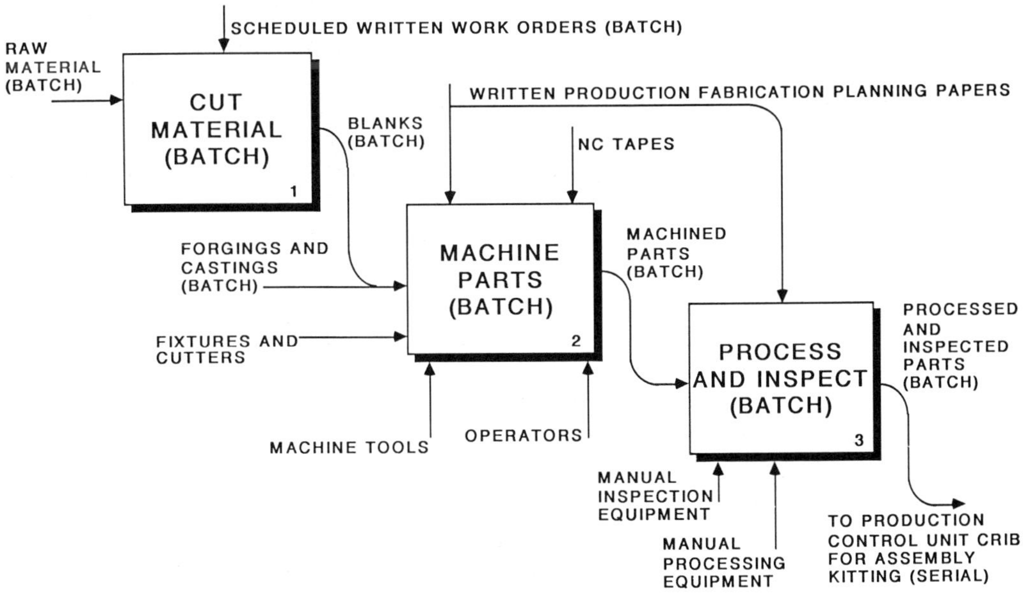

FIGURE 8–19 IDEF Model of As-Is Machining Activity (Courtesy of LTV Aerospace and Defense Company).

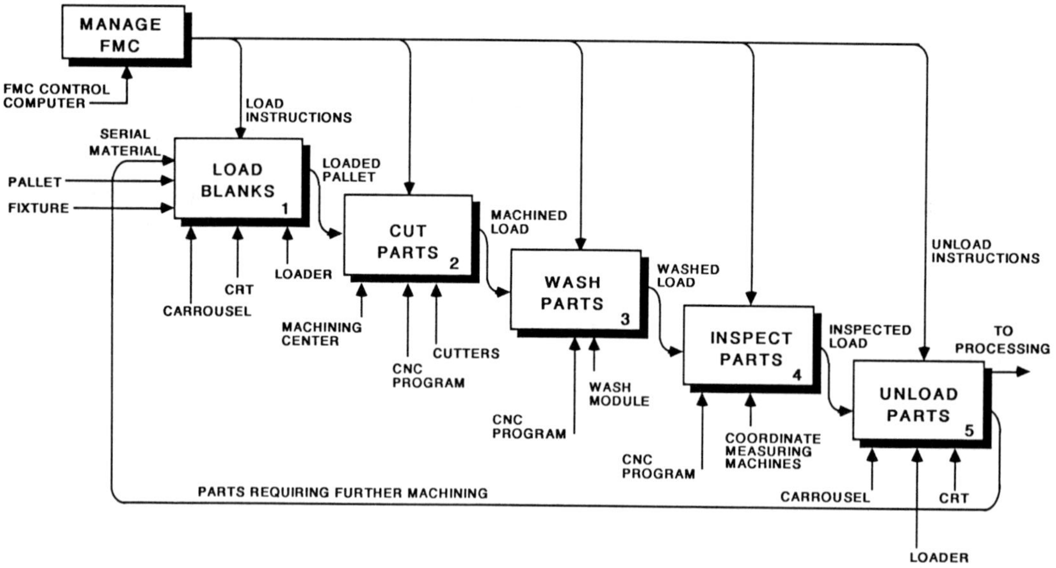

FIGURE 8–20 IDEF Diagram of the FMS Automated Machining Activity (Courtesy of LTV Aerospace and Defense Company).

may number up to sixteen on each pallet, are then moved automatically onto one of two carrousels that each hold up to ten pallets. The load/unload stations are co-located in pairs adjacent to the respective carrousels. Part pallets enter and leave the FMC via the carrousels and are then loaded or reloaded at the designated load/unload station. The loaded pallets remain on the rotating storage carrousels until the cell is ready to machine the particular part load included on a designated pallet.

When a pallet is required by the cell, an automatic robocarrier cart (discussed in detail later) is dispatched to retrieve the pallet from the carrousel and to deliver it to a machining center. Each center has a total queue capacity of five pallets: two in the wait station, one in work, and two in the complete state awaiting transport by an assigned robocarrier cart.

Once part machining is completed, the pallet is transported via an assigned robocarrier cart to a wash station, where chips are removed automatically. After the pallet load has been processed through the wash station, the next available robocarrier picks up the pallet with part(s) and transports it to the assigned inspection station, where the CMM automatically verifies part geometry for each configuration represented on the pallet. The pallet of inspected parts is then routed to the assigned carrousel. Once returned to the carrousel, the completed pallet load is forwarded to a load/unload station where parts are removed or remounted, as required, for additional machining.

Flexible Machining Cell Description

The system highlights of the flexible machining cell are described below, followed by Table 8.1 which lists the components of the cell.

System Highlights

- Complete direct numerical control. The host computer for the system is Digital Equipment Corporation's (DEC) PDP 11/44. A DEC PDP 11/70 computer fulfills scheduling, simulation, and system backup requirements, and two DEC PDP 11/24s are used: one for material handling and one for the CMMs. The FMC is the first VAPD system to utilize full DNC (i.e., no tapes required).

Table 8.1 FMC Components

Elevated computer control room, which houses all but the VAPD business host computer (IBM 3081)
Eight horizontal single-spindle machining centers; a ninth machining center has been planned and the foundation installed, and will be used for part program certification.
Two load/unload carrousels, each with two load/unload positions
One automated wash station
Two automatic coordinate measuring machines
Four automatic wire-guided vehicles (robocarriers)
One central coolant distribution/chip recovery system
One manual wash station
One battery charging spur
One fixture buildup/teardown station to marry and divorce heavy fixtures and risers from machine pallets
Two material review stations for visual inspection of rejected parts
One calibration cube, which is used by the machining and inspection centers to probe ports and bores on the cube to provide a "footprint" of machine; it is used by maintenance for troubleshooting.
An automatic cutter delivery system (which has been planned for use) from Cincinnati Milacron Incorporated (CMI) of Cincinnati, Ohio, utilizing laser-etch bar codes on cutter holders. Once in place, human operators for cutter load/unload will be eliminated from the loop and replaced by robocarriers and a three-axis robotic arm.
Computerized cell loading, scheduling, simulation, and tool-cutting control
Palletized part load/unload at remote automatic work changer stations
Cutting tool setup in the central crib area, where electronic gauging is performed
Cutter-diameter compensation for NC programs
No assigned machine operators

- Automated (hands-off) machining. Drilling, boring, tapping, milling, and profiling all are performed in one setup.
- Remote automated (hands-off) pallet transport and machine load/unload system. This system makes a major contribution to the 3:1 productivity ratio in part processing time.
- Automatic in-line part cleaning and inspection capabilities.

Computer Numerical Control Machining Centers

The computer numerical control (CNC) machining centers are manufactured by Cincinnati Milacron Incorporated. The single-spindle centers perform profile milling, drilling, boring, reaming, and tapping operations. Tool-cutting storage, selection, and ninety tool change capabilities all are automatic, as is the pallet shuttle system on each machine (five stations). Ferrous and nonferrous materials can be machined, with a prismatic work area of up to 36 inches long by 36 inches wide by 32 inches high. The machines are capable of three- and four-axes simultaneous contouring (X, Y, Z, and B) and torque-controlled machining. The part/fixture weight capacity is 5,000 pounds, which represents one of the common denominators used in all aspects of FMC part planning design. The machining centers have an enclosed area for chip/coolant contaminant. In addition, they possess the following features: part surfaces sensing probes, broken tool detectors, torque-controlled machining, and a 40- to 5,000 revolutions per minute spindle.

The centers are under Acramatic 900 series computerized NC with the following capabilities and features: floppy disk local program storage, air-conditioned cabinet, more than 200 totally resident diagnostics, fixture offsets, cutter diameter and length compensation, and circular and helical interpolation.

Pallet Carrousels

The pallet carrousels also are manufactured by Cincinnati Milacron. Each holds ten pallets and provides for their automatic rotation and transfer. Each carrousel has a pickup/delivery (P&D) station for the robocarrier and two rotatable part load/unload stations with computer terminals. The carrousel is under the control of an Allen-Bradley programmable controller.

Coordinate Measuring Machines

Manufactured by Digital Electronics Automation (DEA) of Torino, Italy, the CMMs provide unmanned manufacturing verification of all machined parts. The CMM is of the three-axes gantry type, with a fourth axis provided by a rotary table having 360 indexed positions. The prismatic work capacity is 72 inches long by 72 inches wide by 54 inches high. The machine's positioning accuracies are on the order of ± 0.0008 inch for X and Y axes, and ± 0.0004 inch for the Z axis. Maximum positioning speed is 300 inches per minute.

Measurements are performed by electronic touch probes, with full three-dimensional misalignment compensation. The CMM provides geometric dimensioning and tolerancing of linearity, flatness, roundness, sphericity, cylindricity, conicity, parallelism, squareness, and angularity errors. The machine is capable of off-line programming using Cartesian or polar reference systems with coordinate conversions in a language called "HELP 11" developed by DEA. Measurement reports are generated as needed. CMMs are under the control of the DEC PDP 11/23 computer. (Allen-Bradley programmable controller drives the rotary axis.) The DEC PDP 11/24 host for the two CMMs communicates with a DEC PDP 11/44, the cell host.

Robocarrier System

Figure 8–21 is a photograph of the Eaton-Kenway, AWGV. Figures 8–22, 8–23, and 8–24 present lifting detail, plan view, and elevation schematics, respectively. Through the use of simulation at CMI and VAPD, an optimized layout and traffic system model was developed to aid the high throughput of the FMC. The FMC employs four of these Model XK all-electric (wire in floor) robocarriers, which have a work envelope of 36 inches long by 36 inches wide by 36 inches high. Each robocarrier costs approximately $75,000.

The DEC PDP 11/44 host computer receives instructions daily from the VAPD business host computer, and breaks down the work order into manageable FMC "pass" work orders. The DEC PDP 11/44 in turn commands the DEC PDP 11/24 to move pieces through the FMC. Automatic, real-time control is provided by four (quadrant) zone controllers that interface to the DEC PDP 11/24 computer. This sectionalizing permits modular control and potential nondisruptive expansion. The DEC PDP 11/24 determines the best robocarrier on the basis of queues, travel times, and so on, to carry a given part.

The robocarriers have a sixteen-hour operating time after each recharge of batteries (lead acid, 350 ampere hours at 48 volts, 1,500 pounds per set). The cell comprises four recharging stations with four extra sets of batteries. Robocarriers self-check their battery status, and upon low charge level, automatically deliver themselves to the battery-charging spur as well as report low levels to maintenance. Recharging takes about eight hours.

These robocarriers weigh 3,300 pounds each (with batteries) and travel at 85 to 135 feet per minute with the maximum 5,000 pound load (about 200 feet per minute with 2,200 pound load is standard). The carts are of bidirectional design to navigate efficiently out of a P&D stand; however, the optimal design configuration involves one-way express lanes to maximize machining throughput. The cart path includes 48-inch radius turns, which significantly reduces the overall cell square footage, and 12-foot minimum length lead-in distances to pallet exchange points. The carts navigate turns at 40 feet per minute. They possess a 10

FIGURE 8-21 Robocarrier System (AWGV) (Courtesy of LTV Aerospace and Defense Company).

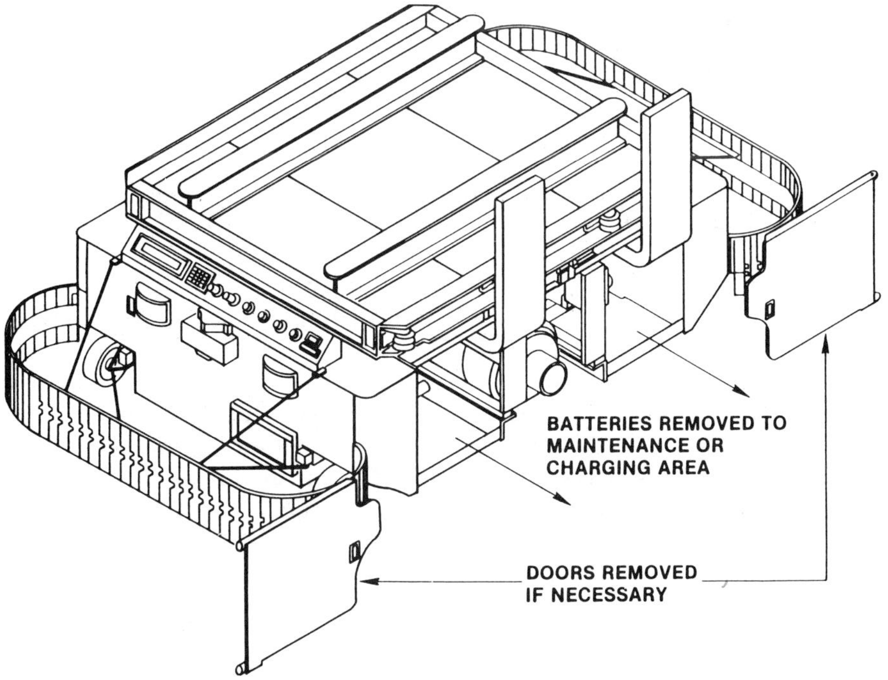

FIGURE 8-22 Robocarrier Lifting Detail (Courtesy of LTV Aerospace and Defense Company).

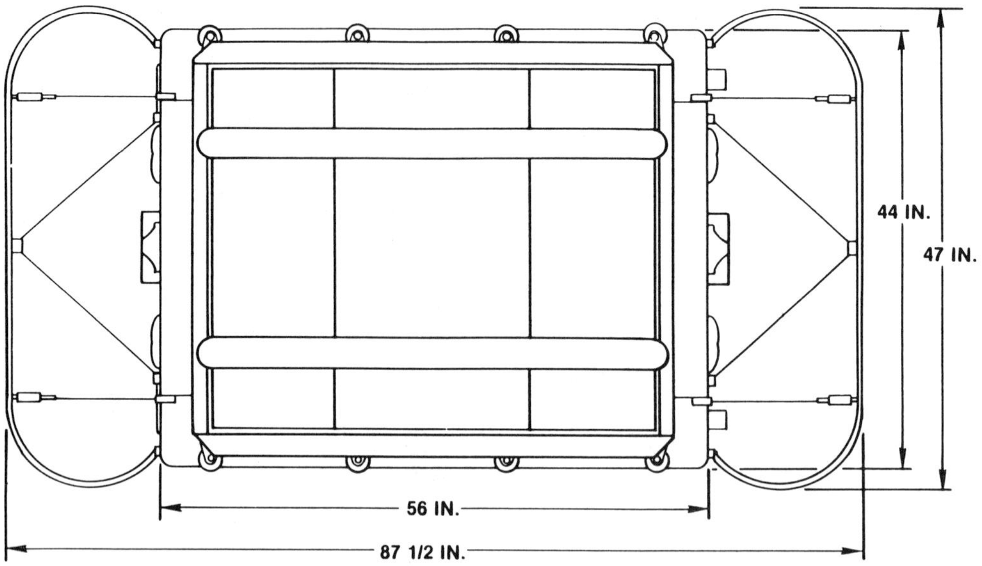

FIGURE 8–23 Robocarrier Plan View (Courtesy of LTV Aerospace and Defense Company).

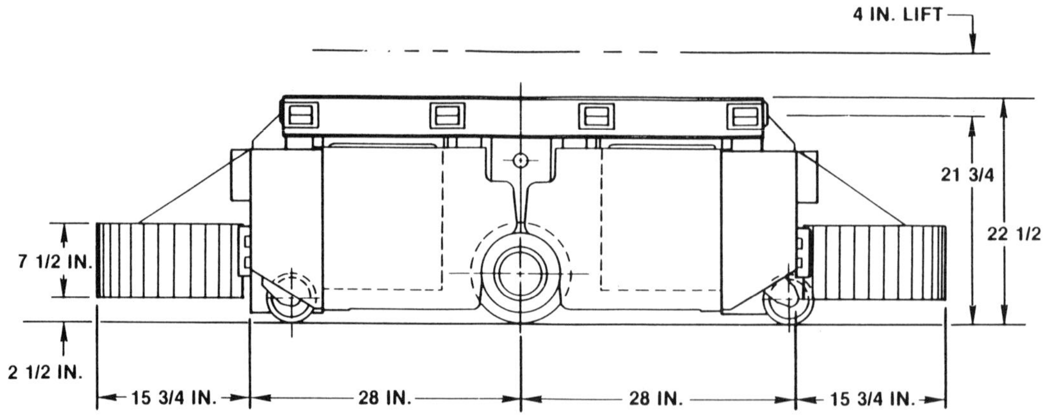

FIGURE 8–24 Robocarrier Elevation (Courtesy of LTV Aerospace and Defense Company).

degree incline tolerance, are articulated to handle bumps and joggles, and can execute a full emergency stop within 12 inches at full speed.

An important innovation in the basic AGV system design was the addition of chip/coolant guards to the top of the robocarriers to prevent machining coolant from contaminating the electronics located below in the carts. Other significant features include a light-emitting diode (LED) array on the front of each cart to permit on-board diagnostics as an aid in troubleshooting, and pendants to permit manual control if needed.

Guidewire

VAPD chose to bury all utilities in the FMC floor slab. The robocarriers follow low-frequency, low-voltage signals, necessitating the 480-volt machine electric power to be at least 4 feet from the guidewire (as a general rule), or 1 foot per 120 volts. The guidewire operates at three signal levels of 48 volts (direct current), 400 milliamperes (maximum). These signals are used:

- FL (4.8 kilohertz). As the cart follows this signal, the RSK (zone controller) switches the wiring into zones to establish the robocarrier path.
- FE (6.7 kilohertz). This is a command signal from the RSK to the robocarrier. It sends status messages and load transfer commands when the robocarrier is on a floor stop or an appropriate floor plate.
- AW (10 kilohertz). This robocarrier-to-RSK signal acknowledges load transfer commands and status requests from the RSK and in turn relays them back to the DEC PDP 11/24 computer.

The FMC's seven-layer under-the-floor utility design was achieved using an IBM CAD/CAM interactive computer graphics system. The guidewire was installed and glued with epoxy into a 12-inch deep concrete floor sawed to a depth of 2 inches. All FMC utilities (electric power, water, air, vacuum, coolant distribution, and so on) are located in the concrete floor. Concrete re-rod must be at least 4 inches away from the guidewire and cross it only on a 90 degree angle. A special aid for troubleshooting and guidewire installation was an electronic "sniffer," which is used to perform preventive maintenance on signal wires and adjust zone amplifiers. Although design and implementation of this configuration was expensive and complex, it was crucial to the success of the FMC.

Automatic Cleaning Module

The module's part work envelope is 36 inches long by 36 inches wide by 36 inches high. Manufactured by Taylor & Gaskin Incorporated, Detroit, Michigan, it accomplishes automatic part transfer, rollover, wash/spray, and discharge. Lubrication and coolant-level maintenance also are performed automatically. Cleaning solution is pumped from the 850-gallon stainless steel reservoir at 400 gallons per minute and 45 pounds per square inch to flush chips from the fixturing. Chips are then delivered to the central chip collection system through a central flume. An electrically heated air system is used to dry parts. The cleaning module is controlled by an Allen-Bradley programmable controller.

Central Coolant Distribution/Chip Recovery System

The chip/coolant system was manufactured by Henry Filters Incorporated, Bowling Green, Ohio. The system is unmanned and integrated to all milling machines. Its functions include removing tramp oil from coolant to promote solution longevity, and segregating/handling ferrous and nonferrous chips. It will accommodate expansion with its modular, below-grade, flume-type configuration. Operational statistics of the system include a 3,000 gallon-per-minute pumping capacity and removal rate, a 13,247 gallon dual reservoir, and 498 linear feet of flume. It also includes an automatic, self-cleaning suction filter (no consumable media). The system is under the control of an Allen-Bradley programmable controller.

Central Computer System

The master machining schedule is created by the VAPD business host computer. The FMC's DEC PDP 11/44 receives a 20-day window's worth of work orders daily. Work orders are examined, selected, and scheduled into the FMC. From the 20 days' worth of work orders, the FMC host selects enough work orders for a 24-hour production run. After the work orders to be run have been selected, they are simulated for maximized throughput before being authorized for actual work the following day.

Figure 8–25 is a diagram of FMC information flow. Figure 8–26 summarizes the functional hierarchy for VAPD's flexible machining systems.

Training

Training for the AGVS in this FMC involved two efforts. First, maintenance technicians (one mechanical and one electronic per shift) were trained during physical construction of the FMC

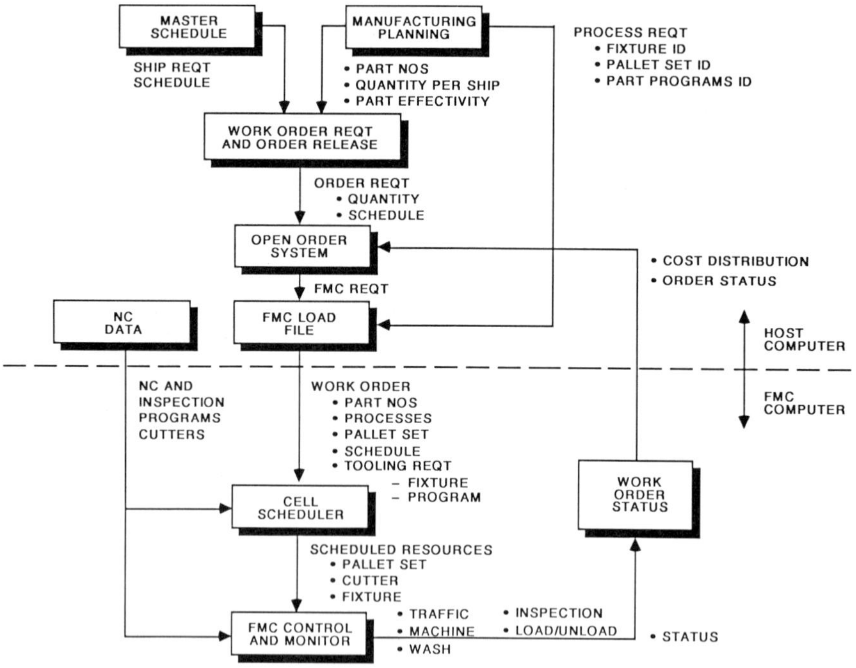

FIGURE 8–25 FMS Information Flow (Courtesy of LTV Aerospace and Defense Company).

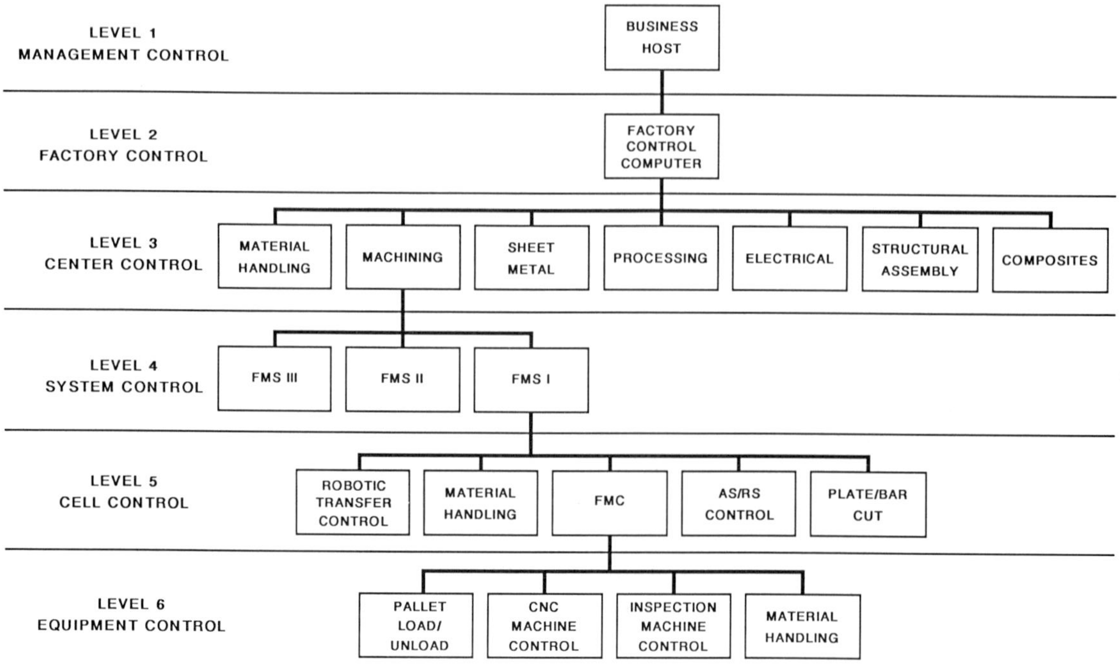

FIGURE 8–26 Functional Hierarchy for Flexible Machining Systems (Courtesy of LTV Aerospace and Defense Company).

so they could observe wiring paths and the like. These trainees included typical assembly parts, service personnel, and specialized electronic troubleshooters. The latter also received training to interface with the DEC PDP 11/44. Control-room-operator (computer) personnel training included instruction on controlling or modifying the AGVS system model manually if required.

Conclusions

In the evolutionary process of developing the FMC, a holistic systems engineering approach was first used to establish a baseline of aerospace machining tasks. On the basis of issues and concerns defined through analysis of the as-is model, a streamlined, flexible model was constructed to represent future machining cell functions in the FMC.

Although the FMC effectively reduces costs of machining detail parts, it exists as an island of automation. Therefore, you must guard against automating the as-is and perpetuating the mistakes of the past. In many respects, the ultimate goal of this state-of-the-art modular manufacturing element is to banish the remnants of outdated methods and practices that were geared to batch manufacturing operations of high-rate production.

Incorporating the AGVs into this highly automated machining environment proved challenging, but it emerged as a key to the success of the FMC and LTV-VAPD's factory-of-the-future goals.

CASE-IN-POINT

FMS INSTALLATION FOR COMPLETE PRODUCTION OF CONVEYOR SPIRALS AND DRIVE SHAFTS FOR COMBINE HARVESTERS PLANRING, USA FARMINGTON HILLS, MICHIGAN*

Description of the Flexible Manufacturing System

The flexible production line is designed for complete production of conveyor spirals and drive shafts for combine harvesters. The product range consists of:

- Spiral shafts (five different types)
- Drive shafts (six different types)

The two product groups consist of three subgroups

(a) End shafts
(b) Center tubes
(c) Conveyor spirals

The driveshafts consist only of groups a and b. The center tubes and the conveyor spiral are supplied ready for assembly by the customer, taking into account the parameters laid down in the "production specifications."

A total of eleven different workpiece types, of which five have welded transport spirals, are manufactured completely in a flexible manufacturing system, including welding the spirals to the shafts.

Total production quantity is 420 complete shafts and spirals per hour in two-shift operation, with 80 percent utilization of the equipment.

Machining Tasks

- Machining of the two shaft ends (average throughput 765 pieces per hour) by end machining, turning, milling, drilling in a manufacturing system with a total of thirty-

*Permission granted by Planring U.S.A., Inc.

four machines; workpiece diameter 48–120mm, weight 0.25–20kg, length 65–580mm; a total of twenty-one different workpiece shapes. Connection of the two shaft ends with externally provided adaptors on three electron beam welding machines and one laser welding machine; workpiece diameter 48–120mm, weight 15–60kg, total length 1653–4028mm; a total of twelve different workpiece shapes (smooth-shafts).

- Welding the transport spirals onto the center pieces, eight workpiece shapes; spiral diameter 125–250mm, total weight 7–37kg, length 1428–3866mm; executed on 6 laser welding machines.
- Connection of the two shaft ends to the welded spirals; eight workpiece shapes, shaft diameters 48–57mm, total weights 8–40kg, total length 1800–4071mm; executed on two laser welding machines.

Required Machines and Automation

- 3 cells for automatic loading of pallets with raw parts
- 6 manufacturing cells for end machining
- 1 manufacturing cell for deep hole drilling
- 17 manufacturing cells for turning
- 8 manufacturing cells for slot milling, 3 of these equipped as double machines with drilling units
- 1 manufacturing cell for slot milling as a single machine
- 1 washing machine
- 3 manufacturing cells for electron beam welding of the end pieces
- 1 manufacturing cell for drilling
- 1 manufacturing cell for laser welding of the end pieces
- 2 manufacturing groups for laser welding of spirals, consisting in each case of 3 laser welding machines for welding the spiral
- 1 laser welding machine for the end pieces
- 3 flexible, computer-controlled transport systems as rack servicing units, together with racks for storing the stacks of workpiece pallets
- 2 process computers for FMS control

Separately installed manufacturing cells:

- 1 special machine for milling the finished shafts, equipped with loading crane and the buffer

The complete system is divided into two parts which are interconnected by conveyor belts which act as buffers.

Section 1: Mechanical Machining (pinion machining). Subsystem 1 consists of thirty-seven independent manufacturing cells (machines, NC gantry loaders, and workpiece storage area),—that is, each cell is controlled by a standard cell control and can be operated independently of the host computer for fully automatic machining after it has been set up and provided with workpieces in the storage area.

This results in considerable advantages in instituting the system, for any necessary changes in the technical tasks, resulting from its flexibility with respect to machining methods and capacity requirements.

Section 2: Welding. 1 washing machine, 4 independent manufacturing groups for welding of ends, and 2 manufacturing groups, each consisting of 4 laser welding machines for the transport spiral and ends are connected to the system in Subsystem 2.

Shaft Machining

The raw materials for this part of manufacture are forged parts, which can be divided into two groups:

- Calibrated round material with forged flanges
- Round-forged stepped-shafts with forged flanges

These parts are received in steel containers (dimensions 1000 × 1000 × 800mm) in front of the pallet loading positions. The batch weights must not exceed 2000kg.

Each container is marked with the corresponding part number—that is, parts stored in the container. Only identical parts are stored in one container.

Assignment for loading the pallet loading positions is carried out by the host computer:

1. The container is lifted by a forklift truck onto the tipping device of the inlet side of the sorting device.
2. The specified part number is entered by the operator and acknowledged by the computer.
3. The tipping device empties the container onto the inlet side of the sorting device.

The inlet side has a maximum capacity of 4 tons. When this weight is reached, the tipping device is stopped.

If the weight of the parts on the inlet side drops to a minimum, an optical signal is activated to draw the operators attention to the absence of supplies.

To permit specific pallet loading per part, the quantity on the inlet conveyor belt can be dimensioned with the aid of the part weight. Superfluous parts are automatically diverted to another container.

Example. The actual sorting is carried out in a vibrating cylinder which is equipped with the necessary diverters.

The sorted parts pass to a feed trough which leads to a separator, passing through a part detection device.

Before a faulty part is signaled, the feed is stopped and the faulty part is extracted automatically.

The parts passing through the separator are correctly positioned for the loading crane. The separator releases a part only if the previous part has been picked up by the crane. Furthermore, the feed trough has a minimum/maximum path (electrically protected) which controls the input of parts to the trough. Loading onto the pallets is carried by an NC loading crane which was programmed by the part acknowledgment by the operator during loading. The pallet is filled with the correct number of parts per roll (each pallet has 5–7 rows). The conversion time with the vibrating cylinder empty and the crane empty is approximately 45 minutes.

The filled pallet stacks are now recorded by the computer and are controlled and monitored by the computer until emptied and buffered for new loading. The pallet stacks are inserted and removed from the racks by computer-controlled rack serving units. The three-story rack rows (one double row, one single row) are used for assignment to the waiting areas and for buffering.

Each compartment of the rack has a code field similar to a chess board, for example Row A, Story 2, Position 29, Code A-229, and so on.

This means that a location of a specific pallet (part) can be determined from the computer at any time. Furthermore, it is possible to obtain information of the current state of manufacturing of the parts in this position. In addition, each waiting area before the various machines can be selected to check which pallets (parts) are in the area and how full the area is.

If the computer fails, system operations can be continued with manual control on the basis of the last printed production lists. As each pallet has a clearly visible part code, distribution of parts from the three-story rack to the various waiting areas can be carried out with the rack serving units (which have a driver's cabin).

Machining of the shafts is divided into three main groups:

- End machining
- Turning
- Slot milling and boring

One deep-hole drilling and one special drilling machine are additionally used for only one specific part.

First Operation: Deep-Hole Drilling. The forged part is placed on a prism and held and clamped on both sides by clamping bells. (The front side, in which the hole is to be bored, must be flat and free of burrs.) The hole is drilled with a single-lip drill.

Breakage of the drill is measured and indicated by means of a pressure-measuring device and the increased torque.

No conversion is necessary, as this machine handles only one workpiece.

Second Operation: End Machining. The machine is loaded and unloaded by a crane. The workpiece is placed on a prism, and is axially oriented by an end stop integrated in the machine.

Centric, Three-point Clamping—Simultaneous end machining on both ends—that is, centering and flat sinking.

Centering Drill with (Infrared) Breakage Monitor—two machining spindles, each with three tools (double tool storage). The tool is changed automatically either when it breaks or after its lifetime has been exceeded (counter).

Each machine handles the complete workpiece program, which means that only a selection switch needs to be actuated to convert it. In addition, the tool clamping elements automatically move to the correct distance. This means that it is only necessary to adjust the end stop manually and to provide the appropriate tool heads. Conversion time is 15–20 minutes.

Third Operation: Turning. The workpiece is moved between the centering tips of the CNC turning machine and clamped in position.

The workpiece is driven by end blades which engage the unmachined ends of the workpiece, the pressure being provided by the tailstop. The machining sequence is controlled by a program read in from paper tape. Up to three diameters of a workpiece can be measured by a measuring station integrated in the gantry and corrected by an integrated cutter compensation. As each blade has 2 twelve-position revolver tool holders, it is possible to install several tools of each kind (3–4 tools). This permits continuous machining up to batch size of 560–640 parts.

Breakage of the tool is detected indirectly by the measuring device and the rejected parts are thrown out via a trough. The empirically determined tool lifetimes are recorded continuously with a decrementing counter and the revolver is switched to the second, third, or fourth tool of the same kind when the lifetime is reached. Tool breakage must cause no damage to the machine.

The following activities are necessary to convert from one workpiece type to another:

1. Read in the new program from paper tape.
2. If necessary, change or adjust the end drive and tail stock.
3. Change the tools.
4. Set the measuring station to the new diameters.
5. Program the loading device by means of selection switches.
6. Adjust or possibly change the gripper in the portal.

These conversion activities are signaled by a workpiece number display in the machine and take approximately half an hour.

Fourth Operation: Key Milling. All keys and grooves can be machined on these CNC double spindle key milling machines, some of which have horizontal or vertical drilling units. Loading and unloading of the machines is carried out by means of a gantry, and workpieces are placed on prisms.

Each milling spindle has a six-position tool magazine, again permitting three or four tools of the same kind to be provided for each workpiece and permitting continuous machining of batch sizes of up to 560–640 parts.

Each machine has the complete pinion parts program stored permanently, which means that conversion from one pinion type to another is carried out with a selection switch. Each milling tool is checked for breakage after the end of each milling cycle.

Rejected parts (due to breakage of the milling tool) are thrown out via a trough. The empirically determined tool lifetimes are recorded continuously by decrementing counter, and the spare milling tool is fetched automatically and clamped into position when the lifetime has elapsed or if the tool breaks. The carbide-tipped milling tools can be sharpened up to five times. The resultant actual dimensions diameter and length are determined on a tool-setting device and printed on self-adhesive labels. These labels are then attached to the corresponding tool holder.

When the tool magazines are filled with the corresponding tool holders, the printed actual values must be entered into the control, together with the tool position 1–6. This programming is absolutely necessary for all fitted slots to keep the rejection factor as low as possible and bypass the continuous measurement of slot dimensions.

Conversion activities are indicated by a workpiece number display on the machine and take approximately half an hour.

Fifth Operation: Special Drilling Machine. Drilling of two cross-holes is carried out only on one single part and is carried out on a four-spindle drilling machine with two drilling operations and two burr-removal operations. Loading and unloading is carried out with a gantry which places the parts on prisms of the clamping device.

The twist drills are checked by means of sensors for breakage. A decrementing counter indicates the empirically determined tool lifetime and stops the machine when the lifetime is reached or if the drill breaks. As this is a single-purpose machine, no conversion is necessary.

All sharpening or presetting of tools must be carried out at a central location. To guarantee correct assignments to the various machines, transport of the tool holders with cutting tools, including perforated tapes and a corresponding master tool, must be carried out in special tool cartridges.

The sharpened carbide-tipped milling tools are adjusted on a special tool adjusted unit (adapter SK40) and deviations from the nominal dimensions are printed on labels. These labels are stamped to the tool holder and the dimensional deviations must be entered into the corresponding milling program when the tool is used.

The turning and end-machining tools are set to their nominal dimensions, as they are fitted with adjustable clamps. These tools holders (heads) are clamped into the corresponding adapters or mountings of the machine without correction of the execution program.

All presetting of tools is carried out on three units: two optical setting units with a 150mm screen and magnification factor of twenty for turning, milling, and end-machining tools. The third unit is a pure length adjustment unit and is used for presetting of deep and special drilling tools.

A sufficient number of program tapes (up to three per program) are supplied with the system and stored at a central location, from which they can be placed in the appropriate cartridges.

Assembly and Welding

The basis for this part of manufacture is:

- Shafts (manufactured as described previously)
- Center tubes (provided by customer)
- Conveyor spirals (provided by customer)

These also can again be divided into groups:

(a) Drive shafts
(b) Conveyor spirals

Two different manufacturing facilities are provided for the two groups:

1. For welding the ends to the drive shafts (electron beam welding) and to the conveyor spirals (laser welding), the pinions must be free of grease. This is carried out in an immersion and steam degreasing unit.
2. The center tubes (center parts) must be straightened before welding. This is carried out on a six-roller straightening machine.

Degreasing

This unit is integrated in the FMS, degreasing of the parts being carried out together with the pallets.

A complete stack of pallets is taken from the rack by a conveyor belt and moved into the degreasing unit. The treatment chamber is hermetically sealed to prevent escape of unwanted materials.

Degreasing is carried out in two steps:

1. Immersion (coarse cleaning)
2. Steam degreasing

The two tanks are filled with perchloroethylene; immersion is carried out cold and steam degreasing warm at 120°C.

The extracted air passes through an active carbon filter and the separated per concentrate, which is trapped by this filter, is automatically returned to the tank. If the amount of oil in the steam degreasing bath becomes excessive (40 percent oil), the treatment temperature rises and the system should be cleaned, at the latest, when it reaches 140–145°C.

For this purpose, the unit must be shut down (during the third shift or the weekend) and the mixture of perchloroethylene and oil in the steam degreasing bath is cleaned by distillation. The remaining oil/dirt mixture is pumped out. After this, the chamber can be filled from the stand-by tank and the system can be restarted after heating up for one hour.

Roller Straightening

This unit is not integrated in the flexible manufacturing system as the tubes, which have a maximum length of 7m, must be cut to length after straightening. The unstraightened tubes are lifted with a crane into a magazine from which they pass singly into the inlet trough of the machine.

The dual inlet rollers feed the tubes through the triple roller frame and straighten them. The straightened tubes are tipped from the outlet trough onto a conveyor belt.

Here, a transfer transport facility provided by the customer can immediately transport the tubes to the cutting system or they can be lifted onto a transport trolley.

Drive Shafts

The ends are welded to all shafts in the electron beam welding machines and the laser beam welding machines.

Electron Beam Welding

The ends are transferred from the pallets of the waiting areas via gantry to the pinion mountings in the machine. The center tube is transferred from the V buffer with separator via transverse transport device to the machine and placed in the clamping facility of the machine by a gantry.

After all three components (two ends, one tube) have been assembled on the evacuation chamber door, this door swings through 180° into the welding chamber and a welded shaft is lifted off by the portal and placed on a transfer belt for movement to the outlet magazine.

Laser Beam Welding of Pinions to Center Tubes

The pinions are transferred from the pallets of the two waiting areas by a portal into the pinion mountings on a separate lifting/stepping conveyor for tubes and pinions.

The supporting tubes (center parts) are transferred from a V buffer with separator to a transfer belt and moved in front of the lifting/stepping conveyor. This takes a tube from the transverse transport belt and moves it together with the pinions to the clamping device of the machine.

Here, the pinions are fixed by clamping mountings. The lifting/stepping conveyor of the pinions is lowered and the two slide units with the clamped pinions are moved together, pushing the pinions into the tube. This provides sufficient centering of the pinions and tube and mounting devices.

The actual laser welding requires approximately 8 seconds, which means that a maximum floor-to-floor time of 25 seconds must be expected.

Conversion of this unit consists of changing the mounting elements on the lifting/stepping conveyor and the clamping mountings in the slide units, together with adjustment of the end stops of these units. The lasers are moved by motors.

Assembly and Laser Welding of the Spirals

The supporting tubes on the spirals are placed on separate conveyor belts and placed by means of separators into the tube insertion device and welding machine.

After the tube has been inserted into the spiral, and aligned radially and axially, the spiral is pressed against the tube and welded into position. Any superfluous spiral length is cut off.

The conveyor spirals welded in this manner are placed on a transfer transport device which feeds the pinion welding machine. To avoid collision during transverse transport, mutual interlocks are provided between the transfer device and the transverse transport.

CASE-IN-POINT

*FMS INSTALLATION FOR A MAJOR MANUFACTURER OF COMBINES PLANRING, USA, FARMINGTON HILLS, MICHIGAN**

Description of Flexible Manufacturing System

After almost one year of planning, PLANRING received an order for a flexible manufacturing system from a major manufacturer of combines. The system was designed to produce

*Permission granted by Planring U.S.A., Inc.

drive shafts and augers in forty-three flexible cells each equipped with robot gantry loaders almost fully automated. The system is arranged in two major sections: shaft end production and shaft assembly and welding.

Shaft End Production

At the start of the line, forgings are filled into hopper feeders and physically inspected. Undersized parts are rejected, and good parts loaded automatically onto pallets using gantry robots. The pallets are then transferred to a double-row shelf system for storage. The computer assigns available storage space to the pallets and memorizes all parts' data required for routing the pallets to the various machine operations. The machining of shaft ends from forgings to finished parts is performed on four groups of machines:

- Six mill to length and centering machines
- Seventeen turning machines
- Ten keyway milling machines
- Two drilling machines

All machines are fed by a central coolant supply system. Chips are removed by means of automatic chip conveyors located underneath the machining lines. The finished shaft ends are put back onto the pallets, sent through an automatic washing machine, and stored in the shelf system until required for assembly. A total of 520 shelf spaces are available, with each space holding up to fifteen pallets, with 2000 pallets circulating within the entire FMS. The system can produce up to 760 shaft ends per hour of twenty-one different types in any desired product mix. These parts range from: 58mm to 585mm length; 16mm to 122mm diameter; 2kg to 21kg weight.

Shaft Assembly and Welding

The end user provides the center tubes and the spirals in "ready-for-welding" condition. These materials are brought to various magazines in front of the welding and assembly cells. The shaft ends are transferred automatically from the shelves to the end welding cells.

The shafts are produced from a center tube and two shaft ends. These parts are welded together on special rotary welding machines. Depending on their carbon content, some shafts are even welded on E.B. welding machines.

The augers are produced from a center tube over which a spiral is stretched automatically, tack welded into position, and stitch welded over its length. In a subsequent operation, two shaft ends are welded to each auger using machines similar to those used for the shafts.

The assembly and welding section consists of:

- Three rotary end welding machines
- Three electron beam welding machines
- Eight spiral assembly and welding machines

All machines are equipped with gantry robots to allow fully automatic parts' handling.

A number of automatic tool monitoring, measuring, and gauging stations ensure constant quality throughout the FMS. The assembly section can produce eleven different types of shafts and augers and a total of up to 410 parts per hour. These parts range from:

- 1195mm to 4013mm length
- 50mm to 250mm diameter
- 8.8kg to 57kg weight

The FMS can be operated in three modes of operation:

1. *Fully Automatic.* Upon request by each cell, the computer distributes shop orders to individual cells or machining lines.
2. *Semi-Automatic.* Used for alternative material flow strategies during run-off or setup periods, or to direct pallets into or out of the system.
3. *Manual.* The pallet stacks and the machines are marked visibly with a part identification code thus permitting manual assignment of parts to the correct machines.

The computer prints lists containing information on the system's actual state at short intervals thus providing support for the dispatcher.

The FMS is the first of its kind to be delivered to this customer, hence extensive training of personnel is of great importance. At this point, Planring had provided 3.6 human/years of training and planned to continue until a total of approximately 6 years of training time would be spent on the project. Planring set up the FMS at the customer's plant and planned to stay with the project for 18 months after the installation has been put into service. Approximately 12 human/years of service engineers' time was supplied for installation, continued training, and transfer of know-how at the customer's plant.

CASE-IN-POINT

*ADVANCED ROBOTIC APPLICATION FLEXIBLE ASSEMBLY SYSTEM FAIRCHILD REPUBLIC CORPORATION, FARMINGDALE, NEW YORK**

Description of Flexible Assembly System
Robotic Cell Clustering Concepts

An ICAM 9401 Gantry robotic cell is used as a basic element or "building block" for clusters of cells. This robotic assembly cell utilizes the following:

- Gantry robot
- Cassette "kit" storage trays
- Automated assembly fixture
- Drivmatic riveter

Using all of these elements, the cell builds subassemblies. Clusters of cloned cells are capable of building large scale subassemblies.

The advantages of cloning and clustering Gantry ICAM 9401 robotic cells are that it gives the company the following:

- Cost Advantages
  - (a) Several cells can be concurrently operated by one computer
  - (b) Commonality of cell design—the only differences are in tooling and Gantry size
  - (c) Software has been developed already
- Technical Advantages
  - (a) CAD/CAM interface
  - (b) Cells in cluster are programmable via MCL
  - (c) System is flexible

*Permission granted by Charles Rosbeck Fairchild Republic Corporation, and Michael Hitchcock of the Manufacturing Technology Division of the ATR Force Wright Aeronautical Laboratories, Wright-Patterson AFB.

Replacement tooling for alternate tasks can be prepared off-line and inserted into Gantry with minimal down time.

- Easy to program
  (a) MCL Macro commands are easy to construct and use
- Example of Macro call commands:
  (a) Load TL/3—generates motions needed for Gantry robot to traverse to tooling table and lock end effector 3 onto the end of its arm. (Figure 8–27 shows a rib assembly cell, and figure 8–28 shows a bulkhead assembly cell.)

Gantry Robotic Cell

The Gantry Robotic Cell was selected because of the following considerations:

- Size of cell components
- Part fabrications
- Future expansion
- Other manufacturing processes

Design Concept

The design concept was to:

- Provide adaptability to diverse applications (flexibility).
- Provide the cell with pre-packaged kits containing the detail parts of an assembly.
- Provide cells with MCL program associated with the assembly.
- Use an assembly fixture capable of being programmed to accept a family of assemblies.
- Fixture details by means of a quick-setting adhesive for riveting.
- Have robot position details at drivmatic for final riveting.

The principal subsystem/functions are:

Subsystem	Function
Robot	Provides accuracy and dexterity required to service the entire cell envelope
End effectors	Provides the performance required to optimize specific tailored assembly tasks
End-of-arm tooling	Provides quick change of end effectors while cell is operating
Tooling table	Provides an area to both store and change end effectors
Automatic assembly fixture	Provides a working area capable of .001 in positioning accuracy
Adhesive arm	Lays bonding adhesive onto a positioned sub-assembly detail
Automatic drivmatic	Provides drilling and riveting operations to previously bonded parts
Tray storage rack	Provides cassette-type storage of previously loaded trays and easy accessibility for both robot and production control
Trays	Provides a holding and locating system for detail parts
Staging table	Provides easy access for robot to manipulate details located in the trays
DG MV 4000	Provides the capabilities of controlling the entire cell
GCA CIMROC	Commands the functions of the Gantry Robot and the end effector
MODULYNX	Commands all the functions of the assembly table
AB PLC 2/15	Controls the drivmatic and the Automated Rivet System
Loctite 205	Controls the dispensing of adhesive to the dispensing head

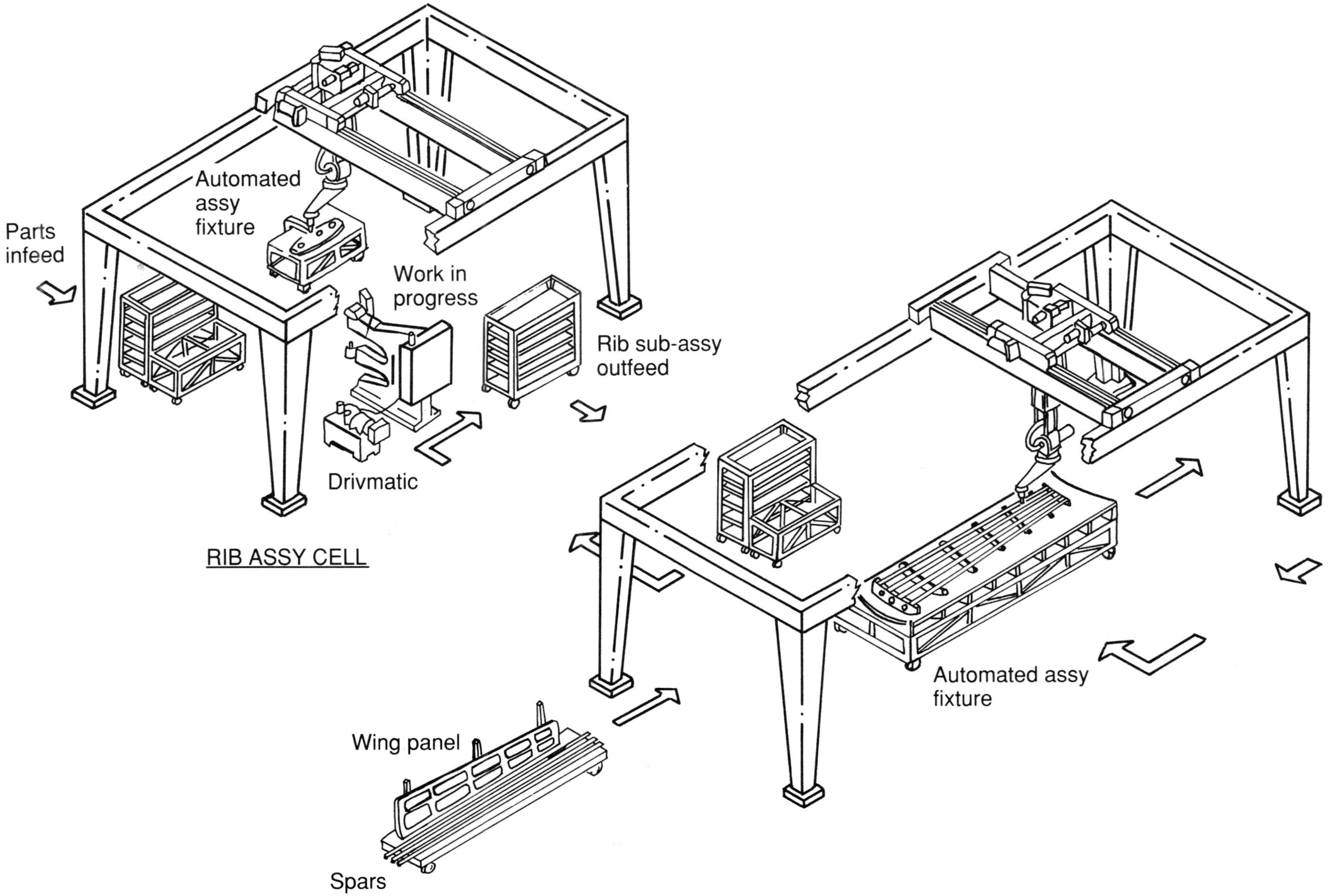

FIGURE 8–27 Rib Assembly Cell (permission granted by Charles Rosbeck, Fairchild Republic Corp., and Michael Hitchcock of the Manufacturing Technology Division of the Air Force, Wright Aeronautical Laboratories, Wright-Patterson AFB and Planring, USA, Inc.).

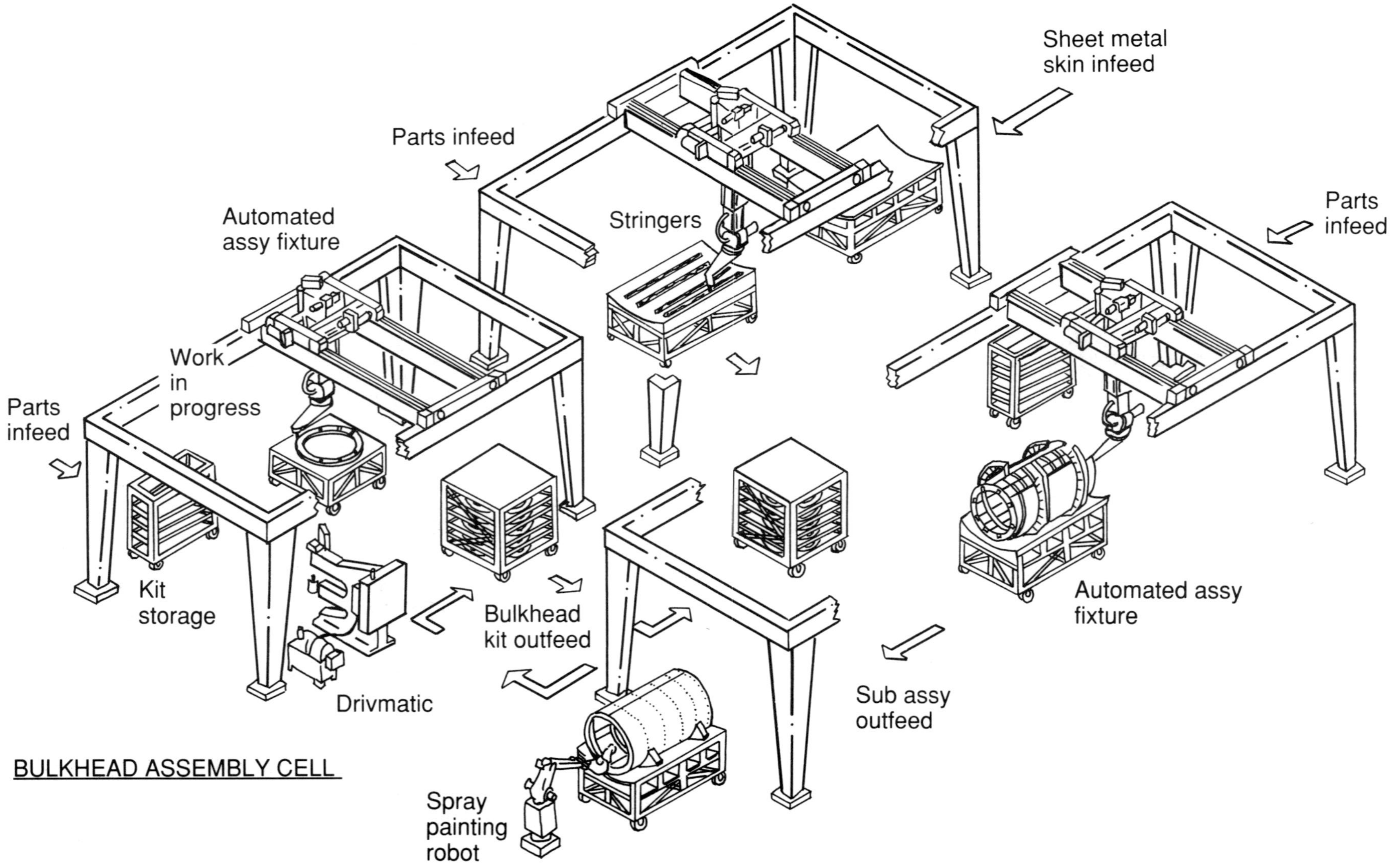

FIGURE 8-28 Bulkhead Assembly Cell (Courtesy of Fairchild Republic Corp./Wright Aeronautical Laboratories).

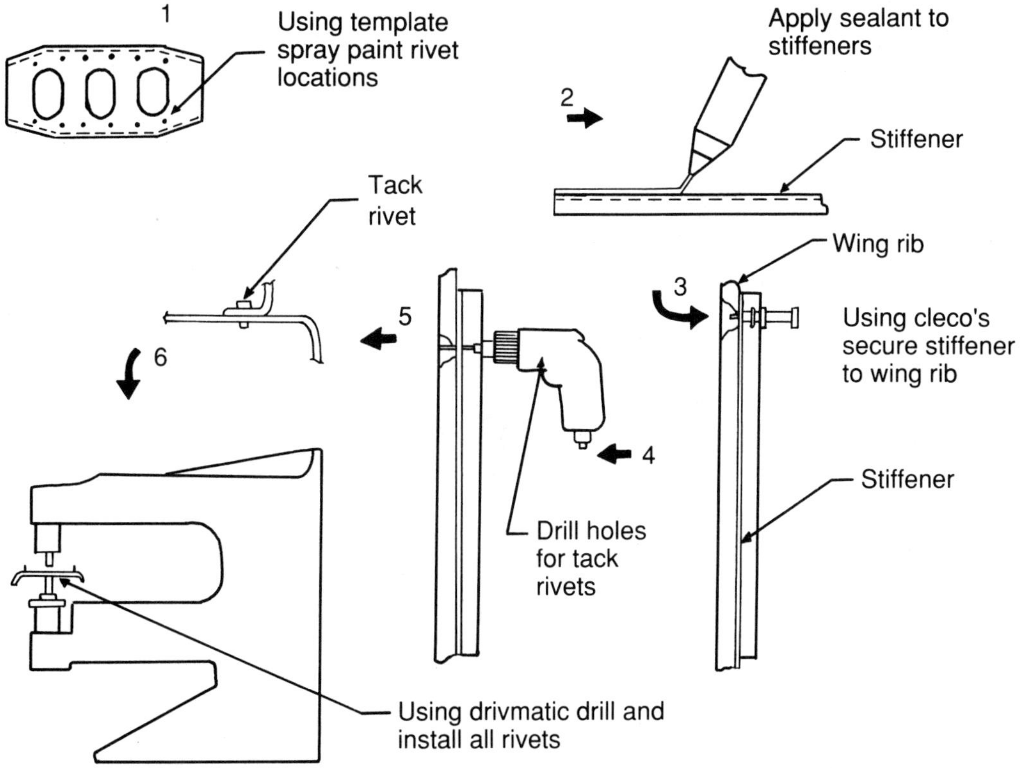

FIGURE 8–29 Schematic Presentation of Conventional Subassembly Operations (Courtesy of Fairchild Republic Corp./Wright Aeronautical Laboratories).

Figure 8–29 shows a schematic presentation of conventional sub-assembly operations. Figure 8–30 shows an overall flow diagram of this system. Figure 8–31 shows a storyboard photo depicting the robotic sequence of events, programmed in MCL, that assemble detail parts. The following is a step-by-step description of how the advanced robotic system operates.

Step 1 (figure 8–32). Overall view of the robotic cell with a robotics engineer flow charting the assembly software process, and another robotics engineer operating software code through the data general terminal.

Step 2 (figure 8–33). This photo shows the flexible assembly/adhesive application fixture with adhesive arm in the elevated position.

Step 3 (figure 8–34). This photo shows the flexible assembly/adhesive application fixture with adhesive arm in the deployed position.

Step 4 (figure 8–35). This photo shows the workstation table in foreground with incoming details module in the background.

Step 5 (figure 8–36). This photo shows the Drivmatic drill/riveting machine in foreground controlled by Allen-Bradley PLC attached to back of machine frame. Automated rivet feed hoppers are in the background to the left.

Step 6 (figure 8–37). This photo shows the detail part kit trays, stacked in incoming details module. The operator calls up program and does preliminary safety checks.

Step 7 (figure 8–38). The robot retrieves the forklift end effector from the end-effector dock. The drivmatic automated drill/riveter is in the background.

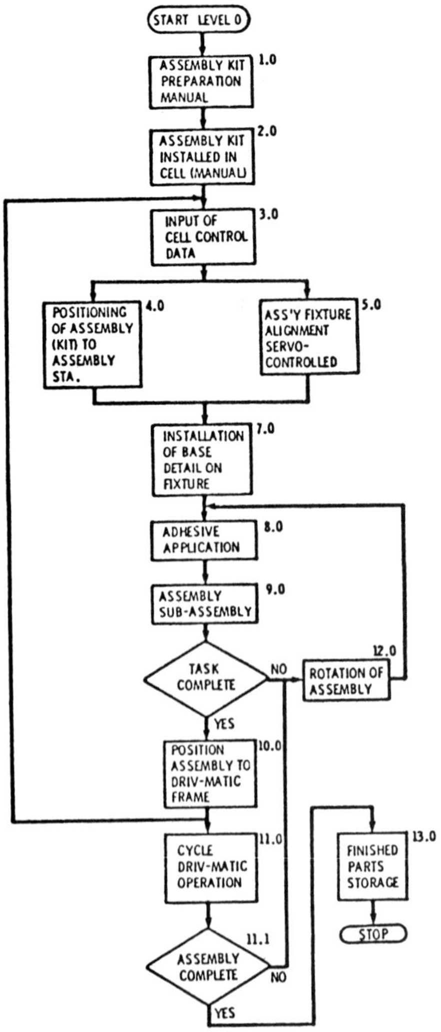

FIGURE 8–30 **Overall Flow Diagram of the System** (Courtesy of Fairchild Republic Corp./Wright Aeronautical Laboratories).

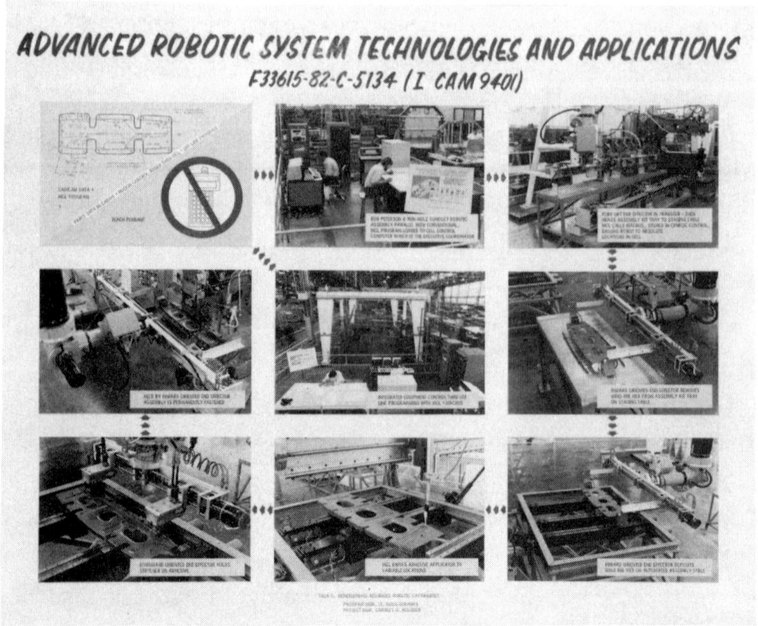

FIGURE 8–31 **Storyboard Photo Depicting the Robotic Sequence of Events** (Courtesy of Fairchild Republic Corp./Wright Aeronautical Laboratories).

FIGURE 8–32 **Overall View of Robotic Cell** (Courtesy of Fairchild Republic Corp./ Wright Aeronautical Laboratories).

FIGURE 8–33 **Flexible Assembly/Adhesive Application Fixture-Elevated Position** (Courtesy of Fairchild Republic Corp./Wright Aeronautical Laboratories).

FIGURE 8-34 Flexible Assembly/Adhesive Application Fixture-Deployed Position (Courtesy of Fairchild Republic Corp./Wright Aeronautical Laboratories).

FIGURE 8-35 Work Staging Table with Incoming Details Module (Courtesy of Fairchild Republic Corp./Wright Aeronautical Laboratories).

FIGURE 8–36 Drivmatic Drill/Riveting Machine (Courtesy of Fairchild Republic Corp./ Wright Aeronautical Laboratories).

FIGURE 8–37 Detail Part Kit Trays, Staked in Incoming Details Module (Courtesy of Fairchild Republic Corp./Wright Aeronautical Laboratories).

FIGURE 8–38 Forklift End Effector Retrieved by Robot from the End Effector Dock (Courtesy of Fairchild Republic Corp./Wright Aeronautical Laboratories).

The robot fetches a kit tray from a specific shelf in the kit storage rack and places it on the staging table.

The forklift end effector is designed specifically for manipulating kit trays. It features:

- A dual pneumatic actuator clamping system generating 301 pounds of holding force.
  - (a) Provides securing requirements of wing rib details in kit tray, which is being removed from incoming details module by forklift end effector held by robot (figure 8–39).
  - (b) Provides positioning accuracy. Foreground of photo shows robot carrying forklift end effector, having removed detail parts kit tray from the incoming details module, and placed the tray on the work staging table. Background contains flexible assembly/adhesive application fixture, with web detail shown, located on variably positioned pins (figure 8–40).

Step 8 (figure 8–41). This photo consists of a close-up of detail parts, wing rib web and stiffeners, in kit tray resting on work staging table.

Step 9 (figure 8–42). This photo shows the robot, having retrieved inward oriented end effector from end-effector dock, positioning end effector to clamp on wing rib web prior to lifting the web from the detail part kit tray.

Step 10 (figure 8–43). This photo shows the wing rib web detail being placed on locating pins of flexible assembly fixture by inward oriented end effector held by robot.

The robot using double grippers removes rib detail from the kit tray located on the staging table and locates that rib onto the positioning pins located on the assembly table.

The double gripper end effector (downward oriented) is designed specifically to pick up and place parts with flanges (figure 8–44). Its features are:

FIGURE 8–39 Wing Rib Details (Courtesy of Fairchild Republic Corp./Wright Aeronautical Laboratories).

FIGURE 8–40 Robot Carrying Forklift End Effector (Courtesy of Fairchild Republic Corp./Wright Aeronautical Laboratories).

FIGURE 8–41 Closeup of Detail Parts (Courtesy of Fairchild Republic Corp./Wright Aeronautical Laboratories).

FIGURE 8–42 Robot Having Retrieved Inward Oriented End Effector (Courtesy of Fairchild Republic Corp./Wright Aeronautical Laboratories).

FIGURE 8–43 Wing Rib Web Detail Placed on Fixture (Courtesy of Fairchild Republic Corp./Wright Aeronautical Laboratories).

- 2 degrees of freedom (X-Y)
- 2 synchronously movable grippers
- 30″ track
- Position feedback (LVDT)
- Pneumatic actuator (101 pound grip force)
- Interchangeable grippers

Foreground of figure 8–44: Robot has moved back (out of view). Wing rib web detail is shown on locating pins of flexible assembly adhesive application fixture. Adhesive application arm is shown in the elevated position.

Background left: Incoming details module with detail part kit trays and detail parts.

The automated assembly fixture is designed to provide a working area to accomplish all parts mating and bonding operations. Its features are:

- Modulynx controller
- Single RS232 computer interface link
- Three motorized locating pins
- 6 axes of freedom (X-Y)
- Positioning accuracy of .001 inch
- Tilt actuated adhesive applicator arm
  - Tilts to home position when parts are being positioned
  - Positions the adhesive gun (applicator)
- Loctite Model 205 Controller
  - Controls adhesive flow
  - Controls adhesive head

FIGURE 8–44 Wing Rib Web on Locating Pins of Flexible Assembly Adhesive Application Fixture (Courtesy of Fairchild Republic Corp./Wright Aeronautical Laboratories).

The assembly fixture tilt applicator arm moves from home position, and the adhesive applicator applies a series of drops along the contoured path which will be later occupied by the stiffener. Tilt arm then returns to home position.

Step 11 (figure 8–45). This photo consists of the wing rib detail, shown on pins of flexible assembly/adhesive application fixture. The adhesive applicator arm is in the deployed position with the adhesive applicator applying adhesive to the web.

Step 12 (figure 8–46). In this photo the inward oriented end effector has been placed back in the end-effector dock. The downward oriented end effector has been selected by the program to retrieve a wing rib stiffener from the detail parts kit tray on the staging table. The stiffener is shown being held in contact with the wing rib web, which has had adhesive applied to it with a light pressure for 5 to 30 seconds in the flexible assembly/adhesive application fixture.

The robot returns to the tooling table and exchanges and end effector for an (inward) end effector and returns to the assembly table. The robot then retrieves the track glued assembly from the assembly table and moves to the drivmatic.

Step 13 (figure 8–47). This photo indicates that the downward oriented end effector has been returned to the dock. The inward oriented end effector, selected through MCL program, has removed adhesively assembled details from flexible assembly/adhesive application fixture, holding the web detail at both ends. The sub-assembly is shown in the jaws of the drivmatic having permanent attach rivets installed.

The robot places the completed assembly on a parts output table.

The robot exchanges its end effector for a forklift end effector and returns tray to its proper shelf location.

FIGURE 8–45 Wing Rib Web Detail Adhesive Applicator Applying Adhesive to the Web (Courtesy of Fairchild Republic Corp./Wright Aeronautical Laboratories).

FIGURE 8–46 Inward-Oriented End Effector Placed Back in the End Effector Dock (Courtesy of Fairchild Republic Corp./Wright Aeronautical Laboratories).

Step 14 (figure 8–48). This last photograph, in this sequence of events, consists of the following:

- Foreground: T-46A aircraft in stage of assembly.
- Background: Robotic sheetmetal assembly cell.

Results

The following comparative figures show the time saved by robotic technology versus conventional methods on all applications handled by this flexible manufacturing system:

Conventional Methods: 1.913 hrs. to drill and rivet
Robot: 0.642 hrs. to drill and rivet
Savings = 66.4%

Figure 8–49 shows savings in bar chart form on one type of assembly, the SF340 Wing Ribs.

The remaining figures are illustrative of the various components of this FMS Application.

- Figure 8–50—Logic trees
- Figure 8–51—Schematic-control room/work envelope
- Figure 8–52—Cell controller hardware
- Figure 8–53—Robotic system architecture
- Figure 8–54—Cell controller data flow
- Figure 8–55—MCL data flow

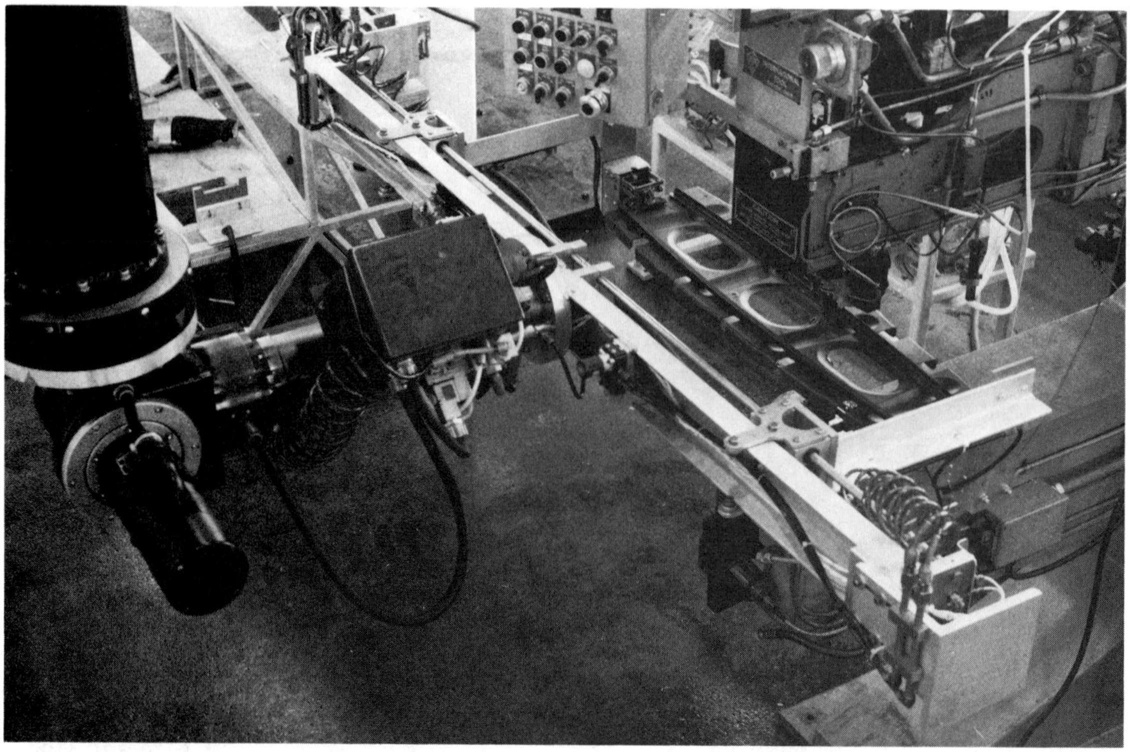

FIGURE 8–47 Downward-Oriented End Effector Returned to Dock (Courtesy of Fairchild Republic Corp./Wright Aeronautical Laboratories).

FIGURE 8–48 Foreground: T-46A Aircraft in Stage of Assembly; Background: Robotic Sheetmetal Assembly Cell (Courtesy of Fairchild Republic Corp./Wright Aeronautical Laboratories).

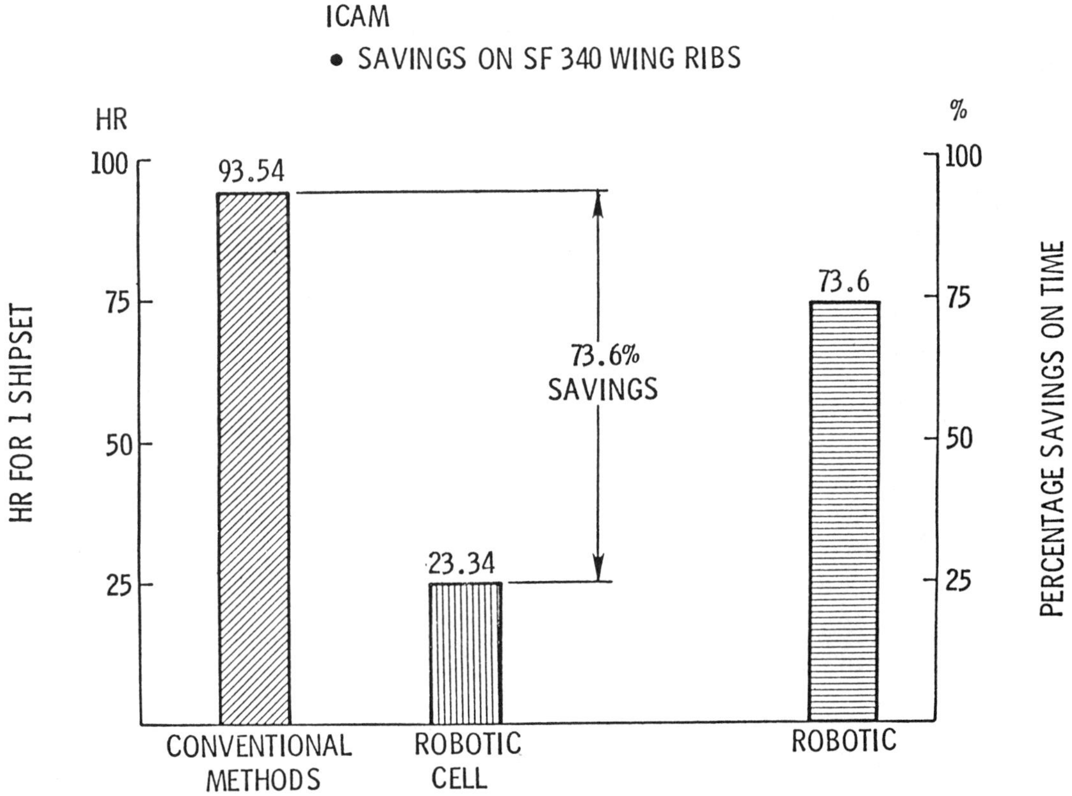

FIGURE 8–49 Savings Analysis—FS340 Wing Ribs (Courtesy of Fairchild Republic Corp./Wright Aeronautical Laboratories).

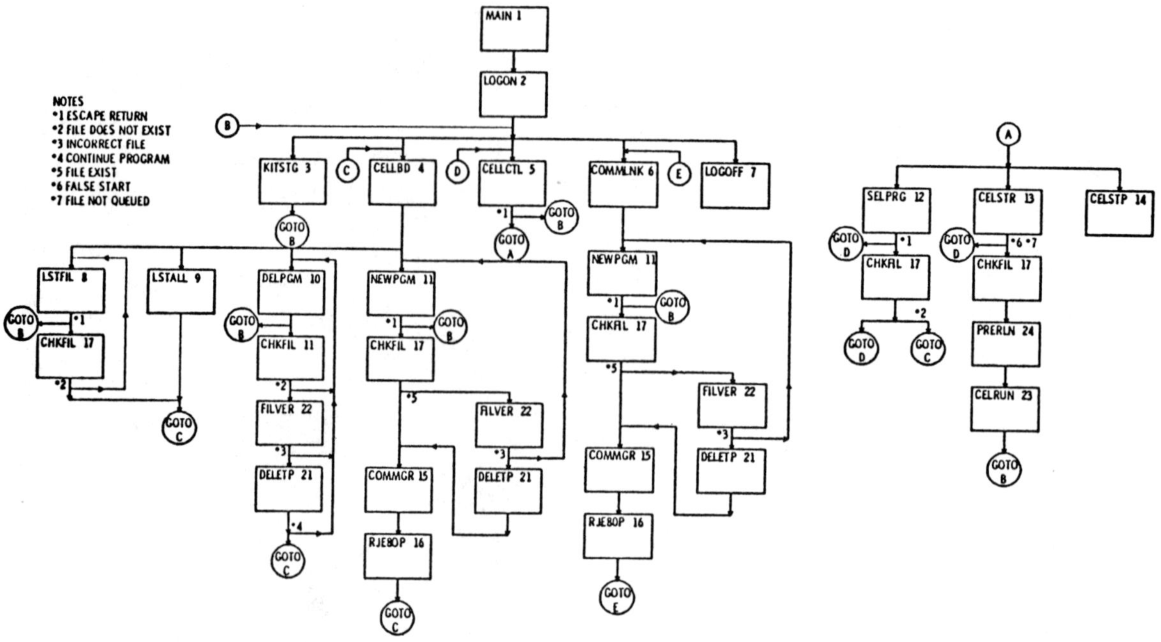

FIGURE 8–50 **Logic Trees** (Courtesy of Fairchild Republic Corp./Wright Aeronautical Laboratories).

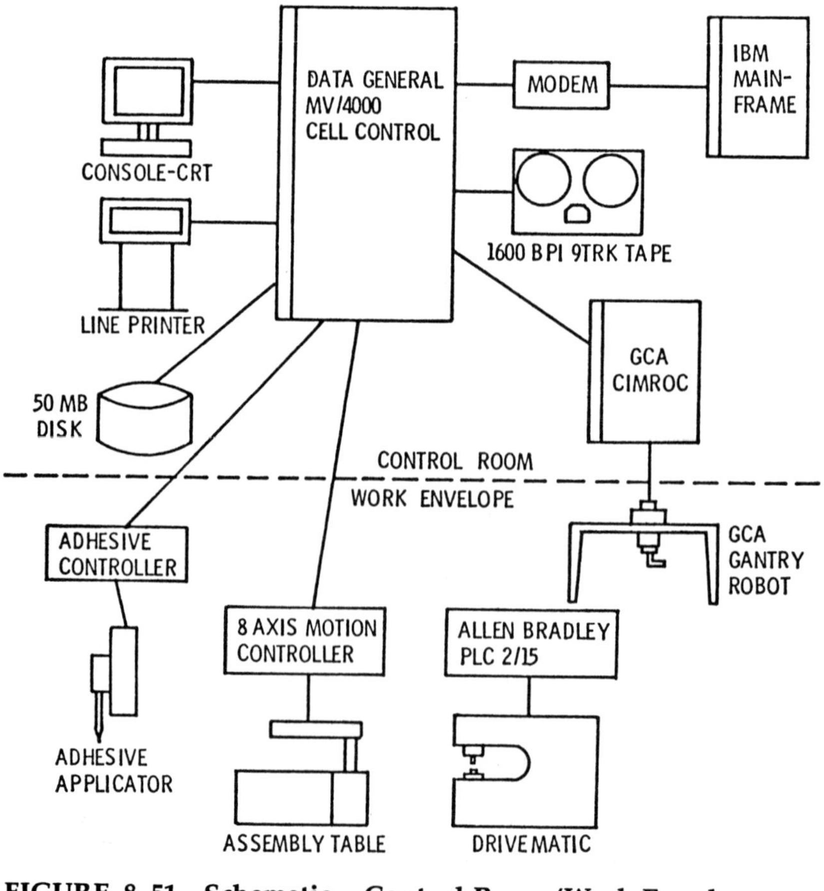

FIGURE 8–51 **Schematic—Control Room/Work Envelope** (Courtesy of Fairchild Republic Corp./Wright Aeronautical Laboratories).

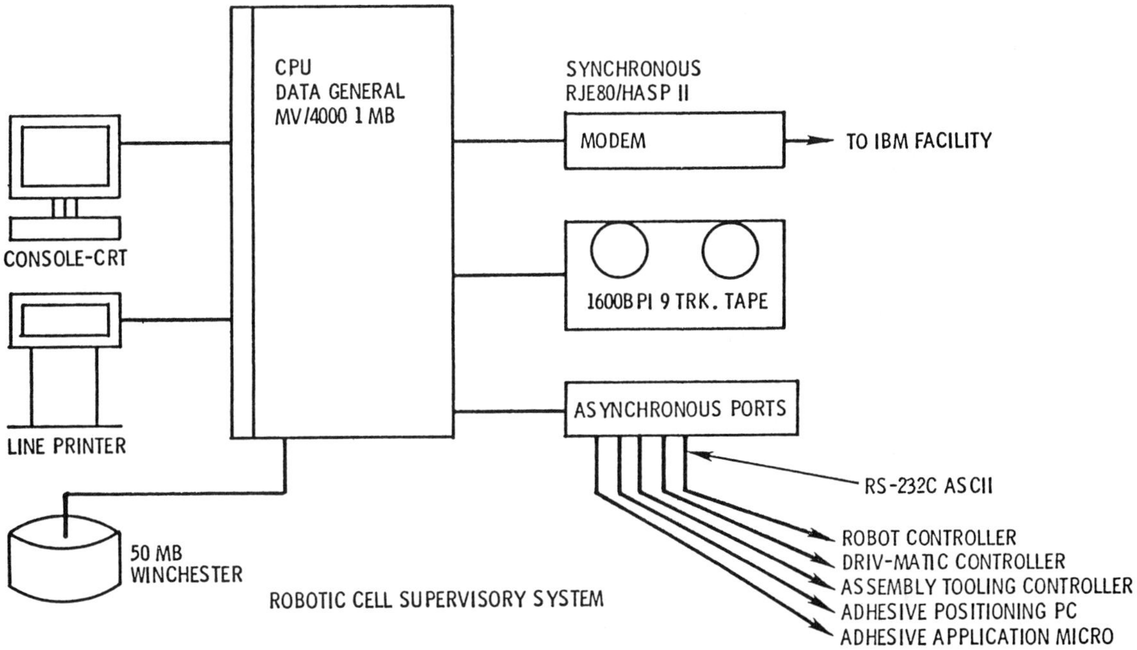

FIGURE 8–52 **Cell Controller Hardware** (Courtesy of Fairchild Republic Corp./Wright Aeronautical Laboratories).

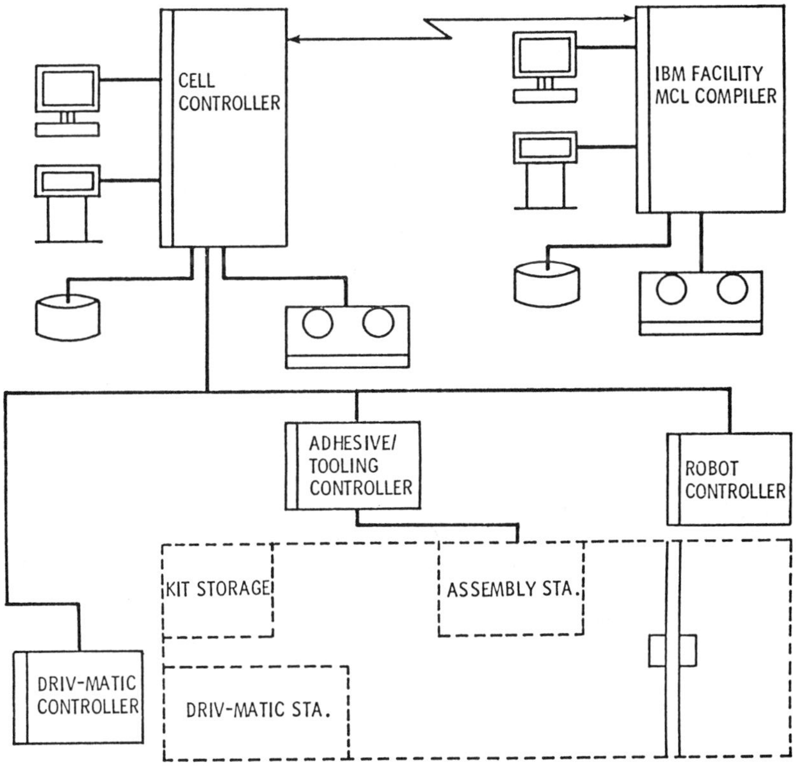

FIGURE 8–53 **Robotic System Architecture** (Courtesy of Fairchild Republic Corp./Wright Aeronautical Laboratories).

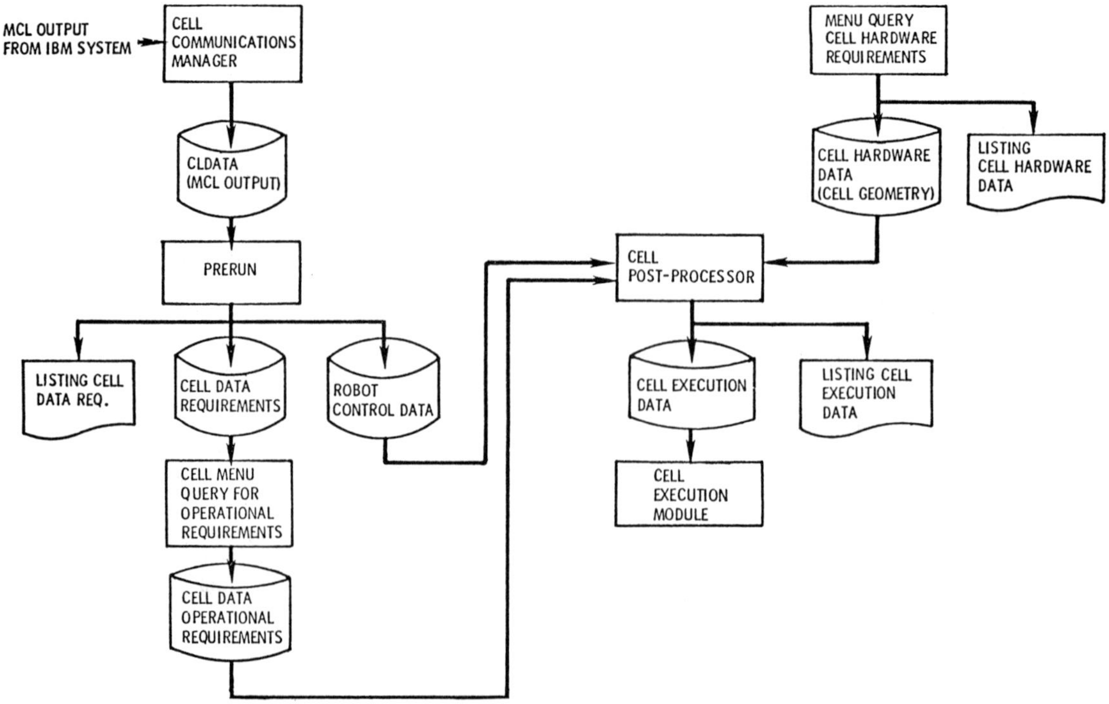

FIGURE 8–54 Cell Controller Data Flow (Courtesy of Fairchild Republic Corp./Wright Aeronautical Laboratories).

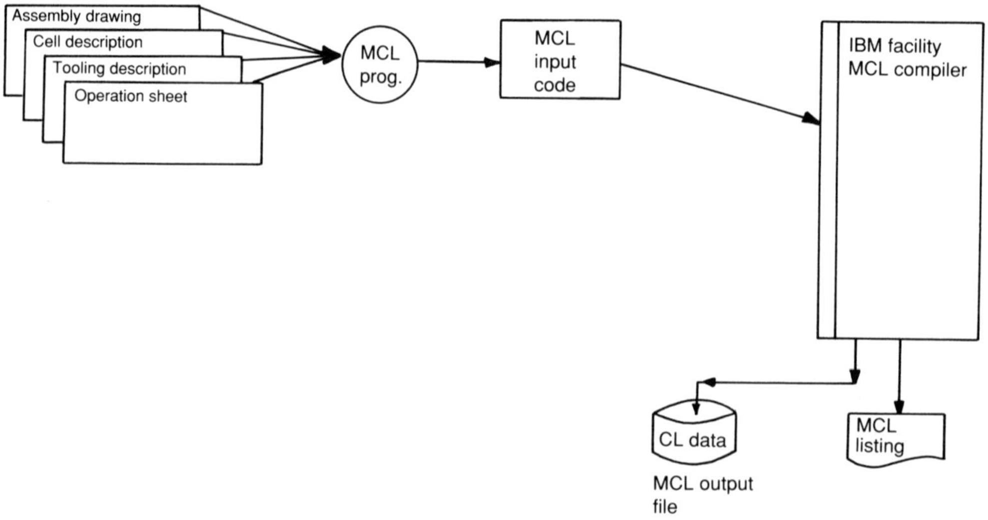

FIGURE 8–55 MCL Data Flow (Courtesy of Fairchild Republic Corp./Wright Aeronautical Laboratories).

chapter 9

Developing Quality Assurance Techniques in Consonance with the Manufacturing Environment

PART A. QUALITY DESIGN AND STATISTICAL TECHNIQUES

TRENDS IN CORPORATE QUALITY POLICY

American industry is going through a most challenging period, one that requires optimum manufacturing efficiencies with the highest standard of quality excellence to be competitive in the international marketplace. In the area of quality control, the most comprehensive expression of a company's commitment to quality is contained in the corporate policy on that subject. Corporate quality policy provides the overall basis of the standards to be followed in the design, engineering, manufacturing, and marketing of all company products. The policy specifies a process by which the quality of the company's product is verified at every critical point from concept to customer usage.

Tips for Successful Automation: Traits of Successful Quality Programs

To be successful, a quality program must embrace certain key elements, such as:

- A *goal* to provide customers with products that equal or exceed standards and customer specifications.
- A *practice* defined in program guidelines. The quality guidelines should reflect a company's conviction that fault prevention, rather than detection, is the key to quality. The ultimate measure of a vendor's success in pursuing an effective, aggressive quality program is the prevention of nonconforming hardware.
- The *process* of a company's program is achieved by program monitoring. A product quality assurance committee of operating managers should review, evaluate, and initiate the necessary operating changes to assure product quality. Through careful and

continuing monitoring of each product and thorough internal communications, this committee can play a key role in the program's success.

- The *feedback* of the quality program should be performed by real-time systems, or other techniques that speed up information flow to individuals that can correct production deficiencies immediately, or through automated means.
- The *ongoing task* of the quality program is fostered by continual training and retraining of individuals to broaden and strengthen their operating and technical skills. The effectiveness of a company's quality assurance program relies heavily on management's commitment to train and educate its people. It is essential that employees at every level and functional area learn to apply quality techniques.

In a corporate environment that stresses quality and the role of the individual employee, a company can implement its specific quality program.

ESTABLISHING THE VALUE OF A MANUFACTURED PRODUCT

A customer should expect product value with some relationship to the cost of the product. This implies that a company has an obligation to satisfy the needs of the marketplace, not only in terms of functional value, but overall intrinsic value as well. Therefore, the value of any manufactured product is dependent on two factors:

- The integrity of the materials that go into the process, and
- The integrity of the process itself.

The first of these factors mandates the best materials from a company's suppliers. A company should expect the same commitment to quality of its suppliers that it demands of itself.

The second factor requires the maximum possible control of variability—that is the essence of manufacturing quality.

Together, they enable a company to supply customers with products free of abnormal, unwanted variations—*products that are made right the first time.*

PREVENTATIVE DESIGN CONTROL TECHNIQUES

To minimize quality problems it is of utmost importance that quality assurance personnel engage in preventative design control techniques in coordination with design engineers and manufacturing engineers. Design engineers perform their role by calling up the design using CAD/CAM software, reviewing the part in question, and modifying or changing the design specifications as required.

Quality can be built into the product through a number of steps including design assurance—drawing and design review, specification review, test plan and procedure review, and other analytical review procedures that enable product quality to commence from a sound base point.

Quality engineers also can apply preventative design assurance methods to software that may involve a safety factor, and automatic test equipment to assure computer programs, data, and documentation will meet customer contractual requirements.

Two important requirements are being added to the concept of what makes a good design; manufacturability and testability. For example, some of the companies in the electronics industry require the design engineer to also produce an acceptable test program before the design can be approved.

CAD, as a design tool, most impacts quality assurance in the area of simulation (discussed in more detail in Chapter 8). *Simulation* is testing before the product is manufactured or even prototyped. Simulation packages are of two basic types:

- Simulation of Products. Ensuring that the product will do what it is supposed to do.
- Simulation of Process. Ensuring that the planned manufacturing process will work.

Product simulation in mechanical design mostly takes the form of structural analysis such as finite element analysis and modeling (refer to Chapter 2). Modeling programs can even check to see if parts will fit together.

There are a number of excellent product simulation packages in the electronics industry. Most of the electronic design automation vendors provide some kind of logic simulation. Many include timing analysis and display, as well as design rule checking. Some packages can recycle design simulation data for test generation, including providing IC test programs for different brands of testers.

Process simulation packages improve quality by providing a way to test the manufacturing process without using raw material and valuable operator and machine time. They also perform a checking function on the capability of various automated equipment before going into production.

Generally, design for manufacturability will entail more initial design time, but companies using this approach have found that it reduces prototyping time and speeds manufacturing operations.

HOW THE JAPANESE CONTROL QUALITY THROUGH QUALITY FUNCTION DEPLOYMENT*

In Japan, companywide quality control (CWQC) is characterized by: (1) a philosophy, (2) a specification, (3) a mechanism to deploy customer desires vertically and horizontally throughout the company. The mechanism has been named *quality function deployment* and serves as the "operational definition" of CWQC. The philosophy and specification of CWQC are covered briefly below.

1. Many U.S. companies have developed quality philosophies in recent years. A company philosophy cannot be copied or adopted from someone else; it must be developed to reflect the individual company culture and management vision. When developed in this way, it can be deployed vertically and horizontally throughout the company via policy and operating practice to change the thinking of all employees.

2. In Japan, the specification for CWQC (Japan Industrial Standard Z8101-1981) is followed very uniformly by most companies. In U.S. companies, quality control and CWQC mean different things to different people, and much effort is lost in debating the virtues of individual quality programs. This tends to paralyze management from acting on quality

*Permission granted by The American Supplier Institute. "The American Supplier Institute (ASI) in Dearborn, Michigan, is a non-profit organization dedicated to the promotion of Quality Management, and the improvement of product quality and industrial productivity. Dr. Genichi Taguchi is Executive Director of ASI, which provides instruction and implementation consulting services with regard to Taguchi Method and Quality Function Deployment applications. The ASI also provides instruction and implementation consulting services on Statistical Process Control methods based on the teachings of Walter A. Shewhart, W. Edwards Deming, and Joseph M. Juran."

issues since there is little consensus within the industry as to what quality control and CWQC mean.

3. The main objective of any manufacturing company is to bring new (and carry over) products to the market sooner than the competition with lower cost and improved quality.

Six key terms associated with QFD:

Quality Function Deployment—an overall concept that provides a means of translating customer requirements into the appropriate technical requirements for each stage of product development and production—that is, marketing strategies, planning, product design and engineering, prototype evaluation, production process development, production, and sales. This concept is further broken down into "product quality deployment" and "deployment of the quality function" described later.

Voice of the Customer—the customers' requirements expressed in their own terms.

Counterpart Characteristics—an expression of the voice of the customer in technical language that specifies customer-required quality; counterpart characteristics are critical, final product-control characteristics.

Product Quality Deployment—activities needed to translate the voice of the customer into counterpart characteristics.

Deployment of the Quality Function—activities needed to assure that customer-required quality is achieved; the assignment of specific quality responsibilities to specific departments. (The phrase "quality function" does not refer to the quality department, but rather to any activity needed to assure that quality is achieved, no matter which department performs the activity.)

Quality Tables—a series of matrices used to translate the voice of the customer into final product control characteristics.

To understand quality function deployment (QFD), you must first understand that the approach to quality is fundamentally different in United States and Japanese companies. In Japanese companies, the customer's voice drives all activities, while in many U.S. companies, it is the executive's voice or the engineer's voice that prevails. Furthermore, compared with many U.S. companies, Japanese companies concentrate more on what the customer likes; U.S. companies pay more attention to fixing what the customer doesn't like. That is, the Japanese put more effort into designing quality at the product development stage, while U.S. companies put a greater emphasis on problem solving.

In QFD, all operations of the company are driven by the "voice of the customer"; QFD therefore represents a change from manufacturing-process quality control to product development quality control.

QFD brings several benefits to companies willing to undertake the study and training required to put the system in place:

- Product objectives based on customer requirements are not misinterpreted at subsequent stages.
- Particular marketing strategies or "sales points" do not become lost or blurred during the translation process from marketing through planning and on to execution.
- Important production control points are not overlooked; everything necessary to achieve the desired outcome is understood and in place.
- Tremendous efficiency is achieved because misinterpretation—of program objectives, marketing strategy, and critical control points—and need for change are minimized (figure 9–1).

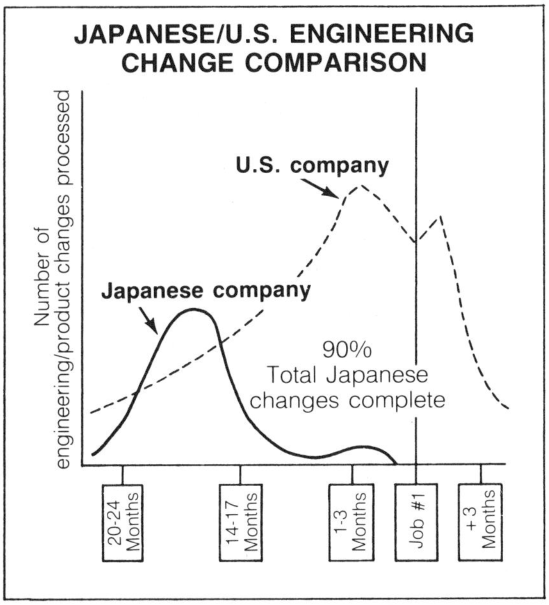

FIGURE 9–1 Japanese/U.S. Engineering Change Comparison (Courtesy of American Supplier Institute).

Four Key Documents in a QFD System

The QFD system concept is based on four key documents.

1. ***Customer Requirement Planning Matrix***—translates the voice of the customer into counterpart control characteristics—that is, it provides a way of turning general customer requirements, drawn from market evaluations, comparisons with competition, and marketing plans, into specified final product control characteristics.

2. ***Final Product Characteristic Deployment Matrix***—translates the output of the planning matrix—that is, the final product control characteristics, into critical component characteristics. Thus, it moves one step farther back in the design and assembly process.

3. ***Process Plan and Quality Control Charts***—identifies critical product and process parameters as well as control or check points for each of those parameters.

4. ***Production Operating Instructions***—based on the critical product and process parameters; these instructions identify operations to be performed by plant personnel to assure that important parameters are achieved.

The overall QFD system based on these documents traces a continuous flow of information from customer requirements to plant operating instructions. Thus it provides "a clear operational definition"—a common purpose, priorities, and focus of attention. Let us examine each document in greater detail.

The purpose of the planning matrix is to translate customer requirements into important final product control characteristics that are to be deployed through the design, development, processing, and production control system.

Putting this matrix together involves eight steps. Step 1 is to state the product requirements in customer terms, as shown by the example in figure 9–2. The primary requirements, which are the very basic customer wants, are normally expanded into secondary and tertiary requirements to obtain a more definitive list. This information usually comes from a variety of sources: marketing research data, dealer input, sales department wants, "buff" magazines, special customer opinion surveys, and so on. This first step is the most critical

Primary	Secondary	Tertiary
Product is Dependable	Reliable, Trouble-Free	• Always starts • Absence of troubles that: —Stop the vehicle —Result in an inoperative function —Result in inconvenience
	Lasts a long time	• Absence of unexpected component wear-out • Absence of unexpected appearance deterioration
	Easily and quickly serviced	• Can be serviced quickly • Parts are readily available • Service is effective

FIGURE 9–2 **Step 1: State Requirements in Customer Terms** (Courtesy of American Supplier Institute).

part of the process and it is usually the most difficult because it requires obtaining and expressing what the customer truly wants—and not what we think he or she expects.

The customer requirements make up the vertical axis of the upper left portion of the planning matrix. Step 2 in developing the matrix is to list—across the top horizontal row of the matrix—the final product control characteristics that you believe should be assured to meet the customer stated product requirements (figure 9–3). These characteristics are the product requirements that relate directly to the customer requirements, and must be selectively deployed throughout the design, manufacturing, assembly, and service process to manifest themselves in the final product performance and customer acceptance.

Consequently, these characteristics must be capable of being expressed in measurable terms since the characteristic output is to be controlled and compared with objective targets. Normally, there is an abundance of these characteristics. The question is: Do you have the right ones in terms of customer requirements?

After labeling vertical and horizontal axes of the upper portion of the planning matrix, we take Step 3: Developing a relationship matrix between the customer requirements and the final product control characteristics (figure 9–4). Since there are varying degrees of correlation between customer requirements and product characteristics, a set of symbols is used to identify the significance of the relationships, as shown in the figure.

The benefit of filling out this relationship matrix using appropriate symbols is that it quickly indicates whether the final product control characteristics adequately cover the customer requirements or expectations. Absence of symbols (or a majority of "weak relationships" signs) indicates that some customer requirements are not addressed or show a weak relationship to final product control characteristics. Therefore the design has little probability of meeting those particular needs expressed by the customer. A very important aspect of this relationship matrix is its ability to identify conflicting design requirements. In the case of conflicting requirements, the design must be optimized to meet individual target values. (Taguchi methods are used to do this, as described later.)

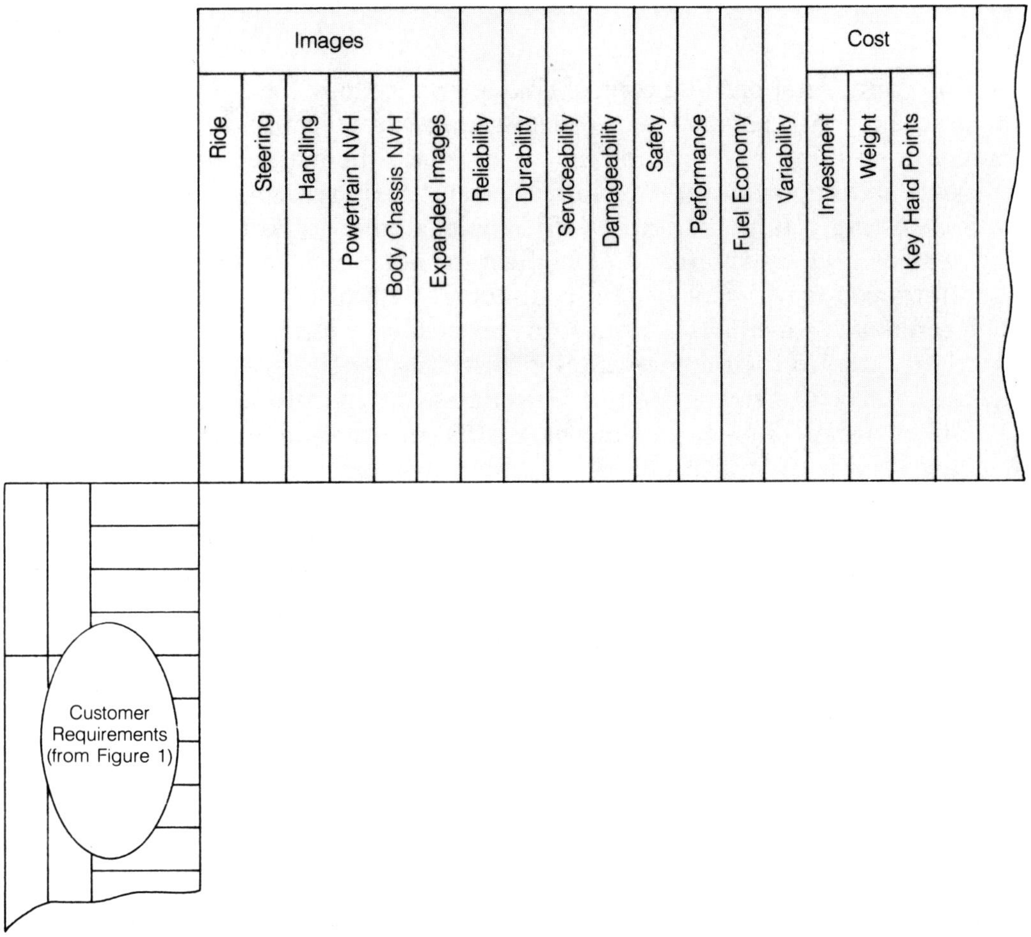

FIGURE 9–3 Step 2: List Existing Final Product Control Characteristics (Courtesy of American Supplier Institute).

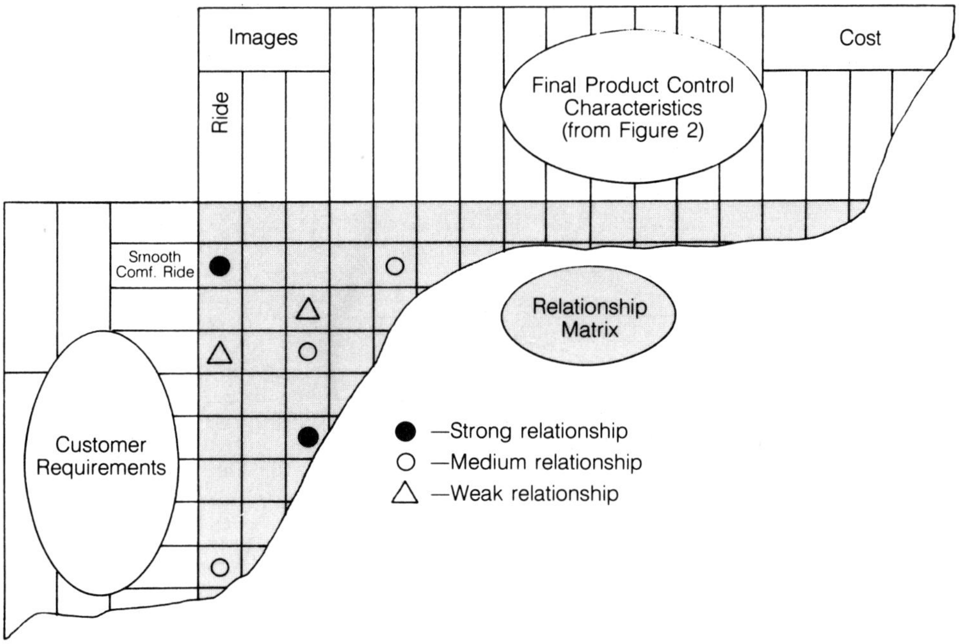

FIGURE 9–4 Step 3: Develop Relationship Matrix between Customer Requirements and Final Product Control Characteristics (Courtesy of American Supplier Institute).

At this point, final product control characteristics may have to be modified or supplemented to assure that all customer requirements are addressed adequately.

Step 4 in building the planning matrix is to add the market evaluation, which covers customer-expressed importance ratings for the listed requirements, and competitive evaluation data for existing products (figure 9–5). The customer importance rating, which represents the areas of greatest interest and highest expectations/satisfaction expressed by the customer, allows you to set priorities for product areas that require improvement.

The competitive evaluation data shows how your product is viewed by the customer and your relationship to your competition in fulfilling a particular customer need. In short, it highlights the absolute strengths and weaknesses of the products that require improvement. The information comes from surveys, media information, and feedback from marketing, sales, and service organizations. The competitive evaluation data should include both your product and your major competition as shown in figure 9–7. In most cases this evaluation is based on a rating system to clearly indicate market strengths and weaknesses.

Step 5 provides for posting of final product control characteristic competitive evaluations (figure 9–6). The data utilized are usually obtained from tests or evaluations conducted within your company; the information covers your own and your competitors' products. Again, these data should be expressed in objective, measurable terms, or in case of subjective data, translated into a numerical performance rating based on consensus evaluation.

Market evaluations are compared with the product control characteristic evaluations to determine areas of inconsistency between what the customer says and your own evaluations. (See lower part of figure 9–6, Step 5B). For example, if the market data indicates that a competitive product best satisfies a customer requirement, whereas your evaluation indicates that your product is better for the particular product control characteristic that relates to that

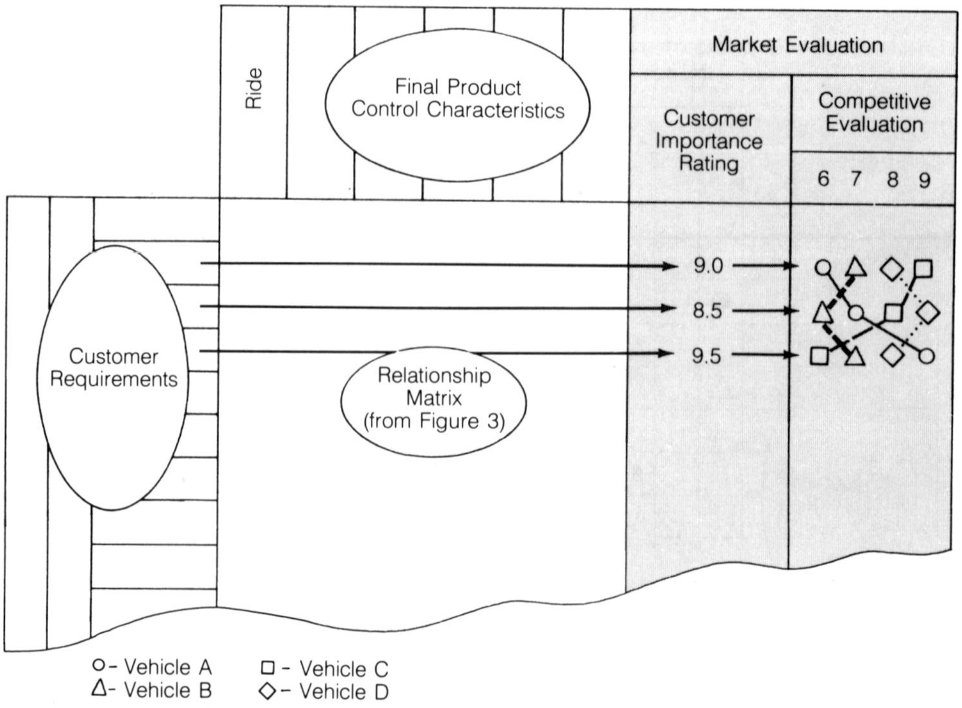

FIGURE 9–5 Step 4: Enter Market Evaluations (Courtesy of American Supplier Institute).

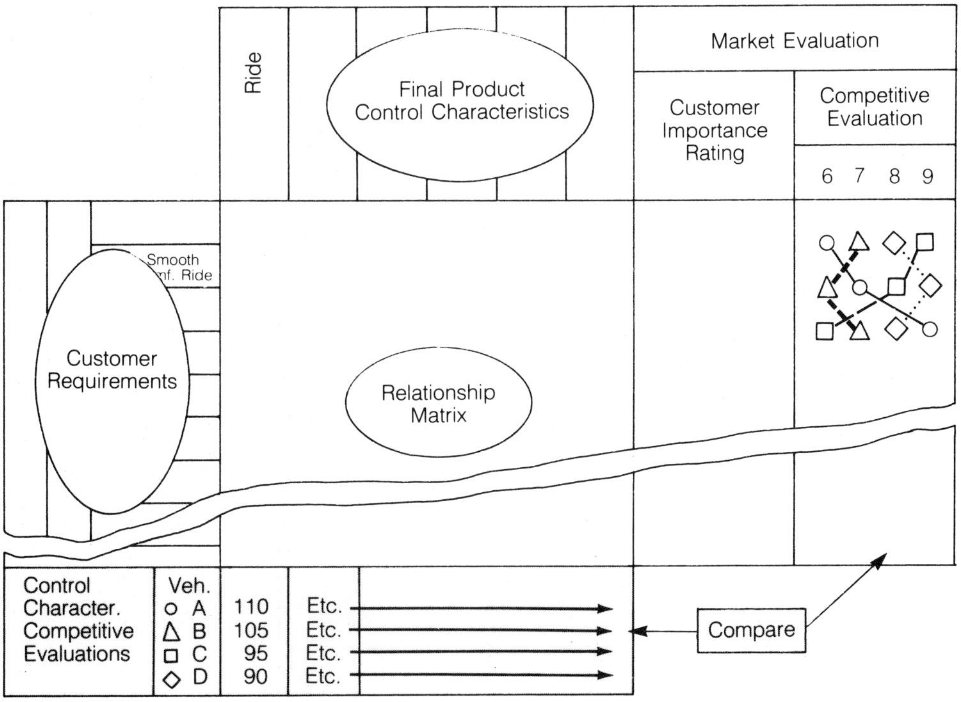

FIGURE 9–6 **Step 5A: Enter Final Product Control Characteristic Competitive Evaluation** (Courtesy of American Supplier Institute).
Step 5B: Compare Control Characteristic Competitive Evaluations to Market Competitive Evaluations (Courtesy of American Supplier Institute).

need, then there is something wrong with the internal evaluation of that characteristic—or the wrong characteristic was chosen to meet that need.

In Step 6, we use the right-hand column of the planning matrix to list key selling points for a new product (figure 9–7). These selling points are advertisable characteristics to be emphasized in a particular market segment. Based on the results of such a comparison—such as best-in-class ride or interior roominess—the product marketing, merchandising, and promotion strategies are established.

The selling points are based on:

- Degree of their importance in meeting the priority customer needs.
- Your company's past performance in these areas.
- Associated costs to incorporate them in the product. All company activities should be involved in the selling point selection process to assure that what is claimed is delivered.

Step 7 is the development of targets for each of the final product control characteristics (figure 9–8). These targets are based on the agreed upon selling points, the customer importance index, and the current product strengths and weaknesses. As mentioned previously, these targets should be measurable values that can be evaluated on the final product.

Performance to these targets is measured at each of these stages of the product development and prove-out process:

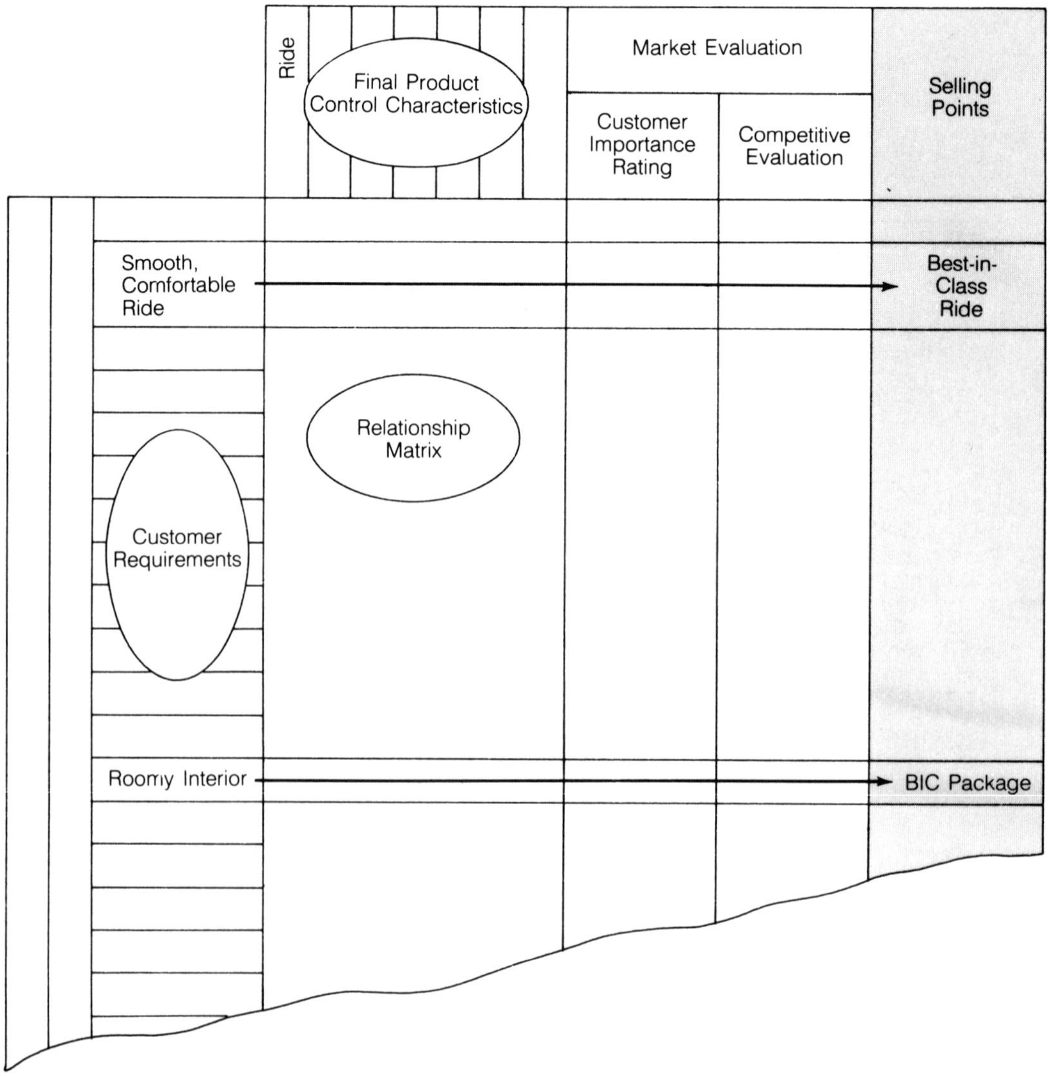

FIGURE 9–7 Step 6: Determine Selling Points for New Product—Advertisable Characteristics Pertinent to the Market Segment (Courtesy of American Supplier Institute).

- Evaluations of mechanical and functional prototype builds
- Engineering sign-offs
- Units produced with production tooling just prior to production, and on first units at start of production

Step 8 in the development of the planning matrix involves the selection of the product control characteristics that are to be deployed through the remainder of the QFD process—that is, from planning through production (figure 9–9). This selection is based on customer importance, selling points, competitive evaluations, and the difficulty of achieving the characteristic target.

Characteristics that have a strong relationship to meeting customer needs, are important to the customer, have poor competitive performance, or are determined to be strong selling points must be deployed or translated into the language of each discipline or function

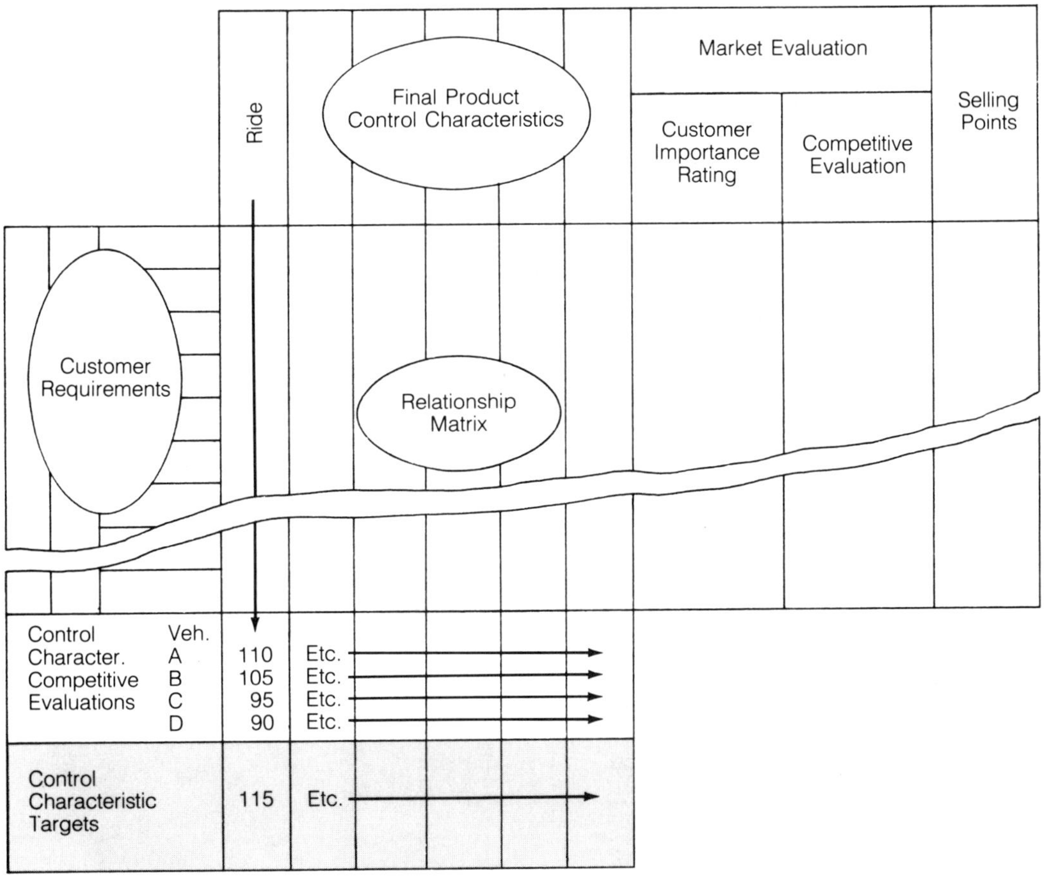

FIGURE 9–8 Step 7: Develop Targets for Final Product Control Characteristics (Courtesy of American Supplier Institute).

in terms of requirements, actions, and controls to assure that the voice of the customer is maintained accurately through every step of product and process development and continuing on to production, marketing, and sales.

Conversely, those characteristics that are not as critical in meeting customer requirements, are not strong selling points, or have targets that are easily achievable need not be subjected to rigorous deployment processes.

Once the planning matrix is completed, we move on to the *hardware deployment matrix,* the second key QFD document. In this matrix, the final product control characteristics are carried from the overall vehicle level down to the subsystem/component level (figure 9–10). This is the first step of the procedure to materialize customers' needs.

In this phase of deployment, both the customer requirements and the final product control characteristics are listed in greater detail. The finished component characteristics affecting the final product characteristics are identified, and a matrix is developed to show whether, and to what extent, the relationships between component and product characteristics are critical. It is the critical finished component characteristics that are to be deployed further and monitored in the production planning and control system.

The component deployment chart (figure 9–11) expands the list of the components to include the finished component characteristic and the essential critical or control characteristics of the component. These characteristics relate directly to meeting the desired target value

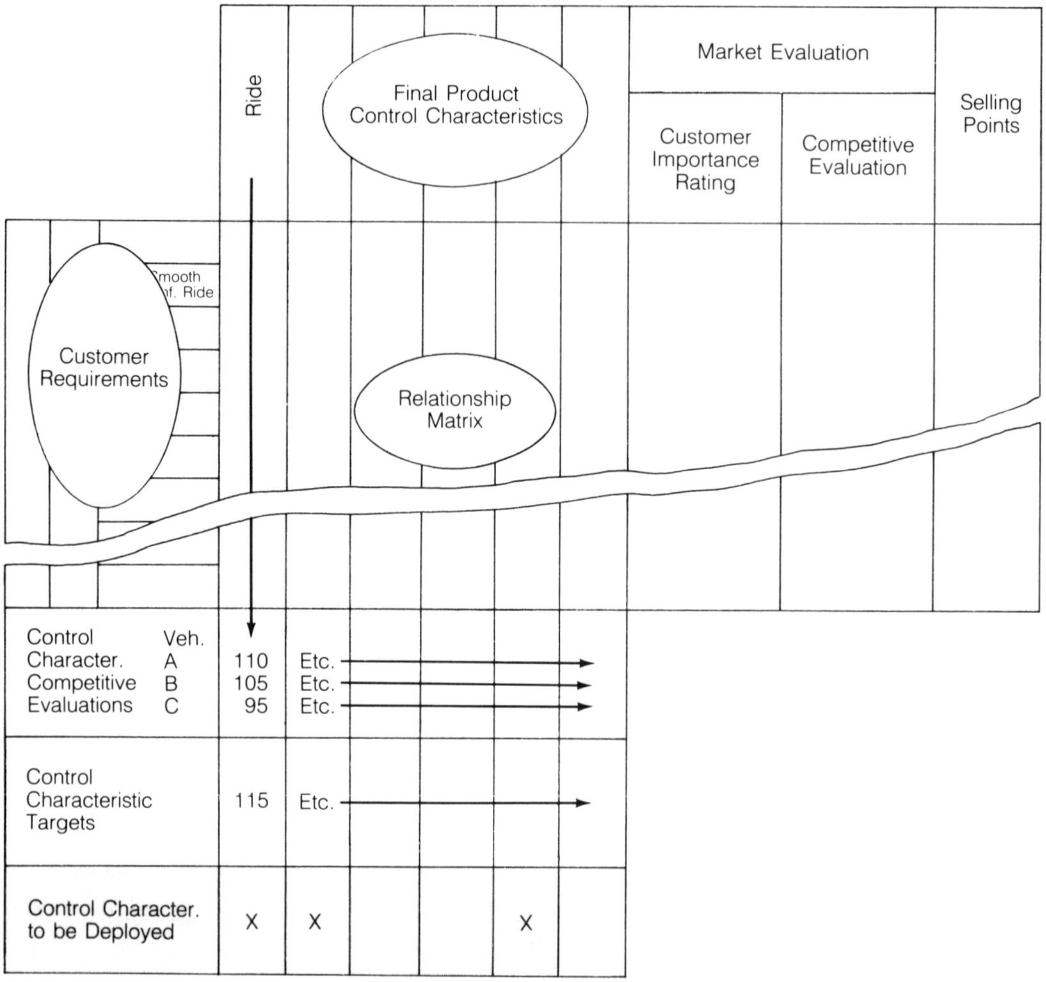

FIGURE 9–9 Step 8: Select Control Characteristics to Be Deployed, Based on Customer Importance, Selling Points, and Competitive Evaluations (Courtesy of American Supplier Institute).

of the final product control characteristic, and to customer expectations. These critical component part characteristics are then "deployed" through the *process plan and control chart.*

The process plan and control chart is the third major document in the QFD system; it represents the transition from development to execution of the production phase. The process plan chart (figure 9–12) develops a relationship matrix between the process used to produce the components and the related critical component part characteristics. As with previous matrices, symbols are utilized to highlight strong/weak relationships.

There is a powerful distinction made in this phase of deployment. If a critical product component parameter, such as spring diameter, surface, or rate is created or affected directly in a given step of a process, that parameter is identified as a control point. These points establish the data and strategy for the product quality control plan and are essential for achieving product characteristics that meet the high priority customer requirements. If critical process parameters, such as time, temperature, and pressure must be monitored to assure the component parameters are achieved, these parameters are designated as checkpoints and become the basis for operating instructions and the process control strategy.

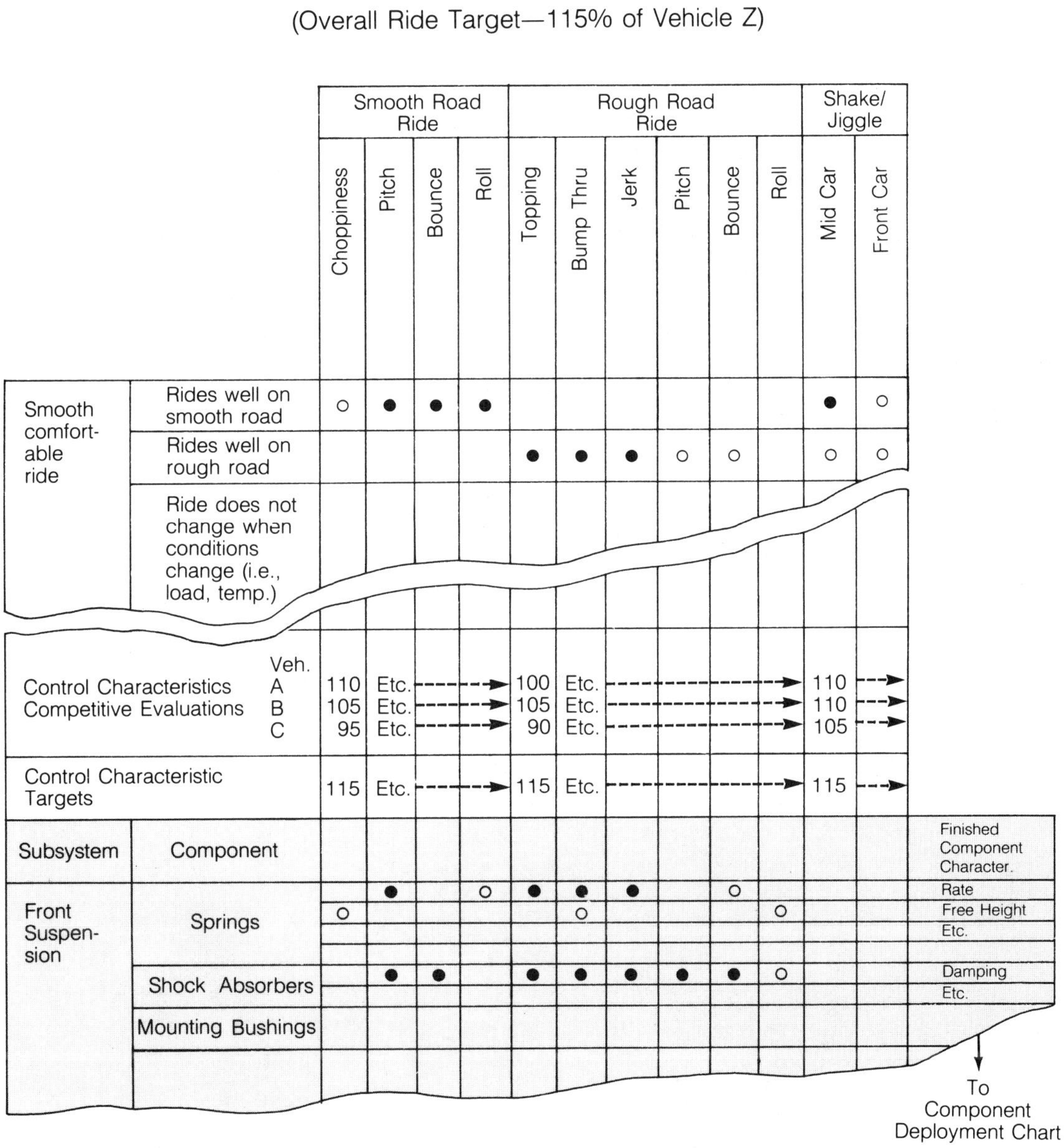

FIGURE 9–10 Step 9: Finished Component Characteristic Deployment Matrix from Planning Matrix (Courtesy of American Supplier Institute).

The quality control plan chart (figure 9–13) includes a schematic of the process flow and the control method, sample size/frequency, and checking method for each of the listed control points. (This information also is used for developing the production operating instructions, as described below.)

We now have provided the means for controlling key parts and assemblies (control points) and ongoing monitoring of important process parameters (check points). Designation of these points is essential if we are to assure that the finished component characteristics most directly related to customer satisfaction will be achieved.

The *operating instructions sheet* is the fourth and final key QFD document. It basically defines the operator requirements as determined previously by the actual process requirements; the process plan charts checkpoints, and the quality control plan charts control points.

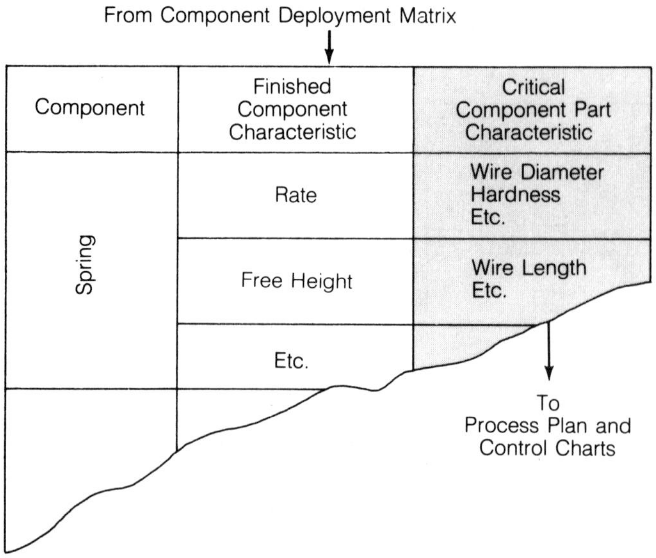

FIGURE 9–11 Component Deployment Chart (Courtesy of American Supplier Institute).

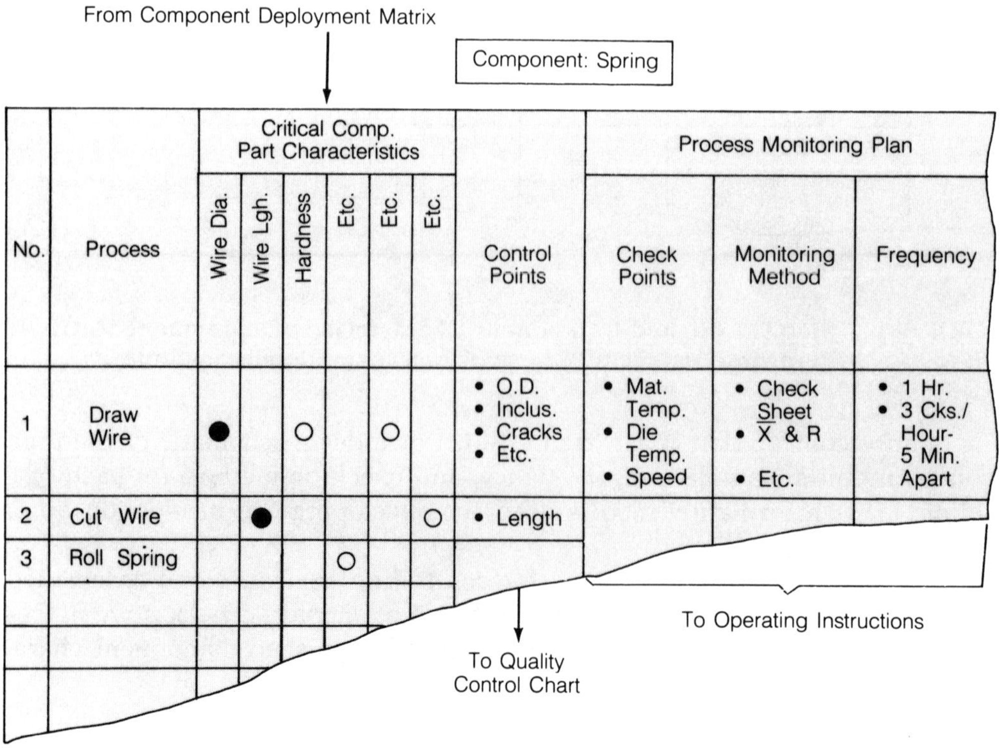

FIGURE 9–12 Process Plan Chart (Courtesy of American Supplier Institute).

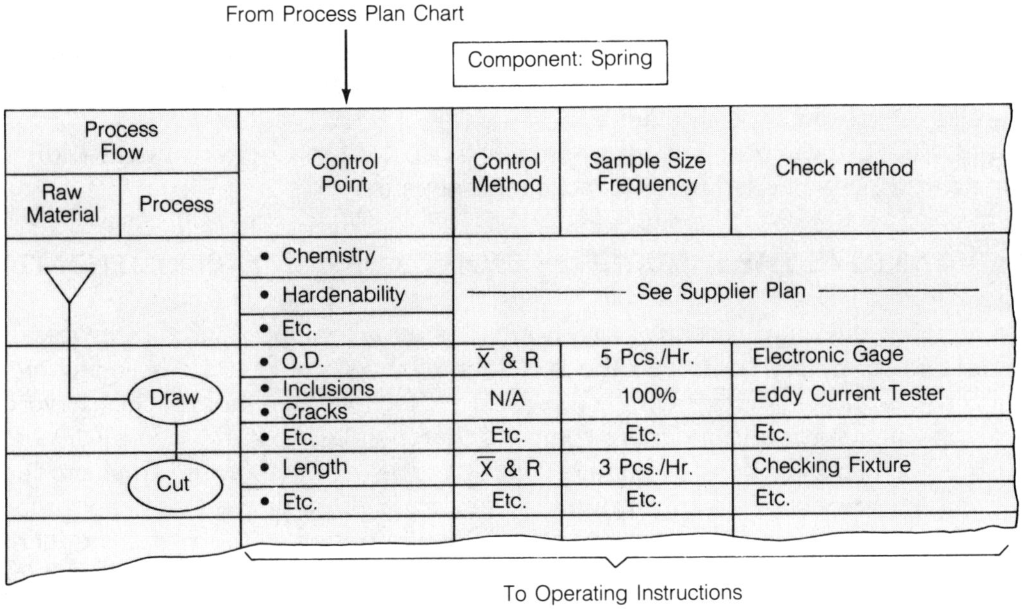

FIGURE 9–13 Quality Control Plan Chart (Courtesy of American Supplier Institute).

Many variations in the operating instructions can be anticipated based on individual process situations. It is important that this document, which relates to the checkpoints and control points, clearly conveys to the operator:

1. What parts are involved
2. How many he or she should check, and what tool to use
3. How to make the check

These three points must be tied directly back to the key final product control characteristics and customer requirements so that the operators understand the relationship between their functions and the achievement of customer satisfaction.

Advantages of QFD Systems

Implementing QFD is a challenge for any company. It requires more work in the early planning stages. QFD also makes it more difficult to change direction once a project is underway because all of the interrelated elements of the system must be revised. The system also can be complicated to administer. (How do you communicate the concept to all affected activities? Who maintains the documents throughout the entire development cycle? How do you incorporate the product/process changes in the QFD documents as changes occur during development and production?)

Despite these potential problems, QFD offers many advantages. It is a customer driven system for evaluating the relationships between:

1. Customer requirements and characteristics to be used to develop and control the product.
2. Customer (market) and company evaluations of competitive products.

QFD also provides means of translating customer requirements into critical product control characteristics, component characteristics, process control characteristics, and operating instructions. There is a continuous flow to assure that all programmed elements are related directly to customer requirements.

Finally, QFD promotes a team approach by fostering a better understanding of requirements and what is needed to meet those requirements.

UNDERSTANDING THE TAGUCHI METHOD OF DESIGN EXPERIMENTATION*

In general, the quality control function relies on a combination of in-process inspections and statistical quality control methods that employ control charts. These techniques are used to tightly control manufacturing processes. However, these approaches have become strained as products have become more complex, thereby placing greater emphasis on reducing the influence that variations in manufacturing processes have on the final product. As a result, companies are now concentrating on achieving high quality on complex products by starting at the front end of the product cycle, with the design of both the product and its manufacturing process. This new methodology is referred to as *design quality* or *off-line quality control.* In broad terms, off-line quality control includes all quality engineering activities carried out before a product goes into full-scale production. It is not enough to have a product that works well when manufactured according to the exact design specifications. Instead, the product also must be easy to manufacture and insensitive to variability on the factory floor.

Doctor Genichi Taguchi: A Profile

Dr. Genichi Taguchi, Executive Director of the American Supplier Institute, was awarded the Williard F. Rockwell, Jr. medal in 1986. This medal is the International Technology Institute's highest award. It is awarded to a maximum of three individuals each year—no more than one from any continent—for excellence in technology.

Recipients of the award must be persons of extremely high stature in the international technology community and electable to the world level of the Hall of Fame for Engineering, Science, and Technology.

Dr. Taguchi's major contribution has involved combining engineering and statistical methods to achieve rapid improvements in costs and quality by optimizing product design and manufacturing processes. Many regard this methodology to represent one of the major advances in the history of manufacturing industries.

Taguchi Method: A 3-Step Design Process

The Taguchi Method is a different approach to the more conventional methodologies because it is highly efficient in its ability to extract relatively large amounts of information from small-scale experiments. This translates to less prototype material cost, less time for running the experiment, less testing time, and less time for analyzing the results. The primary tenant of the Taguchi Method is to improve quality without incurring capital and material cost increases.

The Taguchi Method is an engineering tool used prior to putting a product or process into production. By comparison, American quality control tends to use tools to control quality during production. Although quality control during production is essential, the Taguchi

*Permission granted by American Supplier Institute, Inc.

Method seeks to minimize the necessity for on-line quality control activity by reducing variability in products and processes, thus increasing their ability to overcome the many uncontrollable changing conditions experienced during production or use.

The Taguchi Method refers to quality assurances' pre-production responsibility to develop products and processes with the smallest possible variance, and quality control's responsibility to keep processes on the target or nominal value of the specification—off-line QC and on-line QC respectively. To express these responsibilities in statistical process control terminology, quality control maintains the X-bar chart and keeps it "mean-centered," while quality assurance works to reduce the R-bar by minimizing variation.

The Taguchi Method therefore makes a fundamental departure from conventional U.S. quality efforts in its strategy for quality improvement. U.S. manufacturers attempt to reduce variation, and therefore improve quality by removing the *cause* of variation. The Taguchi Method reduces variation by removing the *effect* of the cause.

The Taguchi Method advocates a three-stage design process:

1. System design
2. Parameter design
3. Tolerance design

System design is concerned with the development of any new product or process. The engineer designs a system with a specific function. In the conventional approach, system design constitutes the total design process, whereas in the Taguchi Method, system design is only the first of three steps.

The second step, parameter design, attempts to "optimize" the performance of the system through experimentation. Different parameters in the systems are modified for experimentation using the least expensive materials. Unlike conventional U.S. statistical methods that are concerned almost exclusively with average performance, optimization is concerned with both the *level* of performance (the average or mean) and the *robustness* of the product to extraneous variables (the variability). From the quality standpoint, parameter design is the most important aspect of the design stage. When parameter design is completed, the system is at its highest level of performance at the lowest cost. The system then can be compared with its requirements, for example, an engineering specification, to see if further development work is needed.

Parameter design is enhanced by another feature of the Taguchi Method. Conventional experimentation attempts to control extraneous factors that might affect the experimental results. The Taguchi Method introduced the extraneous factors in a controlled manner. Dr. Taguchi calls these factors "noise" and classifies them into three types. The first, inner noise, includes factors within the product such as deterioration; the second, outer noise, includes uncontrollable factors external to the product or process such as temperature or power supply; the third, between product noise or manufacturing variation.

The final step is tolerance design. Only if the product is not acceptable at its optimum level, is tolerance design considered. In tolerance design, unlike parameter design, higher grade materials or more expensive processes with tighter tolerances will be considered. The Taguchi Method has specific methods for finding optimum tolerances. In typical U.S. development work, if the system is not adequate following the initial system design, upgrading of materials and tightening of tolerances is considered next without optimizing the system using parameter design. How often has upgrading in this manner been unnecessary?

Products or processes can even be optimized during system design. Since the Taguchi Method can be used in simulation experiments with the performance of a system, it can be described fairly well with mathematical formulas. Simulation with the Taguchi Method offers a low-cost way to discover which factors are signal factors, which are control factors, and what the optimum level is for each factor in the design.

An important concept in the Taguchi Method is the signal-to-noise (S/N) ratio. The S/N ratio is one of Dr. Taguchi's techniques that allows analysis that takes into account the variability caused by a factor or independent variable. In conventional statistics, factors are usually analyzed to determine their effect on the mean (average) performance of some quality characteristic or functional parameter. The Taguchi Method demonstrates, however, that some factors have an effect on changing the mean, while others have an effect on changing the *variability*. The role of factors affecting variability is simply missed by conventional experimental methods. For example, in a TV power circuit, the output voltage is a critical quality characteristic. It must be at a specified level (115 VDC) with as little variation around the target value as possible. In an experiment, several components were changed to observe their effect on the level and variability of the circuit output. Two resistors, A and B, were found to have a significant effect on the circuit's performance. Resistor A had a pronounced effect on the level of output of the circuit without affecting the amount of variation. Resistor B, on the other hand, had a pronounced effect on the amount of variability, while not having as strong an effect on the level of the circuit output as Resistor A. Changing the resistance value of A does not effect the S/N ratio—A is called an adjustment factor. On the other hand, some level of resistance B has a higher S/N ratio than other level. The Taguchi Method strategy with this product would be to set Resistor B (the adjustment factor) at a value that minimizes the overall variation in the circuit output, then adjust Resistor A (the signal factor) to obtain the target voltage of 115 VDC, without adversely affecting the variability. This is an example of parameter design. The most unique application of S/N ratio is its power to evaluate dynamic characteristics.

The S/N ratio also can be used to evaluate test equipment, gauges, and to develop meaningful specifications and tests for new products. The S/N ratio is a powerful feature of the Taguchi Method.

Another strength of the Taguchi Method is its ability to deal with a variety of data types. Most companies generally break data down into two general categories: variable data and attribute data. The Taguchi Method can deal efficiently with attribute data and what is called category data. Category data involves ranking of experimental results in three or more categories such as poor, acceptable, and good. Further, with variable data, Dr. Taguchi makes a meaningful distinction among the smaller-the-better, bigger-the-better, and nominal-the-best dynamic characteristics. Each of these types of data have special characteristics and are treated differently in the mathematical analysis.

Taguchi Methodology: How the Technique Works

Taguchi Methodology includes very unique and easy-to-use orthogonal arrays and linear graphs to analyze the interaction of variables and reduce their influence.

Orthogonal arrays, which orthogonal tables originate from, are useful for efficient data collection. In experimental designs, orthogonal means "balanced," or "separable," or "not mixed."

The capability and flexibility of assigning numerous numbers of variables of varying characteristics is a great feature of orthogonal tables, and the methods to analyze experimental results are much simpler. However, the most important reason to use orthogonal tables is the "repeatability or reliability of conclusions," which are drawn from such experiments.

To minimize the number of experiments, a conventional type of experiment, which may be called the "one-factor-by-one" method may be used. To explain this method, let's use a chemical reaction as an example. If we want to investigate the superiority and inferiority of two kinds of catalysts, there are some other variables that must be accounted for, such as reaction temperature, concentration of raw material, reaction pressure, and so on. In the case of one-factor-by-one method, these variables must be kept constant, such as keeping the temperature at 120°F, the concentration at 60 percent, and the pressure at 150 psi. However, the conclusion obtained from this experiment is true only if those other conditions are kept exactly identical, which may not be possible. If the experiment had incorporated the use of orthogonal tables, the difference of the two kinds of catalysts is determined as an average effect while the conditions of other variables vary. In such experiments, a variable having a consistent effect no matter how the conditions of other variables change, will become significant when the analysis of variance is conducted. That means a large effect obtained from an orthogonal array does not vary if there are some variations in the levels of other variables. On the other hand, if the difference between the two kinds of catalysts varies greatly once the levels of other variables change, the effect of catalyst tends to be small. The reliability, or the high reproducibility of experimental effects, is by far the greatest benefit to be derived from using orthogonal arrays. It is said that since orthogonal arrays were applied in experiments, the results obtained from a small-scale laboratory experiment have become adapted satisfactorily to the actual manufacturing field. That is because a variable with consistent effect under various conditions of the other variables has a good possibility of reproducing its effect if a condition of the manufacturing scale changes.

MAINTAINING QUALITY DURING PRODUCTION THROUGH STATISTICAL PROCESS CONTROL*

Even with product and process designs that have been planned effectively (using for example, quality function deployment and design experimentation), ongoing control during manufacturing is necessary. It is now becoming more accepted by manufacturing personnel that quality control must be an integral part of the production process. One of the methods used to reduce defects and increase productivity is statistical process control (SPC). Whereas traditional software applications such as production and scrap counts, machine utilization and monitoring, preventive maintenance, material usage and MRP, employee time and activity reporting, electronic releases, and capacity planning were treated as totally separate "productivity" tools, they now are being expanded to become part of a quality control program, and in turn, part of the framework of a fully integrated manufacturing system.

There is a natural overlap in the physical design of a manufacturing system and SPC system. The overlapping area lies in the method in which input or data is received into a computer system, or the way that data collection is accomplished. Most advanced manufacturing systems provide data collection on the plant floor for immediate error correction, validation, and closed loop feedback. However, many SPC systems are set up off the plant floor because SPC hardware suppliers still used office grade equipment, which is not conducive to the harsher manufacturing environment.

Today, with the industrial grade data collection devices available, a manufacturer can create a plant floor data collection system and benefit from both improved product quality, as well as coordinated and integrated manufacturing methods. A closed loop production floor data collection system can analyze and index variable and attribute data on a real-time basis, electronically compute required control charts, and provide immediate status feedback to a plant terminal's display. In cases where a trend or shift is detected, pre-programmed suggestions on corrective action can be routed to the plant terminal nearest the process. This allows for corrective action that can be put to immediate use in the reduction of defects. In

*Permission granted by Ford Motor Company.

addition, these corrective actions also are indexed and entered into a data base where they can be recalled in a report format to reveal corrections made to a process over multiple shifts or runs.

The implementation of this type of plantwide network can actually reduce or eliminate the net costs of gathering and analyzing quality information when you consider the other data the plant floor operator terminals can provide. Since all quality and production data is built into an intelligent data base with pre-programmed prompts and suggestions, the system performs equally well on second and third shifts in addition to the first shift, when most office personnel are available.

The discussion that follows gives more conceptual background on statistical process control as part of the manufacturing management system.*

Why Prevention Is Always Better Than Detection

Typically, companies have depended on production to make the product and on quality control to inspect the final product and screen out items not meeting specifications. This usually results in work being checked and rechecked in an effort to catch errors in production. This may be referred to as a strategy of detection. However, this approach is wasteful because it causes time and materials to be invested in products that are not always usable. In essence, after-the-fact inspection is expensive and unreliable since the wasteful production has already occurred.

Instead, a much more effective approach is to avoid waste by not producing unusable output in the first place—a strategy of prevention. To accomplish this objective it is necessary to understand and implement a soundly conceived program embodying the elements of a statistical process control system.

Four Basic Elements of a Process Control System

A process control system can be described as a feedback system. There are four basic elements in this system (figure 9–14). They are:

1. The Process. The process is the entire combination of people, equipment, input materials, methods, and environment that work together to produce output. The total performance of the process—the quality of its output and productive efficiency—depends on the way the process has been designed and built and on the way it is operated. The rest of the process control system is useful only if it contributes to improved performance of the process.

2. Information about Performance. Much information about the actual performance of the process can be learned by studying the process output. In a broad sense, process output includes not only the products that are produced, but also any intermediate "outputs" that describe the operating state of the process, such as temperatures, cycle times, and so on. If this information is gathered and interpreted correctly, it can show whether action is necessary to correct the process or the just-produced output. If timely and appropriate actions are not taken however, any information-gathering effort is wasted.

3. Action on the Process. Action on the process is *future-oriented* because it is taken when necessary to *prevent* the production of out-of-specification products. This action might consist of changes in the operations, operator training, changes to the incoming materials, and so on, or the more basic elements of the process itself such as the equipment, which may

*From *Continuing Process Control and Process Capability Improvement,* Ford Motor Company Publication, 1985, pp. 2–8, copyright © Ford Motor Company, Adapted with Permission.

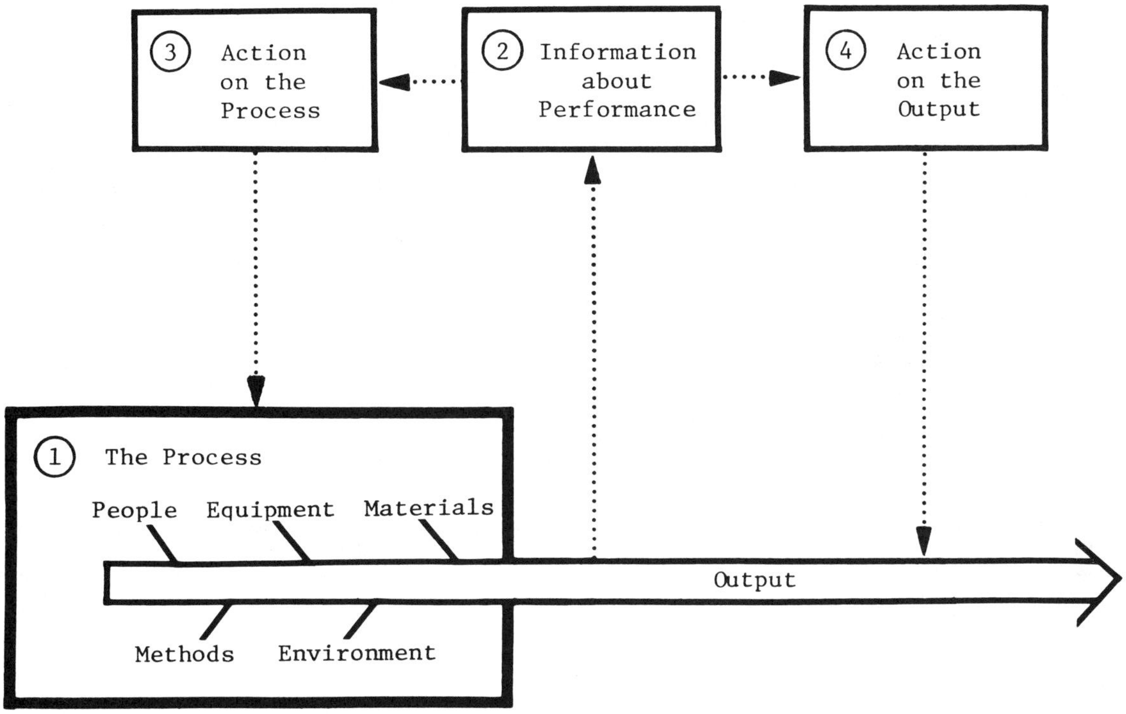

FIGURE 9-14 A Process Control System (Courtesy of Ford Motor Company from Ford Motor Company Publication, Continuing Process Control and Process Capability Improvement, 1985, pp. 2–8; copyright Ford Motor Company. Adapted with permission).

need rehabilitation, or the design of the process as a whole, which may be vulnerable to changes in shop temperature or humidity. The effect of actions should be monitored, and further analysis and action should be taken if necessary.

4. Action on the Output. Action on the output is *past-oriented* because it involves *detecting* out-of-specification output already produced. Unfortunately, if current output does not meet customer requirements consistently, it may be necessary to sort all products and to scrap or rework any nonconforming items. This must continue until the necessary corrective action on the process has been taken and verified, or until the product specifications have been changed.

Inspection followed by action only on the output is a poor substitute for effective first-time process performance. Therefore, the discussions that follow focus on gathering process information and analyzing it so that action can be taken to correct the process itself.

Common and Special Causes of Variation

To use process control measurement data effectively, it is important to understand the concept of variation.

No two products or characteristics are exactly alike because any process contains many sources of variability. The differences among products may be large, or they may be almost unmeasurably small, but they are always present. The diameter of a machined shaft, for example, would be susceptible to potential variation from the machine (clearances, bearing wear); tool (strength, rate of wear); material (diameter, hardness); operator (part feed, accuracy of centering); maintenance (lubrication, replacement of worn parts); and environment (temperature, constancy of power supply).

Some sources of variation in the process cause very short-run piece-to-piece differences, such as backlash and clearances within a machine and its fixturing. Other sources of variation tend to cause changes in the output only over a longer period of time, either gradually as with tool or machine wear, stepwise as with procedural changes, or irregularly as with environmental changes such as power surges. Therefore, the time period and conditions over which measurements are made will affect the amount of the total variation that will be present.

From the standpoint of minimum requirements, the issue of variation is often simplified: Parts within specification tolerances are acceptable, parts beyond specification tolerances are not acceptable; reports on time are acceptable, late reports are not acceptable. However, to manage any process and reduce variation, the variation must be traced back to its sources. The first step is to make the distinction between common and special causes of variation (figure 9–15).

Common causes refer to the many sources of variation within a process that is in statistical control. They behave like a constant system of chance causes. Although individual measured values are all different, as a group they tend to form a pattern that can be described as a distribution. This distribution can be characterized by:

- Location (typical value)
- Spread (amount by which the smaller values differ from the larger ones)
- Shape (the pattern of variation, whether it is symmetrical, peaked, etc.)

Special causes (often called assignable causes), refer to any factors causing variation that cannot be explained adequately by any single distribution of the process output, as would be the case if the process were in statistical control. Unless all the special causes of variation are identified and corrected, they will continue to affect the process output in unpredictable ways.

The Connection between Local Actions and Actions on the System

Dr. W.E. Deming emphasizes an important connection between the two types of variation discussed previously and the types of action necessary to reduce them (figure 9–16).

Special causes of variation can be detected by simple statistical techniques. These causes of variation are not common to all the operations involved. The discovery of a special cause of variation and its removal are usually the responsibility of someone who is connected directly with the operation, although management sometimes is in a better position to correct the variation. The resolution of a special cause of variation usually requires *local action.*

The extent of *common causes* of variation can be indicated by simple statistical techniques, but the causes themselves need more detailed analysis to be isolated. These common causes of variation are usually the responsibility of management to correct, although other people connected directly with the operation sometimes are in a better position to identify these causes and pass them on to management for correction. Overall, though, the resolution of common causes of variation usually requires *actions on the system.*

Dr. J.M. Juran first noted that only a relatively small proportion of all process troubles—industrial experience suggests about 15 percent—is correctable locally by personnel connected directly with the operation. The majority—the other 85 percent—is correctable only by management action on the system. Confusion about the type of action to take is very

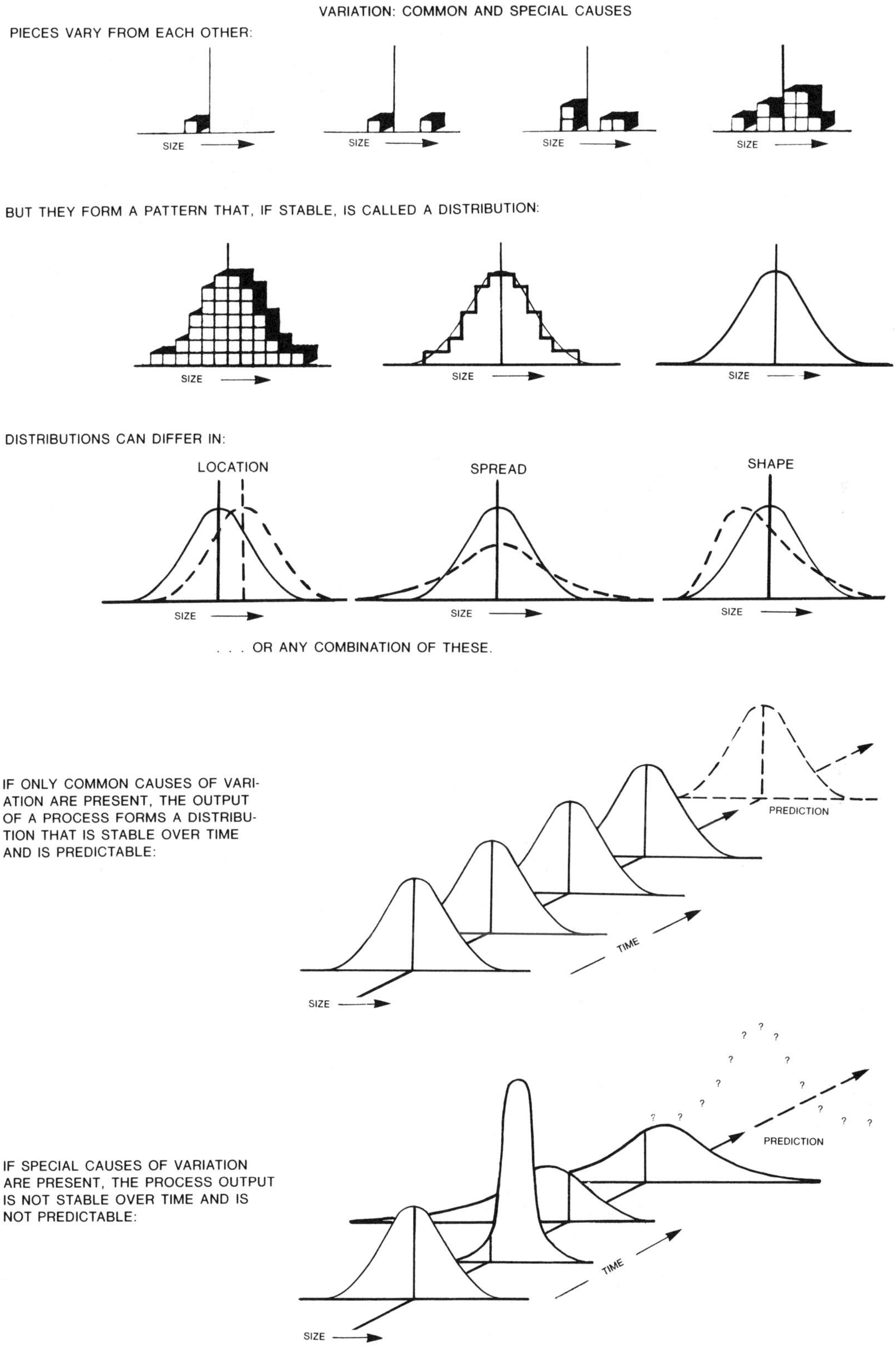

FIGURE 9-15 **Variation: Common and Special Causes** (Courtesy of Ford Motor Company from Ford Motor Company Publication, Continuing Process Control and Process Capability Improvement, 1985, pp. 2–8; copyright Ford Motor Company. Adopted with permission).

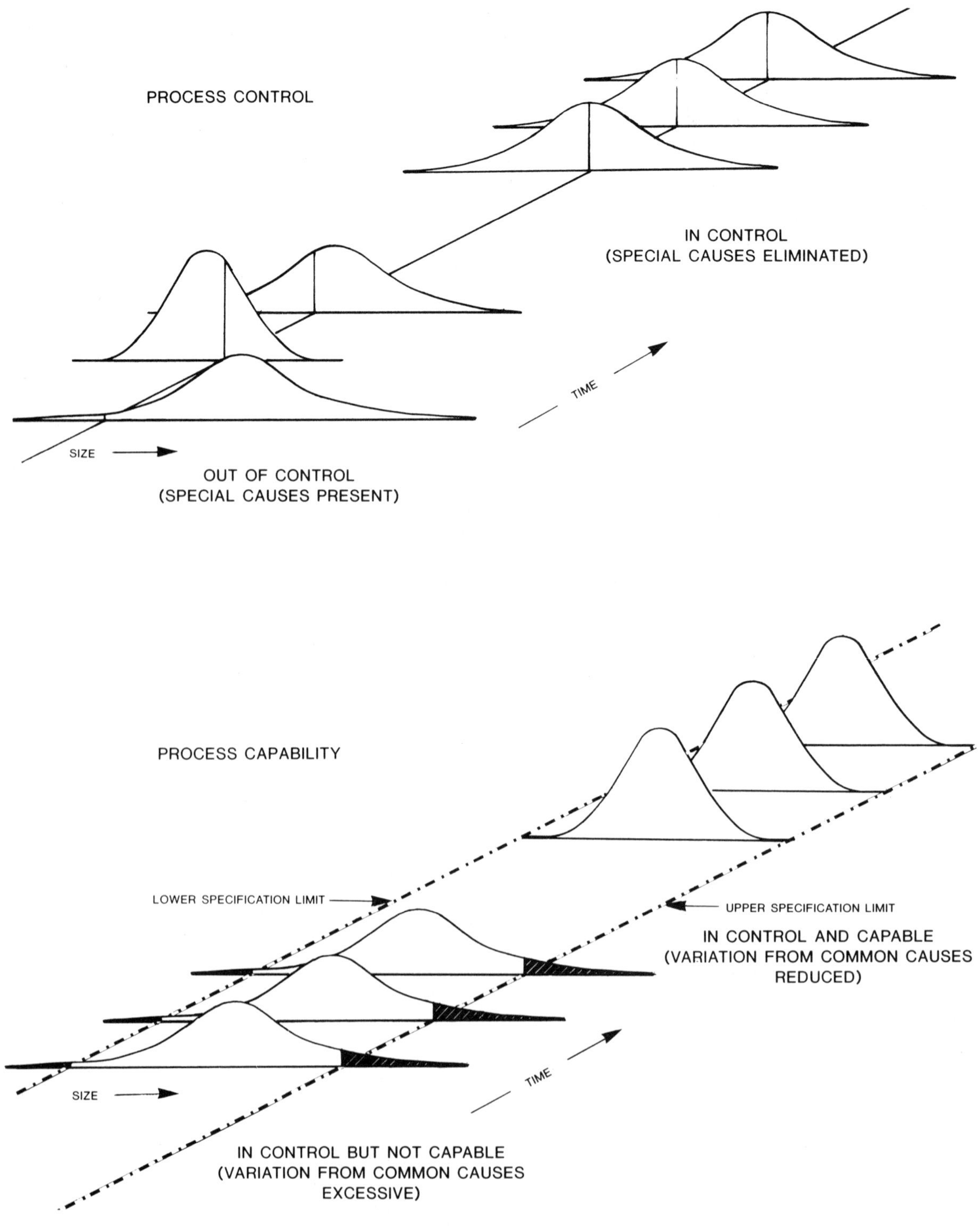

FIGURE 9–16 **Process Control** (Courtesy of Ford Motor Company from Ford Motor Company Publication, Continuing Process Control and Process Capability Improvement, 1985, pp. 2–8; copyright Ford Motor Company. Adopted with permission).

costly to the organization in terms of wasted effort, delayed resolution of trouble, and aggravated problems.

The Relationship between Process Control and Process Capability

The goal of a process control system is to make economically sound decisions about actions affecting the process. This means balancing the risks of taking action when action is not necessary (overcontrol) versus failing to take action when action is necessary (undercontrol). These risks must be handled, however, in the context of the two sources of variation mentioned previously: special causes and common causes.

A process is said to be operating in statistical control when the only source of variation is common causes. A state of statistical control is not a natural state for a manufacturing process. It is instead arrived at by elimination—one by one—by determined effort of special causes of excessive variation. The initial function of a process control system is to provide a statistical signal when special causes of variation are present, and to avoid giving false signals when they are not present. This will enable appropriate action that can eliminate those special causes and prevent their reappearance.

Process capability is determined by the total variation that comes from common causes—the minimum variation that can be achieved after all special causes have been eliminated. Thus capability represents the performance of the process itself, as demonstrated when the process is being operated in a state of statistical control. Capability is often thought of in terms of the proportion of output that will be within product specification tolerances. But a more current view espoused by Drs. Taguchi and Deming, among others, looks at closeness to a most desired target. Since a process in statistical control can be described by a predictable distribution, the proportion of out-of-specification parts or the variation from target can be estimated from this distribution. As long as the process remains in statistical control, it will continue to yield predictable performance. Management actions to reduce the variation from common causes (for example, through use of design of experimentations) are required to improve the process' ability to meet specifications consistently.

In short the process must first be brought into statistical control by detecting and eliminating special causes of variation. Then its performance is predictable, and its capability to meet customer expectations can be assessed. This is the basis for continuing improvement.

CONTROL CHARTS: TOOLS FOR PROCESS CONTROL

Control charts developed by W. A. Shewhard effectively direct attention toward special causes of variation when they appear and reflect the extent of common cause variation that must be reduced by management action (figure 9–17).

Several types of control charts have been developed to analyze both variables and attributes. However, all control charts have the same two basic uses which are:

- *As a judgment*—to give evidence whether a process has been operating in a state of statistical control, and to signal the presence of special causes of variation so that corrective action can be taken.
- *As an operation*—to maintain the state of statistical control by extending the control limits as a basis for real-time decisions.

Process improvement using control charts is an iterative procedure, repeating the fundamental phases of collection, control, and capability. First, data is gathered according to a

Upper Control Limit

Process Average

Lower Control Limit

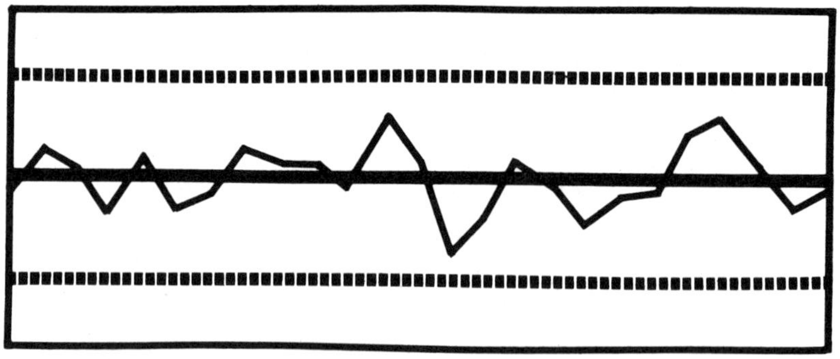

FIGURE 9–17 **Control Charts** (Courtesy of Ford Motor Company from Ford Motor Company Publication, Continuing Process Control and Process Capability Improvement, 1985, pp. 2–8; copyright Ford Motor Company. Adopted with permission).

careful plan. These data are then used to calculate control limits, which are the basis of interpreting the data for statistical control. When the process is in statistical control, it can be interpreted for process capability. To monitor improvements in control and capability, the cycle begins again as more data is gathered, interpreted, and used as the basis for action.

1. Collection. The process is run, and data for the characteristic being studied is gathered and converted to a form that can be plotted on a graph. This data might be the measured value of a dimension of a machined piece, the number of flaws in a bolt of vinyl, railcar transit times, number of bookkeeping errors, and so on.

2. Control. Trial control limits are calculated based on data from the output of the process. They reflect the amount of variation that could be expected if only variation from common causes was present. They are drawn on the chart as a guide to analysis. Control limits are not specification limits or objectives, but are reflections of the natural variability of the process.

The data is then compared with the control limits to see whether the variation is stable and appears to come only from common causes. If special causes of variation are evident, operation of the process is studied to determine what is affecting the process. After actions (usually local) have been taken, further data is collected, control limits are recalculated if necessary, and any additional special causes are studied and corrected.

3. Capability. After all special causes have been corrected and the process is running in statistical control, the process capability can be assessed. If the variation from common causes is excessive, the process cannot produce output that meets customer needs consistently. The process itself must be investigated (using, for example, design of experiments), and management action must be taken to improve the system.

For continuing process improvement, repeat these three phases. Gather more data as appropriate and work to reduce process variation by operating the process in statistical control and continually improving its capability.

Tips for Successful Application: Why Use Control Charts?

Some of the important benefits that can come from using control charts are summarized below.

- Control charts are simple and effective tools to achieve statistical control. They lend themselves to being maintained at the job station by the operator. They give the people closest to the operation reliable information on when action should be taken and when action should *not* be taken.
- When a process is in statistical control, its performance to specification will be predictable. Thus both producer and customer can rely on consistent quality levels, and both can rely on stable costs of achieving that quality level.
- After a process is in statistical control, its performance can be improved further to reduce variation. The expected effects of proposed improvements in the system can be anticipated, and the actual effects of even relatively subtle changes can be identified through the control chart data. Such process improvements include:
  (a) Increased percentage of output that meets customer expectations (improve quality)
  (b) Decreased output requiring scrap or rework (improve cost per good unit produced)
  (c) Increased total yield of acceptable output through the process (improve effective capacity)
- Control charts provide a common language for communications about the performance of a process between the two or three shifts that operate a process; line production (operator, supervisor) and support activities (maintenance, material control, process engineering, quality control); different stations in the process; supplier and user; and the manufacturing/assembly plant and the design engineering activity.
- Control charts, by distinguishing special from common causes of variation, give a good indication of whether any problems are likely to be correctable locally or will require management action. This minimizes the confusion, frustration, and excessive cost of misdirected problem-solving efforts.

*PART B. USING COORDINATE MEASURING SYSTEMS TO IMPROVE QUALITY CONTROL**

WHY IMPLEMENT A HIGH PERFORMANCE COORDINATE MEASURING SYSTEM?

The purpose of a coordinate measuring system (CMS) is to evaluate and monitor the performance of machine tools.

A CMS is used to categorize machine tools according to their controllability and therefore monitor and control their performance in process. To determine CMS from other systems, a CMS may be called capable of such services as "high performance" CMS. The words *high performance* primarily connote the ability to provide high accuracy and highly reliable data. This ability depends on a combination of machine accuracy, software power, and management systems integration.

COMPONENTS OF A COORDINATE MEASURING SYSTEM

The various components that make up a modern coordinate measuring system are explained in the next few sections.

*Reprinted courtesy of the Society of Manufacturing Engineers. Copyright 1984, from Autofact 6 Conference Proceedings.

Coordinate Measuring Machine

The coordinate measuring machine (CMM) is the focal point of the system, but without refined electronic probes, computers, and software, it would be incapable of keeping up with the speed of current manufacturing processes and doing the sophisticated analysis of 3D geometry required to evaluate machine tool performance. In basic terms, the primary function of the CMM is to act as a data-collection device. Whether positioned manually, by joy sticks, or under CNC, a reliable report of its position at the instant the probe contacts the surface of the part is its primary task. The reliability of this information depends on the resolution and accuracy of the CMM's position sensors and on the straightness and squareness of its ways. Although of secondary importance, the speed with which the CMM can collect data can be significant in overall performance.

Probe Head

The probing mechanism is a significant CMS component since its performance will affect the reliability of the position information provided by the CMM. Other than optical probing systems, there are two fundamentally different types of electromechanical probe heads: touch sensitive and position sensitive. Both alternatives are capable of fulfilling most measurement requirements, but each has special strengths. Touch sensitive is stronger when dealing with classical geometry and position sensitive is stronger when dealing with compound curved surface geometry.

The following performance characteristics are important in both cases:

- Lobing
- Probing force
- Repeatability
- Maximum probe extension length
- Maximum probe cluster weight
- Automatic probe changing/orienting

Position sensitive probe heads are capable of scanning over features and collecting data at rates of up to twenty points per second. As such, they are far more efficient in nonclassical measurements than even vector-drive touch sensitive systems. Since certain probe head designs excel significantly over others, it is important to test them in actual measurement exercises.

Styli

Probe tips or styli play a significant role in a CMS. Now almost exclusively spherical, the following characteristics are important:

- Sphericity
- Hardness
- Lightness
- Ball diameter should be at least twice the shaft diameter

Fixtures

Fixtures and tools are important in the CMS from the standpoint of efficiency. Although they generally are simple and rarely accurate, good fixturing for parts, especially in robot-loaded systems, should be emphasized.

Digital Readout/Controller

The digital readout (DRO) of a CMM displays its position instantaneously and sends that position to the computer as raw data for geometric analysis. In CNC systems the DRO expands to become a controller serving as a command module for the motor drives.

Joy Stick Control Console

In motor driven systems, the joy stick control console represents the human/machine interface for manual machine positioning. Good joy sticks are proportional and smart.

On-Line Computer Hardware

The computer continues to assume an ever more important role in the CMS as speeds increase and prices decrease. The "computer" consists of a central processing unit (CPU) plus a host of peripherals that provide a variety of capabilities.

The major components are:

- CPU where match and data processing occur
- CRT providing visual communication with the operator
- Keyboard providing operator access
- Printer providing permanent alphanumeric and graphic reports
- Plotter providing permanent, multicolor pictorial reports
- Mass storage providing permanent and semi-permanent means of recording operating software, part programs, and measurement results
- Data I/O for receiving raw data from the DRO/Controller for sending positioning commands to the controller (CNC only), for handling general communications with peripherals, and for linking the CMS with a corporate computer network through a COM LINK

The main computer can be a stand-alone device dedicated to one CMM or a multitasking system supporting several CMMs with dedicated terminals. Each approach has relative strengths and weaknesses. As long as communications networking is available it becomes moot whether processing is local or central.

On-Line CMS Software

On-line CMS software is now recognized as the central element in a CMS. Software is the interface between the CMS and the human user, providing access to the world of 3D geometry. On-line software can reside in a local or a central multitasking computer, but is called on-line because it interacts directly with the CMM and the operator. On-line software can provide some or all of the following:

Primary Functions

- *Manual* mode operation for impromptu measurements
- *Automatic* mode operation for computer-prompted or motor-driven inspection

Secondary Functions

- On-line and/or off-line part program entry
- Part program editing

- On-line statistical data processing
- Off-line computer communication

Communications Interface

This is the link between the CMS and the rest of the corporate organization, in particular the engineering and the manufacturing groups. This link permits input to the on-line computer of:

- Archived part programs
- CAD-generated part programs
- Nominal data
- Scheduling commands

and output to the off-line computer of:

- Raw data
- Processed geometric data
- Statistical data
- Part programs for archiving
- CMS utilization data
- Material disposition data

Off-Line CMS Software

Software resident in the off-line computer can duplicate many on-line capabilities such as:

- Off-line part program entry
- Part program editing, listing, and cataloging
- Statistical data processing

and provide a number of specialized off-line capabilities such as:

- Post processing of CAD-generated part programs
- Machine tool corrective-action feedback
- Machine tool capability analysis
- CMS scheduling
- Robotic part-handling management
- Engineering data base review from the standpoint of inspectability
- Management information processing

The growth of the CMM into the CMS has enhanced its capabilities but also has created significant new management challenges and made the job of selecting a CMS more difficult. The following material describes the capabilities that characterize the High performance CMS.

TRAITS OF HIGH PERFORMANCE CMS SYSTEMS

The areas that impact CMS performance are:

- Measuring machine accuracy
- Probe system accuracy
- Overall system speed
- Software power
- Software ergonomics
- System reliability

The following sections cover the meaning of and need for each of these in detail.

Accuracy

The need for high accuracy can be explained as follows: All machining processes are subject to variations, of which some are random and some are trending. Measuring processes are subject to both types of variations as well. Regardless of tolerances, if the long-term variations of the measuring process are not significantly lower than the short-term variations of the machining process, the measuring process is incapable of delivering data of any value whatsoever for process control. To establish requirements for the capability of a measuring machine in a particular manufacturing environment you must establish a confidence limit hierarchy. If you assume moderate factors of five for the machining process and of two for the measuring process you would have:

Machined Part:	Tightest tolerance bandwidth = T
Machine Tool:	Maximum short-term variation = STV(MT) = T/5
Measuring Machine:	Maximum long-term variation = LTV(MM) = STV(MT)2 = T/10

Thus, if the diameter of a bore is 1.000″ ± .005″, T would be 0.010″ and short-term machine tool variations would be required that did not exceed 0.002″. Similarly, if you were to measure the same bore again and again long-term measuring machine variations would be required to produce diameter measurement fluctuations not exceeding 0.001″.

If random machine tool fluctuations exceed one-fifth of a part's tolerance bandwidth, the process cannot be controlled. Furthermore, if the results produced by a measuring machine suffer long-term fluctuations exceeding one-half the control "limits" of this machine tool, the measuring machine is inadequate as a performance monitor. Hence, the basis for the familiar 10-to-1 rule has been established.

However, the requirement above for measuring machine performance was based—and must be based—on the long term. What does this imply for short-term requirements? To have any hope of achieving long-term stability, surely short-term or random measuring machine result fluctuations should be less than one-half of the long-term requirement.

Thus, the short-term accuracy of measurement results must be better than one-twentieth of a feature's tolerance bandwidth. Since the highest accuracy you can expect from a digital measuring device is twice its resolution, the resolution required to achieve these results is one-fortieth of the tolerance bandwidth. In review, the moderate factors that have been chosen lead to the following hierarchy:

Machined Part:	Tightest tolerance bandwidth = T
Machine Tool:	Maximum short-term variation = STV(MT) = T/5
Measuring Machine:	Maximum long-term variation = LTV(MM) = T/10
Measuring Machine:	Maximum short-term variation = STV(MM) = T/20
Measuring Machine:	Minimum resolution = R(MM) = T/40

A few examples will demonstrate the rationale of the conclusions noted above:

Tolerance Bandwidth of Machines Part Characteristics	STV(MM)	R(MM)
0.001″(±.005″)	0.0005″	0.000250″
0.002″(±.001″)	0.000100″	0.000050″
0.001″(±.0005″)	0.000050″	0.000025″

You may recall that the confidence factors that had been chosen were "moderate." Conservative factors might lead to a doubled total or a factor of 40 for the maximum short-term variation in measuring machine accuracy as a function of tolerance bandwidths.

The table above provides a basis for selecting a CMM to do the required job. The factors influencing the accuracy of a measuring machine consist of five components:

- Measuring machine geometry
- Position sensor accuracy
- Probe system repeatability
- Probe system lobing
- Software algorithms

The only way to safely evaluate the performance of so complex a system is to demonstrate its ability in real measurements. Only calibrated gauges can be used for this purpose, namely objects about which a great deal is known with high confidence.

Key Point: Do not use a production part

Overall System Speed

System speed means many things. It means floor-to-floor measuring time, setup time, and program entry and debug time. It does not simply mean CMM slew rates or acceleration rates, although these do affect system speed. System speed is important because it impacts the manufacturing cycle and to a certain extent even the engineering cycle. System speed surely affects payback.

Software Power

On-line software is responsible for converting raw X, Y, Z point data into 3D features such as planes, spheres, and cylinders. It also is responsible for evaluating relationships between machined part features, such as distances, angles, intersections, and bisectors, not to mention creating the skewed part coordinate systems (PCS) in which these features are expressed. It also is responsible for providing the means to tolerance measured features. In short, regardless of CMM accuracy, it determines the extent to which a machined part can be measured in accordance with a drawing. The best way to evaluate the power of CMS software is to put it to the test on actual measurement challenges. Software power consists not only of the ability to measure a feature like a cone, but of the ability to measure it at any angle, with any taper and of tolerancing its multiple characteristics. These characteristics are:

- Location of the apex
- Location of the axis pierce point
- Attitude of the axis
- Diameter at a nominal point on the axis
- Location of a point on the axis at a nominal diameter
- Included angle
- Taper
- Conicity
- Concentricity
- Coaxiality
- Parallelism
- Perpendicularity
- Angularity

Powerful software also will automatically store measured features and permit them to be recalled into new coordinate systems at any time, or used to build new coordinate systems and shift or realign old ones.

Software Ergonomics

Software is the essence of CMS. Human access is enhanced by ergonomic or human engineering, and is just as essential as geometric power. Good ergonomics make it easy to manipulate the software power to measure features in exactly the way the drawing specifies. Good ergonomics are characterized by:

- Single keystroke function access
- Simple, smart, and informative CRT displays
- A la carte function architecture
- Plain English logic sequences
- Easy escape and entry paths
- Easy error recovery
- Broad-based operator warnings and assistance

These factors greatly influence the efficiency of a CMS and the ease or difficulty of its management.

System Reliability

The purpose of a CMS is to provide reliable data on which to base expensive decisions.

IMPLEMENTING A HIGH PERFORMANCE CMS

Implementing a high performance CMS is more difficult than selecting one. A host of management objectives need to be met including:

- Internal selling of CMS concepts to engineering and manufacturing personnel to gain their understanding and support
- Management personnel selection and training
- Operating personnel selection and training
- System integration planning
- Establishment of part program writing, test, and documentation procedures
- Establishment of calibration procedures
- Host–computer interfacing preparation
- Material flow planning
- Physical plant preparation

In the early days of coordinate measuring machines the flow of information and material in a manufacturing operation was extremely random and disassociated.

With the advent of the high accuracy, high performance CMS, the relationship between manufacturing and inspection has changed dramatically.

Quick feedback of reliable data now makes good manufacturing decisions a matter of course. Integrating computer-aided inspection with CAD means better cooperation between engineering and inspection and improved performance of the whole team. Communication, trust, and control of manufacturing variables becomes common, leading to decisions that are inexpensive to implement instead of consequences that are expensive to correct.

TYPES OF COORDINATE MEASURING MACHINES*

As discussed previously, the focal point of a CMS is the CMM. This equipment (and its component parts) are used to help inspection keep pace with modern production requirements. It replaces long, complex, and inefficient conventional inspection methods with simple procedures that are up to twenty times faster and more accurate. A CMM reduces or eliminates NC machine down time, cuts scrap and rework, and can be operated by workers of lesser skill after brief training.

Typical uses for a CMM include quick inspection of the first workpiece produced by NC machining; random sampling of production runs; preventative maintenance checks of dies, jigs, and fixtures; and final, qualifying or receiving inspection of machined parts and castings. A CMM can check the dimensional and geometric accuracy of everything from engine blocks and circuit boards to sheetmetal parts and TV picture tubes. Because it's not dedicated to one function, a CMM measures virtually any part at any state of production with precision and substantial time savings.

One of the leading companies engaged in the design and manufacture of dimensional measurement equipment is the Sheffield Measurement Division of Warner & Swassey, A Cross & Trecker Company. It's products are sold under the Cordax® designation. Cordax CMMs provide instant measurement results without complicated setup and operating procedures. They combine surface plate, micrometer, and vernier-type inspection methods into one easy-to-operate machine. The Cordax line consists of machines that range from manual CMMs to a totally computer-controlled inspection center.

*Permission granted by the Sheffield Measurement Division, Warner & Swassey/A Cross & Trecker Company.

Apollo Series: Offering Speed Plus Accuracy

The Apollo series Cordax coordinate measurement system offers five to eleven micron accuracy, with excellent throughput (figure 9–18). It features an exclusive wraparound ring bridge that completely surrounds the measuring area in a rigid symmetrical frame for more precise repeatability. Automatic accuracy enhancement electronically corrects roll, pitch, yaw, vertical and horizontal straightness, scale error, and squareness deviation in all three axis, and even compensates for probe extensions. This system is very fast for its size and accuracy: 10″ per second velocity, with 80″ per second^2 acceleration and deceleration. Other features include a precise balanced central drive to eliminate "walking" and settling problems, free and clear access to the measuring bed for fast part handling, and manual, joystick, or direct computer models in a variety of sizes.

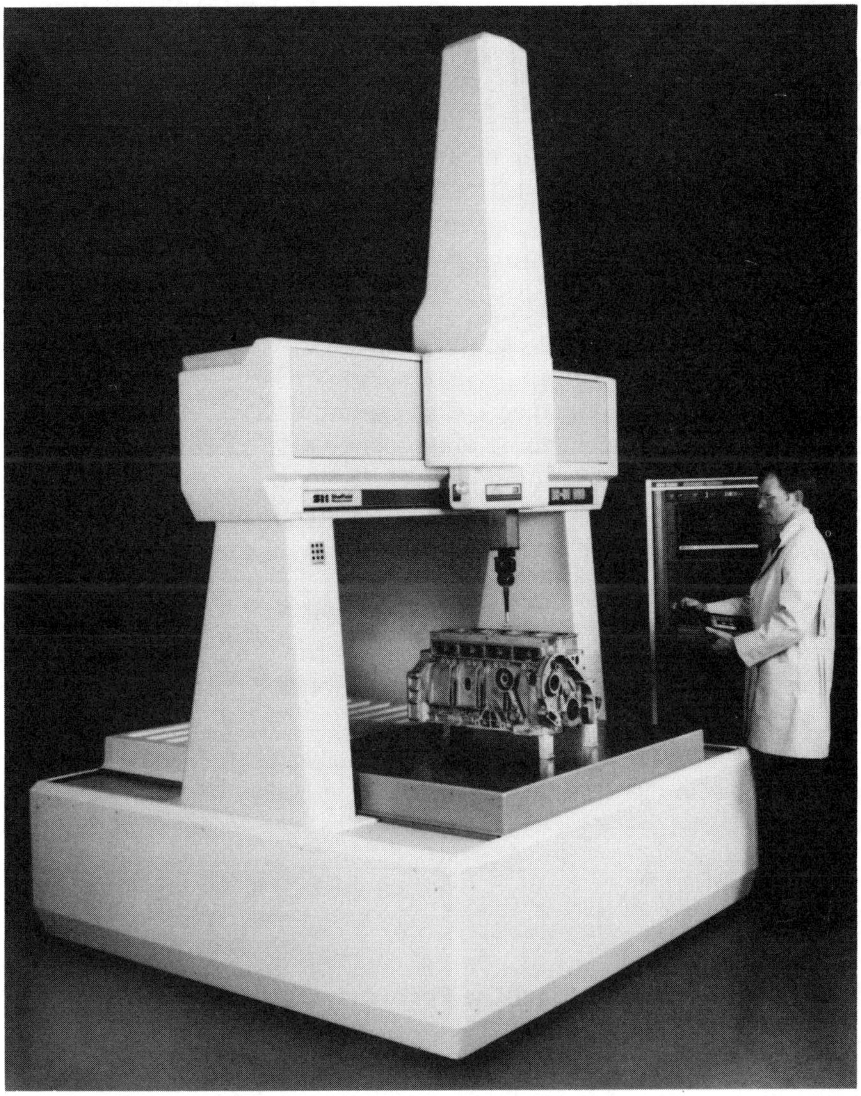

FIGURE 9–18 Apollo Series (Model RS50) (Courtesy of Sheffield Measurement Division, Warner & Swassey/A Cross & Trecker Company).

1800 Series: Achieving Excellent Repeatability

The Cordax 1800 Series CMMs consist of both manual and direct computer control (DCC) models. They give an operator easy starts, easy stops, faster setups, and more control with excellent repeatability. Due to ease of handling, there is less resistance and inertia so manual inspections are less tiring for the operator. Therefore DCC inspections are more precise.

The 1800 Series CMMs supply an immediate evaluation of machine tools and their operators, resulting in less down time and scrap. The units' open cantilever design allows workpiece inspection from three sides, with adequate access from the rear. The DCC model is controlled through an operator friendly interface (OFI) that lets an operator select and implement measurement commands from a CRT menu. It does not require programming language or down time for want of a parts program. Operator training is minimal, since training entails "talking" to the CMM.

2000 Series: A System for Large Workpieces

Cordax 2000 Series horizontal arm CMM provides coordinate measurement on large, bulky workpieces that cannot normally be inspected on a vertical arm machine. These machines are particularly well adapted for measuring large gear cases and engine block-type parts, where high precision alignment and geometry measurement are required. Their standard rotary table permits an operator to measure five sides of a workpiece with a single setup, greatly speeding inspection and minimizing part handling.

EXAMPLES OF COORDINATE MEASURING MACHINE SOFTWARE*

The purpose of a coordinate measuring machine is to provide high speed, reliable data on which to base expensive decisions. The essence of a coordinate measuring machine is software. Regardless of the configuration, bearing design, or resolution, software is the key to getting the job done. Software endows the machine with the intelligence to accomplish objectives like go/no-go true position tolerancing, measurement of parallelism, or determination of concentricity.

Computer evaluation of machined part geometry can be complex. One company that has developed an extensive range of CMM software is the GEOMET Systems Software Division of Helmel Engineering Products, Inc. The company's proprietary products include the following software packages:

- GEOMET®
- GEONET
- GEOSTAT™
- GEOVISION™
- GEOTOOL™

GEOMET®: Software to Enhance Machine Performance

GEOMET is a sophisticated computer-assisted software package for the enhancement of manual, 3D coordinate measuring machine performance. GEOMET provides:

*Material based on information from Geomet Systems Software division of Helmel Engineering Products, Inc. Used with permission.

- *Power* to analyze machine part geometry in keeping with ANSI Y14.5M concepts.
- *Efficient Operator Access* to complex 3D geometry through the dedicated GEOMET keyboard. Each key is dedicated to a specific inspection function, generally designated by an icon. An icon might be a picture of a cylinder or two intersecting lines symbolizing a spatial intersection. Smart, minimenus support the keyboard, posing only pertinent questions and providing proper guidance.
- *Universality* permitting all measuring machines to speak the same language, use interchangeable part programs, and thus form CMM networks.
- *Speed* to provide efficient throughput and responsiveness to operators and programmers.

In addition, a newly developed GEOMET product has all the features of the present product and will extend part program capability across the visual/computer controlled CMM spectrum as well.

GEONET: Software to Integrate CMMs

Coordinate measuring machines are found in receiving inspection departments, in-process inspection areas, and metrology departments, both at corporate headquarters and at outlying divisions. GEONET integrates all of these various CMMs among the many inspection groups in a corporation by providing complete:

- Interchangeability of operators
- Interchangeability of part programs
- Uniformity of inspection algorithms
- Uniformity of inspection reports

The benefits of GEOMETizing interrelated CMMs are even greater when the network includes vendors. In these instances, part drawings, inevitably open to interpretation, are replaced by universal part programs guaranteeing completely controlled inspection procedures.

> ***Key Point:*** By transferring receiving inspection to the vendor site, parts can be shipped directly to the assembly line, thus enhancing the implementation of just-in-time manufacturing.

CMS Performance Verification

A universal CMS language is not enough to ensure that one or any GEOMETizing CMM can be trusted to produce reliable results. The CMMs in a network must therefore pass a common accuracy test to be viable team members. The best approach is to build a calibration artifact that mimics typical part geometries but consists of gauge blocks, and ring and ball gauges. Next a companion GEOMET program is written to measure the artifact exhaustively. Once a CMM has been calibrated using classical methods, the artifact is measured with the companion GEOMET program to establish a performance fingerprint. The artifact and companion GEOMET program are then sent to every other CMM in the network. Passing the artifact test qualifies a CMM as acceptable. Regular fingerprints of in-house and vendor CMMs can be made to provide ongoing verification of their condition.

Network Communications

Part programs may be moved among machines using micro-floppy (3 1/2″) disks. In a more sophisticated method, part programs and measurement results can be sent over telephone lines, together with other pertinent communications. This is particularly appropriate in buyer/vendor relationships, where a rapid exchange of information can assist with process control, as well as help to accelerate the release of products for just-in-time shipping.

GEOSTAT™: A Tool for Statistical Analysis

GEOSTAT is a data-base software system for statistical analysis of dimensional and other manufacturing data. It is a member of the GEOMET software family in the area of statistical process control (SPC). This software provides direct access to measurement results generated by the GEOMET CMM software packages, as well as related data from other inspection systems. GEOSTAT becomes an integral part of the inspection system and can be made callable from the GEOMET 200 and 300 start-up menus. It operates directly on the inspection data collected during GEOMET part program runs and allows the compilation of such data into the SPC file system without elaborate user intervention. Written for the Hewlett Packard 200 and 300 computer systems, GEOSTAT adds powerful statistical analysis capabilities to the inspection operation.

Software Capabilities

Chart Generation. A complete selection of commonly used SPC charts can be generated. Included in the list of available displays are:

Xbar/Range, Xbar Sigma, NP and C Control Charts. Xbar/Range and Xbar/Sigma each are pairs of related charts in which the average of the values in each subgroup and the range or standard deviation, respectively, of the values in each subgroup are plotted as a function of order of production. The NP chart plots the total number of defective pieces in each subgroup and the C chart plots the total number of defects in each subgroup as a function of the order of production.

P and U Control Charts. The P chart plots the number of defective pieces in each subgroup. A piece is counted as defective only once, even if it has more than one defective characteristic. The U chart plots the number of defects in each subgroup, thereby counting each piece multiple times if it has multiple defects.

Data for these charts may be grouped by operator name, time interval, in constant size. Subgroups, or in time sequence subgroups of varying size.

Polar Plot. This chart allows the selection of two dimensions whose measured values will be plotted as X,Y points on a graph with the origin of the graph at the midpoints of the tolerance bands. The in-tolerance zone is drawn as a circle whose diameter is the width of the tolerance bands.

Capability Analysis and Histogram. This technique allows the user to analyze the variability of a production process and to determine how it relates to the product specifications. The analysis measures the performance of a process if no assignable causes are present. Along with the histogram, a selection of indices of process performance are computed. Typical uses of this chart are predicting if design specifications can be met, assigning equipment to production, and making adjustments during manufacture. The ability to superimpose a normal distribution of equal area on the histogram plot allows the user to visually detect overlapping distributions and seriously nonnormal distributions.

Regression Plot. Linear least squares regression can be performed on selected data sets. The routine calculates and plots the best fit line along with the raw data points and an indication of the uncertainty of the fit. Data may be fitted as a time series or one dimension may be plotted as a function of a second.

In all of these instances, the plots may be directed to the CRT display of the computer to a pen plotter, or may be output to a dot matrix printer in either the standard or compressed format. For all of these plots, a wide variety of data selection, formating, and labeling options may be selected to present the data analysis in the format most suitable for the application at hand. The charts may be generated interactively, one at a time, if the application is a routine one calling for a standard selection of SPC functions, the batch charting option for creation of a command file for the automatic production of the needed series of charts, with no need for operator intervention.

Data-Base Storage. Measurement results are stored in a simple data-base format for maximum flexibility in selecting data for analysis. This allows the user to focus only on those characteristics that affect the production process without the need to wade through large volumes of incidental data. Specifically, the user has complete control over data-base contents, choosing to store and analyze all of the inspection results or only that subset known from experience will provide the most sensitive indicators of process performance. Stored data is automatically tagged as to date and time, and may, as well, be tagged with user-defined manufacturing information. Data produced by off-CMM inspection operations may be combined manually in the same data base. Specific data items may be flagged for temporary deletion. Process control information may be retrieved *in toto* or selectively on the basis of time, date, or user-defined manufacturing information.

Automatic Data Transfer. Once the data file formats have been established, data transfer from the GEOMET inspection system to GEOSTAT is largely automatic, requiring only an indication of the file name(s) to be transferred. Proper formating and placement of the data are ensured and there is no chance for operator-induced error.

Manual Data Entry. Inspection information from sources other than GEOMET part programs may be incorporated into the analysis by means of sophisticated, user-friendly input routines that ensure correct formating and placement. Both manually and automatically entered data may be edited as necessary.

HELP Function. This feature may be toggled on or off, providing the learner with automatic prompting for acceptable responses, data entry formats, and so on.

File Maintenance. A complete set of utilities is provided for creation, editing listing, and other manipulations on all data and command files.

Error Trapping. Extensive user protection is incorporated throughout the software to prevent system crashes due to incorrect or unreasonable inputs.

Data Communications. In many cases, either for reasons of security or operating efficiency, it is desirable to conduct the SPC analysis of production data on a computer system other than that used for operation of the CMM. To suit such instances, a variety of data transfer options can be added to the basic GEOSTAT system, allowing one- or two-way communication of measurement results and other information between the GEOMET installation and a remote GEOSTAT facility. Transfers can be accomplished over data communications links or by physical transportation of data disks.

Similar data transmission options are available to connect GEOMET and GEOSTAT systems to IBM PCs and PC look-alikes.

Hardware Configuration

GEOSTAT runs on all currently standard GEOMET 100, 200, and 300 hardware configurations—that is, HP 200 and 300 Series desktop computers with a minimum of 1.25 Mbytes of memory, the HP disk-based BASIC language, any HP Raster Interface Standard printer (for example, HP82906A impact printer or HP2225A think-jet printer), and HP CS-80 hard disk/3.5″ flexible disk drive with a hard-disk capacity of 10 Mbytes or greater. A graphics plotter, for example the HP7470A, is optional but recommended.

GEOVISION™ and CMM Programming Alternatives

Efficient use of CMM depends not only on the analytical power and human engineering of their software systems, but to a great extent on the ease with which part programs can be written. CMM part programs are merely sequences of measurement instructions stored on magnetic media. But since part programs can be carefully thought out, tested, and perfected—and when run, allow no deviations from their rigid formats—they guarantee that all parts are measured in the same way and in the same sequence. Thus, in addition to more than tripling the speed of the inspection process, they can be thought of as creating *standard inspection procedures* for each part.

Until recently there have been only two ways to enter part programs into a CMM computer:

1. Off-line entry, using keystrokes only and requiring no support from a CMM
2. On-line entry, sometimes referred to as "self teach," in which the steps of an actual measurement of the part are stored as a program

On-line entry is almost essential for CNC CMMs, since it is by far the most reliable method for defining the probe path of a motor controlled machine. Manual machines, on the other hand, require no probe path instructions, and are best programmed off-line.

Recently a third method for CMM part programming has emerged: *computer aided inspection.* CAI involves generating a CMM program using the graphics capabilities of a CAD terminal. The method is basically a computer simulation of on-line (self teach) programming, in which images of the part and the probe cluster replace their actual counterparts.

CAD/CAM/CAI Concepts

CAD-based CMM part programming and its link to GEOMET are best understood by reviewing CAD/CAM systems in general. (Refer to Chapter 2 for more detailed information.) The software components of a modern, integrated CAD system are:

- CAD software to create computer models of physical parts
- CAE software to study the physical behavior of an object using a CAD model in a computer simulation of the operating environment
- CAM software to create cutter path programs using a CAD model to manufacture a physical part in a machine tool
- CAI software to create inspection programs based on a CAD model to inspect the physical part in a CMM

A computer-integrated manufacturing system includes CAD workstations, machine tools, and measuring machines. The generic machine tool programs are CAD specific. That means they are only meaningful in the CAD environment in which they were created, and cannot be understood or downloaded to machine control units (MCUs) without modification. To run on a specific MCU, a generic CAM program must be translated into the native language of the target MCU. The translation procedure is called *post processing.* (Refer to Chapter 11 for more detailed information.) Since a single CAM-generated program can be run through a large number of post processors, it can be made to run on a large variety of machine tools.

CAI programs generated for CMMs are CAD-specific also, and must be post processed to run on different CMM controllers. Unfortunately, the set of commands needed to manipulate a CMM is vast compared with that needed for machine tools. Thus, CAI systems and their associated post processors are much more complex than their CAM counterparts.

The GEOVISION post processor is based on Computervision's CAI product AUTOMEASURE™. Detailed descriptions of AUTOMEASURE and GEOVISION follow.

GEOVISION and AUTOMEASURE

Computervision's CAI system runs in a CADDS 4X environment. Providing access to all CADDS 4X developed models, it includes a wide range of CMM programming tools such as:

- Feature measurement commands
- Feature construction commands
- Part coordinate system creation and manipulation commands
- ANSI Y14.5 tolerancing

Of particular interest is the library of standard probe tips for Renishaw, DEA, and Metrotast that can be graphically assembled into clusters and moved around the image of the part to create the desired probe paths.

Once a program is complete, a graphics simulated run can be initiated, enabling the programmer to scrutinize the probe path for previously undetected collisions, and to do any editing required. The final step is to convert the program into a neutral data file (NDF) and store it in the host until ready to run.

An off-line GEOMET system is ideal for program entry using standard GEOMET techniques or when incorporating GEOVISION, and for post processing AUTOMEASURE neutral data files without impacting inspection operations on the floor. GEOVISION can reside in on-line GEOMET systems as well.

When a particular NDF is needed for the first time, the GEOVISION/HP300 is made to emulate a Computervision terminal, extracting the NDF as ASCII characters, and storing it on the 300's Winchester disk. Once the NDF is transferred, GEOVISION converts it into a full-fledged GEOMET program. Once in the GEOMET format, the program can be listed and edited like any other GEOMET program, and finally stored back on the Winchester or on a micro-floppy disk, ready to be moved to an on-line GEOMET system and run.

GEOVISION Benefits

With the advent of GEOVISION, there are now two ways of creating GEOMET part programs. Although both methods will continue to be viable, the benefits of the CAD link are:

- By integrating the CMM programming activity into CAD, Inspection, Engineering and Manufacturing are brought into closer cooperation.

- Inspection and manufacturing programs are derived from the same corporate data base.
- Part programs can be written and tested long before the first article is produced.
- Programs can be written and tested without consuming precious CMM time.
- Measurement maps and results maps can be produced efficiently.

The last item is particularly important for manual CMMs, where a graphic description of the measuring sequence is required for smooth operation. Thus, a simple modification of a perspective view of a part, including numbered pointers, can be generated and then plotted for inclusion in the inspection support folder. This can be amplified further by providing setup drawings showing part fixtures and probe cluster configurations.

GEOTOOL™: Full Fledged On-Machine Inspection

A related software package is under development by GEOMET systems, referred to as GEOTOOL. This is a 3D geometry software system for on-machine inspection.

In-process inspection for process control has become accepted as essential in the drive for higher productivity and quality. However, true in-process measurement and control, already well established in cylindrical grinding machines, is only in its infancy in 3-, 4-, and 5-axes milling and boring machine tools (MTs). Such machines, whether manual or computer CNC, depend mainly on external measuring devices such as CMS.

Currently, some developmental effort is taking shape in regard to on-machine inspection (OMI) on both manual and CNC MTs. In the CNC case, this is accomplished with machine control unit (MCU)-based software systems of relatively limited capability and MT "hardened" electronic probes. In manual MTs, traditional dial indicator techniques are still in use, unsupported by software. Providing MTs with inspection software sophistication is the province of GEOTOOL, a derivative of the universal CMS language GEOMET, modified to suit the specialized requirements of the MT environment.

Why On-Machine Inspection Works

In spite of the growing use of OMI, two fundamental questions continue to require resolution: First, is it viable to inspect a part with the same machine that produces it? If yes, then second, what are the potential problems and benefits? The answer to the first question is a definite *yes:*

> A machine tool, just like a CMS, is an XYZ positioning device. If possessed of adequate accuracy and provided with analytical software and probes, it must, like a CMS, be capable of inspecting parts.

GEOTOOL is just such a software system. Derived from its sister product, GEOMET, GEOTOOL has wide-ranging analytical, ANSI Y14.5M-based, measurement and tolerancing routines. Furthermore, GEOTOOL software does not rely on the leadscrew of an MT, but on auxiliary linear scales and DRO with high speed data latching and computer communication ports. GEOTOOL already interfaces with many common DROs, and is compatible with a variety of common MCUs. In the latter case it operates with tool changer-based electronic probes and compensates for one- and two-axes rotary tables.

GEOTOOL systems are based on the Hewlett Packard 300 Series computer with graphics CRT, 1.3 Mbytes of RAM, dedicated GEOTOOL keyboard, a combination of 10 Mbyte hard disk and 540 Kbyte micro-floppy disk drive, an 80-column dot matrix printer, and an MT interface. Available in a mobile cart from which it can be plugged into any GEOTOOL prepared MT, the system temporarily converts such MTs into manual CMSs. Future developments envision the possibility of a centrally located GEOTOOL processor capable of hard-wired communication with a group of MTs outfitted with simple raw data collection electronics.

Using probe paths generated manually or by a programmable MCU, GEOTOOL simply processes the XYZ position data received from the associated DRO when triggered by the electronic probe. Desired collisions (probings) are registered as measurement points; unexpected collisions must be recognized as such and must bring the system to an emergency stop, followed by a request for help.

A CAD link develops OMI part programs directly from the engineering data base. These programs can then be downloaded to GEOTOOL via CAD-specific post processors. SPC routines can be accessed through GEOTOOL to provide machine performance modification commands directly to the MCU.

Benefits of On-Machine Inspection

The benefits of on-machine inspection are:

1. Inspection turnaround is instantaneous.
2. Assurance of quality becomes closely integrated with the manufacturing process.
3. Parts exceeding the capacity of existing measuring machines can now be inspected right on the MT which makes them, with unheard of efficiency compared with current dial indicator methods.
4. Rough machining operations can be checked on the finishing machine prior to investing further work in a part, with no interruption in work flow.
5. Where possible, corrective action can be taken immediately on defective features without ever removing the part from its fixture or from the MT.
6. Machining operations can be fingerprinted with OMI through comparison with external measurements using secondary devices such as CMSs.
7. Although similar to CMS part programs, OMI programs can be oriented toward analyzing machining problems, instead of simply addressing drawing call-outs.
8. In certain cases, lengthy machine setup can be replaced by part coordinate system determination followed by appropriate tool-path modification.

PART C. TECHNIQUES FOR IMPROVING QUALITY ASSURANCE THROUGH INDUSTRIAL VISION SYSTEMS

THE ACCELERATING USE OF VISION SYSTEMS: AN OVERVIEW*

After decades of research and development, a key element in the automation of American factories—vision systems—is being installed at an accelerated pace. Hundreds of products—everything from cars and submarine propellers to microchips, light bulbs, and frozen pizzas—are now being made more cheaply, more accurately, and more safely with the help of a growing variety of machine vision systems. The vision systems do jobs too tedious for humans to perform consistently such as identifying or inspecting parts on assembly lines or labels on bottles. They inspect dangerous items, such as radioactive materials. They also can

*Permission granted by Mechanical Technology, Inc.

confirm in seconds that hundreds of microscopic connections on semiconductor chips are intact.

In addition, when vision systems are attached to robots and other machines operated by computer programs, the machines become much more capable of recognizing and responding to changes in conditions, much as a human worker would. Therefore, vision systems increase the "intelligence" of machines used in assembly, repair, and maintenance.

In a machine vision system, the camera(s) substitutes for the eyes and the computer(s) for the brain. Such a system can be used to interpret an image either to cause some immediate decision to be made during a production operation, or to build a data base about a series of workpieces so an analytical model can be developed to describe the expected characteristics of the objects. The system must therefore be able to verify object presence or absence, measure image features, and recognize objects.

The ultimate goal of an image recognition process is to reduce to a minimum the ambiguities, or the number of ways in which an image can be interpreted, using the least amount of information possible. At the lowest level, an object's location establishes its identity. Higher up the ladder of difficulty is an identification by its two-dimensional shape, requiring the analysis of subtle variations in light intensity or surface structure.

Components of an Industrial Vision System

Whatever the amount of detail required, most industrial vision systems are comprised of five basic components as described below.

1. Light Source. *The light source illuminates the part to be inspected.* The goal of most designers is to eliminate the need for special lighting, but some systems may require unusual light frequencies, such as infrared or X-rays, to bring out the features that human eyes miss. Highly focused laser light can be helpful in a wide variety of tasks, from locating edges to range finding.

2. Camera. *Captures the image of the part.* Vision systems typically use either vidicon cameras, which resemble consumer video cameras (more popular in earlier systems) or solid-state cameras, employing a charge-coupled device (CCD) or a charge-injected device (CID) image sensors. (These are increasingly popular for vision systems.)

3. Frame Buffer. *Stores the image in the computer memory for processing.* Improvement of lighting and cameras is one way to make pictures more sharply defined, providing more data to the vision system. Resolution is further enhanced by image processing. Most processing redefines the image in digital form—that is, the light levels at various points in the image are represented as 1's and 0's. Computer programs can then highlight important features or eliminate small discrepancies that may occur if, for example, a speck of dust or oil is on sheetmetal.

Some processing programs define everything above a certain light intensity as 1 and everything below it as 0. Such "binary" systems are fast and relatively inexpensive.

They can quickly distinguish a hole from a flat surface in a part, but they are dependent on consistent lighting.

A more sophisticated approach, known as gray scale, divides images into as many as 256 levels of light intensity. However, processing and analyzing gray-scale data require extremely powerful computers.

4. Vision Processor. *Analyzes the image.* Image analysis involves such data as position of features, shape, area, and distribution of sharp changes in light intensity which may indicate edges, holes, or the presence of a different material.

Interpretation of the image involves identifying the object, confirming its quality by checking it against preprogrammed models of what is acceptable, or deciding from the data whether an object is oriented properly for the next stage of the manufacturing process.

Other interpretation programs might identify trends that show when a manufacturing process is drifting away from specifications, or making decisions.

*5. **User Interface.** Permits users to access the system for programming, teaching models, and the like.*

Machine vision systems are nearly always used as part of a total factory environment. The data dictates appropriate actions to either a human operator or a robot. Vision systems can be evaluated according to certain established criteria. The importance of each depends on the specific application.

Functions Industrial Vision Systems Perform

Industrial vision applications use vision systems to perform one or more of the following functions:

1. ***Recognition.*** (What is it?) Determine the identity of a part by recognizing:

- Symbols, such as bar codes or alphanumeric characters
- Physical features, such as shape and size

2. ***Location.*** (Where is it?)

- Locating the position and orientation of a part.
- Relaying this information (a reference point and the rotation of the part) to an external device, typically a robot.

3. ***Flaw Detection.*** (What's wrong with it?) Finding and discriminating irregularities or flaws involved:

- Size
- Shape
- Location

4. ***Gauging.*** (What size is it?)

- Performing precise dimensional measurement of a part and its components
- Verifying minimum and/or maximum tolerance
- Counting the number of parts

5. ***Verification.*** (How good is it?) Ensuring that:

- Specific features or components of a part exist, are correctly shaped, and are located correctly
- Undesired features are absent

How Industrial Vision Systems Work

Industrial vision systems must perform three distinct operations during vision analysis:

Step 1: Acquire the Image

Using specialized lighting and optics, a camera (or cameras) takes a picture of the part. The image is then prepared for analysis by:

- Converting the initial analog signals to digital ones
- Transmitting the encoded (digital) image to the vision processor

Step 2: Analyze the Image

The vision processor analyzes the image to find key features by:

- Examining the part using mathematical algorithms to note physical characteristics
- Determining if the critical features are within specified tolerances for quality

Step 3: Act on the Results

Once the vision processor has analyzed the image, the system relays the data by outputting the results to programmable controllers, robots, process control systems, or mainframe computers.

How to Use a Vision System

User Interface

The most apparent, immediate difference in vision systems can be found in the style and type of user interfaces they support. When evaluating a system, consideration should be given to the needs of the user—an operator, engineer, or both, and the amount of interaction that is desired with the system. This applies to both the software interface and hardware interface.

Software Interface

Commands. Many systems rely on a series of cryptic computer commands that must be learned and memorized by the user. Although time-consuming and awkward, these systems may be preferred by users who wish to program at a very high level.

Menus. Menus simplify the process somewhat, reducing the need to memorize and type long commands. Learning time is still substantial with menu-driven systems; their scroll-through, sequential organization can vary between confusing and tedious.

Icons. Well-designed, easily-recognizable icons are the optimum choice for fast learning and operation. Descriptive icons eliminate confusion, memorization, and lag time as they help and guide the user through each sequence.

Hardware Interface

Display and Keyboard. Practically all vision systems feature some type of display and keyboard for direct user input. Both color and black and white displays are available; color is the best choice for clarity and ease-of-use. Some systems feature touch screens. These are the simplest to use, but can cause problems in the factory environment. Touch screens must be kept clean from dust and grease; operators must remove gloves before using. Keyboards can be standard or membrane-type; the standard providing familiar tactile feedback, the membrane able to withstand spills.

Light Pen. Light pens are hand-held light-sensitive "wands" that activate areas of a display by direct contact. Connected to the system via thin cords, light pens are prone to disconnections and tangled and severed cords. The pens are fairly inexpensive; frequent replacement is typical.

Trackball. Factory-rugged trackballs bring all the flexibility and easy access of a desktop mouse to the factory floor.

Factors to Consider When Automating Your Application

Evaluating Your Needs

Factors that influence the feasibility, complexity, and cost of integrating industrial vision into your manufacturing are:

How complex is the part? The amount and complexity of features dictate how successfully the part can be analyzed. The more complex the image, the more powerful the vision system. To accurately inspect intricate parts or assemblies a system must:

- Utilize sophisticated algorithms to extract large amounts of information from the image.
- Provide enough memory to store all the inspections necessary.

How many different parts do you want to inspect? A vision system that is to find or inspect a large number of diverse objects will further require software with the capability to store the models of all the parts.

Do you require specialized fixturing? How parts are presented to a vision system is critical in determining the features and capabilities you require. A system that must recognize parts in different orientations requires more capabilities than a system that inspects parts presented in a standard, uniform position.

What are your minimum cycle times? Finding a vision system that can meet your minimum required cycle times is fundamental.

Do you require (now or in the future) statistical process control? SPC takes vision inspection beyond basic pass/fail capabilities to the analysis of trends in dimensional variations. By monitoring changes in tolerances, SPC systems can actually predict a breakdown in part quality *before* it occurs.

Tips for Successful Automation:
How to Take the Risk Out of Buying Your First Vision System

Make sure your system is easy-to-use for operators as well as engineers.
A well-designed user interface reduces the time needed for training and programming and ensures that your vision system will be readily accepted by factory personnel. Features such as icons eliminate the time-consuming, error-ridden, and often frustrating process of learning intricate commands and procedures.

Before buying, supply your vendor with sample parts.
Part characteristics that may result in problems include:

- Reflectiveness
- Size
- Potential damage from high intensity light

There are ways to adapt to many of these constraints, but solutions may add cost and equipment to the system. It is best to determine potential difficulties up front, since it is impossible to compare competitive vision systems or receive accurate pricing until these special needs are known.

Specify the operating environment in characteristics that will affect the vision system. Vision systems offer many choices in packaging. To select the right system for the right price, it is important to understand your factory's requirements in terms of temperature, vibration, dust, fumes, ambient light, and so on. NEMA-12 packaging will ensure a system's capability to operate in harsh factory environments. If this level of durability is not necessary, less expensive packaging may be sufficient. Many factories require 19″ rack mount cabinets to meet their standards; for others, panel mounting will suffice. Select a supplier that offers a choice in packaging.

Make sure the system can operate within your required cycle times. Because applications vary to such a great degree, there are no uniform standards by which to measure a system's speed. The only valid comparison of speed is based on how fast a system can execute your application.

If possible, automate your least complex vision application first. The experience you gain from your first installation—a working knowledge of system capabilities and limitations, a supportive relationship with a vision equipment supplier, training, and so on—will make the automation of future, more complex applications much easier.

Selecting an Industrial Vision Supplier

In addition to providing competent service and support and comprehensive training both on the company's products and their technology, the supplier also should provide you with the cost of a complete system that can go beyond the cost of the basic system hardware, and should include:

- Programming
- Installation
- Special lighting, cameras, and optics
- Specialized fixturing
- Peripheral equipment, such as printers or PLCs
- Communications and networking

Attributes of a Prominent Vision Supplier: An Example*

American Industrial Vision Corporation (AIVC) designs and manufactures ARGUS, a modular, high-speed inspection system for dimensional gauging, complex assembly verification, and 3D part inspection.

ARGUS verifies part presence, part absence, dimensions, and tolerances, both on and off-line.

The ARGUS inspection system:

- Uses proprietary mathematical algorithms to detect edges and other critical physical characteristics.
- Processes only important portions of the image, ignoring irrelevant changes in lighting and background.
- Supports a standard open-ended VMEbus hardware architecture that provides reliability, easy-maintenance, and cost effectiveness with 32-bit computer architecture based on the Motorola 68010 family for maximum performance.

*Permission granted by American Cimflex.

- Features easily recognizable icons, color graphics, Vision Ladder Logic, and many other user-friendly devices.
- Supports the industry standard MAP network, which allows the system to communicate with host computers, programmable controllers, robots, and other factory equipment.
- Can be used as a stand-alone for a variety of applications, but is designed to be easily integrated into large, turnkey systems for CIM.

CASE-IN-POINT

AUTOMATED INSPECTION SYSTEMS FOR BEVERAGE CAN ENDS MECHANICAL TECHNOLOGY, INC. END SPECTOR™ SYSTEM

Situation

Quality control methods used in the can manufacturing industry involve a large staff of full-time technicians using contact sensors to perform measurements at several workstations. These techniques result in high labor costs, unavoidable inaccuracies in measurement, and delays of up to 12 hours from the time a part leaves the production line to the time out-of-tolerance can ends are identified.

Can ends are precision components that must satisfy a number of exacting requirements. They must fit precisely on the can body. They must be easy to open and yet be reliable and leak-proof before they are opened.

To prevent defects, manufacturers must measure, among other dimensions, the thickness and width of the metal along the score cut for the can opening, the diameter of the rivet used to hold the pop-top tab, and the diameter and height of the curl where the can end fits onto the can body.

Solution

In seeking to improve efficiency and accuracy in measuring can ends, Mechanical Technology, Inc. (MTI) considered systems that use contact or noncontact gauging. The End Spector™ machine vision system was developed as the most practical way to meet the varied requirements. Machine vision techniques do not involve complex probe arrangements and can be modified easily as can design changes.

This precision measurement system uses five video cameras, structured light sources, a vision processor and a host computer to measure thirteen critical dimensions on each can end in less than 3 minutes. A robotic loader and a precision, motorized stage are used to provide fully automated handling of parts during the inspection process. The system performs noncontact measurement with an accuracy of up to 100 micro-inches, completes the inspection process at a single site, and records and analyzes data via a host computer.

The system also represents a breakthrough in the use of machine vision for differential measurement. With a unique method involving two high resolution video cameras, special structured-light projectors, and a series of specialized image-processing operations, the system performs measurements of metal thickness with a resolution of 30 micro-inches.

A further advantage of the machine vision inspection systems is that it can be programmed to run special lots of parts to obtain selected measurements rather than the full set of thirteen parameters. This is necessary when the system operator wishes to evaluate the effectiveness of a corrective adjustment in the manufacturing process.

To aid in overall productivity analysis, the system's host computer stores inspection results and also contains diagnostic and self-check routines that verify the correct operation of major systems elements.

Results

The End Spector™ system is a unique example of specialized machine vision techniques applied to large-volume inspection of a precision part. It illustrates how various equipment and techniques such as precision positioning, robotics, structured light, machine vision, and image processing can be combined to address a specific functional need in an industrial quality control environment.

How to Use Vision Systems in Assembly Operations*

Vision systems are becoming more evident in assembly operations as operators begin to accept the benefit of this technology in regard to measurement, alignment, and recognition. Vision systems can offer the assembly line operator control of his or her entire assembly process from initial input to end product. Mistakes can be caught immediately and misalignments rectified before the long process of multiplying a costly error has a chance to get started.

This is because vision systems offer the assembly operation the ability to see what is going on at every stage. Usually this is achieved with very little or no slowdown in the assembly operation. Vision systems also offer reliability, accuracy, repeatability, and flexibility.

Speed. Vision systems are fast. There is no comparison between mechanical cycle time and the speed of electronic measurement systems.

Reliability. Vision systems are extremely reliable. Once a microprocessor-based system is programmed to perform a task, it does it every time in the same way.

Accuracy. The accuracy of a vision system is directly proportional to the field of view (FOV) of the camera. Therefore, a system may have very fine or very coarse resolution. This relates directly to accuracy. Accuracy may be maintained, in some cases, by multiplexing several cameras to divide the FOV into smaller pieces. The trend is toward cameras and systems with greater resolution capability.

Repeatability. Vision systems have consistent repeatability because what they see is what is measured every time.

Flexibility. Vision systems are characteristically very flexible, but even they may be designed to do a specific task, and only that task. With a single goal a system can be built that will be very fast and relatively inexpensive. But when this approach is carried to an extreme it is like building hard gauging.

Software is the key to flexibility, and hardware is the key to speed. That's why the best vision systems are an integrated blend of hardware and software with the integration beginning at the conceptual design phase. Hardware is used to do the repetitive operations, and software allows the user to program his or her system to perform a wide variety of tasks.

With proper design, a vision system can be both fast and flexible. It also can be very cost effective.

Uses of Vision Systems in Assembly

Robots, which can also be regarded as handling systems, need guidance. A company can save a considerable amount of money in fixturing, which is generally very specific to the part being handled, by having a vision system tell the handler where the part is and what its orientation is. The vision system can be reprogrammed for each new part, whereas if fixtures

*Material based on information from View Engineering, Inc. Used with permission.

are involved, it's back to the drawing boards and then to the shop, which results in incurring attendant time and cost.

Vision systems also are of value in part identification and verification since they can inform an operator ahead of time when to try to assemble parts that were never meant to be assembled. However, at present, vision systems cannot both identify the part and inspect it in detail dimensionally because of the large FOV generally required to see the whole part, which precludes the accuracy of measurement normally required.

Vision systems also are capable of doing many types of inspection before, during, and after each assembly operation, such as to verify that the parts have actually been assembled and that they have gone together correctly. They can often also guide the assembly operation.

Types of Vision Systems

Vision systems now on the market can generally be divided into two types: edge finders and correlators. Both types are referred to by most manufacturers as *pattern recognition systems,* but there are significant differences between the two and a clear distinction will be useful.

Edge Finder

Edge finders are designed to find edges. These systems look for transitions from black to white, white to black, or from gray level to gray level.

When such a transition occurs, the edge finder system notes where it occurs (by line in the vertical direction, and by clock count in the horizontal direction). When a second transition occurs, it also is noted and the number of picture elements (pixels) between the transitions is noted. If it is desirable to make a measurement, then the system will have been calibrated (the size of a pixel determined) and the physical distance between the two transactions can be computed.

An application of such a measurement might be where the lines that cross on one end are measurement bars that are positioned by the operator, and which can then be programmed to look for transitions and to compute dimensions.

The same sort of measurement bars can be used for go/no-go measurements. In this method, each bar is placed over the position of a segment of a seven segment display as normally used in digital watches. The edge finder system can then be used as a simple pattern recognition system to read the numerals.

In general, edge finders have limited application in flaw detection, but they can be used for the detection of gross problems. Here we refer to flaws as defects which are not simply wrong dimensions. They might be such things as chips missing from glass or ceramic parts, bridging or gaps in traces on PC boards, improperly printed characters, or a myriad of other defects. While edge finders are useful in some cases, the programming may be time-consuming and difficult.

The alignment of parts for the control of handling systems can often be performed effectively by an edge finder. For example, when showing the position of measurement bars placed to direct a pick-and-place machine to pick up an integrated circuit, two bars are placed to find the position and angle of the left hand edge, and a third is placed to determine the position of the top edge. More complex gate sets would be required to find the position and orientation of more complex parts.

Correlation Systems

Correlation systems are actually template matching systems. That is, a complex image is stored in memory and the correlator searches the scene for a match to the image stored in memory. One major difference between a correlator and an edge finder is ease of programming.

To program a correlator, all that is required is to position the reference image in the field of view and press a button to store it in memory. From then on, the system will search whatever scene is presented to the camera for the best match to the reference.

The uses of a correlator somewhat overlap those of an edge finder, but also extend the use of vision systems to other areas. Correlators are more useful for alignment of complex parts because of the ease of programming and because of the capability of searching automatically through complete 360° of rotation without additional programming. This capability also makes the correlation system useful for part identification and sorting.

A correlation system can store many references in memory. Therefore, if it is known that a part presented to the correlator is one of the ten predetermined types, the correlator simply searches each of the ten stored references until a match is found. In this way the correlator not only determines which type of part is present, but also finds its position and orientation.

This information is then available to be used by the handling system to direct a manipulator to pick up the part and put it in the proper bin.

A more sophisticated system might, even at this point, combine an edge finder with a correlator. In the most general application of a combined system, the correlator identifies the part and finds its position and orientation. From the part identification it then directs the edge finder which gate set to use. From the position and orientation information, the measurement gate set is positioned properly over the image of the part to make the desired measurements.

> ***Key Point:*** In the use of such a combined system, usually a rather large field of view is required by the correlator to allow for the entire range of possible part displacement and to get enough information about the part in the field of view. If the edge finder shares the optics and camera with the correlator, then the resolution will be poor. This means that precision measurements will not be possible in most cases.

An Application in the Watch Industry

In the competitive watch industry, labor costs are becoming more and more important and on-line vision systems are being considered seriously for many assembly operations. This application shows how multiplexing four cameras can be used to check the progress of a sub-assembly after each operation. By time sharing one set of electronics (multiplexed) among four cameras, costs are reduced dramatically without any slow down in throughput.

At one station, two cameras are used to check that the pin is vertical and square to the base. This is done by establishing several points along one edge of the pin with each camera. By proper camera placement and curve fitting techniques, the position of the edge can be established as if the resolution (pixel size) of the camera were much better than it actually is. In many cases such as this one, sub-pixel accuracy can be realized by clever processing of the data. This is another manifestation of the flexibility that can be provided by a well designed vision system.

At a following station the concentricity of the gears is to be established. This is a critical parameter in watch operation and is very difficult to monitor by conventional means.

And, at the last station, the sub-assembly is checked for the presence of two parts (the position is not critical), and the location of the contact spring is verified.

Vision systems can be used in watch assembly from beginning to end. Even the final verification of time-keeping accuracy can be performed. It is an easy matter to see that a vision system can be used to either read the hands of an analog watch or the display of a digital watch. The elapsed time as read on the watch can be compared with the elapsed time as measured by the accurate crystal clock in the vision system.

The adaptability of vision systems to the assembly process is well exemplified by the watch industry. From the alignment of pins, through the concentricity of gears and final verification of the operation of the watch, a single type of vision will do the job.

Through multiplexing, the cost is kept low and the speed of the system maintains the throughput of the assembly line.

The following case-in-point illustrations are representative of various types of vision systems that companies are using for assembly inspection applications in today's industrial manufacturing environment.

CASE-IN-POINT

AUTOMATED ASSEMBLY LINE FOR REAR BRAKE DRUMS
VOLKSWAGEN AG
*AMERICAN CIMFLEX, INC.**

Situation

Volkswagen AG, as part of their investment in the new Rabbit, have installed an automated assembly line for rear brake drums at their plant in Braunschweig, West Germany. The line, based on the modular transfer system manufactured by Robert Bosch GMBH, uses manual, semi-automatic, and automatic workstations to perform the complete assembly operation on the drum.

Brake assemblies are a critical safety component on a car, and rigorous quality assurance based on 100 percent inspection is enforced.

Solution

The company, in its desire to further improve these standards, and to cope with the increased production made possible by the new line, installed a vision system designed at the University of Manchester. It uses knowledge model-based software and is ideally suited to inspect complex assemblies and components. A model derived from a drawing or CAD system can focus the attention of the computer on those areas of the part critical to the inspection process.

The inspection station, constructed as a module on the transfer line, uses three high-resolution cameras to view the part. In addition, the station provides a 180° rotation of the part to allow imaging of all four side views together with a plan view. Operation of the station is controlled from the vision computer, making the image analysis and control independent of the remainder of the line.

Correct component illumination is critical to the operation of any vision system in providing good contrast between the features of interest and the background. In this system, three-color structured lighting, matched to filters on the cameras, is used to highlight different parts of the assembly in each view. Red diffuse lighting is used for the plan view, while green tangential lighting is used to highlight surface texture on the brake shoes.

In operation, a brake assembly mounted on a transfer line pallet is moved into the station. The plan view is then analyzed to find the exact location and orientation of the shoes, which may move relative to the backing plate. Specific checks are then made on each item in the assembly to check for the presence of correct compounds, correct directional tolerances between components, and for possible damage during assembly. Image processing techniques

*Permission granted by American Cimflex.

range from edge and hole finding to locating position, Fourier analysis for components such as springs, to texture analysis for showing surface roughness due to grinding faults.

Results

This inspection system, installed by Volkswagen AG, is completely operational and an integral part of the routine production process. As a result, the company has achieved its desired objective of maintaining rigorously high standards of control over brake assembly production.

CASE-IN-POINT

DASH PANEL INSPECTION SYSTEM
FORD MOTOR COMPANY
*THE AMERICAN INDUSTRIAL VISION CORPORATION**

Situation

On a fully automated assembly line, one of the major problems to overcome is the disposition of less than ideal parts as they are presented to an automated work cell, and then verifying that good parts have been assembled properly. Error detection before assembly of any product is fundamental to all manufacturers, particularly from an economic standpoint.

Solution

American Industrial Vision Corporation (AIVC) delivered an economically feasible solution to Ford Motor Company in the form of an eight camera vision system for its Escort instrument panel assembly line. The cameras are divided into two stations with four cameras on each station. Each station operates asynchronously through a single /SIGHT** industrial vision controller.

Operational Sequence

The first station checks the instrument slip retaining holes to ensure that the holes are not flashed over, which would inhibit the robot from inserting the clips, and also ensure that there is enough plastic to form a proper hole. The latter is a more severe error that renders the part unusable, requiring removal of the part from the assembly line. The second station verifies that the robot has indeed inserted the clips in all of the holes correctly and additionally, if required, tracks which clips have not been inserted, sending that information upstream to the robot.

CASE-IN-POINT

C-WASHER INSPECTION SYSTEM
FORD MOTOR COMPANY
THE AMERICAN INDUSTRIAL VISION CORPORATION

Situation

Vision system reliability is always a requirement, especially in hostile, complex-viewing environments. Rear axle C-washer inspection is one such example. Limited field of view, cluttered viewing area, and variable lighting combine to create a very harsh inspection task. In addition, the following customer requirements had to be met:

*Permission granted by American Cimflex.
**/SIGHT was a custom vision system that was the predecessor of ARGUS, although these applications also would be typical of the ARGUS vision system.

100 Percent Reliability Rating. Rear axles shipped without C-washers are a serious threat to the consumer. Failure of such a defective axle may lead to questions of alleged manufacturer negligence.

Axle Variability. Inspection operations on eight distinct axle types require a high degree of system flexibility. The option of adding new types in future model years also must be addressed.

Inaccurate Fixturing. The saddle fixtures used for axle transport do not provide a reliable base for any type of noncontact gauging. Axle movement of ±5 inches in both X and Y dimensions, gearcase rotation of ±5° and 360° C-washer rotation all add up to highly unrepeatable visual reference points.

Part Contamination. Oils, grease, shadows, and glare result in a very noisy image. Due to these conditions, conventional vision techniques were not applicable.

Solution

AIVC proposed a /SIGHT industrial vision inspection system with special ultraviolet capability, A /SIGHT system consisting of a single /SIGHT controller, multiple cameras, and an AIVC designed inspection tunnel was designed, built, and tested for customer approval.

Operational Sequence

C-washers are coated with UV metal marking dye. After installation, the axle enters the inspection tunnel. Special lighting fixtures illuminate the C-washer area with long-wave UV light (365nm). The resultant image is spectrally purified by camera-mounted cutoff filters.

The inspection tunnel serves two purposes:

1. Preventing outside lighting from interfering with the inspection process
2. Limiting the effects of UV exposure to surrounding operators

Photoelectric sensors located in the tunnel detect the presence of an axle on the saddle fixture and signal the /SIGHT system to start an inspection.

The /SIGHT controller snaps frames of the viewing area and tests for both right and left C-washers. An addition test is performed to check for the presence of "failure bolts" on the axle gearcase (indicating the axle is faulty for reasons other than missing C-washers). Good axles are allowed to pass. If one of both C-washers are absent and/or a failure bolt is present, alarms are sounded indicating the appropriate status.

Conclusion

Since this and the preceding cases are typical industrial applications, you may reasonably conjecture that at least over the foreseeable future, most fully automated assembly line systems will at one time or another become dependent on industrial vision technology to ensure their proper operation and to provide functional feedback for their continuous operation when dealing with inconsistent parts.

HOW LASER INSPECTION SYSTEMS ACHIEVE 100 PERCENT INSPECTION*

Non contact laser-gauging systems are finding increasingly widespread use in industry—both by manufacturers and users. This advanced technology has been able to help both large and small companies improve quality and productivity. One of the most compelling reasons for the installation of a laser-gauging system is to achieve 100 percent inspection

*Permission granted by Automation Systems, Inc.

thereby eliminating defective parts flowing into the assembly operations. With the continuing spread of automation, and increasing production speeds, this innovative approach can avoid costly machine jams and "crashes," tooling damage, product rejects, and unnecessary down time. While 100 percent inspection is sometimes available from an OEM, it is usually prohibitively costly because it is based on a visual and/or gauge inspection of every part. As a result, companies have found it more advantageous to install noncontact forms of gauging, principally by laser inspection. In addition to inspecting incoming vendor items with this method, manufacturers also are employing laser equipment to provide a reasonable cost, 100 percent inspection of materials flowing through their internal plant operation.

Advantages of Laser Inspection

The advantages of laser inspection are as follows:

- Noncontact at high speeds
- 100 percent inspection of all components
- No customer returns
- Labor cost savings (no manual inspection)
- Improved quality of product
- Microprocessor-based equipment
- Statistical software for production run analysis

There are many types of industrial applications that can benefit from the use of laser inspection systems. Some of these are:

- Gauging
- Gauging/parts qualification
- Identification operations

Gauging

This involves laser systems used to inspect a variety of parts for length, height, and thickness, as well as camber (bow), and part orientation. These systems can gauge parts on- or off-line at production speeds without contacting the part. Inspection tolerances depend on feed rates and part configuration.

Principles of operation. A 2-milliwatt helium/neon gas laser generates a collimated beam of light which is expanded and directed onto a part to be inspected.

The laser light shadowed by the part is imaged onto a linear diode array which contains pixel elements with known centers. The pixel produces a pulse when lit; no pulse when darkened. By counting the number of lit or darkened pixels and comparing them to a present number (entered via BCD or key pad) the actual part dimension can be measured. Accept/reject signals indicate whether the part is acceptable or to be rejected from the system.

Gauging/parts qualification

This type of system inspects parts for over and under length, over and under height, diameter, over and under width, thread presence, and count.

The system gauges and sorts, by dimension, which can qualify the parts prior to automatic assembly.

Principle of Operation. A 2-milliwatt helium/neon gas laser generates a collimated beam of light which is optically processed, focused, and directed onto the part. In most cases, the inspected products are received from an automatic feed system and gravity is fed down a custom-designed presentation V rail, which is on a 30° incline. For length inspection, the laser source is split into three beams. The center beam is stationary and acts as a reference for the other two. Both the upper and lower beams can be adjusted independently by micrometer dials that define the distance between over- and undersize inspection with a single pass. The beams then act as an optical length gauge. The interruption of the beams by the part generates a sequence of events that indicates the length of the product. Photodetectors convert the information to electrical signals for logic processing. Electronic decisions are made, and when the part is found to be out of tolerance an air solenoid ejects it. Acceptable parts proceed uninterrupted.

For width inspections, two additional vertical laser beams are added. One vertical angled beam with photo multiplier tube receiver is added for thread inspection. Two horizontal beams are added when height measurements are required. An expanded beam and diode-array scanning receivers are added when diameter inspection is required.

Identification Operations

This type laser system inspects a wide variety of threaded parts for the presence or absence of threads and thread quality. This application will inspect for internal threads on such parts as nuts, bolts, shafts, plugs, plates, links, studs, and so on.

Principle of operation. A 2-milliwatt helium/neon gas laser generates a collimated beam of light which is optically processed, focused, and directed onto the part. In most cases the inspected products are received from an automatic feed system and gravity is fed down a custom-designed presentation rail which is on a 30° incline and tilted back 15°.

Examples of Laser Inspection Systems

The following examples are representative of advances in laser inspection systems technology designed and manufactured by Automation System Inc.

Bearing Inspection System (figure 9–19)

This laser inspection system was designed and manufactured for flaw detection of self-lubricating type bearings. The system can locate flaws of .005″ in diameter as the part is rotated at sixty parts per minute. Flaws such as nicks, digs, metal slivers, and incomplete coating on the ID surface are located by this noncontact method.

The bearings are initially fed by a vibratory bowl to a gravitational track where they are laser inspected for height, thus ensuring that the camber is present. Rejects are ejected at this inspection station while good parts are allowed to continue to the spin station. A standard ASI Model 615 laser instrument scans 100 percent of the ID as the part is rotated at the spin station. The reflected light of the ID is captured by a receiver and processed to determine if an attenuation of the reflected light occurs. Attenuated laser light indicates that a void is present on the ID surface and accept/reject gates separate good from defective products.

The system is microprocessor-based and contains a self-learn mode of operation for the height gauging. The dimension of a know part is dialed into the system. That part is

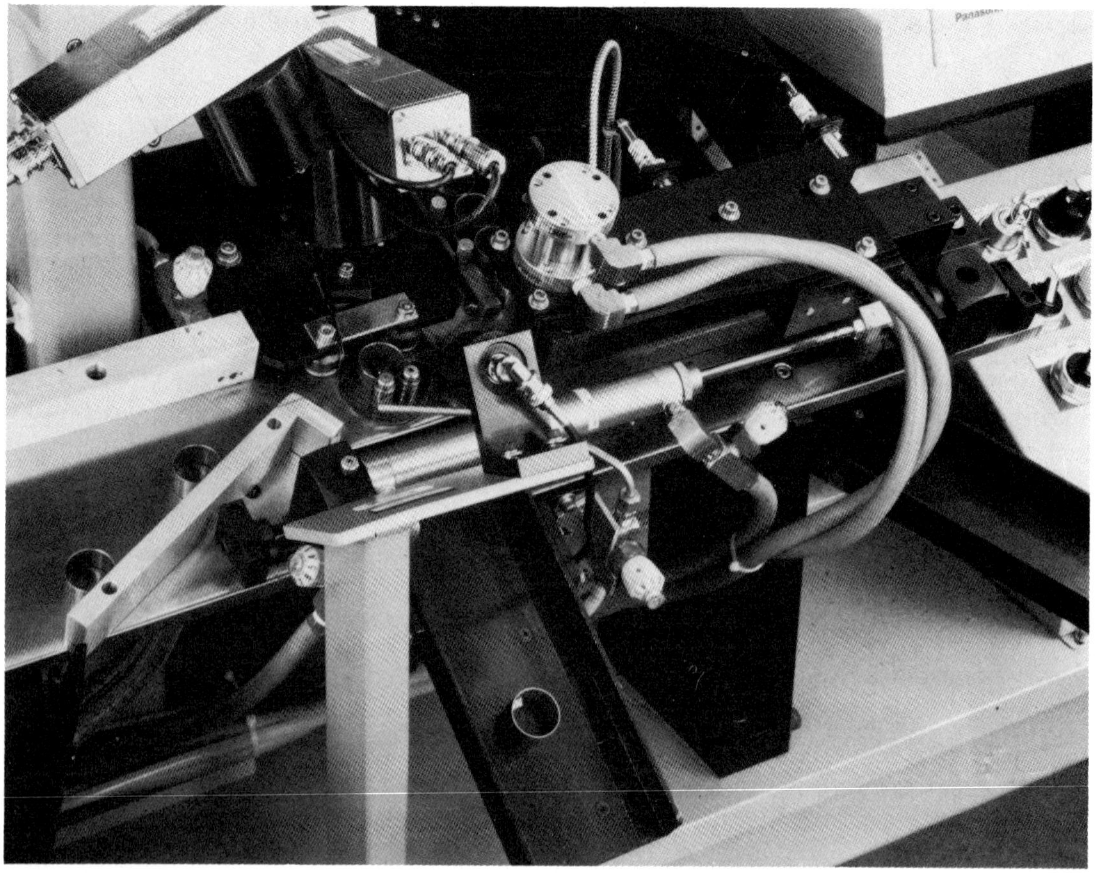

FIGURE 9–19 Bearing Inspection System (Courtesy of Automation Systems Inc.).

manually fed through the inspection area, calibrating the system to the part. The part tolerances (min/max) are dialed into the system and production parts now can be inspected automatically.

Laser Gauge for Pump Vanes

This application is a turnkey system designed and manufactured for inspection of vanes (slippers) for automotive power steering and transmission pumps at a rate of 100 parts per minute.

The vanes are bulk loaded into a vibratory parts feeder bowl and fed down a gravity tract to ASI's Model 325 laser inspection gauge. The inspections include squareness, parallelism, length, and width to a tolerance of one-ten thousandth of an inch. The inspections are accomplished with eight laser beams that shadow the part onto eight photo diode array receivers. The laser light received is optically magnified onto the diode array pixels (contraction for picture elements). The number of blocked or dark pixels are counted and compared with a preset window of acceptance (tolerance). Dimensional tolerances of parts outside of this range are automatically ejected out of the system pneumatically at this point in time. Part thickness also is gauged by ASI's "Triangulator" to .001″ by triangulating two laser beams (top and bottom) onto two arrays and comparing the results.

The equipment contains a clean optics circuit which allows up to 90 percent attention of the laser light before cleaning of the lens is required. This allows the equipment to operate in a hostile environment without error in gauging the parts. An eddy current station also is added for detection of a hard–soft condition of the vanes.

The dimensional parameters are entered into the system via a CRT keyboard and are easily modified by this user-friendly system. The equipment is microprocessor-based with an RS232 available output. This allows the laser inspection system to be linked to a manufacturing computer or to a statistical process control package such as ASI's Model 730 Printer/Microprocessor.

At present, this system is the only known method of gauging these vanes to the tolerance specified on a noncontact basis at production speeds.

Rotational Station for Laser Inspection

This device is designed for rotating cylindrical parts for laser scanning applications. Components such as piston rings, valve seat inserts, and bearings can be rotated while inspecting for surface flaw defects or gauging. ASI's standard model 615 (flaw) or 655 (gauging) performs the inspection to .001″ (.025mm), resolution to .0001″ (.0025mm).

The basic variable speed drive applies spring loaded force on the O.D. of the part to be inspected. The part is supported in the fixture and rotates on carbide balls/drive wheels as it spins in the inspection station. The fixture is designed to minimize radial movement of the part and to release the part at the accept or reject chute.

The fixtures are pinned onto the dial plate and are interchangeable for different part diameters. Changeover time is less than 15 minutes.

The parts are being inspected at 60-100 parts per minute depending on the O.D. size. The inspection speed limitation is in the feeding and not in the laser inspection. The laser inspection is accomplished with a high frequency scanning beam that impinges off of a specular surface. The reflective signal is received by a light detector and processed. Discontinuities such as scratches, porosity and marks are viewed as attenuators of light and are causes for reject.

CASE-IN-POINT:

APPLICATION OF IN-LINE LASER INSPECTION TECHNOLOGY
AUTOMATION SYSTEMS, INC.

Situation

In recent years the manufacture of electronic products has moved rapidly toward the concept of modularization. Various electronic modules are used in building-block form to produce a wide range of electronic products. This has necessitated a rapid increase in the demand for electronic connectors.

To meet this need, the connector industry has developed new technologies capable of producing higher quality innovative connectors at lower costs. Connectors now are being assembled at the rate of hundreds per minute at very high quality levels. Since the electronics industry has mandated zero defects—at these increased production rates—the manufacturers of electronic connectors have moved toward in-line 100 percent inspection. This eliminates defective connectors as they are produced; provides a monitoring capability of the assembly process; and can automatically shut down a process if it is producing an excessive number of rejects.

In addition, the high levels of manufacturing automation within the electronics industry (such as robot loading of connectors into printed circuit boards) has placed demands on connector manufacturers to supply trouble-free devices (no bent pins).

Since customers return defective components to a connector manufacturer, which can be very costly, a laser inspection system was designed to solve the problem at production speeds.

Reasons for Choosing Lasers as the "Eyes" of the System for Connector Inspection

A helium/neon laser operating at 6300ang provides the unique properties of a continuous wave light beam to examine areas of the contacts for dimensional characteristics. High beam brightness, beam collimation (low-beam spread), small beam diameter, and monochromatic light permit optical illumination with only two milliwatts of laser energy. The beam collimation and small beam diameter provide high resolution at high speeds for part inspection. Monochromatic light permits the optical system to operate in most ambient environments.

For the reasons above, lasers were chosen as the "eyes" of the system for connector inspection.

DIP Sockets

The initial inspections required on DIP sockets were for contact hole presence/count and overall width of the molded socket. A system was required to inspect for the above prior to contact insertion into the socket at 250″ (9.8mm) per minute.

The molded sockets are bulk loaded into a vibratory feeder. The feeder orients the socket and feeds it end to end to an inspection rail. The rail is tilted forward by 30° and canted 15° toward one side. This enables the socket to feed in a stable fashion by gravity through the inspection area. Part separation is guaranteed by the correct interface angle between feed bowl and the rail. The part is laser inspected while in motion on the track.

As the gate beam is interrupted, the inspection process commences. The vertical laser beams on both channels generate analog waveforms on photodetectors mounted under the rail, as the beams pass through the openings in the socket holes and slots in the rail. The analog information is digitized and compared with pre-established limits as set by BCD (binary coded decimal) switches. The digitized pulses that are counted verify that the holes are present and the correct number are present (none blocked by plastic flashing).

One additional vertical beam on a translation stage verifies light presence by establishing a maximum acceptance for socket width to .002″ (.05mm).

Unacceptable parts exit the rail by means of an air-jet blast on the inspection rail. The acceptable product is allowed to continue through to the end of the rail. The qualified parts then continue on to the contact insertion station. Machine jams caused by hole blockage at the insertion station are eliminated thereby increasing overall productivity. Missing contacts are eliminated thereby improving the quality of the product delivered to the end user.

The next task was to solve sprung contacts to .005″ (.13mm) after insertion with certain socket designs. These problems caused poor electrical contact between the DIP semiconductor and the socket contact. The sockets are gravity fed down a vertical rail and inspected in the following manner: Gate beams are interrupted and the inspection process commences. Vertical laser beams are allowed to be transmitted alongside the contacts, between the contacts and the molded socket wall. The pulse width of the transmitted light intensity is measured below the rail on photodetectors. The pulse width and count is compared with preset threshold levels to determine if the part is acceptable. This method also would determine contact presence and count.

The next problem to resolve was pin straightness to .005″ (.13mm) on each row of contacts. Bent pins caused pin bending when trying to automatically insert sockets into printed circuit boards. Pin bending results in a poor electrical connection at the specific point.

The sockets are gravity fed down a vertical rail in a "dead bug" orientation and inspected as follows: Two angled laser beams are triangulated off the pins at an appropriate point and reflected into a photomultiplier receiver for processing. The received analog signal is processed, digitized, and compared with threshold levels for inspection of bending to ±.005″ (.13mm). It is important to maintain the socket in a stable position while it is transferred through the inspection area. Pin presence and count also are verified.

The need arose to have the ability to inspect for sprung and crushed contacts on sockets which have different centerlines. This was to be accomplished without adjustment of the laser beam position when centerlines varied on the sockets. Also, bent pins in any true position (TP) direction to ±.003″ (.08mm) was required. At the same time, requests were made to be able to inspect at speeds greater than 500″ (19.7mm) per minute.

The sockets are fed in a dead bug position on a driven belt to maintain part separation of .005″ (.13mm) or greater and maintain part stability upon speed up during inspection.

One horizontal laser beam is used to gate the pins, which commences part inspection for pin count. A second horizontal laser beam at an appropriate angle performs pin count by measuring the number of pulses received on the photodetector. The number of times that the beam is blocked by the pins is stored in the systems microprocessor. This data is compared with information entered via a keypad into the user-friendly microprocessor. Also, by measuring the time differential between the gate start/stop and the number of pulses/pulse width, the true position (distance between each pin) can be calculated. If it is not equal, then the pin is bent from its true position.

Another 2KHZ scanning laser emanates from the bottom of the socket to measure open, crushed, or missing contacts. If the contact is down (blocking hole) light will reflect back to the photo multiplier tube (PMT) receiver which will verify the reject, and eject the part out of the inspection area. An open or missing contact will transmit a greater amount of light through the contact to the PMT receiver above the presentation area, thus verifying this reject. The PMTs measure this reflected light intensity and compare the data with preset threshold levels.

The scanning laser is used to overscan across the socket contacts allowing centerline variations without laser beam adjustments.

The same technology that ASI developed for DIP socket inspection was then utilized to solve connector inspection problems.

Connectors

The ribbon-type connectors are fed end to end to the inspection area after contact insertion. The rail receives the part and allows it to be gravity fed through the laser inspection unit at a rate of 250 parts per minute.

As the gate beams are interrupted, vertical laser beams measure for plugged holes (crushed contact). Angled laser beams inspect for bent/missing pins to .005″ (.13mm) true position by triangulating off the pins on both rows.

The gap between the split contacts can be measured by a scanning laser, if contact twisting is a production problem.

The manufacturer guarantees positive pin alignment which ensures a positive electrical connection when the ribbon cable is pressed into the pins.

The telephone-type connectors are fed end to end, contact up, after contact insertion. The rail receives the part and allows it to be gravity fed through the inspection area at a rate of 250 parts per minute.

As the gate beam is interrupted, vertical laser beams impinge on the contact surface. The reflected light is received in a PMT and its intensity is compared with preset threshold levels. A sprung or missing contact by ±.003″ (.08mm) will reflect less light or no light.

Rejects are ejected out of the system by air; good parts are allowed to travel to the packaging system. *NOTE:* Sprung or missing contacts cause poor electrical contact in a telephone circuit.

The printed circuit (P.C.) card edge connectors are fed end to end, pins up, after contact insertion. The rail receives the part and allows it to be gravity fed through the inspection area at a rate of 250 parts per minute.

As the gate beam is interrupted, vertical angled beams triangulate off of the P.C. contacts at an appropriate angle. They inspect for missing, sprung, or damaged contacts to ±.003″ (.08mm). PMT receivers are utilized and measure the light intensity.

A horizontal angled beam is used to count the pin tails and measure TP to ±.003″ (.08mm). The detector receives light or no light, and those pulses are compared with preset pulse counts. In essence, the microprocessor measures the gap between the pins and verifies TP reject/accept criteria.

This inspection guarantees that the P.C. card can be plugged into the connector and a good electrical contact will be obtained. This inspection also verifies pin presence and ensures that a connector can be inserted into a printed circuit board, providing trouble-free electrical contacts.

Innovations to Watch: Advancements in Laser Inspection Technology

The following products are now being reviewed for laser inspection based on previously developed technology.

Headers

Headers can easily be inspected for TP of the pins on both sides of the connector. Horizontal laser beams impinging at an appropriate angle can measure the spacing between the pins to determine straightness. Also pin height to ±.002″ (.05mm) can be detected with a scanning horizontal laser beam. Another horizontal expanded laser beam with a diode array receiver monitors the body thickness and accounts for the bow factor of the molded body.

Insulation Displacement Connectors

Many of the principles discussed previously are utilized in the inspection of IDC type of connectors. Vertical laser beams impinge off of the recess contact to ensure presence and proper insertion depth. The blade area is inspected by a scanning laser beam for proper gap dimension to ±.002″ (.05mm).

Other potential areas are measurement of contact dimensions while the contact is on a carrier strip and inspection of semiconductors, such as DIPS and DUAL QUAD PACKS for pin TP.

The application work performed on the DIPS and connectors has resulted in a proven production system for high-speed inspection. Connector manufacturers are shipping zero defect product to their customers.

The process is limited by the number of parameters that can be measured in a single pass. The present system measures only critical parameters.

Further work in the area of triangulation of laser beams for thickness measurements is underdevelopment. Also, as handling systems increase in output, the inspection speeds must be increased without loss of measurement resolution.

Types of Laser-Based Systems

The beam configurations generated by various existing laser-based systems include the following:

1. Static spot
2. Circular spot
3. Linear spot
4. Multiple spot

Because the geometry of the laser beam determines the inspection system capability and its ultimate use, laser systems are classified according to the type of their beam.

In operation, the transmitted laser beam interacts in various ways with the target being inspected. Through this interaction the beam at the receiver input will be encoded by various target parameters such as its dimensions, surface quality, and so on. The receiver extracts or decodes this information and uses the resultant data for making decisions in inspection. In particular, consider the several ways that an incident laser beam can be encoded and produce different detected signals at the receiver by its target. It can be in any of the following ways:

Sequential interrupted. The detected signal is of a digital nature—that is, it is either on/off or go/no-go type. It is used in simple length sorting equipment.

Reflected. This results in an analog detected signal whose amplitude and duration are used in the analysis of the inspected part. This approach is used basically in areas of surface flaw detection.

Shadowed. This produces a digital signal which is used for more accurate dimensional gauging, for example, where the sequentially interrupted method becomes inappropriate.

Transmitted. When detected, this is either a digital or an analog signal and is basically used in applications where the material to be inspected is clear or transparent.

The following specific examples illustrate each one of the systems:

Figure 9–20 shows a system that generates a static spot beam which is split into three beams to gauge the length of a part. The detected beams will result in on/off signals that will contain part length information. This unit typically will be used to gauge dimensions of parts like pins, bolts, and plates. In addition to normal gauging functions, the unit can be made to sort undersized and oversized parts. With certain laser inspection systems, production rates as high as 600 parts per minute to within tolerances of ±0.001″ are possible. A micrometer dial setting enables the unit to handle parts from 0.25 to 2.0″ in overall length.

Figure 9–21 is an example of a laser system based on the circular scanning principle. The generated beam traces a circular pattern on parts to be inspected. This flying laser spot scan rotates at about 30 rps. In addition, the system is designed so that the position of the laser spot can be accurately defined through the use of an optical encoder. The resultant system favors the inspection of parts like bearings, gaskets, nuts, and washers.

In a shadowed system (figure 9–22), parallel light beams are generated enabling the detector to "see" a "shadow" of the object being inspected. Certain units are capable of

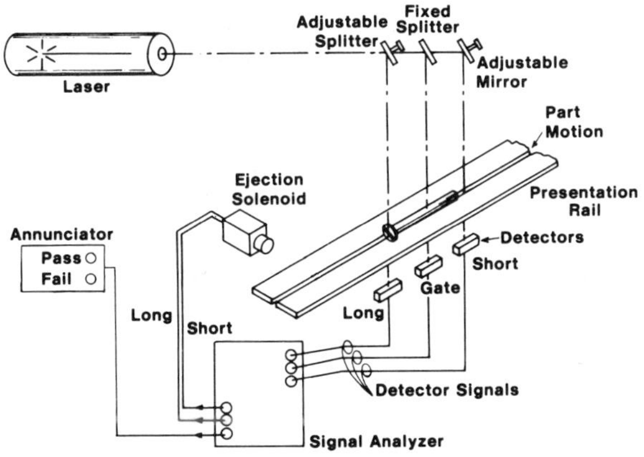

FIGURE 9–20 Laser-Based System-Static Spot (Courtesy of Automation Systems Inc.).

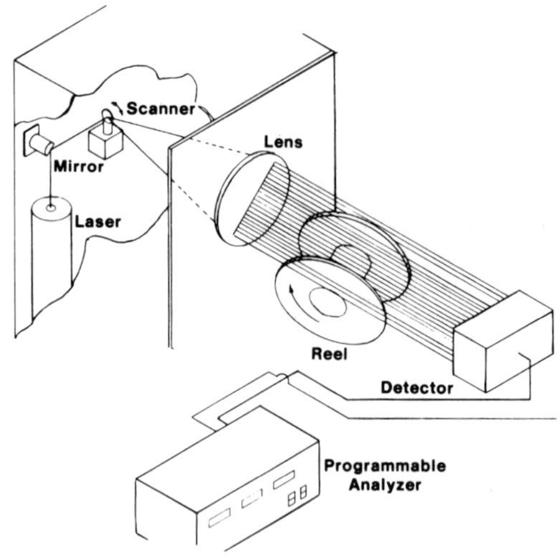

FIGURE 9–21 Laser-Based System-Circular Spot (Courtesy of Automation Systems Inc.).

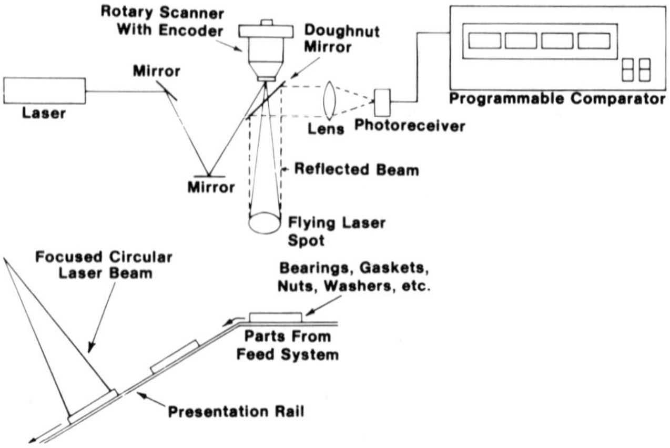

FIGURE 9–22 Laser-Based System-Linear Spot (Courtesy of Automation Systems Inc.).

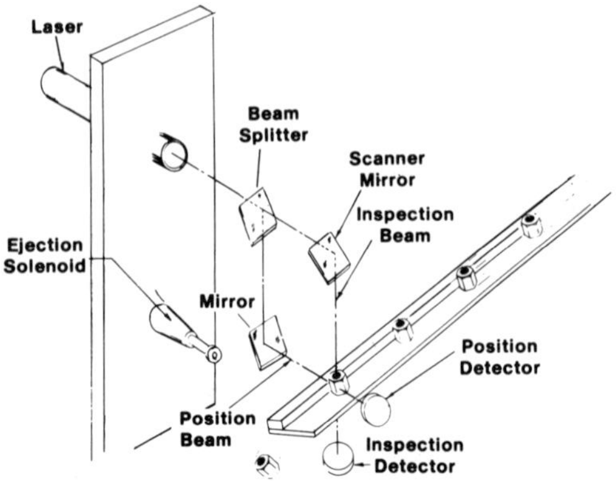

FIGURE 9–23 Laser-Based System-Multiple Spot (Courtesy of Automation Systems Inc.).

scanning a focused spot of laser beam along a precise linear path at the rate of 3000 per second with dimensional tolerances held within 0.0004″. The system permits the detection of parallel variations in flange dimensions of a part like a tape reel.

Using transmitted signals (figure 9–23), parts can be inspected for surface defects and missing operations as well as their dimensional accuracy. For example, threaded fasteners can be searched for surface problems, missing internal or external threads, damaged threads, or even mixed sizes of threads. The laser beam is scanned in a vertical plane so that it "walks" down the threads. At the same time, the horizontal beam determines the location of the part under inspection.

CASE-IN-POINT

*ROTOR STACKING PROCESS CELL MANUFACTURING TECHNOLOGY MECHANICAL TECHNOLOGY, INC.**

System Description

A rotor stacking process cell (RSPC) was designed by MTI for two air logistics centers (ALCs) that serve as engine repair facilities. Under this program, an automated system was developed that increases gas turbine rotor measurement and assembly efficiency at the ALCs.

This program involved the design and implementation of an RSPC that ensures the straightest possible rotor assemblies from the available ALC parts inventory. The RSPC eliminates the number of engines rejected for excessive vibration attributed to inaccurate rotor assembly. By combining state-of-the-art electro-optical/laser gauging with conventional noncontact sensors and probes, the RSPC evaluates critical incoming rotor parts and determines their required disposition (scrap, partial rework, no rework) prior to their entry into the engine rework cycle.

This unique system accommodates added variables required for reworked parts and provides a rapid, accurate, and detailed inspection of key rotor parts. Using this inspection data, the RSPC determines which parts provide a proper snap fit. The system's MTI-developed combinational stacking algorithm routines determine the proper orientation of the rotor parts so that the assembled rotor has a minimum center line deviation. It therefore indicates how such a straight stacking approach can reduce vibration levels.

Results of Program

The RSPC has been designed to:

- Increase engine throughput/productivity
- Reduce engine reject rate
- Reduce cost of overhaul and test cell qualification time
- Allow maximum use of existing ALC rework facilities
- Increase availability of engines for flight service
- Increase surge capabilities
- Reduce field rejects of replacement modules
- Produce a higher quality engine
- Better utilize experienced personnel

Other Advantages

Major components (disks, spacers, hubs, shafts) are inspected using a laser coordinate measuring machine to determine component compliance with technical order requirements.

*Permission granted by Mechanical Technology, Inc.

Computer analysis of the measurements permits the identification of the proper combination of components and their orientation relative to each other. This process ensures minimal deviation from the assembled center line, and as a result, minimal rotor vibration.

HOW MOBILE QUALITY WORK CELLS OPERATE ON THE PRODUCTION FLOOR*

The purpose of a mobile quality work cell (QWC) is to get quality control "at the source" on the production floor. A quality work center of this type can be set up to function with most brands of electronic gauges, and to provide direct gauge input for SPC with real-time corrective action. The QWC is designed specifically for multiple cavity/index SPC applications. A typical mobile unit can accept input from over thirty-two different gauge ports, and the CRT screen display proves full control chart displays (figure 9–24).

One of the most complete mobile quality work cells has been designed and manufactured by LDJ Electronics, Inc. This QWC is a mobile, industrial-grade quality workstation. The station includes a work surface to hold the gauges, CRT, electronic communications, electronic protection circuitry, and gauge interfaces (figure 9–25).

Parts and process specifications are downloaded from the host computer. Measured data prompts are displayed on the CRT and data is sent directly to the host via a push button or foot switches. Statistical results are sent back to the CRT in seconds, allowing the operator to make real-time adjustment on the process.

The QWC works in conjunction with the LDJ® monitor real-time statistical process control and SPC preprocessor systems. Using the SPC preprocessor, specifications and quality

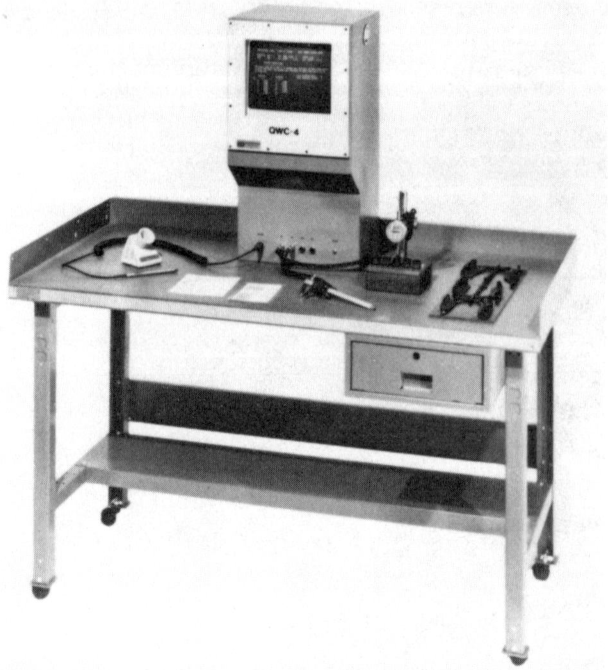

FIGURE 9–24 Mobil Quality Work Cell (Courtesy of LDJ Computer, Inc. and Systems Specialist, Inc.).

*Permission granted by LDJ Computer, Inc. and Systems Specialist, Inc.

FIGURE 9–25 LDJ Monitor—Production Monitoring (Courtesy of LDJ Computer, Inc. and Systems Specialist, Inc.).

control data can be downloaded directly into the QWC. The data base contains the preprocessor setup instructions and specifications for the jobs. The standard QWC supports four gauges, measuring eight different characteristics. By daisy chaining, the QWC-12 can support twelve gauges and ninety-six different characteristics. The QWC supports optional radio frequency communications, which eliminates the need to be hard wired on the factory floor, making the system truly portable.

Bar code entry is supported allowing the operator to bar code part numbers, employee numbers, job numbers, QC characteristics, and others. Standard templates can be made and attached to the QWC, making data entry easy for the operator. The QWC is designed to be located at the machine, making "quality at the source," without additional labor a reality.

HOW TO MOVE QUALITY CONTROL TO THE SHOP FLOOR WITH REAL-TIME MONITORING*

In its most basic form, real-time monitoring consists of quality information being disseminated from the production floor through the use of gauges on machine tools, or data captured on charts.

Gauges provide no permanent record. Although charts provide a permanent record, they do not offer a user an ability to obtain an instantaneous reading of what is occurring. In either case, neither of these in-process inspection methods provides machine-readable output for statistical analysis or contains process control capabilities. Neither gauges nor charts can initiate prompt remedial or corrective action.

*Permission granted by LDJ Computer, Inc. and Systems Specialist, Inc. as well as Penn Video, a subsidiary of Ball Corporation.

Programmable controllers are a further advancement in the evolutionary process leading to real-time monitoring. Programmable controllers (PCs) are attached to individual pieces of equipment, take readings, perform simple comparisons and computations, and initiate corrective action such as changing a temperature or closing a valve. Although PCs can perform various functions at a very high speed, their computational and communications abilities are very limited.

Programmable controllers also are called *event status controllers,* which is more descriptive of their function.

Objective of Real-Time Monitoring

The primary objective of real-time monitoring is to get the quality control activity out of the QC laboratory and on the shop floor where it can monitor conditions as they occur. As we all know from experience, it is very costly to identify problems after bad parts have been produced or unnecessary down time has been incurred.

Real-time monitoring is concerned with:

- Detection of "off," "pause," and other machine status
- Monitoring inputs of raw materials and process variables
- Monitoring outputs
  - (a) Cycle counts, cycle times
  - (b) Quality of products as they are produced
- Sensing out-of-control conditions and then either:
  - (a) Reporting conditions in real time, or
  - (b) Initiating corrective action (sending signals to control devices)

Real-time monitoring is a very important tool to improve quality, reduce scrap, and eliminate unnecessary down time. With real-time monitoring, information can be disseminated on process conditions in a rapid manner to whomever is in a position to take effective action. In addition, real-time monitoring provides an ability to accumulate history and a set of statistical analysis tools to analyze what is occurring.

Example of Cost-Effective Real-Time Machine Monitoring

Every plant manager envisions a factory that integrates the production control department with a real-time machine and process monitoring system that identifies the exact status of labor efficiency, raw materials, finished inventory, shift production, machine idle time, parts cost, and so on. However, all plants have unique features. Therefore, a factory monitoring system must be flexible to meet the diversified needs of the factory automation marketplace.

The basic needs of manufacturers for a factory monitoring system are:

1. Production and scrap counts
2. Machine utilization and monitoring
3. Preventive maintenance and tooling maintenance
4. Material usage and materials requirements planning

5. Employee time and activity reporting
6. Electronic release and scheduling
7. Capacity planning
8. Production scheduling

An excellent method, developed by LDJ, Inc., to meet this need is the LDJ Monitor™ system, which provides a flexible approach to factory monitoring by using a super microcomputer host in conjunction with an event/status controller on the shop floor.

The host system is able to handle a large number of transactions at high speeds by incorporating an interactive data base. The data base allows data manipulation, report generation, and real-time displays of the large amount of data flowing into the host system.

A factory monitoring system must integrate three main factory automation elements to increase productivity, bottom line profits, and cash flow. The first element is a *real-time display*—a continuously updated color terminal showing machine/part status, completion time, labor and machine efficiency, and so on. It is a window to the factory floor for management. The second element is a *real-time statistical analysis* of processes and parts quality to reduce the production of defective parts. The third element is an *integrated preventive maintenance system* which reduces machine and employee idle time.

The event/status controller is able to monitor and react to various digital and analog inputs for a wide variety of different kinds of machinery and processes. This machinery could include such things as injection molders, blow molders, grinders, screw machines, stamping machines, cold headers, assembly lines, and so on.

Analyzing the cost effectiveness of a factory monitoring system requires a look at specific machines and the savings associated with the various cost elements.

On the following page is an example of a worksheet for identifying time savings generated by a factory monitoring system. Time savings are obtained by reducing the production of defective parts and reducing nonproductive time—that is, idle or machine down time. This example describes a plastic injection molding company and covers their three major molding departments.

For the bushing department, 11.3 minutes are lost through producing defective material, excessive setup time, unplanned tool repair, mold and die failure, and excessive die setup time.

Following the bushing department, Group 1 savings result from reducing defective part production by 2.3 minutes per shift. As shown below, the average dollar value of each part is $.03, and 3600 parts are produced per hour. The calculations for twenty machines in the bushing department bring the total monthly savings to $5,382.00.

In the last calculation, as noted below, Group 2 savings result from reducing the nonproductive, or idle time of the machines. Again, referring to the bushing department, each machine had a burden rate of $25.00 per hour. Since the idle time of each machine is reduced by 11.3 minutes per shift, sixty-five shifts per month, the cost savings per month is $6,120.00.

The trim-mold and gear department savings are calculated in the same manner.

For all departments of this molding factory, the savings is $13,638.00 per month for reducing scrap, and $7,496.00 for reducing idle machine time. The total monthly savings is $21,134.00.

With the significant savings generated by increased quality productivity and decreased down time, paybacks of a few months are not uncommon. With minimum capital

Nonproductive Time Savings

Downtime or Idle Reason	Department: Bushing — Minutes saved per machine per shift	Department: Trim Mold — Minutes saved per machine per shift	Department: Gear — Minutes saved per machine per shift
External			
Operator error			
Operator absence			
Quality control hold			
Producing defective material	2.3	1.8	2.6
Out of raw material			
Wrong material (too much/too little)			
Excessive external setup time	2.0	1.0	2.0
Locate tooling		1.0	
Setup instructions unclear, missing			
Preventive Maintenance			
Locate plant engineer			
Identify/locate repair parts			
Wrong repair parts			
Missing repair parts			
Reduce unplanned tool repair	3.0	1.5	
Reduce unplanned machine repair		0.5	
P.M. procedure unclear, missing			
Machinery Related			
Hydraulic			
Pneumatic			
Limit switch			
Bearing failure			
Gear/tool failure			
Mold/die failure	3.0	2.0	4.0
Transfer/feed problem			
Excessive setup time	1.0	1.0	1.0
Poor setup			
Other			
______________			
______________			
Total	11.3	8.8	9.6

Group 1 Savings Reducing Defective Part Production

		Department		
Molding Department		**Bushing**	**Trim Mold**	**Gear**
A	Dollar value	$ .03	$ .20	$ 1.00
B	Parts produced per hour	3600	3000	692
C	Minutes producing defective parts	2.3	1.8	2.6
D	Defective parts/OCC $= \frac{B \times C}{60}$	138	90	30
E	Occurrences per month	65	43	43
F	Savings per month (F = A × D × E)	$269	$774	$1290
G	Number of machines	20	9	1
	Total savings per month for all machines (H = F × G)	$5382	$6966	$1290

Group 2 Savings Reducing Nonproductive Time

		Department		
Work Center/Part Number		**Bushing**	**Trim Mold**	**Gear**
A	Number of machines	20	9	1
B	Burden rate/machine	$25	$20	$35
C	Minutes per shift per month	11.3	8.8	9.6
D	Number of shifts	65	43	43
E	Hours per month per machine $= \frac{C \times D}{60}$	12.24	6.30	6.88
	Savings per month (F = A × B × E)	$6120	$1135	$241

investments, a monitor system can be leased and the analysis of most plants will show a substantial positive cash flow. In this example of the plastic injection molding factory, the lease payment is approximately $4,174.00 per month leaving a positive cash flow of $16,960.00 per month and a yearly cash generation of $203,520.00.

TYPES OF INSPECTION EQUIPMENT AND TECHNOLOGIES*

There have been many new types of inspection equipment and technologies introduced into the manufacturing environment. Some of these innovative advances are discussed in the next sections.

*Permission granted by Grumman Corporation.

Ultrasonics Detect Problems in Critical Process

Automatic bridge ultrasonic scanners are used by inspection for checking critical processes such as electron beam weldments for defects as small as 0.003″. These ultrasonic testers also can handle raw material slabs up to 55 feet long and weighing as much as 8000 pounds, while rapidly finding more typically sized defects in the 0.050″ range.

Flawless electron beam welds are as strong as the original material. Nondestructive evaluation is an important part of the manufacturing process. To verify what cannot be inspected visually, critical electron beam weldments are ultrasonically scanned at several depths simultaneously using a dual manipulator ultrasonic bridge scanner.

Radiography: Another Technique for Finding Internal Flaws

Radiography is useful in applications not suited to ultrasonics, such as castings. It also is an extra test for critical processes like electron beam welding. It even shows internal parts of components for checking dimensions and performance.

A feature of high-resolution radiography is the use of range-finding laser beams to precisely line up the X-ray head and measure its distance from the part to within ten thousandths of an inch. Special collimators have been developed for on-site radiographic examination of welded or brazed joints. This device eliminates dangerous radiation scatter that previously required the clearing of production personnel from work areas. With this improvement one company was able to realize an 80 percent savings in labor power due to the elimination of the job clearing requirement.

Automated radiography is used for checking long weldments. It consists of an automatic X-ray and film reading system. A key technical feature is the use of a computer to automatically deblur the X-ray films. In some instances, these systems have reduced labor power requirements by as much as 60 percent.

Penetrant Inspection Locates Surface Flaws

Penetrant inspection is used to detect surface flaws under ultraviolet light.

Strategically set up close to machine shop areas, penetrant inspection facilities are able to detect flaws as small as 0.050″ reliably, even in parts 30 feet long.

Other Equipment Technologies

A part monitoring test technique called *acoustic emission* finds flaws by detecting the acoustic waves given off by propagating defects. Fluid components can be measured in parts per billion using mass spectrometry. Computerized spectographs can analyze the elements in metal stock and forgings. An atomic absorption spectrometer can analyze samples in solution with an instantaneous direct reading printout. Probes used in hot furnaces verify temperatures. Microphotography identifies oils and other lubricants.

Composites Inspection: A Typical Application Involving Several Technologies

Control of composites quality begins with the layup of the filament strips and continues with monitoring autoclave heatups and cure rates. A contour following ultrasonic squiter is used to inspect laminates and bonded assemblies. Once the initial path is traced, this device—state-of-the-art in industry—reads out a continuous computerized scan of the part.

In addition, tangential filmless X-radiography is used to detect voids in composites radii that cannot be tested ultrasonically.

CASE-IN-POINT

REFLEX INDUSTRIAL AUTOMATION SYSTEMS, LTD.
*AMERICAN CIMFLEX, INC.**

Situation

American Cimflex, working with its United Kingdom licenses, Reflex Industrial Automation Systems, Ltd., decided that a need existed for a new swagging application that would reduce breakage and lower the proportion of failing tests encountered in quality control.

Solution

In the heat exchanger production process, once a laced tube stack has been positioned and the robot cycle initiated, the operator is free to lace the next stack. Meanwhile, the cell is working concurrently with the lacing phase of the production. By demanning the swagging task, a robotic cell has been created that improves repeatability and consistency.

The process is designed so that the robot controller monitors and controls the complete work cell. For example, during the roller expanding process the mandral is withdrawn only when a preset maximum torque value is reached. If the maximum torque is not achieved before the maximum preset penetration is reached, then the tool is withdrawn and the actual maximum is compared with a minimum preset torque. If the minimum is not reached, the system sets off an alarm, and the robot stops and points at the appropriate tube. The alarm condition may indicate either a broken or worn tool, which the operator can then correct. Because of the large number of analog and digital input/output channels available on the robot controller, a separate programmable controller is not required.

Results

By robotizing this application it became possible to lower the amount of re-roller expanding and re-testing that was normally required. Over expansion and consequent end-plate growth also have been reduced or eliminated, resulting in cost reduction benefits to customers employing this approach. The MERLIN systems developed by American Cimflex can create complete work cells, as was illustrated in this swagging application.

FIVE TYPES OF ADVANCED APPLICATIONS OF AUTOMATION IN THE CALIBRATION LABORATORY**

As industry accelerates its introduction of advanced manufacturing technology into the production environment, there is a corresponding need to also install automatic test equipment. This necessitates establishing a complete calibration laboratory capable of measuring, inspecting, and testing new inspection devices. In addition to its primary activities, the calibration laboratory also must exercise cost control through developing new methods of calibration, new calibration systems, and new standards of higher accuracy and broader ranges. By implementing these cost control techniques, a company should be able to obtain significant productivity gains, while also upgrading its quality levels.

A company's approach to calibration automation can be guided by these basic rules:

*Permission granted by American Cimflex.
**Permission granted by Grumman Corporation.

1. Automate where it is an economic or technologic advantage.
2. Utilize "turnkey" systems wherever possible.
3. Work closely with system manufacturers.
4. Maintain diligent technical oversight and control over systems operation to assure a high quality level.
5. Innovate and be creative—but be practical.

There are many calibration automation operating systems presently in use for calibrating the following:

- Pressure transducers
- Thermocouples
- Digital and analog meters
- Voltage calibrators
- Current calibrators
- Resistance decades
- Resistance bridges
- Potentiometers
- Temperature indicators
- Current shunts
- Voltage dividers
- Amplifiers
- Oscillators
- Function generators
- Pulse generators

In addition, further advances in automated calibration systems include the types of instruments discussed in the next few sections.

How the Surface Plate Calibration System Detects Costly Measurement Errors

Calibration of surface plate has always been a challenge to people in quality control. Most are calibrated on site due to their size and weight. Instead of conventional methods, an automated technique can be employed utilizing a laser interferometer interfaced to a minicomputer. Thus, data can be acquired and analyzed in a real-time mode. Costly measurement errors are noticed quickly and corrected promptly, thus greatly reducing the need for repeat calibrations. Data plots are provided both in numeric and contour form for the surface plate.

Laser interferometer measuring techniques also can be utilized for calibration of precision linear devices such as the 100″ Schwein barometer. Other advanced systems applications consist of automated laser calibration to multi-axes machine tools and measuring machines.

Load Cell Calibration Systems Test Static and Fatigue

Strain gauge-type load cells are used for static and fatigue testing. A systems configuration consisting of automating the calibration of load cells prompts the system operator to perform a specified test sequence while the system is automatically acquiring and processing the data.

The computer directs the operator to select a standard of comparable range and sensitivity to the test article and to mount the units in the test fixture. A force range of 0 to 10,000 pounds can be applied in either tension or compression mode. A series of zero load measurements and resistance calibration equivalent value (RCEV) tests are performed automatically to assure that the circuits are electrically operational. By responding to "touch screen" commands, the operator exercises the transducers to prepare them for calibration and then performs two calibration runs from zero to full scale and back to zero. At each test point, the system samples the excitation voltage and output voltage of the transducers. A least squares solution is performed on this information, which provides the sensitivity, deviation, repeatability, and hysteresis data necessary for the user of the transducer.

Capabilities of Automated Microwave Calibration System

This system is used for calibration of microwave components and active microwave devices. The system provides the following measurement capabilities:

- Attenuation
- VSWR
- Frequency
- Power
- Modulation

The system consists of a receiver and signal generator. Each frequency is controlled by a synthesizer. A modulation meter, power meter, and frequency counter provide the modulation, power, and frequency measurement necessary for signal generation calibration. For component calibration, a set of bridges is employed for VSWR measurement.

Companies using this system have been able to reduce calibration times up to 90 percent for broadband microwave components as well as reducing signal generator calibration times by over 50 percent.

Where to Apply Automated Microwave Power Sensor Calibration Systems

This system is used for calibration of power sensors. Power sensors requiring power meters having an analog voltage output also can be calibrated with this system. It also has the capability to perform calibration of thermistor, bolometer, and thermocouple-type sensors. Some companies have reduced calibration time by up to 95 percent using this system.

Automation of Low-Value Resistance Calibration Provides Quick, Accurate Comparisons

Certain calibration techniques are used for basic electrical standards involving the calibration of standard resistors, shunts, thermal converters, and standard cells. This system provides a quick, accurate means of comparison between NBS calibrated reference standards and units submitted for test. For example, a low-resistance comparison configuration resolution of 0.01 parts per million is achieved with uncertainties shown to be approximately 0.5 part per million at the Thomas one-ohm level.

Applying the Automated Oscilloscope Calibration System

Oscilloscopes and their related plug-ins can represent the largest group of instruments in the workload of an electrical calibration laboratory. Using this instrument, a systems operator inputs and acquires data concerned with oscilloscope calibrations.

In summary, automated calibration systems have been developed for many test inspection tasks. However, the challenge remains to introduce automated calibration methods for many nonautomated applications encompassing an assortment of small inspection tools such as micrometers, calipers, thread gauges, hole gages, and so on. It is anticipated that the use of such instruments will decrease due to the emphasis on quality assurance methods built in as integral parts of automated manufacturing and assembly processes discussed in other chapters, rather than traditional inspection methods of products. Despite this trend, a large volume of these tools will probably remain in the calibration workload for the foreseeable future. Some of these not calibrated by automated methods will, in all likelihood, be picked up as existing systems and upgraded in performance.

IDEAS FOR APPLYING COMPUTERIZED QUALITY ASSURANCE SYSTEMS*

As more computerized quality assurance systems reach the marketplace, companies are beginning to achieve significant benefits, advantages, and cost savings for these expenditures. Such as the following:

Benefits of Using an Integrated Dimensional Metrology Analysis System

The integrated dimensional metrology analysis system (IMS) is used to reduce machine tool down time, improve machine tool productivity, and achieve higher quality.

This system is designed to greatly speed up the calibration and troubleshooting of machine tools. It is capable of gathering, analyzing, plotting, and storing machine calibration data provided by a laser measurement system software package.

It includes a CRT display which allows both alphanumeric and graphic output of measurement results. A built-in thermal printer provides hardcopy for permanent record keeping. Calibration data can be stored on small magnetic tape cartridges for later recall, analysis, comparison, or plotting.

IMS is capable of gathering and analyzing all of the data commonly needed for complete machine–tool calibration. This includes linear position error, pitch and yaw error, straightness, squareness, parallelism, and flatness.

Purpose of IMS

IMS provides all inspection requirements that exist in the production process. This includes inspection of:

- Raw materials
- Manufactured product
  - (a) Preprocess
  - (b) Work in process
  - (c) Post process

*Permission granted by Grumman Corporation.

- Production process parameters
  (a) Tools
  (b) Fixtures
  (c) Machine tools
- Verification/calibration of the inspection process parameters
  (a) Inspection fixtures
  (b) Inspection gauges
  (c) Inspection equipment

Post Process Inspection versus In-Process Inspection

Today, companies are becoming more aware that most metrology labor-hours are being spent on the post process inspection of product. This is because companies tend to establish quality control and dimensional inspection activities on post processing which seeks to find defective parts *after* a product run. This approach is costly and ineffective since it results in an inordinate generation of scrap and reworked materials. Instead, there is a strong movement in the industry to prevent defects through improved control of the product in the work-in-process stage, and as with the Taguchi Method—to make the product right the first time—through up-stream improvements at the design stage, as well as in the manufacturing engineering stage. In the latter stage, through improved in-process inspection techniques and dimensional control of the tooling, fixtures, and machine tools, greater efficiencies will accrue as well. The emphasis of an integrated metrology system is toward prevention with the inclusion of the dimensional verification of the product process parameters as well as the inspection of products. Much of this advancement can be traced to new technologies such as computer-controlled coordinate measuring machines, optical techniques, electronic-gauging systems, laser systems, photogrammetry, and fiber-optic systems.

Functions of an Integrated Metrology System

An integrated metrology system consists of a number of functions that are highly interrelated and dependent on one another. These functions fall into three major areas, which are considered to be subsystems of the integrated metrology system. They include the following:

- Inspection of manufacturing process parameters, including inspection of production tools, fixtures, and processes
- Verification of inspection process parameters, including inspection of inspection tools, fixtures, and processes
- Inspection of product, including parts, subassemblies, and completed products

Integrating IMS with Major Functions of the Manufacturing Process

Each of the major subsystems of the integrated metrology system support and are integrated with major functions of the manufacturing process. These include:

- The manufacturing planning and build function—defining all parameters dealing with the manufacturing process
- The QC/inspection planning and build function—defining all parameters that support the quality assurance function
- The machining and fabrication of product function—the actual process of making the finished product

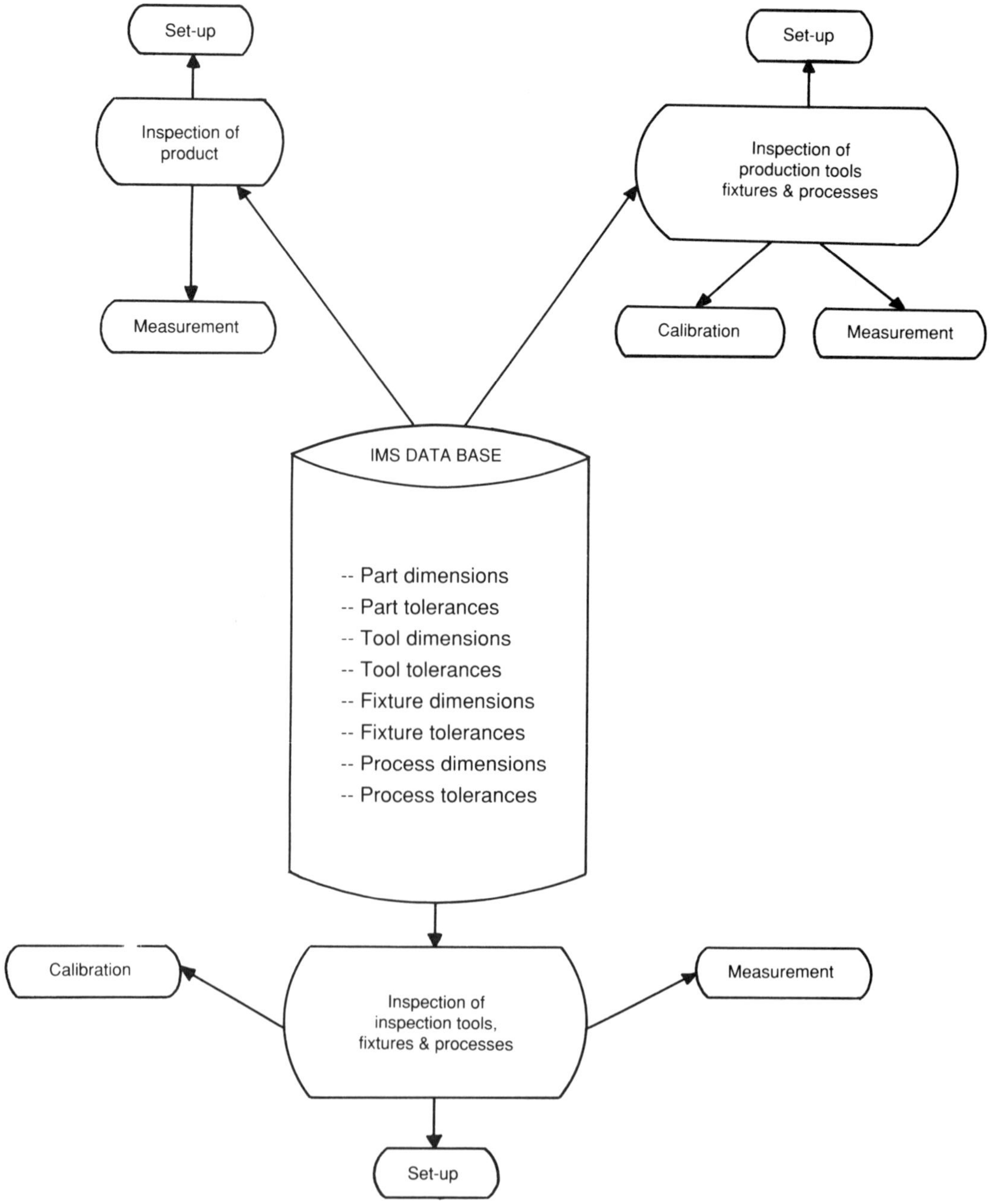

FIGURE 9–26 Integrated Metrology System (Courtesy of Grumman Corporation).

Each of these major subsystems is interconnected through an information flow and communications exchange between all functional activities.

Within each subsystem, communication and information flow are needed to control the individual metrology activities. This involves the storage and access to information that is needed for each functional activity. The data link that ties all of this data required for the integrated system is shown in figure 9–26.

In summary, the key to the development of an integrated information flow and communications network is its ability to be compatible with the development of CIM (refer to Chapter 11). To accomplish this task, it is necessary to marry computer-based systems with manufacturing-related hardware and software. Once all of these activities are interfaced, the IMS can become an integral component of the "factory of the future."

Applying Computer-Aided Tolerance Analysis*

Tolerances are fundamental to manufacturing. The choice of design tolerances affects the build-ability of the product. The selection of manufacturing processes and the sequence of processes affects tolerance stacking. Tolerance stacking leads to scrap and rework thereby increasing the cost of the product. A computer-aided tolerance analysis (CATA) system addresses this problem.

The effect of tolerances is so important that there are three major groups in a manufacturing organization that are concerned with tolerances. These are design, process engineering, and quality control.

The Importance of Tolerances

Modern manufacturing of a multicomponent product is based on the concept of interchangeable parts. Interchangeable parts are made possible by recognizing the inherent variation from the nominal dimension caused by the manufacturing process. By allowing a tolerance about the nominal dimension, the variation can be planned into the design of the product thus allowing the production economies of the interchangeable parts concept. Tolerances are fundamental to modern product design and manufacture.

As a general rule, the tighter the tolerance, the more difficult. Hence, the part is more expensive to make. The expense is incurred by either requiring precision equipment that can manufacture the part to the desired tolerance, and/or by adding more operations to the manufacture of the part to alleviate tolerance buildup created by the previous operations. Tolerance buildup is caused by the natural additive affect of tolerances.

Additive Property of Tolerance Buildup

Tolerance buildup also affects the assembly of a product. Tolerance buildup may make assembly of the product impossible such as where a subassembly will not fit into the allocated space. Tolerance buildup results in parts not being manufactured to blueprint specifications, therefore it is a major concern in manufacturing.

Manufacturing and Tolerances

As previously stated, the three separate groups concerned with tolerances are: product design, process engineering, and quality control. It is the responsibility of product design to assign tolerances to the product which ensure that the functional requirements for the product will be met.

It is the responsibility of the process engineer to determine the processes and the process sequence required to create the part from raw material. The process engineer must select the most economical operations available to create the part and at the same time be concerned with the inherent tolerance buildup due to routing a part through a multi-operation sequence.

Finally, it is the responsibility of quality control to monitor the progress of the part created by the chosen processes. Quality control is determining if the parts are being made to blueprint specifications. A major concern is detecting whether or not the parts are being

*Reprinted courtesy of the Society of Manufacturing Engineers. Copyright 1984, from Autofact 6 Conference Proceedings.

made within tolerance. If the parts are being made out of tolerance, it is the responsibility of quality control to determine the cause of the deviation.

Tolerances play a major role in manufacturing. To control tolerances, the three manufacturing groups that deal with tolerances need a method to analyze tolerances and tolerance buildup. Such a tool is CATA.

Understanding Computer-Aided Tolerance Analysis

CATA is a method of analyzing the effects of tolerance buildup due to the manufacturing process. It is based on the concept of "tolerance charts." Tolerance charts are a manual graphical method of presenting the dimensions of a workpiece or assembly at all stages of its manufacture. Its advantages are many:

1. Aids in developing the proper manufacturing sequence and working tolerances for each operation
2. Ensures that the end product will have the desired dimensions and tolerances
3. Is an instrument of communication when manufacturing specifications cannot be met economically
4. Determines if a part can be made to specifications by the available machinery, given that the accuracy of the machine is known
5. Reduces the number of dimensional and routing errors likely to occur in processing complex parts
6. Examines alternate methods of dimensioning the part for process purposes and examines alternate process sequences
7. Guarantees sufficient stock for removal for each operation
8. Reduces scrap and rework
9. Reduces tooling, gauging, and inspection costs
10. Reduces the amount of selective assembly

CATA enhances the advantages of tolerance charts by removing the tedium—the learning and manual calculations required by the tolerance chart technique—thereby minimizing the time required for effective tolerance analysis and control.

Scope and Overview of CATA

CATA was developed under the premise that tolerance analysis and control is influential through the entire scope of manufacturing. Since CAM has been introduced mostly in larger organizations, CATA has been developed for the small shop that does not desire to invest in expensive CAD/CAM software. CATA has stand-alone capabilities including a limited CAD package for data input. However, the analysis portion was developed independently of a CAD system, thus allowing piecewise integration into existing CAD/CAM systems.

CATA software can be described in terms of input, procedures, and outputs. Following are inputs to the procedure:

- Part drawing
- Raw material drawing
- Rough routing information
- Blueprint and raw material dimensions and tolerances

- Stock removal amounts for each type of operation
- Minimum process tolerances

The first four inputs use computer graphics. The two remaining inputs are retrieved from a data base. The tolerance stacking procedure is "automatic" and does not require any user interaction. Its main tasks are:

- Calculate working dimensions from assigned stock removal and final dimensions
- Assign working dimensions and tolerances
- Calculate rough stock removal and tolerances
- Calculate resultant dimensions and tolerances
- Check data and adjust chart.

The CATA output is a plot of the tolerance chart, or a computer generated tolerance chart.

Applications of Data

Design. CATA can be used as a CAE function in design. Coupled with a CAD system, CATA will allow the designer to examine the feasibility of manufacturing the design.

Process Planning. CATA is most appropriately a process planning tool. The process planner still must determine the manufacturing operations and the sequence of operations necessary to create the part. CATA gives the process engineer a tool to aid in determining the most economic process. CATA is a tool for the process engineer to communicate with design and quality control. CATA also can be used as an educational tool for future process engineers.

Quality Control. CATA can be used to perform sensitivity analysis to determine the critical operations of the manufacturing sequence. Control charts can be set up for these operations. Also, in absence of the other sectors using CATA, the QC engineer can use CATA with the existing operation sequence and determine the problem operations.

A computer algorithm with CATA employing graphic techniques, was developed for the analysis of process tolerances. Given part print information, raw material information, and rough routing information, analysis of the process tolerance buildup is completed. The results of the computer-aided tolerance analysis is presented in the form of a computer-generated tolerance chart. The manual graphic technique, *tolerance charts,* is the basis of CATA. Information is entered into the algorithm through the use of computer graphics.

Conducting Detail Parts Inspection with a Quality Information Management System

Companies are implementing automated quality information management systems (QIMS) to improve productivity, quality, and capabilities. The influx of automated manufacturing equipment and robotic cells has not only increased the production rate of parts, but also has reduced the dependence on operator skill to produce parts of consistent good quality.

QIMS provides automated data gathering and analysis of inspection results and manufacturing work center performance to determine, through its product acceptance logic, the part numbers that are to be modified, full, mandatory, or problem inspection candidates.

An automated system of this type will include these features: standardized inspection characteristics and dimensional data, defect codes, problem flags, part number and work order inspection history, and hardness test criteria.

The benefits that can be derived from this system are:

- Increased inspection department throughput (lots per hour) while maintaining or improving the standards of parts quality through the analysis of part number and work center inspection history.
- Requests for blueprints are minimized by automatic determination of the inspection level requiring prints.
- Computer generation of blueprint lists and lists for configurating inspection work modules.
- Providing quality management with timely automated reports and graphical trend data to monitor productivity, effectiveness, and quality from both the part and process point of view.
- Rapid identification and resolution of all significant problems, especially those associated with high speed/high volume manufacturing processes.

A typical QIMS begins when detail parts and their work orders are received from various manufacturing work centers for inspection. These data are then inputted on a CRT, and the system automatically generates blueprint lists for in-process inspectors, or provides inspectors with dimensional data on visual display screens.

> ***Key Point:*** The system can be streamlined by providing selective inspection data for parts that have a low defect rate, or manufactured from work centers with low defect rates and have been accepted on their most recent work order releases. In these instances, the part number need only be inspected for certain characteristics rather than requiring complete blueprint dimensional inspection.

Next, the inspector checks the parts' history, visually inspects the part, and records the results of the inspection via CRT entry. This enables the system to compare inspection results against part specifications, and spot any unusual product problems. During this procedure, inspection data is reviewed continuously and various reports are created, or updated screen displays are generated by inspector, work order number, work center error rate, and part number history. This information is then reviewed by inspection management which evaluates productivity and work center performance.

How Grumman Aerospace Corporation Conducts Quality Operations Planning*

The Grumman Aerospace Corporation engages in extensive activities involved in quality operations planning (QOP). These activities involve the application of various inspection techniques at the production floor level. The use of QOP in various production/inspection related situations are described below.

Inspection Planning for Detail Parts on the Grumman Aerospace Corporation Partsmaker Machine Tool

After the manufacturing–engineering department has selected which parts are to be fabricated on the partsmaker, an NC tape is prepared for the part. QOP determines whether the engineering drawing is fully dimensional or lofted. It then creates a mathematical model

*Permission granted by Grumman Corporation.

of the part geometry on the quality control CAD/CAM file by using the engineering drawing and supplemented data shown on the methods tool order.

Lofted Drawings. On lofted drawings, the CAD software digitizes the part. Upon completion of the graphic drawing, the model of the part geometry is transferred to the CAD/CAM release file. Data stored in this file is used to create a tolerance inspection mylar overlay (TIMO) model in the CAD/CAM file. Upon completion of the TIMO model, QOP transfers the model to the partsmaker release file.

Manufacturing accesses the QOP partsmaker release file and schedules the part to be plotted. Manufacturing plots the QOP partsmaker TIMO and marries it to the work order for release to production.

Machine Part Inspection Planning for NC Inspection Tapes and Mylar Overlays for In-Process Inspection

After QOP determines the NC inspection technique to use, NC inspection tapes are generated to support the machined part first-piece inspection effort when tolerances are greater than or equal to a specific figure. An NC program is then developed to create the inspection probe motion for verification of part contour and characteristics/features. The program is part of the overall inspection package and is used to drive a milling machine with an indicator.

Inspection Mylar Overlay

The inspection mylar overlays (IMOs) are generated to facilitate the visual inspection of a machined part for first-piece and recurring inspection effort. The overlays locate the periphery contours, breakouts, thicknesses, and so on.

Quality inspection-planning numerical-control planners access controlled data bases to design customized quality mylars to facilitate the inspection process. A discrepant material report is prepared if a problem surfaces in the production process.

The NC inspection tapes are generated in CAD/CAM by applying probe motion to an engineering/manufacturing model through the selection of significant geometry/part features. At times, NC tapes are generated using conventional, automatically programmed tool statements. In these instances, programmed output is in the form of either a punch tape or a program residing on the DNC system.

Mathematical models, either existing or created in CAD/CAM are modified, and additional inspection criteria are incorporated. After completion of models, an IMO is created.

The quality operations planning NC inspection tapes are used for first-piece machine part inspection. They are used to "qualify/validate" the manufacturing NC tape. Once approved, the manufacturing NC tape is accepted by quality control and stored in a "controlled" library on a data base, such as DNC.

Prove-out of Numerically Controlled Machine Tapes

To prove out numerically controlled machine tapes prior to production runs, computerized three-axes measuring machines programmed by QOP personnel assures quality by providing instant and accurate read-out using a Hansford coordinate measuring machine. Thus, the engineering and manufacturing system is comprised of three interrelated components: CAD, CAM, and CAI.

An advanced Zeis/Mauser computer-controlled coordinate measuring machine with a range of 200 by 80 inches and a 3D digitizing capability is used for prove-out or qualification of NC machine tapes, tool inspection, machine tool qualification and calibration, and machined part inspection. This machine replaces inspection checking fixtures, supports distributed numerical-control activities and provides fast, real-time measurement and analysis capability.

CASE-IN-POINT

COMPUTERIZED QUALITY ASSURANCE SYSTEM
*CONTINENTAL SYSTEMS TECHNOLOGY, INC.**

As industry moves toward greater involvement in automated inspection applications on the factory floor, companies are developing advanced computerized quality assurance systems. This case-in-point illustration contains all the basic components of this type of system.

Purpose of the Quality Assurance Systems

Quality assurance systems automatically capture raw test data from quality workstations, move it from the plant computer into a data structure, and generate inquiries, reports, and statistical and control charts for user-selectable test groupings and time periods. Raw data is maintained online for three months and then moved into computer-readable backup storage for an additional 21 months. Although regularly scheduled (daily, monthly) reports and charts for preplanned test groupings are required the *most important outputs* are for ad hoc user-specified reports and charts needed for investigation and resolution of quality problems. Some output reports include raw data values, but many others consist of statistical data (averages, control limits, rejects) derived by analysis of the raw data. Preplanned statistical data is maintained for use in correlation with raw data outputs and the wide variety of test groupings is inherent in the system. Since the ad hoc outputs involve any groupings of tests for any time period including both on-line and backup data, the system contains, as part of report or chart generation, inclusion of data-base management software for statistical analysis purposes.

Quality Assurance System Function

The quality assurance system consists of the following functions:

- Inputting and maintaining standards
- Inputting and maintaining inspection sequences
- Downloading of inspection sequence to quality workstations
- User interaction and data collection/validation at UIU microprocessor workstations
- Uploading of test raw data from workstations
- Correction of data and maintenance of backups
- Selective analysis and output of results to users

In developing this interrelated systems configuration, the intention is to provide plant personnel with sound workable approaches for handling quality assurance concerns. The following items highlight some of the beneficial attributes included in the systems design:

- Ease in obtaining desired exception reports
- User-defined data, visual inquiry display, and report formatting
- User-friendly commands
- Graphics for high visibility and trend evaluation

*Permission granted by Continental Systems Technology.

- Color where needed to convey priority or provide contrast
- Pre-programmed core systems for communication, data collection, and data management
- Central support and maintenance of core systems
- Ability to select data from various data bases to structure tailored reports
- Magnetic code to automatically capture basic personnel data
- Optical bar codes to automatically capture product data
- Integrated computer network between the host computer (downloading), and plant computer (uploading)

Standards Input and Maintenance. This function entails input and maintenance by quality supervision of standards to be employed in the analysis of test raw data to identify out-of-specification data or rejects.

Inspection Sequence Input and Maintenance. This function addresses preparation of inspection sequences by quality supervision to be used at the quality workstation to provide visual and aural prompting of inspectors, and control the collection and validation of test raw data. The sequences consist of command lines including necessary standards written in the UIU command sequence language (UIU-CSL) using the plant computer utility editor.

Performance of this function entails interaction with one or more quality workstations located at points in the plant where inspections must be performed. The quality workstation is a particular application of "smart gauges" and a proprietary microcomputer-driven universal interface unit (UIU) developed by Continental Systems Technology, which is employed to prompt inspectors and collect data from analog or digital gauging devices. The quality workstation and its relationship to the plant computer and the smart gauge, which performs a series of dimensional changes checks on cans or ends, is depicted in figure 9–27.

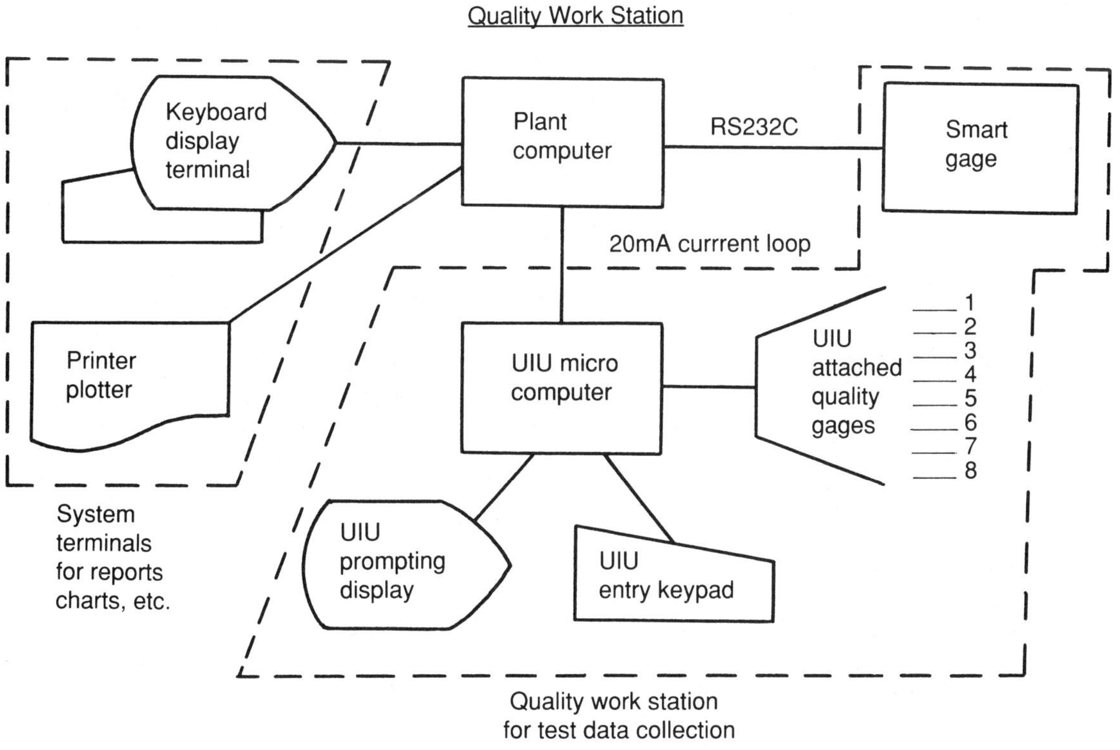

FIGURE 9–27 Quality Work Station for Test Data Collection (Courtesy of Continental Systems Technology).

Downloading of Inspection Sequences to Quality Workstations. This function addresses the downloading of CSL Files to the UIU for storage in the microcomputer RAM. A RAM resident interpretive program in the UIU reads and executes the sequences on request. The communications protocol, message formats, and command summary are contained in the following sections.

User Interaction and Data Collection/Validation at Quality Workstations Using UIU and Smart Gauges. This function entails use of the UIU or smart gauges by the quality inspector to collect raw test data. Although operation of the gauges is *not* part of the required work, this function is described to provide background information related to use of inspection sequences and test operations. The inspection process is as follows:

1. Start UIU or smart gauges and select an inspection sequence via key entry in response to menu or gauge prompting.
2. Following sequence prompts
   (a) Enter test result/reading via the UIU keypad, or
   (b) Position samples in the gauge selected in the inspection sequence and press switch to take readings.
3. Continue keypad entries and/or gauges inputs until all inputs for all samples in the tests required by the inspection sequence have been made.

Uploading of Test Raw Data from Workstations. This function addresses the uploading of data packets to the plant computer. Commands in the inspection sequence control the generation and sending of data packets to the plant computer to append date, time, and additional identity to the data packets received; strip away message framing; and move the test raw data record into the quality inspection data base, suitably keyed or indexed to permit retrieval by predefined or user-definable reporting functions. The data-base management software provides a user capability for defining and maintaining additional nontest data.

Correction of Data and Maintenance of Backups. This function entails correction of data errors by the QA supervisor using a plant computer terminal to enter corrections and mark test raw data records as corrected (for quality audit trail). At specified intervals, the QA supervisor backs up old data and purges it from the on-line file.

Selective Analysis and Output of Results to Users. This function addresses generation of reports, inquiries, and charts by quality assurance personnel. This function includes both the analysis required to derive statistical data from raw data and the generation of outputs required for quality assurance operations. In this functional view the analysis is considered to be part of the overall process of selection, retrieval, analysis, formatting of data, and presentation of output.

Although preplanned, periodical (weekly, monthly) outputs (reports, charts) are necessary, the essence of this function is end-user capability to define and generate ad hoc outputs on demand to address quality problems. *This is a requirement* and generation of all or most reports, displays, and charts with the end-user output function, if sufficiently responsive and efficient, is considered beneficial and an avenue for reducing report programming costs.

Example of a Total Hardware/Software System*

Components of a Factory Inspection Automation Systems

Due to advances in computer-based technology, the movement of materials throughout the production cycle can now be monitored and controlled by automated means. In the

*Permission granted by LDJ Computer, Inc. and Systems Specialist, Inc.

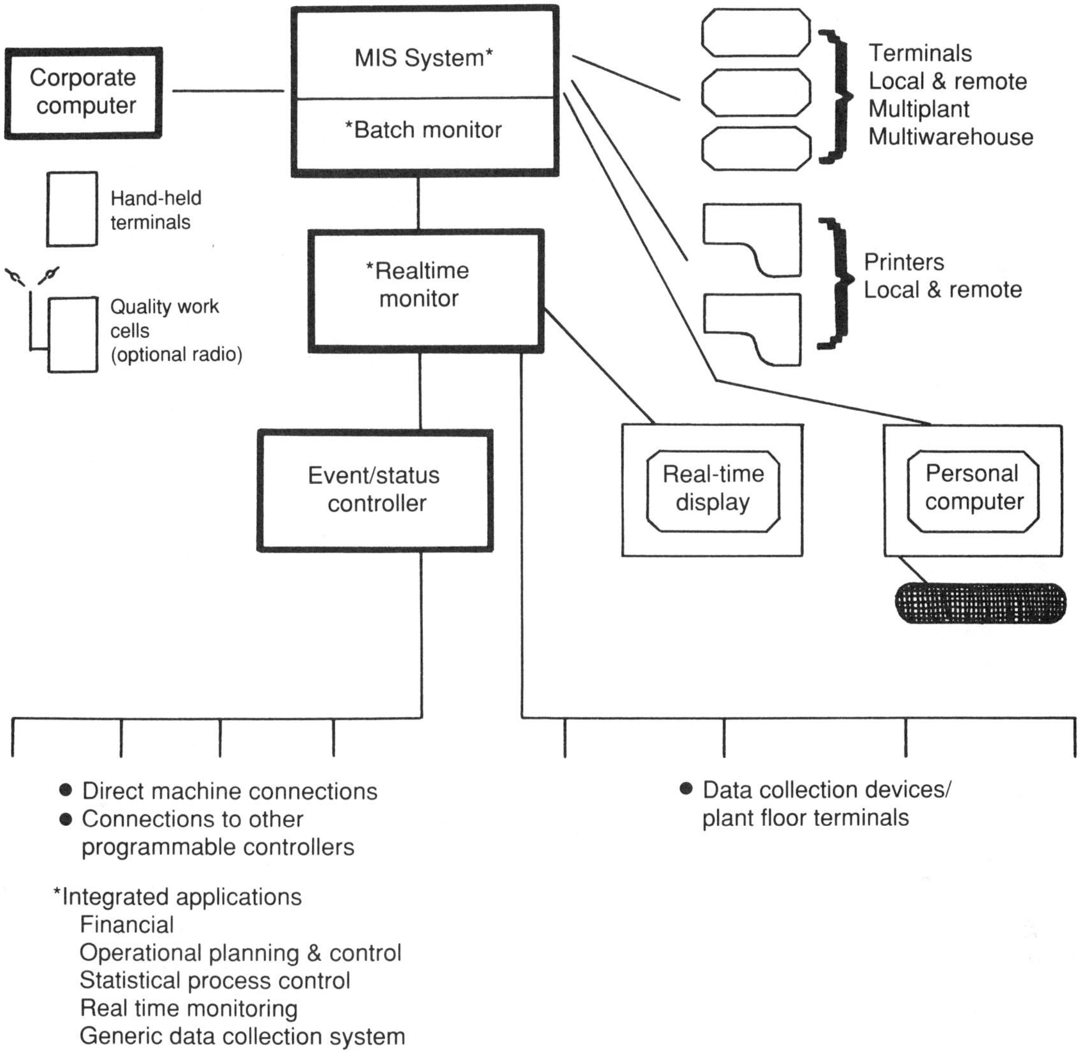

FIGURE 9–28 A Total Hardware/Software System Configuration (Courtesy of LDJ Computer, Inc. and Systems Specialist, Inc.).

case of inspection, a totally automated factory inspection system can be installed consisting of the following hardware/software configuration characteristics. (See figure 9-28.)

Event/Status Controller

A number of event/status controllers are located on the shop floor between various machine tools and the host computer system. Connected to these controllers, may be:

- Other programmable controllers
- Analog and digital sensors
- Contact closures

These controllers are function-limited industrial-grade computers. Their primary functions are to detect machine counts, process parameters, analyze those process parameters, and initiate rudimentary reactions to machine and process irregularities.

The event/status controller sends data and messages back to the real-time host system. In turn, the host system can send reaction commands back to the controller which relays this information to the machine tools.

Statistical Process Control System

Statistical process control functions are performed on the batch host computer. The host computer is integrated with other subsystems for inventory control, machine loading, and maintenance management.

This system stores historical performance information, performs extensive statistical analysis on this information, and generates numerous process control reports including recalculation of control limits.

The system also maintains a log of problems and solutions by machine and job number, which is referred to as a *corrective action system.*

This system also includes a real-time processing module to:

- Immediately flag any statistical violations to operator
- Prompt for corrective action (from operator)
- Display suggested correction action for the violation

These systems include a report writer for use by nonprogrammers.

Quality Work Cell and Statistical Process Control Preprocessor

The quality work cell is a portable unit which is taken to the work center. It contains the following:

- Mobile cart
- Industrial grade CRT
- Data entry pad
- Gauge interface
- Bar code entry
- Optional radio communication

The programmed logic that "drives" the QWC is provided by the SPC preprocessor software contained in the real-time host. It provides the QWC operator with a step-by-step tutorial in performing the QC tests. The system then performs data analysis. This software provides an interface between the QWC and the SPC monitor system on the batch host computer.

Data Collection and Portable Terminals

A variety of data collection terminals can be interfaced to the real-time host computer. These terminals can be customized as to input device (badge reader, key entry, bar code readers, and so on) and output devices (printers, displays). In addition, this system can accommodate hand-held terminals with bar code entry for use in such applications as inventory taking, recording material transfers, and the like. At present, some of the most advanced terminals can store up to 512K of data internally. Terminal substitutions are performed easily as their internal logic (program) is downloaded to them each time they are turned on.

chapter 10

Automated Assembly, Packaging, and Warehousing

PART A. AUTOMATED ASSEMBLY TECHNOLOGY

WHY ROBOTS ARE WELL SUITED FOR PRECISION MECHANICAL ASSEMBLY

Assembling intricate parts is a repetitive task to which robots are well suited. However, putting robots to work on precision assembly is a complex operation that depends on the successful application of several technologies. The key factors to be considered in developing automated assembly operations are:

1. The product must be designed with automated assembly in mind. Existing products may require a redesign to be assembled by the automated production equipment currently available.
2. Components for the product must be available at the right time and with the necessary quality. Automated assembly requires just-in-time management of components.
3. Design technology must be available to handle full automation of the production system.
4. Basic automation machinery should be developed if assembly robots are to be used.
5. Ancillary equipment for the production lines should be standardized. Setup and programming will be simplified by such standardization.
6. Component transport and orientation technology must be developed. Components must be in the right place and position at the right time.
7. End effectors must be designed to suit the performance of specific functions. The interaction between the manipulator and components in small part handling requires a high degree of precision.

8. Automatic inspection and monitoring of operation positions is necessary to ensure assembly progress.
9. Control technology for overall assembly automation is required.

All of these factors must be considered irrespective of the use of assembly robots. However, where assembly robots are used, the need to develop basic machinery is lessened while the development of a design technology for full automation becomes more important because the robots will require the flexibility to handle a variety of production tasks.

Why Speed Is the Major Reason for Using Robots in Assembly Operations

Almost 70 percent of all assembly operations involve simple repetition (pick-and-place) motions and do not require skilled operators. Therefore, the use of robots must be justified on the basis of price and performance: either robot price must be low or robot speed must be increased.

Lowering the cost of the robot itself is often impractical when the manufacturer is faced with customer requirements for improved performance, greater positioning accuracy, more sophisticated and flexible communications capabilities, faster changeovers, reduced idle time, and so on. In addition, ancillary equipment normally accounts for more than 50 percent of the total cost of a robot automation system. Therefore, lowering the cost of the robot itself will have a minimal effect at best.

Thus, the major justification for robot use in assembly operations is increased speed, which will maximize operational efficiency. However, defining and measuring robot speeds is not a simple matter.

It is not sufficient to consider "maximum tip velocity," which may be reached only rarely during actual production operations. A better criterion is the "cycle time" required to complete a specific operation. This provides the basis for a more meaningful comparison of how robots will perform in assembly operations.

In addressing this challenge, engineers should consider the following:

1. The angular speed of the manipulator arm is more crucial than its linear speed.

2. Faster acceleration and deceleration times will decrease cycle time significantly. The well designed production line minimizes robot arm motion. Therefore, acceleration and deceleration are more important than maximum velocity in increasing operation efficiency.

3. Shorter residual oscillation setting times allow the robot to carry out its operations on the product more rapidly. If oscillation time is significant, end-effector action will have to be delayed until the arm is stationary.

4. Faster external communication with the local or central computer ensures that the robot does not have to wait for its next instructions. Communications should be fast enough to reduce waiting times to zero.

5. Simultaneous horizontal and vertical movements should be executed whenever possible. For a typical pick-and-place operation, conventional robot movement consists of three consecutive modes of acceleration and deceleration, referred to as "gate" motion. Certain types of robots, however, avoid the longer cycle time inherent in gate motion with a single continuous "arch motion." With the "arch motion" there is a 20–25 percent improvement in cycle time.

Trends toward Improving Robot Accuracy and Repeatability*

As robots gain further acceptance in the marketplace, there is an increasing demand for improved positioning accuracy and repeatability. This is due not only to the requirements of the precise applications but also because there is a definite trend toward the total automation of assembly lines. When assembly robots are part of a production line that includes vision sensors, tactile sensors, component feeders, conveyor systems, and other complex ancillary equipment, repeatability and absolute accuracy assume increasing importance.

Thus, repeatability is a major factor in the selection of a precision assembly robot. For instance, in the case of such precision operations as watch assembly, it is desirable to have a repeatability of ±25 microns (±.001″) that can be maintained over a long period of time.

The Seiko Group Example: Using Robots for a Delicate Product

A prominent company in this field is the Seiko Group, which manufactures the famous Seiko watches. Within the Seiko Group, the Accusembler family of precision assembly robots has been developed in Japan by Suwa Seikosha for automated assembly of such delicate products as watches (figure 10–1) and a wide variety of small electromechanical components, and Epson, the other well known member of the Suwa Seikosha Epson family, for the automated assembly of printer heads (figure 10–2), floppy disk drive units, and other precision products.

FIGURE 10–1 Watch Face Assembly (Courtesy of Seiko Epson Corporation).

*Permission granted by Kanematsu-Gosho (U.S.A.) Inc.

FIGURE 10–2 Printer Head Wire Assembly (Courtesy of Seiko Epson Corporation).

WHY USING ROBOTS TO ASSEMBLE SURFACE-MOUNTED DEVICES MAKES SENSE

In 1986, about ten percent of all electronic components were surface mounted. By 1990, that percentage will grow to over 25 percent or more than 100 billion devices. It's expected that more than half of all electronic equipment will contain surface-mounted devices (SMDs) by 1990.

Initially, surface mounting was limited to certain passive components and integrated circuits. Now more switches, potentiometers, inductors, ICs, and LEDs are available in surface mounting packages. The trend is for that range of components to widen to include almost anything that can be mounted on a conventional PCB.

How SMDs are Placed on PCBs

The most critical machines of surface-mounted production lines are automatic placement systems with component feeders. There are two types of automatic placement machines: dedicated single-purpose and general-purpose.

The dedicated machines place components being used on large-scale production runs, such as boards for automotive electronics, with placement rates of 50,000 to 100,000 components per hour. Such single-purpose machines are very expensive to buy and set up and are often limited in the type of SMDs they can handle.

General-purpose placement machines, on the other hand, have placement rates of a few thousand units per hour but can accommodate a much wider range of components and can be fed by tape, magazine, or bulk. General-purpose machines can be adapted to handle circuit boards with both SMDs and through-the-hole components. They often can be reprogrammed more easily than dedicated single-purpose machines.

The Right Robots for Pick-and-Place Systems

Precision assembly robots, such as the Accusembler SSR-H and SSR-X series from Suwa Seikosha, are ideally suited for the automatic placement of SMDs. Robots offer both flexibility of application and ease of reprogramming for new PCBs or alternative components. Short-run batch assembly operations can be readily automated with savings in both time and manufacturing costs.

Accusembler precision assembly robots can either be programmed off-line, saving valuable production time, or can be led through the assembly sequences so the robot "learns" the required steps and actions. In addition, the robots may be linked to a CAD/CAM system for direct programming of the operating coordinates.

Using the Assembler in Surface Mounting

A typical surface mounting application of a conveyor rail PCB assembly line is shown in figures 10–3 and 10–4. Two Accusembler robots are used; one to position adhesive dots and solder paste and the other to attach a wide variety of electronic components.

At the left in figure 10–3 is an SSR-H253 robot, a three axes unit with a 250mm reach. The work envelope for this robot is shown in figure 10–4. The robot has two independent

FIGURE 10–3 Typical Surface Mounting Assembly Line with Two Accusembler Robots (Courtesy of Seiko Epson Corporation).

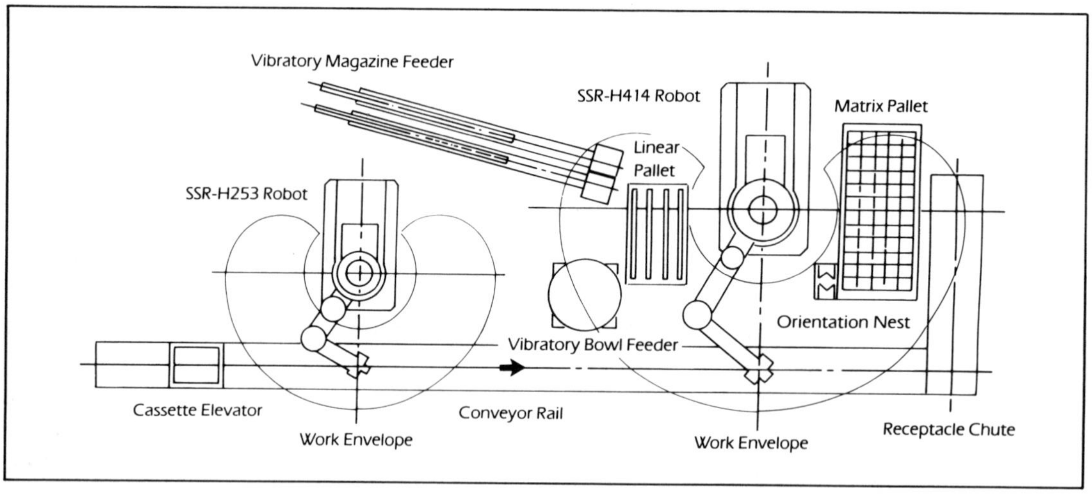

FIGURE 10-4 Diagram of Surface Mounting Assembly Line (Courtesy of Seiko Epson Corporation).

dispensers for adhesives and solder parts and a fixed effector for moving the printed circuit boards along the conveyor rail.

The robot on the right is a four-axes SSR-H414 unit with a 410mm reach. The SSR-H414 has an exceptionally large work envelope as in the figure. This robot has two vacuum chucks for picking up and positioning components and a fixed effector for moving the printed circuit boards.

Surrounding the robot are (from the left) a vibratory magazine feeder for DIP integrated circuits, a linear pallet for miniature molded transistors, and (to the right of the robot) a matrix pallet for flat-pack integrated circuits.

The use of such a large number of different types of component feeders is made possible by the SSR-H robot's unusually large work envelope. Flexible arm robots can pick up components from surrounding feeders, as compared with Cartesian robots, which are restricted to components inside the rectangular work envelope.

Both the vacuum chucks may be rotated to position the components correctly before mounting. For example, the larger vacuum chuck (foreground) picks flat-packs from the matrix pallet and then positions the IC in the orientation nest adjacent to the pallet. The correctly positioned IC is then placed on the printed circuit board.

Although not shown in figure 10–3, additional components could be handled with tape feeders. When component mounting is complete, the PCB is sent to either a receptacle chute or an infrared wave soldering oven.

ADVANTAGES OF THE SUWA SEIKOSHA PROGRAMMABLE EQUIPMENT LANGUAGE ROBOTIC PROGRAMMING LANGUAGES

As discussed in Chapter 6, programmability is a critical aspect of robot applications. Most robot manufacturers have developed unique languages that are optimized particularly to handle robot motions. Although this means that the various robot operation languages are not generally compatible, efforts are underway to develop a standard intermediate language to aid compatibility.

Suwa Seikosha has developed a particularly powerful programming language, Suwa Seikosha Programmable Equipment Language (SPEL) for the Accusembler robots and their ancillary equipment. Virtually no programming experience is necessary to use this language.

The SPEL language offers:

1. *Ease of Learning.* The syntax and structure of SPEL are similar to BASIC. Powerful macros permit major program changes with only a few key strokes.
2. *Applicable to Simple and Complex Tasks.* SPEL is easy to use for simple robot motions, but offers a range of instructions to cover the most complex manipulations.
3. *Flexible Applications.* Up to four different manipulation programs may be stored in the Accusembler robot's microcomputer, each selectable by simply setting a manual switch.
4. *On- or Off-line Programming.* The robots may be programmed at the production line with the direct input of coordinates.
5. *Communication Compatibility.* A parallel, 16- or 32-channel input and 16- or 32-channel output port interface enables external control.

Innovations to Watch: The Future of Robotic Precision Assembly

The application of robots to precision assembly is still in its infancy and many problems and issues still remain unsolved. As applications increase in complexity, several basic issues require consideration.

Sensors

Although there have been considerable advancements in vision sensor technology (see Chapters 6 and 9) it is widely agreed in the industry that the vision sensors must be reduced in price and have a better processing time before they will enjoy large-scale acceptance. Also, it is believed that dedicated, special-purpose vision sensors will probably be more cost effective than general purpose units.

Mobility

Demand for fully mobile robots is increasing, particularly in severe operating conditions such as nuclear reactor stations. In the precision assembly area, mobile robots could handle several tasks sequentially, thus reducing the number of robots required. For example, one robot might palletize components and then move to an assembly site to put the components together. The requirements for a mobile robot are considerably more complex than for fixed units. The cost of such robots will have to come down considerably for assembly operations.

Clean Room Compatibility

The semiconductor industry is interested in using robots for "clean room" assembly of large-scale integrated circuits and other components. However, robots require oil for their moving mechanical parts, which would violate many clean room requirements. True clean room application of robots will have to wait for better particle-free lubrication agents and rust protection agents.

Communications

In the future, precision assembly robots will frequently be part of an overall automated factory. The automated systems will be connected to a LAN for production control. The LAN will demand standardized interfacing and faster, more reliable communications capabilities. Therefore, the robotic systems will be developed to interface with LAN and be integrated into a local factory automation system.

An Example of a Flexible Automation Assembly Robot System*

In addition to the automated assembly robot systems discussed previously, advances in automation technology now make it possible to combine the proven performance of hard automation with robotic technology into a software controlled and managed assembly system. This flexible automation approach can improve manufacturing productivity and reduce operating costs.

The Westinghouse Olivetti SIGMA robot is representative of this type of automated systems. The robot features end-of-arm manipulation, easy access to the work envelope, versatility in application, and state-of-the-art controls and sensing.

The SIGMA robot is constructed with a sturdy I Beam support structure that provides an open working area that is accessible from all four sides. This open frame structure provides for part delivery systems access from any direction or angle, reduces setup and maintenance, and provides maximum materials handling flexibility.

In addition, this robot is designed with advanced sensing interfaces that allow quick recovery from irregular handling situations, often without interrupting the work cycle.

This robotic system was designed specifically for application versatility. It offers:

- A variety of standard manipulating modules
- Special fixtures
- A versatile programming system
- A high number of inputs and outputs for auxiliary equipment

A typical SIGMA robot system consists of a steel frame with a pair of precision overhead rails, one to five operating arms, a multiprocessor control unit, a teach pendant, and a CRT and printer. It operates up to 16 servo axes. Each axis can be operated simultaneously and independent of the others. The control of 16 axes is not limited to the robot arms, but can include X-Y positioning tables, rotary index tables, and the like.

SIGMA robots are ideal for component and subassembly manufacturing operations where fast, repetitive manipulation is required. The SIGMA design allows quick setup and changeover, so it is particularly well suited for quick batch runs of different styles and configurations.

In addition, the SIGMA robot can:

- Function as a work cell supervisor. A programmable logic controller is usually not necessary because standard I/O includes 160 input channels and 128 output channels for ancillary devices. Also, the controller features multitasking program execution, allowing simultaneous control of the robot arms and external work cell devices.

*Permission granted by Westinghouse Electric Corporation.

- Respond to analog sensory devices such as force and displacement transducers. This means it can perform in-process inspection operations such as part present, part seating, part orientation, correct size of part (selective assembly), and so on. It also can interface with vision systems and other external computer controlled equipment via RS232 communication.
- Be configured with up to five arms on a gantry structure, with the work envelope contained within the framework of the structure. This multi-arm capability has been used to incorporate a YAG laser material-handling system which maximizes the laser on time.

Fully Automatic Operation

The SIGMA robot is designed for entirely automatic operation. When the program for an assembly operation is entered into the controller and the components are in the working envelope, the SIGMA robot will complete the programmed work cycle without operator interaction.

Each operating arm can make continuous path motion in X, Y, and Z directions utilizing linear and circular interpolation. Rotary motion about a vertical axis and a horizontal axis also is available.

The arms can utilize various sensing methods. Hall effect sensors provide force and displacement data for configuration, size, and part presentation verification. Vision sensing may be utilized with solid-state cameras mounted on the operating arms to determine displacement and orientation of randomly placed workpieces.

These sensing systems identify the workpiece position on the work surface in relation to the programmed arm motion, and transmit discrete positioning data to the controller. That data is interpreted and compared with the programmed work cycle by the controller, and adjustments are made to the operating arm motion to compensate for any displacement of the workpiece on the work surface.

Programming can be done on-line or off-line. Off-line programming allows uninterrupted operation of the robot.

Operating Arms

The operating arms, suspended from the carriage modules, are the work handling components of the robot. By suspending the arms from the carriage, the system designers enhance the open access construction of the frame leaving the assembly area free of obstacles that could hinder production, setup, and maintenance personnel. This design also keeps all robot operating components within the safety of the frame structure, permitting more efficient utilization of valuable floor space.

A series of optional applications for various assembly modular components may be added to the operating arms as required, such as:

1. *Vertical Axis Revolving Modules*—provide rotation of the workpiece about the vertical axis by either servo control or pneumatic actuation.
2. *Horizontal Axis Revolving Modules*—provide rotation from the normal horizontal working plane by either servo control or pneumatic actuation. These modules make it possible to pick up workpieces supplied by feeders, conveyors, and chutes.
3. *Gripper Modules*—handle the workpieces for assembly. These are internal and external grippers, grippers that conform to the workpiece, and special grippers that fulfill special application requirements.

PART B. HOW AUTOMATION AFFECTS PACKAGING, PALLETIZING, AND UNITIZING

WHEN TO USE ROBOTS IN PALLETIZING AND UNITIZING

Today, robots are able to perform any palletizing or unitizing task, no matter what the size, shape, or weight of the product. Robots are best for low-speed, low case weight applications, and complex operations with almost an infinite variety of patterns. They are most commonly used in two material handling tasks: (1) case palletizing, and (2) unitizing of parts in bins or in specially designed pallets.

In unitizing of unpackaged parts, robots are able to precisely position parts of any size, shape, or weight in a bin or pallet, and also can insert dunnage.

Since robots excel at handling many different palletizing patterns, an unlimited number of programs can be put on a cassette. The robot can switch from one pattern to another automatically, for example, on receiving a signal from a bar code reader or other device. This assumes, of course, that the gripper can handle different sizes and styles of cases.

Robots also are adept at serving multiple stations. Using the reach of their arm or traveling on a gantry, they can retrieve cases from multiple infeeding conveyors. More typically, they stack cases of different products from a single in-feeding conveyor onto several palletizing stations. Not only is this extremely efficient for low-speed lines, it also makes optimal use of floor space.

Robots are naturally well suited for working with other automated equipment. Most applications are synchronized through bar codes or other means with some type of conveyor infeed. Robots can work with other robots, such as robots stacking cases on waiting AGVs.

There is really no limit to the size of cases that a robot can stack. They could conceivably stack 6-foot tall, lightweight cases if necessary, which is certainly beyond the capability of other palletizing machinery. Cases with high centers of gravity are another application in which robots have a natural advantage over other palletizing machinery.

Robots do have limitations though. First, they are slow. Six transfers per minute is the most you can expect. So unless the operation picks up multiple cases, that means only six cases per minute.

To pick up multiple cases, it requires use of a vacuum gripper. Even with several vacuum pads per case, the robot is limited to case weights of about 50 pounds or less, unless it has extremely strong and air-tight cases. For cases over that weight, a mechanical gripper is necessary—that means only one case per transfer.

Thus, robots will never compete with the high-speed case palletizers because high-speed case palletizer machines can handle 20 to 100 or more cases per minute, no matter what the weight of the product. They can form double-high loads 10 feet high or taller.

Without costly hardware, high-speed case palletizers are unable to form more than three or four patterns. They generally require a great deal more floor space than a robot—200 square feet or more.

How to Choose between Mechanical and Vacuum Grippers

It is possible to use either vacuum or mechanical grippers for robotic case palletization. Both require careful design to handle an assortment of cases.

The advantage of using vacuum grippers is that the robot can pick up more than one case at a time. However, 50 pounds is typically the upper limit on case weight. This is because of the porosity (tiny holes) and irregular surfaces of paperboard cases.

Mechanical grippers allow handling of cases weighing several hundred pounds. They also can perform other tasks, such as filling cases or positioning pallets. Mechanical grippers that grasp cases from the bottom minimize the chances of case failure.

For certain applications it is advantageous to use a combination mechanical–vacuum gripper. For example, a mechanical gripper can grasp and position a pallet; a vacuum gripper then picks up and transfers cases.

Why Palletizing Requires Customized Software

The programmability of robots and the gripper design, are what allows robots to handle a wide variety of products and stack in many different patterns. Because there is such a variety of cases and patterns, software must be customized.

Some vendors provide generalized software. The user programs the product's size, gripping points, and the path the robot takes to and from the pallet. Other vendors provide no software. You must teach the robot by leading it through the required motions with a control pendant.

Depending on the experience of the operator, programming or teaching could take as little as a few minutes or an hour or more.

Regardless of the method of programming, store routines on a tape or disk. Don't keep the only record in the robot's controller.

Why Parts Unitizing Software is Always Customized

Robotic case palletizing is often less complicated than unitizing uncased parts. There are similarities between many applications: Cases are the same shape, and they can be stacked in the same patterns.

In unitizing, every application is different. No two parts are alike, containers are different, and dunnage stacked between layers of parts is different. However, a robot is the only way to automate most unitizing jobs. When automating, it becomes a question of which robot to use.

Cartesian, cylindrical, and jointed-arm robots are commonly used for unitization. Spherical coordinate robots are used less, partly because they cannot reach into bins or reach all points of a pallet positioned below the robot.

Because of the unique nature of most unitizing applications, the gripper as well as the software probably will have to be customized. Programming has to be extremely precise, and the system must be designed so that parts, containers, and dunnage are conveyed accurately in and out of the robot's workstation.

Three Types of Grippers for Unitizing

Three different types of grippers are typical in robotic unitizing: mechanical, vacuum, and magnetic. Most applications use either mechanical or magnetic grippers for handling parts; vacuum grippers generally handle lightweight dunnage. Very often a robot will have a dual gripper: one gripper to handle the part, and another to handle the dunnage.

Mechanical grippers are the most common gripper for unitizing. They are able to handle any shape of part of almost any weight. Vacuum grippers are rarely used because they require parts with extremely flat surfaces. Magnetic grippers can, of course, handle only ferrous metals.

Both vacuum and magnetic grippers are weak in shear. This means that although they can lift parts, they have little strength while swinging them horizontally.

What to Do about Dunnage

Some form of dunnage is necessary in parts unitizing applications. Most parts are too irregular in shape to form stable stacked loads.

Dunnage simply may be sheets of corrugated paper, or can be a plastic or metal pallet designed to cradle a specific family of parts. Such pallets could cost hundreds of dollars a piece to design and fabricate.

If a part is being placed into a bin or on a pallet, and will require no further automated handling, it may not be necessary to design dunnage that will cradle the part. If more automated handling takes place further down the line, or if the parts cannot be allowed to slide around at all, some customized dunnage is required.

Think about the design of your dunnage early in planning.

1. Design dunnage so that the same gripper that handles parts also can handle the dunnage.
2. Look for common features of parts. One design may accommodate several parts.
3. Be sure that the robot's gripper has enough clearance while placing parts in the dunnage.
4. If parts are to be shipped to another department, determine if they have any special requirements.

CASE-IN-POINT

MAJOR MANUFACTURER OF OFFICE SUPPLIES

Situation

A major manufacturer of office supplies realized that to meet the challenge of aggressive foreign competition without losing market share, it would have to materially improve its packing and shipping of spare parts and scheduled maintenance kits to field service personnel. As a result, the company decided to institute a modernization program in its distribution center with heavy emphasis on mechanization of its packaging operation.

Solution

After analyzing the characteristics of its distribution center's activities, the company embarked on a multiphase modernization program.

The initial phase involved the installation of five new packaging lines. These five "islands of automation" ranged from standard to custom-designed packaging units, consisting of the following:

- A vibratory feeder that automatically counts and combines small parts into arrays for automatic bagging
- A special machine designed by the company's engineers, which packages tubular items such as lamps and shafts
- A modified book packaging machine that boxes circuit boards and other low-profile items
- A dedicated packaging line for scheduled maintenance kits where boxes are automatically set up, sealed, marked (by ink jet), and palletized by a robot palletizer
- Twin, height-adjustable packaging lines that form cartons for final packaging of shipments to regional distribution centers

Subsequent phases involved the integration of the packaging cells that had been created in the prepack and overpack areas, and to further automate operations that were already partially automated.

The packaging cells were physically integrated by fifteen AGVs traveling throughout the facility on over three miles of guide wire. This advancement was coupled with improvements in automatic identification and data transmission technology which enabled complete tracking of materials movement throughout all warehousing activities.

Further automation included introduction of robots and automatic order picking devices for assembling scheduled maintenance kits and installation of robotic palletizers to pick small cartons from the twin packaging lines into large corrugated containers.

Results

Although full integration of all packaging cells by use of AGVs and other similar methods improvements are more difficult to evaluate, the company was able to identify the following dramatic results due to the mechanization of its packaging operation:

- Cycle times about three times faster than they were before mechanization (4 days instead of 12 days)
- 25 percent reduction in personnel engaged in the packaging operation
- 37 percent overall reduction in personnel directly and indirectly supporting the packaging operations
- A reduction in the number of packaging workstations from twenty-five to eight with no reduction in throughput capability

TECHNIQUES FOR AUTOMATING PACKAGING

A Checklist for Achieving Cost-Effective Packages

Companies are becoming increasingly aware of the need to consider packaging design and functionality in the overall cost-effective equation of operating a warehouse facility. This recognition applies to both standard and custom-made packages.

The following checklist can be used as a guide in determining whether your company has considered all factors of this important concern.

Product

1. Have you given your box maker a description of your product and its use? The exact dimensions, weight, and physical characteristics?
2. Is the product likely to settle or shift?
3. Is it perishable, fragile, or hazardous in any way?
4. Will it need extra protection against vibration, impact, moisture, air, heat, or cold?
5. Will it be shipped fully assembled?
6. Will more than one unit be packed in a box?
7. Will accessories, parts, or literature be included with the product?
8. Have you provided your box maker with a complete sample of your product as it will be packed?

Packing Operation

1. Is your box inventory adequately geared to re-order lead time?
2. Is your box inventory arranged to efficiently feed your packing lines?
3. Is your inventory of boxes stored properly?
4. Will you be setting up the boxes on automatic equipment? If so, what type? size? method of closure?
5. Will your product be packaged automatically? If so, what type of equipment?
6. If more than one unit or part goes into each box, have you determined the sequence?
7. Will inner packing (shells, liners, pads, partitions) be inserted by hand?
8. Is your closure system (tape, stitches, glue) compatible with the box, packing line speed, customer needs, and recycling considerations?
9. Will the box be imprinted or labeled?
10. Will a master pack be used for a multiple of boxes to maintain cleanliness or appearance?

Storage

1. Have you determined the gross weight of the filled box?
2. Does the product itself help support weight in stacking?
3. Will the bottom box have to support the full weight in warehouse stacking?
4. Will boxes be handled by lift trucks that use clamps, finger lifts, or special attachments?
5. Will filled boxes be palletized? (The size of pallet and pallet pattern may justify a change in box design or dimensions if only to reduce or eliminate overhang.)
6. Would a change in box style or size make more efficient use of warehouse space?
7. Will filled boxes be subject to unusual conditions during storage (high humidity, extreme temperatures, etc.)?

Shipping

1. Have you reviewed the appropriate rules of the transportation service you intend to use (rail, truck, air, parcel post, etc.)?
2. Is your container authorized for shipment of your product?
3. If the package is not authorized, have you requested appropriate test shipment authorization from the carrier?
4. Have you determined the actual inside dimensions of the transportation vehicle so you can establish how your filled boxes will be stacked or braced?

Customer

1. Does your customer have any special receiving, storage, or handling requirements that will affect box design?
2. Will the box be used as part of a mass display?
3. Is the box intended as a display shipper?
4. Will it contain a separate product display?

5. Will it be used as a carry-home package, requiring a carrying device?
6. Does it need an easy opening feature?
7. Can surface design, symbols, or colors relate to promotional materials or to other products of the same corporate family?

Advances in Weighing and Packaging Technology

Using the checklist above as a guide, an evaluation can be made of various equipment, systems, and materials that are proving to be cost effective in weighing and packaging operations. Following are advances in this technology that are making inroads in warehousing operations:

1. *Electronic scales that can be programmed to store critical data about hundreds of parts and tied to a host computer to update production records.* With improved weighing accuracy, these systems can be used as an inventory control checkpoint. Various check weighing and counting scales can store part numbers and weights in computer memory, print inventory tickets and reports, and direct the reading and printing of labels. Through a direct communications link to a host computer these scales can forward batches of data to update inventory records or ask for part number and weight information not stored in the scale's memory. Also, several advanced scales can be linked together.

2. *In-line automated case forming and packing equipment offers throughput improvements over manual packing methods.* Systems can remove a knocked down case from a stack, erect it, accumulate and pack the product, close and seal the case, and discharge it to a conveyor.

The most advanced systems use microprocessor-based controls that vary speeds and patterns and can detect improperly formed or filled cases, halting the equipment operation if necessary.

There is a variety of automatic and semi-automatic case sealing systems for filling operations. Most operate in-line and handle either fixed, or random, size cases. Electronic controls improve throughput and control material use.

3. *On-site label printers that can either be tied to scales or operated separately.* These printers enable on-demand generation of labels for identifying and tracking finished parts.

Automated methods ranging from reciprocating coders to ink-jet printers are replacing stenciling and hand-stamping techniques. Among the contact printing methods, roller coders and reciprocating coders are common in-line techniques.

4. *The three common methods for securing unit loads.* Stretch wrap, strapping, and shrink wrap are recent improvements in some types of plastic strapping that have improved strength levels to nearly those of steel strapping.

Stretch wrap film has been improved with the use of new, thinner, stretch-and-shrink wrap films that meet strength and toughness requirements. In addition, prestretch grades can increase film yield as much as 200 percent.

Shrink wrap, which must be heated to secure loads, also is being used to save packaging costs. In some applications, shrink wrap can utilize loads so well that pallet or slip sheet bases can be eliminated.

The Right Time to Use a Computer-Adjustable Case Erector/Bottom Sealer

For a packaging line that uses a variety of case sizes, erecting and sealing those cases with mechanical equipment can be costly and time-consuming when changes in case sizes must be made. To handle a flexible packaging line of this type, a computer adjustable case erector/bottom sealer can be installed. This equipment has push-button controls that allow packages to change case sizes more accurately and in less time than can be done manually with tools.

The computer-adjustable machine eliminates the need for an operator or mechanic to adjust mechanical parts.

Automating the Job

The computer controls of this equipment can permit as many as 250 different case sizes to be preprogrammed into the computer-driven board.

This type of machine is controlled by a dedicated microcomputer that controls every mechanical function of the machine except the loading of knockdown cases into the hopper.

In addition, a computer-adjustable top and bottom sealer can be incorporated which features a dedicated microcomputer control system that permits push-button changeover between case sizes in 30 seconds.

The glue system can be set to operate the machine as a top and bottom sealer, a top sealer only, or a bottom sealer only. It offers the same adjustable capabilities of a random sealer.

Tips for Using a Computer Controlled Handling and Carrousel System

One advanced technique that has proven to be most efficient in packaging is a computer-controlled handling system centered around a carrousel system. This type of operation usually involves multiple horizontal carrousels integrated with packaging stations powered by a roller conveyor system. The system transports and stores and tracks parts in bar-coded tote containers, as well as corrugated packaging materials. The handling hardware is controlled by a computer system that receives information from workers and equipment, allocates parts for customer orders, and tightly controls the flow of materials through storage and final packaging.

Steps in Advanced Packaging Operation

Incoming materials are delivered to the packaging area by fork truck. After inspection, parts are placed in tote pans. Each tote contains a unique bar code label on each side.

The inspector uses a keyboard to enter the identification number of the load into computer memory, then performs a visual inspection for quality and quantity of parts. A printer automatically generates packing instructions before the inspector adds shipping documents for that batch of materials.

The worker then uses a hand-held scanner to read bar codes on the tote, the paperwork, and his or her employee badge. When the badge is scanned, the control system starts the roller conveyor automatically.

The conveyor transports each tote past a stationary scanner, which verifies the identity of the tote before reaching the input automatic load/unload device. The device lifts the tote from the conveyor and places it into a storage location on the main carrousel.

Upon demand of the computer control system, tote pans are removed from the carrousel by the output automatic load/unload device, located at the end of the carrousel. The device lifts the tote from its storage location, then places it on an output conveyor. Totes move on the conveyor to the packaging area.

At the packaging stations, workers form individual customer orders for shipment and package them in corrugated cartons.

After placing the packaged order into a tote, the packer uses a hand-held scanner to identify the job for the computer control system. The worker scans bar codes on the tote, paperwork, and his or her employee identification badge. After the scan of the badge, the output conveyor starts automatically.

Packaged orders move to a staging area, where they accumulate on pallets until shipment time.

PART C. SUCCESSFUL IDEAS FOR AUTOMATING WAREHOUSES

BENEFITS OF FOUR-MODULE, COMPUTERIZED WAREHOUSE AND SHIPPING CONTROL SYSTEMS

Operating an efficient warehouse facility in today's competitive environment requires installing a computerized warehouse and shipping control system (WCS). With this system in-place, you can expect to accomplish many significant benefits, such as:

- Improved space utilization, based on computer selection of storage locations
- Improved efficiency of storage and retrieval functions due to computer routing and scheduling of all activities
- Accurate inventory tracking for reductions in safety stocks, minimized likelihood of stockouts and product loss, and fewer customer returns resulting from picking and shipping error
- Enchanced employee performance due to storage and retrieval efficiencies, workload management, and precise productivity reporting
- Shipping cost reductions, based on computerized order consolidation and truck building for maximized use of cube and weight capacity

A number of software packages available now satisfy this need. Most can be customized to fit the exacting requirements of each individual company. The next few sections highlight the sequence of events of the various software modules for a well conceived integrated warehouse control system, which are:

1. Receiving module
2. Storage module
3. Order processing module
4. Truck loading module

Receiving Module: The Key to Accurate Inventory

Accurate inventory management starts on the receiving dock where an incoming item should be labeled and routed correctly. Validation is usually based on the use of bar coded labels; passive transpondence, which uses electronic waves to read microchips affixed to the product or location for extremely powerful operator-independent control over data; or voice recognition technology in which the workers actually talk to the system to confirm task completion.

Generally, a bar coded label is generated for rapid identification of SKUs throughout the warehouse process—documentation that eliminates the need for subsequent manual entry of product data, and thereby minimizes the possibility of error. This "travel ticket" authorizes the items transport through the warehouse facility, and should include storage instructions, unit of measure, count, product description, and so on.

With this data in place, the system begins tracking the elements of the order through a logical multistage receiving process—a process which may include a quality assurance inspection, repackaging, or assembly.

Storage Module: Why Management Strategy Is Fundamental

One of the most important elements in warehouse and shipping control is computer-directed storage. Two concepts are key to this activity. They are:

- *Management Operating Strategy*—a unique set of guidelines that define the optimal operation of the facility. This may involve the decision to institute a policy of using only forward picking locations for new products unless FIFO or lot control guidelines interfere. Or, it may be decided to keep all fast pick zones on a horizontal plane for faster storage and retrieval.
- *Storage Parameters*—specific to stockkeeping units, and are defined in light of managements operating strategy. These are the parameters the system uses to make decisions about where to store each item. For example:
  (a) Environmental requirements, such as temperature, humidity, or hazardous material restrictions
  (b) Demand level (pick frequency) for each item
  (c) Type of demand, such as, broken case, full case, or pallet load
  (d) Dimensions, cube, and weight
  (e) Batch or lot control requirements
  (f) The type of storage device required
  (g) Replenishment thresholds, and the advisability of moving some items directly to forward pick locations

The system should be programmed up front to make storage decisions based on all the parameters of importance to a company's operation, and to consider the ramifications of having the computer assess these factors to achieve managements operating strategies.

Once the order has been picked up, transferred to, and deposited in a storage location, the system should demand validation of the transfer by having the worker scan the bar coded labels on both the item itself and on the storage location, which will result in complete storage accuracy.

Order Processing Module: What to Expect from a Productive System

This sequence of events generally originates with the host computer, which transfers the details of each order electronically to the system, initiating a computer-generated pick order.

The system determines the most productive way to pick, pack, and ship the order, taking a wide range of factors into consideration—everything from order type, cube, weight, lines and priority, to route/stop sequence, workload by warehouse area, processing method, and inventory availability. Before any picking instructions are issued, the system should run quick simulations for all the possibilities. It should be programmed to ensure that the inventory is in ample supply at each location concerned, and that adequate personnel and equipment are on hand to execute the activities to minimize the time and effort involved, including aisle changes and to present the picked items in logical sequence for subsequent routing to shipping or manufacturing.

In addition, the system should analyze:

- The advantages of batch versus single-order picking
- The opportunities for consolidating orders as well as stores and retrieves
- The impact of fast or slow pick zones on efficiency
- The need for allocating items that are in short supply, by making partial shipments or placing some requests on back order
- The possibility of substituting similar products for items in short supply

All these factors should be reflected in the picking instructions generated by the computerized system.

Once those instructions have been issued via bar coded label printout or radio communications, a worker is dispatched to fulfill them. Material-handling devices can range from carrousels and conveyors to stacker cranes.

The key element in the picking operation is validation. Each worker must confirm that the right SKU has been selected from the location prescribed, using bar coded scanning or other validation technologies.

Truck Loading Module: Six Opportunities to Cut Costs

The truck loading module starts in the order picking process. This step should present items properly in route stop sequence or reverse route stop sequence. This module can enable a company to achieve significant freight and handling savings in several ways by:

- Automatically consolidating orders bound for the same destination.
- Overseeing in-motion weighing, particularly important wherever product pricing is a function of weight.
- Overseeing sortation equipment, assuring delivery of all orders to the right loading docks.
- Calculating and displaying the cubes and weights that would be achieved by consolidating various combinations of orders in a given load.
- Selecting the best trailer for a given group of orders based on weight and cube.
- Building the load to maximize cube and distribution weight in accordance with appropriate state regulations.

UNDERSTANDING SOCIAL AND BUSINESS INFLUENCES ON WAREHOUSING AND DISTRIBUTION

A number of social and business factors impact the ways goods are produced, warehoused, and distributed. Some of these factors are:

- *More women in the work force* needing more apparel, goods, and services, and having less time in which to shop for them
- *Trend toward upscale lifestyles* creating a demand for greater variety and improved quality in everything worn, consumed, or driven
- *Transportation deregulation* has changed distribution methods and patterns
- *Just-in-time handling, shipping, and receiving of goods* is overthrowing the cherished notions of how distribution is supposed to work
- *Franchised outlets* for fast foods, clothing, toys, etc. are mushrooming

Keeping the outlets provisioned is a major distribution undertaking. The pressure to track the identity of products and components throughout their life cycles is hastening the implementation of automatic identification in warehousing and distribution. Lot numbers, part numbers, contract numbers must all be maintained and tracked for purposes of regulation, quality control, and marketing.

TRENDS IN MATERIAL-HANDLING TECHNOLOGY

In addition to the impact of social and business factors on warehousing and distribution, several dynamic trends are changing advanced material-handling technology. These trends influence not only equipment and controls, but also the relationships between materials and the operations involving materials—fabrication, processing, assembly, packaging, warehousing, and the ordering of materials. On a macro level, the unmistakable trend in material handling is the march toward automation. This in turn, has expanded the role of material handling to one of material management, and to close the loop, managing materials efficiently and cost effectively can only be accomplished through automation.

As an aid in analyzing material-handling needs, the College-Industry Council on Material-Handling Education has prepared a comprehensive listing of twenty principles of material handling. These principles can be used as a guide in formulating material-handling strategy and requirements. They are listed below.

Tips for Successful Automation: The Twenty Principles of Material Handling

1. *Orientation.* Study the system relationships thoroughly prior to preliminary planning to identify existing methods and problems, physical and economic constraints, and to establish future requirements and goals.
2. *Planning.* Establish a plan to include basic requirements, desirable options, and the consideration of contingencies for all material handling and storage activities.
3. *System.* Integrate those handling and storage activities that are economically viable, into a coordinated system of operation including receiving, inspection, storage, production, assembly, packaging, warehousing, shipping, and transportation.
4. *Unit Load.* Handle the product in as large a unit load as practical.
5. *Space Utilization.* Make effective utilization of all cubic space.
6. *Standardization.* Standardize handling methods and equipment wherever possible.
7. *Ergonomic.* Recognize human capabilities and limitations by designing material-handling equipment and procedures for effective interaction with the people using the system.
8. *Energy.* Include energy consumption of the material-handling systems and material-handling procedures when making comparisons or preparing economic justifications.

9. *Ecology.* Minimize adverse effects on the environment when selecting material-handling equipment and procedures.
10. *Mechanization.* Mechanize the handling process where feasible to increase efficiency and economy in the handling of materials.
11. *Flexibility.* Use methods and equipment that can perform a variety of tasks under a variety of operating conditions.
12. *Simplification.* Simplify handling by eliminating, reducing, or combining unnecessary movements and/or equipment.
13. *Gravity.* Utilize gravity to move material wherever possible, while respecting limitations concerning safety, product damage, and loss.
14. *Safety.* Provide safe material-handling equipment and methods that follow existing safety codes and regulations in addition to accrued experience.
15. *Computerization.* Consider computerization in material-handling and storage systems when circumstances warrant, for improved material and information control.
16. *System Flow.* Integrate data flow with the physical material flow in handling and storage.
17. *Layout.* Prepare an operational sequence and equipment layout for all viable system solutions, then select the alternative system that best integrates efficiency and effectiveness.
18. *Cost.* Compare the economic justification of alternate solutions in equipment and methods on the basis of economic effectiveness as measured by expense per unit handled.
19. *Maintenance.* Prepare a plan for preventive maintenance and scheduled repairs on all material-handling equipment.
20. *Obsolescence.* Prepare a long-range and economically sound policy for replacement of obsolete equipment and methods with special consideration to after-tax life-cycle costs.

MATERIAL-HANDLING EQUIPMENT AND METHODS THAT INCREASE WAREHOUSE EFFICIENCY

Recent advances in various types of material-handling equipment and methods have resulted in a significant increase in warehouse efficiency. These gains have been achieved through improved control of inventories, better responsiveness to customer needs, reduction in warehouse space, virtual elimination of pilferage and damaged materials, and sharply lower labor costs. The following types of material-handling equipment and methods, either individually or in combination, have been the major contributors to this dramatic change in warehousing and distribution:

1. Fork lift trucks
2. Rack system
3. Automated high-rise storage and retrieval system
4. Automated miniload storage and retrieval system
5. Automated guided vehicles
6. Automatically controlled cars
7. Unit load storage and retrieval system
8. Car-in-lane system
9. Man-ride machine

10. Gravity flow racks
11. Carrousel storage and retrieval system
12. Automated selectors and dispensers
13. Sortation and consolidation areas

Why Fork Lift Trucks Remain Popular

Fork lift trucks are still heavily in use, and perform a necessary warehouse activity. However, they have been upgraded considerably from earlier versions and now perform much more than merely transporting materials from one warehouse location to another.

For example, to weigh a load on a lift truck, older systems used a transducer in the hydraulic lift cylinder circuit to interpret pressure as pounds of loading on the forks. Today, simple strain gauge sensors on the sides of the forks do as well, and highly accurate systems use fork lifts with load cable inside them to display the weight to the operator. The rugged on-board display can show pounds or kilograms to totalize the day's loading, deduct the weight of the pallet or container, and even download to a ticket printer. These on-board electronics, which have been added to communicate with the operator, display instructions on CRT screens, print pick orders and labels, and require verification when tasks are completed.

When Using Rack Systems Proves Advantageous*

A rack system offers a company flexibility, convenience, ease of operation, and relatively low cost to purchase and maintain. Although mostly manually operated, rack systems are a necessary and utilitarian component of many warehouse facilities.

A rack system can be used advantageously if the load is very heavy, or heavier than usual, in which case the material handlers should look for a rack position on the first floor of the warehouse.

As part of its modernization program, the Grumman Aerospace Corporation obtained three rack-oriented storage and retrieval systems for large or odd shaped items. They are:

1. Dean Research Corporation's vertical sheet stock storage rack design with narrow openings to store large flat production tool templates. The racks consist of 1.5 inch openings to accept templates up to 10 feet long.
2. Space Saver Corporation's wing racks for hanging intermediate size templates that do not fit in drawers. The swing wing concept saved approximately 30 percent of floor space previously used for this type of storage, and improved tool pick efficiency by 25 percent.
3. Portec Pathfinder™ Wire Guidance System, which enables a vehicle operator to pick parts within narrow rack and shelving storage aisles (figure 10–5).

Racks come in many sizes and shapes: pallet racks, drive-in and drive-through racks, cantilever racks, stacking racks, and gravity flow-through racks.

In addition, racks are used to store pallet loads or other unitized loads. Input and output of loads from storage is often accomplished using a forklift truck, but human-ride machines and AS/RS machines have been successfully integrated with various types of racks to provide an efficient mix of manual and automated handling equipment.

*Permission granted by Grumman Corporation.

FIGURE 10–5 Pathfinder Wire Guidance System (Courtesy of Portec).

How Automated High-Rise Storage and Retrieval Systems Save Space*

Automated storage and retrieval systems (AS/RS) more fully utilize the space set aside for in-plant storage than conventional racking. This is accomplished by reducing the aisle width—the space between racks—to the minimum required by the size of the load, and by increasing the rack height all the way to the ceiling, if possible. As a result, an AS/RS is able to "maximize the cube" of available warehouse space. To accomplish this, a storage/retrieval machine can be designed to travel a precise route between the racks, and move the load to be stored through three directions—back and forth, up and down, and side to side (figure 10–6).

*Permission granted by Eaton-Kenway.

FIGURE 10-6 Automated High-Rise Storage and Retrieval System (Courtesy of Eaton-Kenway, Salt Lake City, Utah).

Basic Equipment

A typical warehouse stock retriever is a twin-masted, top-guided machine that traverses the aisle on a single steel rail similar in design to a length of railroad track. The mast is of a trussed design for maximum strength with minimum weight. In addition to the mast, four other main elements comprise the retriever: a horizontal truck assembly, which holds the two main retriever wheels and the horizontal drive; the vertical lift carriage and drive, which raises the load to the proper level for storage; the shuttle or extractor and its drive, which moves loads from side to side, into or out of the rack storage space; and a top lift-frame assembly, which holds the system electrification and provides for vertical alignment and guidance. Mounted on the front of the retriever is a state-of-the-art, microprocessor-based controller that operates the horizontal, vertical, and shuttle drives guiding the machine through a series of speed ranges to locate the bin and store or retrieve the load.

Systems Demand

Currently, industry is demanding taller AS/R systems up to 100 feet in some cases. They usually are computer controlled with sophisticated peripheral handling systems for in- and out-feed, with the rack structure serving as support for the building's outside walls and roof. In the more sophisticated installations the AS/R system is integrated through the plant's computerized controls network, into the production matrix itself. In this configuration it can respond to the moment-by-moment demands of production by retrieving and storing in-process as well as raw materials, while maintaining a real-time inventory of every item in its rack.

An automatic storage and retrieval system can provide a company with a number of tangible benefits, such as:

1. Saving warehouse space
2. Damage prevention
3. Better stock rotation
4. Labor savings
5. Improved inventory control

1. Saving Warehouse Space

An important justification of AS/RS is its space-saving advantage. Typically such a system will use only one-third of the floor square footage needed to store an equal volume of product in a traditional warehouse. This is accomplished by storing at greater heights and by reducing the aisle space between racks. There are a number of benefits that may be derived from saving space, such as:

- If a totally new facility is being planned, the use of AS/RS will reduce the overall plant size, reducing land acquisition and building construction costs.
- If additional areas for production equipment is needed in an already existing facility, the space often can be gained by consolidating a traditional warehouse area into a space-saving AS/R system.
- In some instances, the life of an existing plant can be extended by taking advantage of the unique characteristics of an AS/R system.

2. Damage Prevention

Another important benefit from the use of AS/RS is the reduction of product damage. This is especially true when finished products must be warehoused for any extended period of time before shipping or use. At this point product damage is more costly than at any other time, since all the value gained from processing, assembling, and packaging has been added to the raw materials. Unfortunately, this is when product damage seems to occur most frequently. As loads are moved into storage, stacked, and restacked during order-picking, or staged in order-building locations prior to loading for shipment, the product is especially vulnerable to damage. Much of this damage is caused by fork lift trucks engaging in this warehousing activity.

This type of damage can be avoided when products move in and out of storage by means of fixed path transporter such as the retriever in an AS/R system. In such a system, the load to be stored moves with mechanical precision to a specific opening in the rack and is retrieved in the same manner. Each stored product has its own unique position in storage and is not stacked or piled in any way. Each bears only its own weight, so unit load distortion caused by bearing the weight of others above it is avoided.

3. *Better Stock Rotation*

The AS/R system is of particular benefit where a FIFO accounting method provides for stock rotation. Since each unit load has its own storage address and is physically independent of all others, the retrieval of any specific unit, no matter where stored, is accomplished easily.

4. *Labor Savings*

The use of automatic storage and retrieval also can result in significant labor savings. With the proper peripheral equipment, the complete storage operations including record-keeping, can be done by one operator.

5. *Improved Inventory Control*

The marriage of the computer with automatic storage and retrieval has resulted in much tighter inventory control. The following steps indicate how this activity can be handled in an automatic storage/retrieval computer controlled warehouse operation.

Unit loads for storage approach the system on automatic conveying equipment and pass an ID station where each load is identified to the computer. This information can be either inputted via CRT terminal, or automatically entered by bar code scan. The bar code label identifying the item would have been affixed at a prior time, usually at the last work or process station. At the ID station the load will then be weighed automatically.

The computer has now recorded in its memory which of the stock-keeping units is represented by this load, and its status. If the item is one of the fast movers then a frequency code on the ID label will be employed to notify the computer of this condition. When all data has been transmitted, the computer will release the load for storage. As the load approaches the rack face, the computer automatically chooses a storage location and directs the load to that aisle for pickup by the storage/retrieval machine. Although some warehouse systems store randomly, most AS/R systems do selective storage.

If the unit load is one of the fast movers the computer has chosen an empty location toward the front of the rack system. This will allow for quicker access when retrieving the load for shipment, thus improving the warehouse throughput. Another factor in the computer's choice of location is the need to balance the storage of all SKUs throughout the rack system so that each aisle has an approximately equal amount of all categories of stored product. This will guarantee that unless all storage aisles become inoperative at once, there always will be some of each category available for retrieval and shipment.

The computer then directs the storage machine to pick up the load, carry it to the designated bin or storage space, and place it in storage. Since the computer keeps track of all empty bins, this command is executed immediately by the storage machine. If for some reason the computer has chosen a storage space already occupied, this will be detected by a photo-cell check before insertion is attempted. When this occurs, the computer flags the anomaly in its memory for later investigation and picks another appropriate location in the same aisle.

If the stored product is one of those requiring quarantine, the computer begins a timed lock-out that will prevent that load from being retrieved for shipment until the quarantine period is over. The storage date of all products is captured by the computer so that when a retrieve command is issued the oldest appropriate stock unit will be chosen, thus affecting the first-in, first-out inventory method.

If the product is retrieved from storage for shipment, the computer can be ordered to retrieve in an optimum sequence. It will choose the oldest in each category first and avoid those items still in quarantine. If, for some reason, it sends a retriever to a bin it thought contained a product but which turned out to be empty, it will note the fact for further investigation and then send the retriever to an alternate location.

The marriage of automatic storage and retrieval with the computer also has resulted in the availability of a variety of real-time, hard-copy reports. These can include inventory by lot number, color, size, weight, quantity, location, or any other attribute used to identify the loads for storage.

Tips For Successful Automation: How to Justify an AS/RS

In justifying installation of an AS/R system, there are a number of factors that should be considered. These are:

1. *A Comparison of Land Acquisition Costs.* In AS/R system will use one-third to one-half of the floor space needed for conventional alternatives. This is an important consideration when evaluating the cost of a new warehouse facility.
2. *Building Costs and Write-Offs.* If the AS/RS structure being considered is a rack-supported building, it is usually possible to classify it as equipment or machinery shelter. This will qualify it for more favorable tax treatment and this savings also should be included in the justification calculations.
3. *Value of Excess Inventory Carrying Costs.* This is a very real item in the cost of warehouse operations. In the traditional warehousing system, these costs, when annualized, can easily reach to 30 percent of the value of the stored materials. It has been demonstrated that AS/RS can eliminate most of the reasons for excess inventory.
4. *Value of Improved Inventory Control to Customer Relations.* The penalty for lack of efficient inventory management is often a costly delay in customer service. The real-time inventory control made possible by a computer-supported AS/RS is therefore a justification factor and must be quantified and included in any bottom line figure.
5. *Cost of Direct Operating Personnel.* In many AS/R systems the number of operating personnel is reduced by two-thirds or more over conventional systems. This personnel would consist of lift truck drivers, order pickers, checkers, packers, expeditors and others who are engaged in the receiving, handling, storing, retrieving, and delivery functions. In documenting the savings due to reduced personnel the cost of maintaining an employee is at least 33 $^{1}/_{3}$ percent higher than his or her base salary, and these costs escalate over successive years.
6. *Machinery and Equipment.* Machinery and equipment for each alternative must be considered when justifying AS/RS. In cases of conventional alternatives, the cost of industrial trucks, pallet rack, pallets, and other equipment must be considered.

CASE-IN-POINT ILLUSTRATIONS

The following case-in-point illustrations are excellent examples of installed automatic storage and retrieval systems. The two companies discussed benefited considerably by gaining better control of their warehouse operation and increased labor productivity.

In the first case, Canadian Tire Corporation, the company achieved its main objective of controlling the movement of a varied product line in an extremely large warehousing complex.

In the second case, the Buick AS/RS installation, the company obtained a number of important secondary benefits in addition to accomplishing its main objective of operating an efficient warehouse facility in a limited physical environment, with a fraction of the space that would be needed in a traditional warehouse.

CASE-IN-POINT

CANADIAN TIRE CORPORATION
AUTOMATED STORAGE AND RETRIEVAL SYSTEM
*THE ANN ARBOR COMPUTER DIVISION OF JERVIS B. WEBB COMPANY**

Situation

The Canadian Tire Corporation was in need of a real-time computer controlled AS/R system to handle a varied product line in its general merchandise distribution center, located near Toronto, Ontario.

Solution

After an extensive analysis and evaluation of the Canadian Tire Corporation's requirements, the Ann Arbor computer division of Jervis B. Webb Company designed and installed an automated AS/R system involving three individual warehouse sections and twenty-eight storage/retrieval machines. This sprawling computer system controls the movement and storage of up to 81,000 pallets of material in one of the largest automated storage and retrieval systems in North America.

The top level of control consists of two warehouse computer systems communicating with three system controllers. Both the warehouse computers and the three system controllers share data storage and decision-making capabilities. Each of the three system controllers acts in a stand-alone mode when a warehouse computer is out of service.

The computer control, similar in the three warehouses begins at an identification station, where pallets of cased products are identified by an operator at a data-entry terminal. Data concerning each pallet, including the part number and cases per pallet, are entered. The warehouse computer verifies this information and selects a storage aisle and bin for pallet storage. The system controller then automatically directs the pallet, by way of conveyor, to a specific storage aisle load station at the AS/RS. Storage aisle selection is based primarily on even distribution of a product across all aisles. Bin decisions are based on storing higher-demand products nearer to the load station.

Picking-lane operators enter retrieval requests by way of data entry units. If the item is in stock and can be accessed by an S/R machine, a command from the system controller to an S/R machine is made. A retrieved pallet is automatically conveyed to the picking lane for which it was requested. A central bypass spur automatically accumulates pallets if necessary.

Results

The automated real-time computerized AS/R system has achieved the following:

- Tight inventory control of its varied product line
- More efficient utilization of warehouse personnel

*Permission of Jervis B. Webb Company.

- Lower operating costs
- Savings in cost of land and warehouse space
- Improved warehouse control
- Dissemination of timely valuable management reports, including:
  (a) *Bin contents report* showing which products are stored in each bin, including the number of cases, whether or not the pallet is full, and the date of storage.
  (b) *Product code report* showing all valid products that may be stored in the warehouse and how many cases of each product are currently stored.
  (c) *Product location report* indicating the number of pallets of a specified product code that are in storage, and their location. The report also indicates the number and position of currently available empty bins.

CASE-IN-POINT

BUICK MOTOR DIVISION OF GENERAL MOTORS CORPORATION AUTOMATED STORAGE AND RETRIEVAL SYSTEM JERVIS B. WEBB COMPANY

Situation

The Buick Motor Division of General Motors Corporation wanted to free up 60,000 square feet of floor space in its engine plant for new assembly and test facilities. Due to its land-locked situation, Buick only had 10,000 square feet in which to fit the 60,000 square feet of storage.

Solution

After analyzing Buick's unusual situation, Jervis B. Webb Company designed and installed a rack-supported structure, immediately adjacent to the engine assembly plant, standing 72 feet high, 65 feet wide, and 150 feet long. Inside, a four-aisle high-rise system operates with four Webb-Triax® storage/retrieval machines.

Loaded pallets are transported to and from the S/R machines by roller conveyor.

The system was built to handle 750 different parts in 2,880 storage openings, each designed to hold a maximum load of 4,000 pounds. Each retriever can handle fifteen dual commands per hour, for a total of sixty dual commands throughout the entire system.

A computerized control system was programmed and installed to enable warehouse personnel to locate and identify parts automatically by simply entering the part number in the computer.

The computer control begins at an identification station, where pallets of incoming parts are identified by an operator. These parts are verified against the computer's product file (a list of every valid part number) and are moved to a sizing station on a powered roller conveyor. There the pallets are weighed and dimensionally checked for proper system handling and storage conditions. Once identified correctly, the warehouse computer directs the pallets to storage in the AS/RS. Incorrectly loaded pallets are automatically directed to a reject area, where they are reloaded or adjusted.

The accepted pallets are continuously tracked from the identification station to the proper storage aisle under the direction of the microprocessor-based system controller. This controller accepts commands from the warehouse computer and then directs the pallets to one of four pick-up stations. There a S/R machine picks up the load for storage.

Each of the four S/R machines carries its own microprocessor controller on board. The S/R machine's controllers communicate with the system controllers to receive commands and report status.

Aisle decisions by the warehouse computer are based on S/R machine availability, balance of product distribution over the four storage aisles, and balance of activity over all four of the S/R machines. Bin decisions are made so that pallets are placed as close to the front of an aisle as possible. In retrieval, the warehouse computer automatically directs the pallet from S/R machines to delivery spurs.

The effects from a system malfunction are minimal. The real-time control permits continuous monitoring of all appropriate conveyor and S/R machines. If a malfunction occurs in any system component, alternate storage locations are accessed. In addition, the system allows semi-automatic operation if the warehouse computer is not functioning. A data entry terminal can direct the AS/RS, and a complete transaction log is maintained and copied daily on magnetic tape for subsequent data-base restoration.

If the conveyors are inoperative, retrievals can be deposited to the rear of the system, where fork-lift operators can pick them up and continue deliveries.

Results

The computer control system has given Buick a much better control on its inventory. In keeping with just-in-time philosophy, Buick is able to carry less inventory, which means a reduction of inventory costs. Buick also is able to reduce parts shortages. Workers no longer need to spend time searching for parts now that inventory is managed effectively by one individual, identifying pallets and requesting retrievals. Because Buick knows where certain parts are, those that are obsolete can be pulled and returned to the source promptly.

The computer control system is a great help to lift-truck operators, who are able to stock the assembly lines in 3–5 hours, leaving them free to restock the storage system or help with shipping and receiving tasks.

The output of management information such as inventory, product, and empty location reports is another important benefit of the computer control system.

Other advantages of the system are: If there is a fault with a part, it may be quarantined within the system until it can be inspected; the AS/RS makes a practice of retrieving old stock before the new; a worker may make a request to the ID station operator for a high priority part, which will then be retrieved before others.

In summary, the newly installed AS/RS has greatly improved Buick's efficiency and accuracy in handling its purchased parts inventory—in a situation where limited warehouse space was a prior major constraint.

Features and Benefits of Automated Miniload Storage and Retrieval System*

A miniload is a fully automated computer controlled storage/retrieval machine designed to bring small parts in bins to an end-of-aisle order picker. The operator stands at a human-engineered workstation while the S/R machine automatically delivers parts bins and puts them away (figure 10–7).

The miniload commonly operates in aisles 40 to 300 feet long and stores bins 12 to 40 feet high. The S/R machine travels horizontally at up to 365 feet per minute.

Real-time inventory control in a miniload system contributes to significant improvements in inventory accuracy, material accessibility, and integration with internal data processing systems. A miniload system will automatically update inventory records, print management reports, and provide inventory status on demand.

*Permission granted by Eaton-Kenway.

FIGURE 10–7 Automated Miniload Storage and Retrieval System (Courtesy of Eaton-Kenway, Salt Lake City, Utah).

Below is a summary of features and benefits of automated miniload storage and retrieval systems:

- *Space Savings*—reduces storage area requirements by using all available ceiling height to greatly improve storage cube utilization.
- *Labor Savings*—eliminates all walking and searching time, thus fewer operators are required. Personnel reductions also can be made in the number of material expeditors, restockers, packers, checkers, inventory auditors, kitters, and supervisors.
- *Inventory Reduction*—through real-time inventory control, backup or safety stocks can be drastically reduced or eliminated.
- *Error Reduction*—picking is done at a comfortable workstation and is disciplined by the computer, thus costly picking errors are reduced or eliminated.
- *Increased Throughput*—the number of picks per hour per operator often doubles or triples with a miniload system since the operator spends time picking, not traveling and searching for parts.
- *Pilferage Reduction*—access to stock is restricted by the secure miniload storage enclosure, and pilferage is discouraged by random, computerized parts storage and real-time accountability.

- *Improved Production*—costly delays in production due to mispacked parts or shortages of materials are reduced or eliminated.

CASE-IN-POINT

*GRUMMAN AEROSPACE CORPORATION**

Situation

Grumman Aerospace, a division of the Grumman Corporation, is a widely diversified company manufacturing a range of products from aluminum canoes and yachts to truck bodies, highly sophisticated aircraft, space components, and avionics systems. The company's Great River, New York manufacturing center produces electronic systems for on-board flight control and ground support equipment.

Before the AS/RS was installed, electronic components were warehoused in open shelves, with an alphanumeric location system. Twenty-two pickers assembled electronic kits using computer-generated work orders that identified location, quantity, and labeling.

Solution

After detailed study of its part storage retrieval and control requirements, the company decided to install a miniload AS/R system. With the new system, picking is done by an operator at each of seven workstations. The AS/RS, under computer control, brings the parts bin forward to the workstation, displays work instructions to the operator, and restores the parts bin to its storage location when a task is completed.

The miniload is a seven-aisle system measuring 35 feet high by 55 feet long.

The AS/RS manufactured by Supreme Equipment and Systems Corporation, is controlled by a PDP 11/44 computer, with two RKO-7 disk drives, linked to the computer holding the master material file (MMF) through telephone modems to division headquarters.

The system is able to handle kitting of more than 24,000 line items in an average week, and occasionally as many as 35,000 line items.

Kitting is done from requirements generated by the host computer MMF, and loaded to the PDP 11/44 computer via magnetic tape feed. For each part, the bin location, part number, and work order number are displayed at the operator's station. Bins are delivered to operators automatically, in an average time of 30 seconds. When the proper number of parts have been removed and bagged, the operator presses a key to signal that the transaction is complete.

A printer generates a label that corresponds to the display readout and the label is attached to the parts bag, which is then placed on a small conveyor above the workstation that takes the parts to a holding aisle for consolidation into the complete kit. The transaction is recorded on a disk for inventory adjustment. When a special part request has to be processed, the automatic retrieval mode can be overridden by the operator.

The parts inventory in Grumman's AS/RS are stored right to left, based on size. Small parts, like resistors and transistors, are stored in the right-hand aisles; larger parts and components in the middle aisles; and complete kits in the two left-hand aisles, at the end of the conveyor. The kits are stored until they are scheduled for manufacturing, at which time they are automatically retrieved and moved to the manufacturing area.

Restocking is done at night, again with computer assistance. Existing parts are restocked simply by calling up the number on the computer display to find the location. For new parts, an operator determines the aisle where the part should be placed and the amount

*Permission granted by Grumman Corporation.

of space needed. The computer then generates a display of a bin with the proper amount of available space. The operator either can accept the space indicated, or request another.

Transaction data stored on the disk memory are transmitted to the host computer for entry into the MMF. Inventories are adjusted and purchase orders generated to restore stock for any part numbers that have fallen below minimum quantities during the day.

Results

By installing this miniload storage and retrieval system, Grumman was able to achieve the following:

- A 24-month payback on the cost of the system
- Significant reduction of stocking and retrieval errors
- Complete, accurate, readily available inventory data
- Ability to assemble parts and kits in quantities that could not be done without an AS/RS
- Able to handle an 80 percent increase in parts stocked, with a corresponding increase in order volume, with 40 percent fewer personnel
- Ability to return 16,000 square feet of space to manufacturing by installing the system in a new addition to the manufacturing building. With space valued at over $20 a square foot, this saving in manufacturing space amounted to more than $320,000 a year.
- Improved security and more accurate inventory control.

In addition to the AS/RS installation in Great River, Grumman also has consolidated its warehouse operation in Bethpage, New York into one building. The miniload AS/RS aisles in this facility are used for a combination of purchased parts in an MMF, manufactured parts in a manufacturing operational data (MOD) system, tools and other supplies, as well as integrating into a conveyor system that delivers storage bins directly to workstations.

Guidelines for Cost-Effective Use of Automated Guided Vehicles*

Automated guided vehicles are an integral component of any fully automated warehouse operation. AGVs are cost-effective since they:

- Offer transport answers from simple to complex
- Can be configured to meet almost any warehouse transport need
- Provide continuous operation at optimum speeds even if they are slower than manned vehicles
- Deliver loads unmanned to delivery points great distances away on or off of the guide path

AGVs basically can be used to run from one location to another with automatic load/unload at one or both ends.

The advantage in layout is flexibility without path restriction or obstruction. This permits setting up racks or an AS/RS, or any other kind of storage where it can be most functionally advantageous without relation to either shipping or receiving.

Today, there are AGVs for virtually every warehouse need. These include:

*Permission granted by Eaton-Kenway.

- Heavy duty AGVs with capacities of 20,000 pounds or more for transporting extremely large loads to and from manufacturing—and from the receiving dock to high-rise storage serviced by stacker cranes.
- AGVs moving unit loads from the dock to pick-up and drop-off stations, narrow aisle fork truck and AS/RS storage in distribution warehouses.
- AGVs to handle retrieved loads to shipping, or to production.
- Highly automated proprietary processes that load finished goods onto AGVs for transport some distance away from AS/RS.

For example, AGVs can be used for a flexible path handling system that moves pallet loads in and out of this distribution warehouse. Truckloads of the same product can be distributed throughout the system. The loads are retrieved in mixed sequence to satisfy shipping demand. The AGVs were chosen to free the floor area for all kinds of mobile traffic.

Guided wire communications (GWC) series vehicles are microprocessor controlled, flexible path, unit load transporters which may operate as stand-alone vehicles or be fully integrated with other automated material-handling and storage systems.

The GWC transporter follows an electronic guidepath in the floor, receiving directions from, and reporting its status to, a central control and communications computer. The GWC vehicles require no operator intervention to load, unload, and transport a wide variety of material load sizes, weights, and configurations. AGVs may come in various models such as Eaton-Kenway types GL, GC, and GS as follows:

- The GL features a lifting top that fine-positions the vehicle against a pickup/deposit (P&D) stand supporting a load, and raises or lowers its top to pick up or deposit the load.
- The optional GC model features a conveyor top that can interface with a variety of powered, static, or gravity conveyors for input or output of material for processing or storage operations.
- The optional GS model features a shuttle top that picks up and deposits a load alongside the vehicle without driving under the P&D stand. The shuttle automatically extends under the load, lifts and transfers it onto the GS vehicle. Reversing the shuttle motions deposits the load.

How Automatically Controlled Cars Can Enhance an AS/RS*

In many instances it is advantageous to use automatically controlled cars as an integral component of an AS/R system. Beumer Corporation is a major supplier of its Autover System which is a computerized, fully automatic supply and retrieval system (figure 10–8). The Autover System, with the so-called Autocars, performs the following functions simultaneously:

- Transportation
- Storage
- Commissioning
- Distribution

*Permission granted by Beumer Corporation.

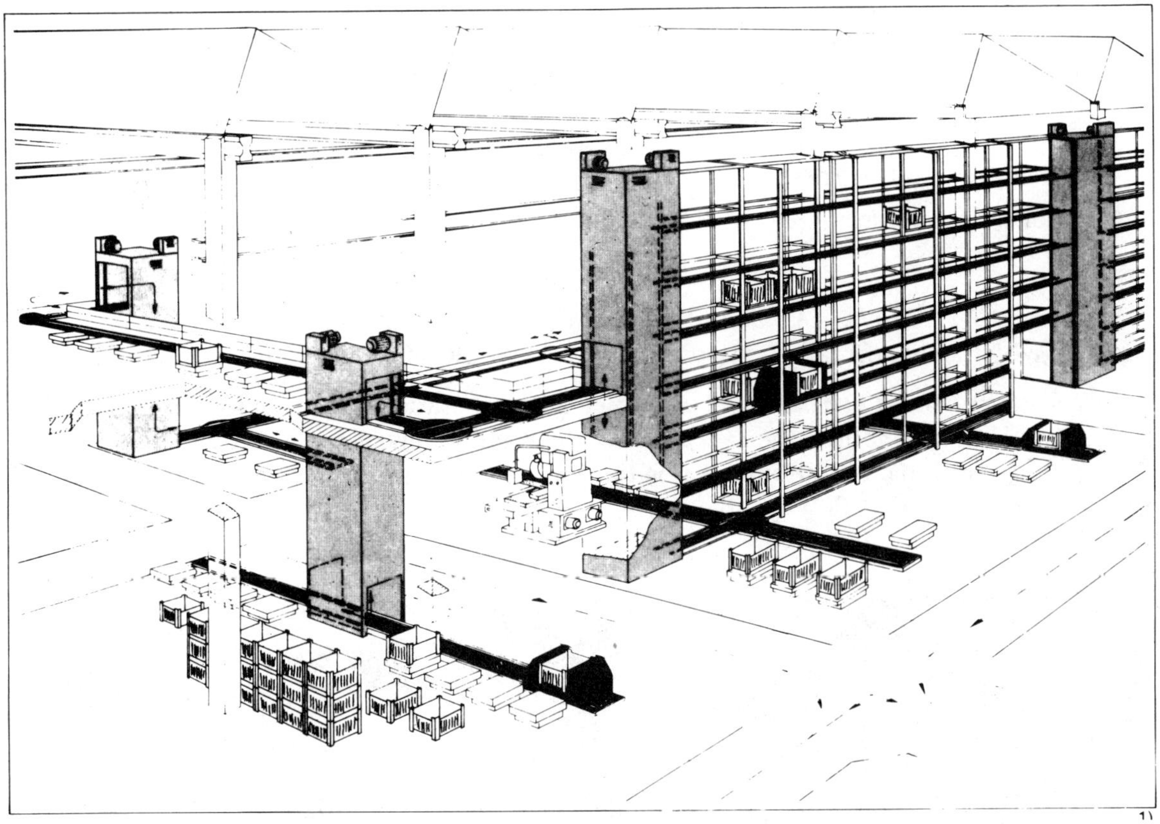

FIGURE 10-8 Fully Automatic Supply and Retrieval System (Courtesy of Beumer Corporation).

The Beumer Autover System has been designed to function most effectively as a sub-system of a computerized integrated and automated manufacturing system (CIAM). This is a logical interface since CIAM provides the basic planning requirements for a supply and retrieval system. From this planning activity, the automatically controlled cars (Autocars) carry out the individual transport commands completely from the storage point to the machine and back independently of each other.

The Beumer Autover System offers a number of advantages, such as:

- High frequency for very extensive plants
- Managing of large transportation and distribution tasks in different levels
- Intermediate transfer of the load unit is not necessary since the Autover System works in the storage, as well as in the commissioning area
- High availability and high throughput through the use of several Autocars, enabling fulfillment of different tasks simultaneously
- High redundance of the Autover System through use of several Autocars
- High adaptability of the entire system to the surroundings
- Clear and quick control of manufacturing through the use of a central system control station

Technical Design of the Autover System

The Autover System consists mainly of the following components:

- Transporting vehicles
- Lift stations and turntables
- Steel construction
- System control

Transporting Vehicles (Autocars). These are transporting and distributing vehicles. The vehicles, which drive on rails, are powered by linear motors (figure 10–9).

Lift Stations and Turntables. Turntables and lift stations are provided to (1) change direction of travel, and (2) allow the cars to enter different levels of the multilevel rail system.

In the lift stations, an automatic positioning mechanism ensures accurate leveling of the rails to an accuracy of ± 0.2mm.

FIGURE 10–9 Transporting and Distributing Vehicle (Courtesy of Beumer Corporation).

Steel Construction. Conventional shelves, rails, pallet deposit points.

Control Technique. The central systems computer controls the processing of commands to individual Autocars. Commands for one particular Autocar are completely independent of commands processed for other cars within the system.

Examples of the System in Use

1. Beumer Autover System in a Storage Facility (figure 10–10). A primary task of this system could be the supply and retrieval of articles to and from other conveying equipment, such as from high storage shelving.

The combination of the Autover System with its Autocars could operate in conjunction with the shelf transport equipment of a high shelf storage unit. The shelf transport

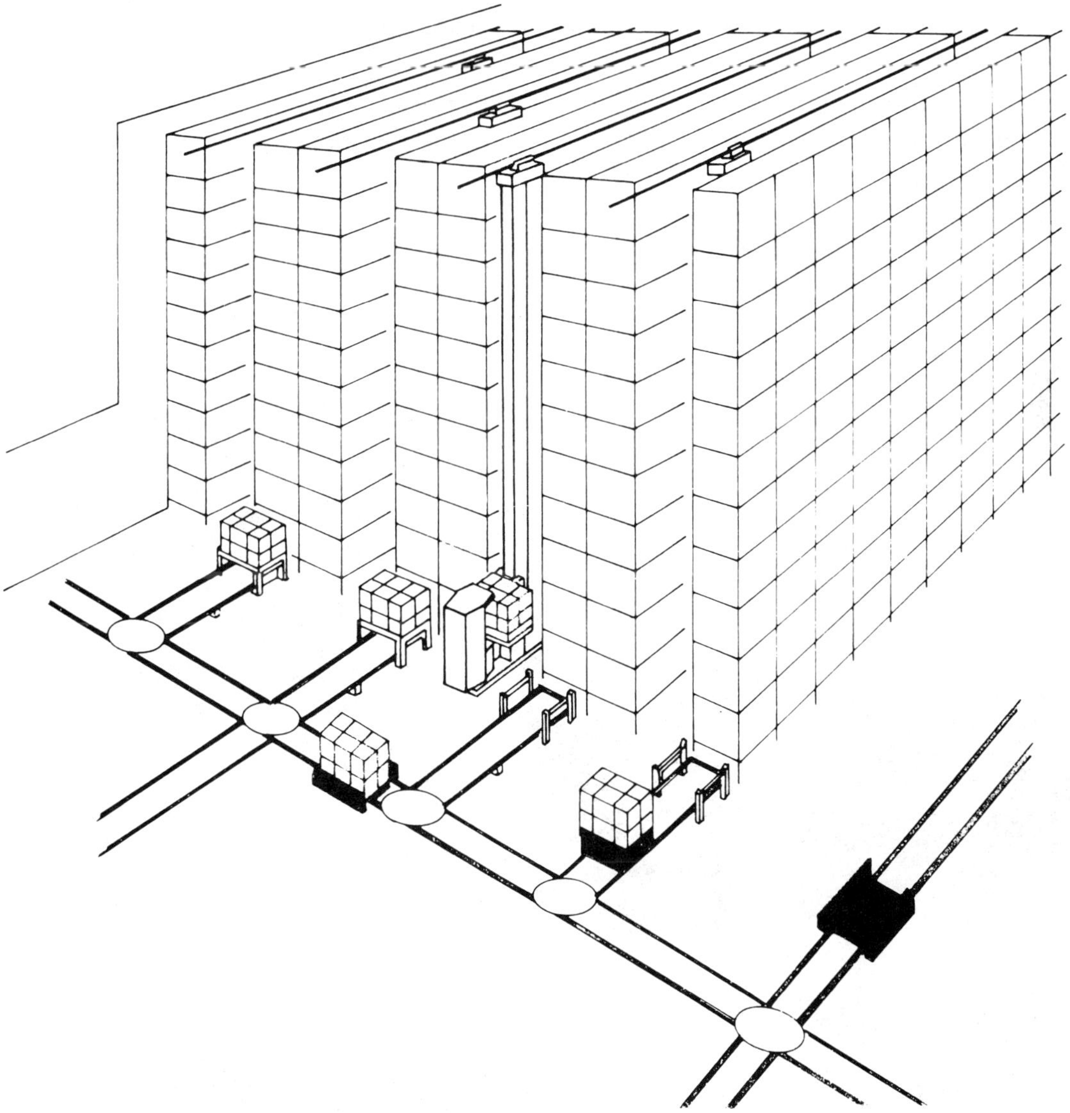

FIGURE 10–10 Autover System in a Storage Facility (Courtesy of Beumer Corporation).

equipment, which operates entirely within the storage facility and cannot leave its work area, can be connected with the rest of the conveying operation quickly and reliably by means of the Autocars. This is especially true if the overall material conveying operation is very large and complex.

2. Beumer Autover System in a Warehouse (figure 10–11). In designing a large warehouse operation, a fast and reliable supply and retrieval of the pallets is of the utmost importance.

In laying out such a warehouse, special emphasis is placed on high redundancy.

The Beumer Autover System can offer a distinct advantage because of its flexibility in the following areas:

- The supply of the full pallets into storage is accomplished by means of two feed tracks, which can be used as emergency retrieval points in the case of a failure of the storage system.
- The pallets to be placed onto the Autocars are stored in readiness also at two points.
- Each storage level is integrated with a feed track, the tracks being connected with each other by means of two lifts operating independently of each other.

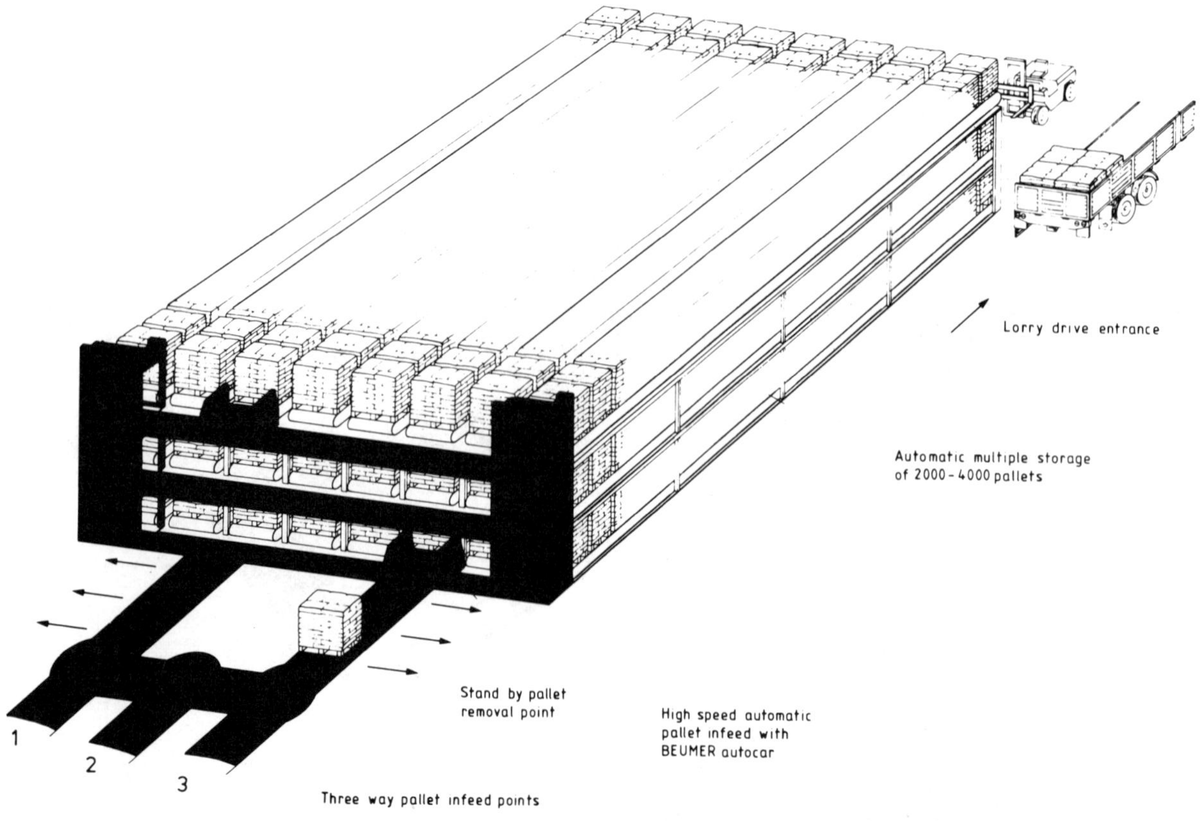

FIGURE 10–11 Autover System in a Warehouse (Courtesy of Beumer Corporation).

- The two lifts transport the Autocars to the appropriate level, from which point material is brought into storage. It also is possible for the Autocars to run a circular course with this setup.
- In case of failure of one Autocar or one lift, it still is possible to run the entire system in emergency operation—that is, at 50 percent capacity using one Autocar or one lift.
- The supply system travels short distances and can approach both ready pallet storage points automatically.
- Increased job safety, as no personnel is necessary for the supply of the storage area.
- All control operations are completely automatic.

How the System Retrieves Material

Retrieval of the pallets from the continuous shelf storage area can be accomplished with the same arrangement of the Autover System as for the supply operation.

Unit Load AS/RS: What It Is and When It Is Useful

A unit load AS/RS is any load configuration handled as a single item (figure 10–12). The unit load storage methods are expressed in a variety of rack configurations, such as:

- Single-wide aisle/single-deep rack
- Single-wide aisle/double-deep rack
- Double-wide aisle/double-deep rack

Unit load AS/RS are used for everything from the storage of automobiles to palletized and containerized loads. They can have completely automatic conveyorized input or be fed by fork lift trucks or AGVs. Today's advanced systems use a variety of special fixture load platforms to handle finished products, to store products on test, or to supply robotic assembly systems.

How Car-in-Lane Systems Combine High Storage Density and Automated Control

Car-in-lane AS/R systems combine high storage density with automated control of stored parts. These AS/R systems can be configured in several ways, but most systems use shuttle cars to carry unit loads into horizontal storage lanes, and transfer cars to move the shuttle cars, with their loads, into different storage lanes. An AS/RS machine can take the place of the transfer car.

Some cars are completely self-contained with on-board power and remote control while others use an umbilical cord for power and control signals inside the rack.

Car-in-lane systems can provide either last-in, first-out (LIFO) or first-in, first-out (FIFO) storage, depending on the layout of the system and throughput demands. For FIFO storage, an additional aisle is needed in system layout to provide access from both ends of the storage lanes.

FIGURE 10–12 Unit Load AS/RS (Courtesy of Integrated Automation, Alameda, California).

Human-Ride Machines: An Automated Storage Alternative

Human-ride machines offer another automated storage alternative with operator intervention. These machines lift an operator to elevated heights, up to about 50 feet where the operator can gain access to part or all of a load placed in storage. Some machines are equipped with shuttle tables, so an operator can easily handle both small parts and heavy unit loads.

Human-ride machines can be combined with gravity-flow racks to offer still another type of AS/RS. Instead of using a shuttle car to advance stored loads down a lane, these systems use gravity-flow racks, which can be controlled in various ways.

Why Gravity Flow Racks Work Well in FIFO Systems

Gravity flow racks have particular application where first-in, first-out order of multiple pallets of the same product is required. The lanes may be as short as four pallets deep, but loading from the rear and picking from the front keeps aisles open for continuous loading and discharge in the same time without lift truck intervention.

Most common flow racks use rollers or wheels and gravity to permit loads to move forward toward the picking front or output end of the rack.

One of the newest gravity flow systems uses troughs for flow rails which support pallet carriers, the edges of which conform to the troughs. When air is introduced into the troughs, it exits through holes under the conforming pallet carriers. The air creates a frictionless film between the two, thus allowing a pallet of virtually any weight to move forward easily.

Gravity flow racks used in conjunction with automated storage/retrieval machines provide some of the fastest manual order picking systems with mezzanines and conveyors at various levels in the racks. Access for the pickers and takeaway for the casegood is provided.

Deciding between Horizontal and Vertical Carrousel Storage and Retrieval Systems

Horizontal

Warehouse applications of horizontal carrousel configurations are always specialized applications. They are used in medium to high-activity load and pick operations. They are most useful in warehousing less-than-full-case lots of small parts. In particular, carrousels are found in companies where there is a high variation/low volume-type operation. This would apply to situations where perhaps only one box per order was being utilized. Based on this requirement, carrousel manufacturers have been able to marry carrousels to transporters and speed throughput in general, and parts picking in particular. It also has brought about as much as 40 to 80 percent savings in warehouse space.

The tangible benefits of increased productivity are as important as the less perceptible benefits of increased accuracy of placement, increased safety, and effects on employee morale because of less fatigue and simplified training.

Bringing the box to the worker has proven especially important in warehousing operations where it is not uncommon to experience time savings of nearly 400 percent in picking operations.

Vertical

A material-handling system that companies are finding to be cost effective in labor savings and increased warehouse productivity is the vertical carrousel storage and retrieval system. This system can be quite versatile depending on whatever level of automation is needed to suit your particular needs, and a potent material-handling vehicle in combination with a computerized material requirements planning and inventory control system. This interface can enable immediate updating of parts inventories on the mainframe computer.

Vertical carrousels are ideal for small parts and supplies storage since they are freestanding, mechanized, circulating shelving systems designed to minimize floor space, maximize cube utilization, and improve productivity by permitting a substantial number of high capacity storage shelves to be loaded and picked from a single work counter.

Why Use of Automated Selectors and Dispensers Is Growing

Many types of machines can automate order picking. These range from full case selectors down to single item dispensers. These methods are extremely productive in high volume operations.

Accuracies in automatic case dispenser systems range in the 98 percent or greater area, and with bar code verification, can reach 100 percent accuracy on order picking.

Automatic replenishment systems can depalletize incoming loads a layer at a time, singulate and orient the cases, and automatically feed them into the correct dispenser lane.

Individual items of an odd shape, as well as uniformly shaped items can be automatically dispensed onto a moving conveyor belt in pick order sequence.

With the pressure from retailers to reduce their inventories and rely more heavily on distribution warehouses to ship them only what they need, the value of case and item dispensing systems and more frequent ordering of small quantities will point the way toward these faster and more accurate order assembly techniques.

How Sortation and Consolidation Areas Are Being Automated

Sortation and consolidation areas are becoming highly automated with moving beam scanners, automatic diverters, and sophisticated conveyor systems that combine to move products into shipping lanes with tremendous speed and accuracy, including an accounting of every item in a real-time mode.

HOW TO MAXIMIZE PRODUCTIVITY WITH COMPUTER-INTEGRATED WAREHOUSING SYSTEMS*

To maximize productivity, materials control and materials handling should function as a single integrated system. With this system's integration it is then possible to track materials throughout the entire warehousing and distribution cycle, from receipt of material to storage, and then to shipment of the product.

In integrated real-time systems, the major elements are the hardware for the physical movement of material, the inventory management system, and an information and control system based on a computer network.

Why Distributed Control Is the Most Effective Approach

The most effective approach for implementing an integrated system that provides total materials control is a distributed network. This is true for several reasons.

One reason is in the nature of automated materials-handling systems, namely that they require distinct levels of equipment control and data processing. A distributed processing network readily meets these requirements. Distributed networks also permit evolutionary change. Functions can operate manually at first, then be automated later on. They also can be automated separately, then interfaced together whenever the time is right.

A distributed network also provides for a distributed data base as well as distributed processing. The data base should be constructed so each processor contains the data locally that is required to make processing decisions.

An S/R machine, for example, should be able to access all locations in the aisle using only the on-board microprocessor and local data. The only outside stimulation required is the command to move to a specific location.

What Happens at Each Level of Equipment Control and Data Processing

An automated materials handling system includes distinct levels of equipment control and data processing.

At the lowest level, each component of the system can be controlled by a dedicated on-board microprocessor. These controls are inexpensive and they offer flexibility and reliability. They also interface well with minicomputer systems and they can be backed up for additional reliability.

*Permission granted by Catalyst U.S.A., Inc.

Today, microprocessors are being used to control storage/retrieval machines, conveyors, robots, guided vehicles, palletizers, weigh scales, and packaging machines. In addition, they can provide self-diagnostics, machine component monitoring, up-time and down-time records, overhaul projections, and materials flow data.

If a system component cannot be controlled by a microprocessor, then provide for that control in the network, and then incorporate it at the appropriate time.

The next level of control is the materials handling computer system, which controls materials flow and materials storage. This system is usually based on a minicomputer and is the intermediary between the mainframe corporate computer, which performs material requirements planning, and the process control provided by the microcomputers on the handling equipment.

The materials-handling computer system accepts plans from the host system and directs material flow and process control. This portion of the system also provides the bulk of the data processing and real-time data collection. It provides management with reports and gives operators the ability to inquire on the status of materials.

The next higher level of control, typically, is the host computer. It provides the capacity needed for extensive data management and file processing.

Material requirements planning (MRP), order entry systems, complex order processing systems, and factory control systems should be at this level. In most cases, those functions create requirements that are much too great for the minicomputers normally used for materials handling and materials control.

The Critical Link between the Host Computer and Materials-Handling Control

One of the primary problem areas in integrated systems is to properly define the interface between the materials-handling control system and the host computer. The crux of the problem lies in materials control.

There are two categories of materials control: (1) inventory control systems, and (2) stock locator systems.

Inventory control systems control materials flow, materials storage, and inventory levels. A comprehensive inventory control system contains all the functions required to control an entire manufacturing facility. Its only interface to the MRP system is to obtain scheduled orders and to transmit inventory transactions, relative to orders, back to the host.

An inventory control system is capable of controlling material from the moment it is received into a facility until it leaves the shipping dock. It can handle all of the functions, from material input to material output.

This type of system, and the level of control provided, is usually implemented when an entirely new facility is built or when an existing facility is completely modified.

A stock locator control system, on the other hand, operates as a slave to an inventory control system. Essentially, it controls the input and output operations of the storage system. It receives material requisitions in the form of pick lists and it transmits material store/issue transactions after they are accomplished.

This type of system usually is implemented when an automated system is incorporated into an existing operation and inventory control is already on a computer system.

The interface between the automated materials-handling systems and the host system may be either batch mode or on-line direct link. Either is applicable to both categories of material control systems.

Systems complexity, however, may be significantly affected by the interaction between the two systems regardless of the type of interface. The level of control from the host to the material control system, and the data which is transmitted, can have serious performance and sophistication requirements.

Basically, an advanced system will eliminate redundant processing and redundant data located in both systems. Also, the system should provide a logical update scheme for data transfer between the computer systems, so the data bases are synchronized but need not be identical during real time.

Tips for Planning the Computer Capacity of an Integrated System

The capacity of an integrated system is a critical design parameter and should reflect future growth as well as current requirements. This applies not only to automated handling systems but to computer control systems as well.

A major consideration, then, in planning the materials-handling control system is the number of transactions that each processor or each computer will be required to perform. These figures must be determined, and hardware with enough capacity for each job must be specified.

A related consideration is the number of inquiry terminals in the system. Too many inquiries simultaneously may saturate the system, slowing response time. This can be critical. System operators will lose respect for the system if they must wait an excessive length of time for the computer to respond to their inquiries. An accurate prediction of required response times should be made.

The Simpler the Real-Time Software System the Better

Another very important consideration is to keep the real-time software system as simple as possible. Most problems in software can be traced to lack of simplicity. So, it's best to perform only the basic material identification, movement, tracking, storage, and material control functions in the materials-handling system computer. Avoid the "bells and whistles." But if you must have them, off-load them to another system.

What to Consider When Planning a Material Identification System

Still another major consideration is material identification. The more automation and integration that you desire, the more critical it is that proper material identification techniques are used.

To control material automatically, materials-handling equipment must be able to "see" and identify materials at decision points. Bar coded labels, magnetic strips, or alphanumeric characters make that possible. Automatic identification equipment then can be used to read the labels and transmit the required data to the computer.

A second consideration is that a part–number scheme and a load identification technique must be established appropriate to the level of automatic operation to be used. To work properly, automated systems require unique numbers for each item to be controlled.

GUIDELINES FOR DETERMINING THE CORRECT LEVEL OF MATERIAL-HANDLING AUTOMATION

Stay with Proven Technology

1. Whenever possible, use concepts, equipment, and controls that have been proven in a similar application and environment.

2. Minimize the risks associated with prototype controls, equipment, and software.
   (a) Carefully examine all the key elements of the system and identify the prototypes or unproven technologies.
   (b) Allow adequate time and budget for prototype testing.
   (c) Do not put too many prototypes in one system.

When to Automate the Loads

1. Do not automate if the physical characteristics of the loads being handled cannot be predicted.
2. Do not try to handle all loads in one system. Exclude exceptionally large, unstable, or unusual load configurations.
3. Do not use an automatic material-handling system as an inspection device for rejecting oversized or unstable loads.

Rigidity Versus Flexibility: Know What to Expect from Equipment

Automated systems are generally quite rigid or fixed in their functions, maximum throughput, and maximum storage capacity. Justification of a system that can handle all possible peak load conditions is usually impractical.

The following are some "flexibility" guidelines in selecting equipment types:

1. Automated storage machines are rigid in their maximum input and output rates.
2. AGV systems are flexible because more machines may be added and guidepaths easily changed, however, AGV justification becomes more difficult as throughput volume goes up.
3. Conveyor systems are rigid from a maximum throughput standpoint, however, they are generally more applicable in higher volume systems than AGVs.
4. Lift trucks and purely manual systems are the ultimate in flexibility, however, they are labor intensive and difficult to control.

How to Establish System Boundaries

Users who take system responsibility may limit the level of automation possible and complicate system interfaces by separately implementing subsystems that should be designed and constructed as one system. Guidelines for establishing systems boundaries are:

1. Do not split responsibility when system design and installation should be performed by one source due to tight load tolerances, complex mechanical interfaces, or complex control interfaces.
2. Use a single source when control functions should be grouped logically in one computer system.
3. Use a single source when one type of equipment—such as AGVs—can be used for multiple functions (such as factory transportation as well as input/output to a high-rise storage system).

Why Simulation Is Essential for Design Verification

The system is too complex or not defined well enough if the design cannot be tested by simulation before it is constructed.

How to Avoid Complicating Real-Time Computer Systems

Real-time computer systems for material-handling automation are intrinsically very complex. Do not complicate them anymore than necessary. Here are some basic rules:

1. Simplify real-time software systems whenever possible by performing only the basic material identification, movement, tracking, storage, and material control functions. When in doubt, *do not* do it in the real-time system computer—off load it to another system.
2. Use standard proven software designs whenever possible. The probability of completing a project on time and in budget is related more to the amount of new, unproven software in the system than any other factor.
3. Do not assign responsibility for the real-time software, which will be installed in one minicomputer, to two different software teams at different locations.
4. Minicomputers have capacity limitations that must be carefully analyzed before a real-time software design is finalized. When in doubt, buy a larger, faster computer or distribute functions to several computers.
5. Do not put MRP, order entry systems, complex order processing systems, or factory production control systems in the material-handling system computer.
6. Resist the temptation to believe that real-time software can be transferred easily from one computer to another. Different operating systems and communication protocols require extensive rewriting of software.

Real-time material-handling system computers are the links that make integrated factory and warehousing operations possible. Following are guidelines for determining the level of computer automation which will facilitate system integration.

1. Whenever possible provide the means for computer tracking of individual loads.
2. Provide the means for communicating real-time material status to higher level factory or distribution center computer systems. However, limit the amount of communication. Highly interactive systems are very costly and difficult to debug.
3. Some real-time computer control in automated material-handling systems is almost always justified. When in doubt, opt for computer control with planned expansion even if current control functions are relatively simple. This greatly facilitates system upgrading and integration with other factory or warehouse systems in the future.

Use automatic identification equipment in material handling-system computers to:

- Reduce the cost of large real-time material tracking systems
- Improve accuracy and speed when inducting loads into a system
- Improve accuracy when integrating manual and automatic elements of a system

How to Overcome the Second System Syndrome

After successfully completing a small automated system, do not be lulled into thinking it's easy to do the same thing on a larger scale. Controls for larger real-time automated systems are usually much more complex than indicated by a simple doubling or tripling in physical size and function. A good rule is that twice as many devices or automatic functions in a system is four to five times more complex from a control standpoint. This also means it will take four to five times as much effort to checkout and put the system into operation.

Manually Operated Equipment versus Total Automation

Although more and more companies are achieving a high order of integrated automated systems, most systems are usually a blend of manually operated conventional equipment, with the predominant role handled by automatic equipment. Simply stated, it costs a disproportionate 90 percent to automate the last 10 percent of the system.

Understanding the Importance of People in System Planning and Implementation

- Do not automate unless you have access to an experienced, capable project team.
- Involve operating management and management information systems (MIS) in system planning. If they are not involved, they will find reason to disown it.

Tips for Establishing the Right Level of Human Involvement in Automated Systems

As stated previously, total automation is very difficult to achieve. Even most highly automated systems are a blend of mechanized and human functions. Rules to follow in determining when to utilize people versus machines are:

1. Minimize the use of people in material induction and tracking functions. They have limited capability to accurately identify and track material at the rates needed for most automated systems.
2. Use people rather than computers for preplanning the day's activity in an automated system.
3. Computers are much more effective in monitoring and controlling an automated system than an operator with an enunciator board. In most cases enunciator boards or system status panels have limited value and are very costly.
4. Use people to make decisions when the computer software required to automate decision making becomes highly complex.
5. Do not make the system operator's job overly complicated by designing software to accommodate every possible exception condition. Handle seldom seen exception conditions by manual intervention.

A Hint for Justifying the System

Improve payback and avoid unnecessary automation by justifying each major subsystem and each system element as well as the total system.

How to Assure Reliability

1. Reliable system operation is possible only with a good maintenance program. Users who are not prepared to make the investment in qualified maintenance personnel, training, and spare parts should limit automation.
2. 100 percent system reliability is impossible in an automated material-handling system, and a system should not be planned if 100 percent up time is a "must." A good rule of thumb is to plan to operate on the basis of 95 percent up time. With proper maintenance all systems should perform at that level or better.
3. A backup minicomputer system is generally needed in all large highly active systems which required 95 percent up time. Semi-automatic backup modes for computer controlled systems are almost never used and their cost is usually not justified.

TECHNIQUES FOR MEASURING PRODUCTIVITY IN MATERIAL HANDLING

Basically, any productivity device to measure material handling efficiency should be used to monitor performance over a period of time to determine trends or to establish a need for corrective action. If the numbers are available, absolute figures can be compared with those from other plants in the same industry, or having similar handling operations. The following are some typical productivity ratios:

1. Material-Handling Labor (MHL) Ratio

$$\text{MHL ratio} = \frac{\text{Personnel assigned to MH duties}}{\text{Total operating personnel}}$$

This ratio represents the number of personnel assigned to material-handling duties in proportion to the entire work force. It can be determined on the basis of head count or payroll costs. Some support activities (maintenance, tool room, production control) are not devoted full time to material handling. An estimate of the percentage of time spent on handling should be used.

A variation of this ratio, called the direct labor material-handling ratio, can be used to measure the percentage of direct labor that is spent on material handling. The required data can be obtained from work sampling or other analysis techniques.

2. Handling Equipment Utilization (HEU)

$$\text{HEU} = \frac{\text{Items (or load weight) moved per hour}}{\text{Theoretical capacity}}$$

The way this ratio is determined will vary from one facility to the next. Therefore, it is meaningful only if used to make relative comparisons over a period of time, within a given operation. To use this ratio properly, you must first decide what is meant by theoretical capacity—or full utilization—and then stick by the definition. For example, some engineers may consider a piece of equipment fully utilized only when it is carrying a full load. On the other hand, others may feel it is properly utilized when empty but heading toward a loading station.

3. Storage Space Utilization (SSU)

$$\text{SSU} = \frac{\text{Storage space occupied}}{\text{Total available storage space}}$$

This measure will be applied most frequently in warehousing and other storage operations. Cubic space should be measured rather than floor area. In collecting the data, keep track of the percentage of bin and rack openings that are empty. Of the ones that are occupied, note whether they are fully or partially utilized, and if practical, try to estimate percentage of utilization.

4. *Aisle Space Percentage (ASP)*

$$\text{ASP} = \frac{\text{Space occupied by aisles}}{\text{total space}}$$

Space is becoming extremely costly in both manufacturing and warehousing. Aisles and traffic patterns should be laid out carefully to use available space most productively. The calculation should generally be based on cubic feet of total space.

Also keep in mind that too low an ASP figure may be as bad as one that is too high. A reasonable number of both traffic and access aisles must be provided to maintain desirable levels of throughput and productivity.

5. *Movement/Operation (MO) Ratio*

$$\text{M/O ratio} = \frac{\text{Number of moves}}{\text{Number of productive operations}}$$

This relationship reflects the overall efficiency of material-handling operations in the plant. It can indicate the number of handling—and rehandling—steps that are involved in receiving, storage, manufacturing, and other departments. Typically, a high ratio will indicate an improvement opportunity in the form of fewer handling steps, simplified operations, or use of mechanized equipment.

6. *Manufacturing Cycle Efficiency (MCE)*

$$\text{MCE} = \frac{\text{Time spent in actual production operations (machine time)}}{\text{Time spent in production department}}$$

This ratio measures how effectively materials flow through the manufacturing process. Delays encountered can be caused by various factors, including poor scheduling and routing, inefficient machine operation, or insufficient local storage. Different figures can be expected from job shop operations and continuous processing systems. As with most of these ratios, the key is to apply and observe them consistently over time.

7. *Damaged Load (DL) Ratio*

$$\text{DL ratio} = \frac{\text{Number of damaged loads}}{\text{Total number of loads}}$$

This ratio shows how effectively and properly crews are handling incoming and outgoing goods and in-process materials. Some programs of sampling should be established to generate damage data. The ratio can be applied to a number of different departments and stages of material flow in the plant.

Tips for Successful Automation: Evaluating Energy Efficiency

More than ever before, every one of us is concerned with the energy efficiency of our equipment and operations, and ratios covering this important factor can be utilized.

It may be possible to determine separate energy efficiency ratios—relating throughput to energy consumption—for individual items of equipment such as conveyors or lift trucks. However, a more meaningful evaluation may be energy utilization for the entire facility, for example, BTUs consumed per cubic feet of warehouse space. These ratios can be kept low with such measures as providing minimum heating and cooling in unmanned areas or automated storage facilities, and using task lighting or lights mounted on mobile equipment in place of permanent overhead lighting.

INCREASING PRODUCTIVITY THROUGH WORK-VOLUME AND WORK-DENSITY ANALYSIS

A first step toward increasing productivity in a facility might be the evaluation of effectiveness and appropriate application of existing handling systems. One technique for accomplishing this task is *work-volume analysis.* It also can be used for determining new system requirements. Work-volume analysis is a procedure in which the characteristics, volume, and handling requirements of the material that is to be transported within an area of the plant are arranged systematically on a chart. By itself, the procedure does not solve any material-handling problems. However, by displaying important handling system data on a work-volume chart, hidden system requirements can be discovered and many benefits realized.

A work-volume analysis normally leads to the establishment of important system parameters. Very often this technique will help to support decisions such as determining:

- General class of equipment to perform a task within a department
- Amount of work to be done in units appropriate to the type of equipment
- Specifications and quantity of handling equipment required
- Placement of equipment and personnel requirements

How to Conduct Work-Volume Analysis

To develop a work-volume chart for a selected class of operations or for an area, detailed data must be collected on each of the types of material that is handled. It is important that all pertinent data is gathered. The actual selection of data to be included on a work-volume chart will depend on the situation under study.

Work-volume charts follow the same sequence that governs the physical flow of materials: receiving, processing, and shipping. Each operation must be examined and identified separately on the work-volume charts because certain operations will require variations in equipment and labor grades. The following steps can be utilized when constructing a work-volume chart:

1. Determine the major class of operations to be studied (i.e., receiving, shipping, processing) and identify the individual operations to be analyzed (material moves, inspections, storage, machining, etc.).
2. Gather data on the material handled at each operation. For each unit handled the size, shape, fragility, material and handling method should be noted. Also the volume, quality inspection procedures, storage method, and characteristics of a unit load are important data. The data should reflect the nature and volume of the work as well as the conditions under which the operations are performed.

3. Examine the data collected and determine if any important information can be calculated from the data. For example, load quality data could be divided by handling unit data to obtain handling units per load.
4. Lay out a matrix (rows and columns) type of chart with columns that are of a width that will facilitate recording the data gathered in Step 2 or calculated in Step 3. Place the headings along the top of the columns.
5. On the left-hand side of the chart identify the areas or operations for which data has been gathered or calculated.
6. Record the data.

When to Use Work-Density Analysis

A work-volume analysis is especially useful for identifying equipment or personnel that are underutilized. A procedure sometimes referred to as *work-density analysis* involves balancing the workloads on the operations that have been charted. There are no known formulas for calculating work density. Each individual operation must be studied and compared with other operations. Some of the general rules to be observed when performing a work-density analysis are:

1. Material-handling personnel are generally most efficient when under constant supervision. Try to concentrate activity to provide effective supervision.
2. Locate storage areas as close to each center of work density as practical.
3. In using conveyor loading and unloading, consider the possibility of automatic handling.
4. Make use of quick-change attachments for trucks and cranes to make them adaptable to different handling operations within an area.
5. Consider compromising the specifications for handling equipment to broaden the usefulness of each unit. This rule applies to all types of equipment—conveyors, containers, trucks, trailers, and cranes.
6. Analyze the work requirements of the material-handling activity as a whole as well as by segments, to improve the work-density factor.

GUIDELINES FOR MAINTAINING PRODUCTIVITY IN THE PICKING OPERATION

One of the key elements in warehousing is the picking operation. Depending on the type of warehouse, 30 to 40 percent of the labor cost can be in this activity. Some of the more basic steps that can be followed to reduce this costly item of expense are:

1. Automate preparation of the packing list, bills of lading, labels, and any other shipping documents.
2. Computerize the inventory control system so that items can be deducted automatically from inventory when picked.
3. Design the storage area so that item placement is arranged strategically for optimum picking.
4. Develop standard packs to speed up the retrieval of items from stock.

5. Design the order system so that a picker can make one pass to pick an order.
6. Properly mark and label items in stock.
7. Install the most efficient equipment around the unique characteristics of your warehousing operation.
8. Design the warehouse layout to minimize the time needed to pick from shelving, the floor, flow racks, pallet racks, or drive-in racks.
9. Avoid equipment breakdowns that have a bearing on picking by providing proper equipment maintenance.
10. Maintain load integrity that can impact the picking operation.
11. Avoid heavy traffic congestion in the picking area.
12. Set up picking standards to measure labor productivity.
13. Establish an effective training program to ensure that pickers are thoroughly cognizant of the methods and procedures involved in this activity.
14. Assign experienced and competent supervision to the storage area to ensure that work is being performed in the most effective manner.

WHY "PAPERLESS" PICKING SYSTEM AND PICK DIRECTOR SYSTEMS CAN CUT LABOR COSTS*

As discussed previously, the picking operation is the highest labor-intensive activity in warehousing. Because of this, a number of vendors have developed "paperless" picking systems, or pick director systems which can sharply reduce labor costs and increase order accuracy to levels approaching 100 percent in stockroom, warehouse, and distribution centers. These various systems all have basic similarities and exclusive features. One of the most advanced systems, SWIFTPICK, has been developed by Integrated Automation.

SWIFTPICK systems consist of miniature "intelligent" computer terminals that are displays, light/switch combinations of a sophisticated communications network, and a computer which controls all these elements. (See figure 10-13.) The displays and lights are mounted in a rugged plastic extrusion that is fastened to the leading edge of flow-rack or shelving. (See figure 10-14.)

How the System Works

When in operation, a picker identifies the order by scanning the bar coded order number on a label or traveler document, or by entering the order ID on a nearby display. The display responds in one of two ways, depending on the configuration in use.

If the zone is equipped with a display at every picking location, the displays prompt pickers with order quantities. All picks in the zone are indicated simultaneously.

If the zone is equipped with a "light-controlled" display, then a display associated with each bay indicates the pick quantity for the first item to be picked in that bay, and a light below the picking location indicates the first item to be picked.

The picker follows the direction on the display and picks the appropriate quantity. When the picker has counted out the quantity indicated, he or she either presses a "pick complete" button on the display or presses the button under the light to indicate that the pick is complete.

*With permission from Integrated Automation; Alameda, California.

FIGURE 10–13 SWIFTPICK Display (Courtesy of Integrated Automation, Almeda, California).

FIGURE 10–14 SWIFTPICK Displays Mounted on Shelving (Courtesy of Integrated Automation, Almeda, California).

If it is impossible to pick the indicated quantity because of insufficient stock, the picker can quickly adjust the quantity using the display. In this manner, higher speed picking is accomplished at the same time as the data collection required for inventory updates and manifest generation, with no paper handling and reduced picking, checking, and data entry labor.

Purpose of the System

The primary purpose of a system of this type is to:

- *Capture* downloads of orders to be picked from a host computer system.
- *Guide* the pickers on the warehouse floor through picking those orders.
- *Track* the progress of each order noting shorts and other exceptions, and time-stamping all transactions.
- *Transmit* uploads of completed orders along with any exception information back to the host computer system.
- "Tie-together" identification number of tote, box, or shipping/picking label with all items that have been picked or put aside to support automated manifesting and billing in subsequent steps in the operation.

SWIFTPICK's Three Types of Picking Systems

A SWIFTPICK system layout is comprised of locations, bays, and zones.

A *location* is the smallest unit in the picking system. By definition, a location may contain only one SKU, although a single SKU may exist in multiple locations. Each location is associated with a display or a light, which is usually mounted just below the physical rack or shelf location.

A *bay* is a collection of locations. Most systems would be comprised of bays which consist of between five and seventy-five locations. The size of the bays and the locations associated with a bay change infrequently, usually only when physical racking changes are made. Although in many instances a bay corresponds exactly with a physical bay of flow rack or shelving, this is not required. Each bay normally contains a bay light which is illuminated whenever a pick is to be made within the bay. This helps the picker locate required picks more efficiently within a zone.

A *zone* consists of multiple bays and is usually defined as the territory which is serviced by a single picker (or more than one picker, at times). It is not uncommon for the size of zones to vary. In fact, changing the number of bays within a zone is the primary means by which the picking supervisor balances the workload for the pickers. In each zone, one or more bar code scanners are used to identify the order number of the carton or tote which is to be processed within the zone. To avoid confusion, only one order may be handled within a zone at a time, although the system will allow orders to be placed on hold so that another order may be processed. This allows priority orders to be rushed through, or temporary shorts to be resolved quickly and economically without forcing a backorder or short shipment.

A *putting* system also can be used for direct "putting." In this method, lights indicate quantities to be distributed into waiting totes or containers of any one SKU.

TECHNIQUES FOR CAPTURING DATA

A major element in a warehouse operation is information captured, transmitted, stored, and retrieved. The scope of data captured in the modern warehouse ranges from automatic identification equipment to vocal and tactile data input devices. The aim of all these related technologies is to reduce the amount of paperwork, time, or complexity in generating and using important alphanumeric information.

Automatic identification technologies cover a wide spectrum of methods, such as:

Image Based Techniques

Type	General Description	Characteristics
Bar codes	Data is encoded in alternating light and dark bars of varying thicknesses. Read by scanner or light pens.	Low cost, high reading accuracy, with less than one error in 3 million reads. Considerable data in little space.
Optical character recognition	Printed alphanumeric codes that can be used alone or with bar codes. Is human readable.	Low cost, variable reading accuracy. Does not require sophisticated computer system to be decoded.
Vision systems	Video system reads many automatic identification codes and identifies parts. Can be tied to a main computer.	Relative high cost. Reads at slow to moderate speeds requires line of sight to read. Distinguishes shapes.

Non-image Based Techniques

Type	General Description	Characteristics
Radio frequency surface acoustic wave	Both techniques encode data on a chip encased in a tag. When in range of a special antenna, chips are decoded by a reader.	Unlike SAW chips, RF ones can be reprogrammed and carry more data. Both are expensive. Readability not affected by dirt or line of sight.
Magnetic strips	A coded message is programmed onto a magnetic film attached to a special card. Noncontact readers decode data.	High data density, reprogrammable, more costly than bar codes, unaffected by dirt or oil.
Voice recognition	A computer-based system that translates operator spoken words into computer data without special codes.	Leaves operator's hands free, relatively high cost, limited vocabulary recognition, must be "trained" to operator's voice to attain high accuracy.

Benefits from these technologies are better control, increased accuracy, more timely real-time information, lower costs, and increased warehouse productivity.

GOALS AND BENEFITS OF DISTRIBUTION RESOURCES PLANNING*

The operational complexities of distribution management have created a requirement for more responses, accurate, and meaningful information regarding customer service and inventory investment. The need to control the desired results from distribution management by projecting requirements, planning activities, and monitoring progress exists at different distribution levels, such as branch warehouses, factories, and vendors.

One of the key objectives of distributed management is to provide the desired customer service levels while minimizing inventory investment and other distribution costs. DRP

*Permission granted by American Software, Inc., 1985, all rights reserved.

software—consisting of planning systems, scheduling systems, and status systems—provides distribution management with the capabilities to obtain these and other desired results.

The user may have a well defined distribution network system. Actual customer demand and anticipated customer demand (sales forecasts) are key influences on replenishment actions that occur at each distribution level. Sales forecasts are required for any location in the distribution network that provides shipments to customers. Safety inventory may or may not exist for each item at each customer shipment location. Safety inventory exists for protecting customer independent demand, or for protecting dependent demand resulting from the requirements at a lower-level warehouse. The quantity of safety inventory required for a specific level of customer service can be reduced as the accuracy of the sales forecast improves.

Numerous benefits are available to the users of a distributed resource planning system. Sales forecasts are more accurate, thereby requiring less safety inventory for a given level of customer service. In this manner, safety inventory is allocated where it is needed, but only to the extent necessary. Safety inventory can be allocated to items, groupings of items, and warehouse locations consistent with providing different customer service levels. Differences in the competitive environment associated with types of items and warehouse locations usually can be identified and reflected in varying customer service requirements. A variety of customer service levels, inventory stocking points and levels, and alternative distribution strategies can be simulated so that the user can implement decisions that will enable the desired customer service levels to be obtained while minimizing inventory investment costs and other distribution costs.

DRP enables the user to combine statistical and expediting methods scientifically so that a given level of inventory investment can provide much greater customer service levels than are possible with either statistical or expediting methods alone. DRP is predicated upon taking replenishment actions based on the relationship of projected inventory levels to projected safety stocks. These safety stocks usually are based on the specified customer service level with appropriate considerations given to replenishment lead times, sales forecasts, and projected forecast errors. If the user monitors the progress of actual customer demand and replenishments, shortages can be anticipated in advance of their occurrence, thereby providing the user with expediting tools such that the shortage will either not occur or its effect will be minimized.

Distribution resource planning (DRP) is an extension of manufacturing resource planning (MRP) to the distribution environment. DRP converts independent, time-phased sales forecasts at each shipment location into dependent, time-phased requirements that are placed on the specific item's source of supply. All of the time-phased replenishment quantities needed at the individual warehouses are aggregated to produce the total network requirements for an item produced at a specific plant or the next higher-level warehouse.

HOW TO APPLY SIMULATION TECHNIQUES*

A simulation model can be used to determine feasibility (functional capabilities) and assess the quality of an automated warehouse system or of related process systems. The simulation model enables the methods engineer to analyze the system concept or design. Simulation permits the engineer to test a system configuration and to determine its ability to satisfy present and projected needs under a variety of operating conditions before hardware exists or a commitment is made to a specific design. It also enables the engineer to test and evaluate control algorithms and management strategies.

*Permission granted by Eaton-Kenway.

Experience has shown that the most successful and productive systems are those evaluated by such models during early design. These systems are installed quickly and are trouble-free from the beginning because the "bugs" were identified by the simulation model and corrected before the hardware existed. Software programming language can be employed to develop discrete event system models customized for each simulation system.

The first step is to define the system to be simulated. A checklist is used to determine user needs and expected benefits from the simulation.

> ***Key Point:*** Accuracy of data to be used in developing the simulation model is extremely important.

Tests comparing simulation results with actual hardware performances and the observed results, are usually within 1 to 3 percent of the simulated predictions. Where hardware performance differs significantly from simulated performance, the problem can usually be traced to a faulty data base for the simulation model.

Next, the definition is reviewed, and elements of the data base verified for accuracy and completeness of the performance characteristics of the various components used in the automated system.

A Checklist for Deciding When to Use a Simulation

The following simulation checklist indicates the type of information used to determine whether to develop a model. If the answer is "yes" on any question, the simulation will probably be beneficial.

1. Are there one or more components in your system that, when loaded with an average workload, have a utilization of 85 percent or more?
2. Are there events in your system that represent merging or synchronization of separate processes?
3. Are there activities in your system that, when completed, trigger other remote activities?
4. Are there pauses in processes in your system that await completion of remote activities, such as inspections of in-process material?
5. Do your processes have variations with wide limits in execution time?
6. Have your process control algorithms been applied in exactly this type of system before?
7. Have you included factors for contention in use of facilities, such as data lines or paths for material flow?
8. Does your system planning include detailed procedures and timing for human/machine relationships?

Four Types of Information Needed to Define the Process

The definition of any process involves four primary elements:

1. Scaled and dimensional drawing or layout of the process
2. Brief narrative of data, material flow, control algorithms, and management strategies

3. Statistical description of system parameters, such as desired inputs, outputs, component performance, and timing of events in the system
4. Specific information that is expected from the simulation

In most cases, all this information can be provided on a single drawing accompanied by appropriate notes and comments. By means of notes added to the system layout, sufficient data can be provided to prepare a simulation model.

The description of an operation should be simple and brief. The methods engineer should be able to develop a simulation model without getting involved in minute details of the process or system. The basic functions can be described simply and include accurate data on matters that affect activity of operators, movement of materials, data transfers, and management strategies.

Figure 10–15 illustrates an order-picking warehouse employing storage/retrieval trucks. Forklifts deliver loads for storage to the aisles. An upper-level conveyor transports retrieval loads to a drop-zone queue conveyor, and forklifts deliver them to one of five picking lanes.

The warehouse stores materials for five order-picking lines. Due to the nature of the products, there are two classes, which are referred to as odd and even. A different class of products is run every week, and during that week most of the storages and retrievals belong to the currently active class of products. Therefore, the warehouse is managed as two four-aisle warehouses with a single overflow aisle: Aisles 1, 3, 5, and 7 for odd products; Aisles 2, 4, 6, and 8 for even products; and Aisle 9 for overflow.

Inputs

Trucks arrive at the dock and are unloaded by forklifts (Point A). Incoming pallet loads are grouped into lots of ten pallet stacks for convenience in checking (Point B). A clerk inspects the loads and checks the shipping invoice, then goes to the dock office and keys the purchase order number and quantity of material received into a CRT terminal (Point C). The warehouse computer verifies the purchase order and automatically selects the storage location in the high rise. ID tickets with aisle storage location are printed in the warehouse office, and the checker-operator places an ID ticket on each load and moves on to the next group of pallets. It takes 30 ± 10 seconds to inspect a pallet, 18 ± 4 seconds to make the terminal transaction, 4 seconds to print a ticket, and 5 seconds to attach a ticket to a loaded pallet.

Storage

When a forklift driver sees a load with an ID ticket attached, the driver picks up the load and delivers it to the indicated aisle, which has a set-down stand on each side (Point D). Forklifts move at an average rate of 300 feet per minute and can pick up a load in 10 ± 5 seconds; a deposit takes the same amount of time.

There are only four storage/retrieval trucks in the system. One aisle is designated for storage of the inactive product class, where loads are placed along the wall adjacent to the set-down stands. When twenty or more loads are accumulated, a storage/retrieval truck is moved to that aisle to store them. Storage/retrieval truck operators also monitor the input set-down stands. Whenever a load is present at input, they pick it up and store it in the location shown on the ID tag (Point E). A storage/retrieval truck takes 15 ± 5 seconds to read an ID tag, 120 ± 30 seconds to complete an average storage cycle (half way down and half way up), 14 seconds to execute shuttle movement, and 12 ± 4 seconds to make the terminal transaction.

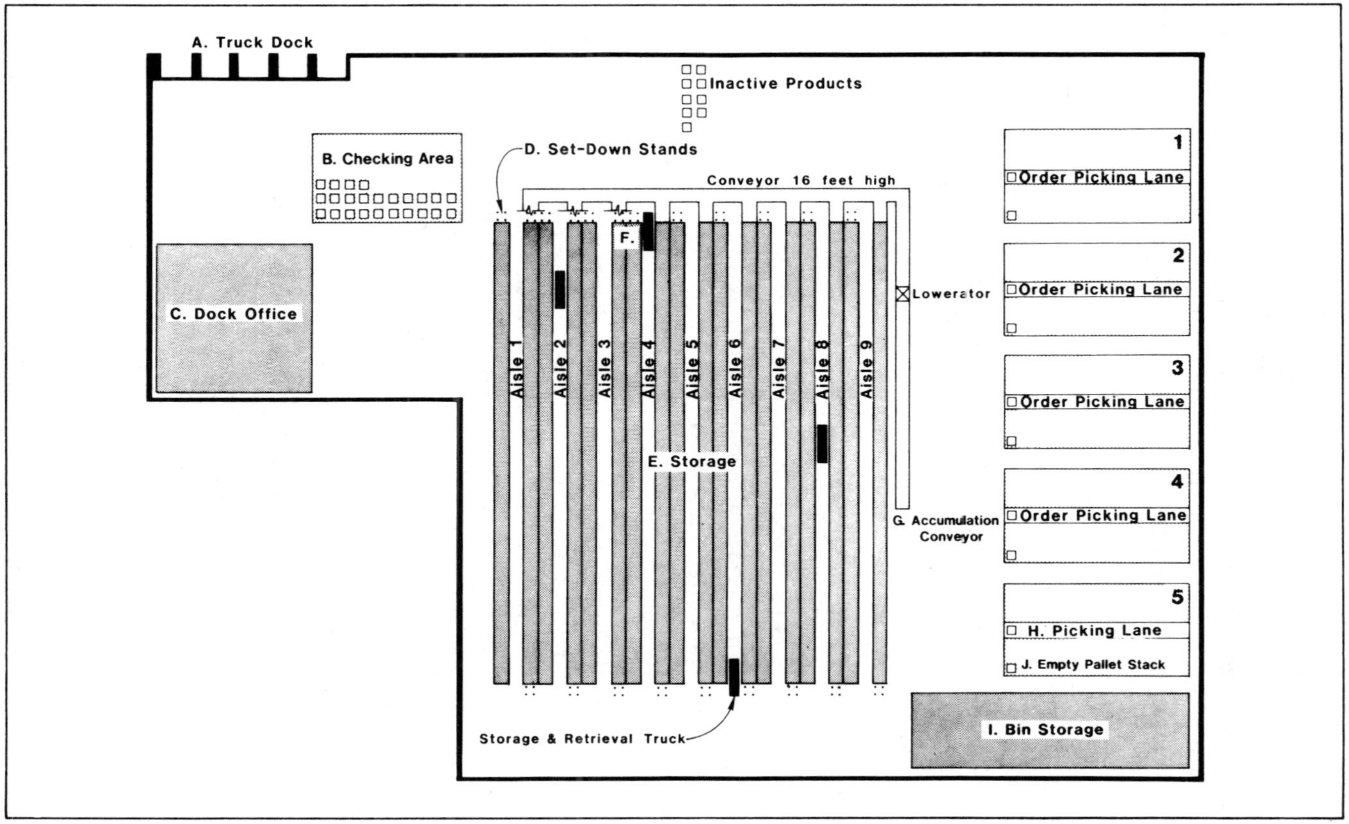

FIGURE 10–15 Order-Picking Warehouse Employing Storage/Retrieval Trucks (Courtesy of Eaton-Kenway, Salt Lake City, Utah).

Retrievals

Storage/retrieval truck operators receive retrieval instructions via an on-board terminal. The terminal prints a move ticket containing storage location ID number and destination; the SRT operator retrieves the load and attaches the move ticket.

Seventy-five percent of retrieval loads are delivered to a head-of-aisle conveyor at the 16-foot level (Point F), and then conveyed (Point G). Local forklift operators remove the loads and deliver them to the picking lines as indicated on the move ticket (Point H).

Ten percent of output loads are delivered to back-of-the-aisle deposit stands, where a single forklift removes the loads and places them in bin storage for special purposes (Point I).

Empty pallets are accumulated at the head of each picking lane. Periodically, the back-of-aisle forklift picks up the empty pallet stacks and delivers them to the truck dock (Point J).

Statistical Representation of Parameters

Specialized software can provide a considerable selection of facilities to simulate a system's characteristics in a model, including discrete value functions, continuous value functions, random numbers, normal distributions, random distributions, and so forth.

Data Requirements from the Simulation Model

A model can be prepared in many ways and with various levels of detail. A user may desire to test the feasibility of a specific approach, such as high-rise storage warehouse versus a conventional pallet-rack facility. This high-level model can be developed at small cost; however, as the concept becomes more detailed, the questions become more specific, and the cost of the model increases.

It is important for the engineer to state the requirements for model response so that the necessary data is derived from the model. You may decide to develop a model that permits analysis at one level, and then by means of added development, more and more details. A phase model may match the internal project program, as progressive steps are taken from feasibility, justification, specification, design, and evaluation.

The storage/retrieval trucks warehouse system discussed previously illustrates the types of questions that a user requires of a model:

1. How many forklifts are required at the dock?
2. How many clerks are required to inspect the loads and operate the CRT terminals?
3. How many storage/retrieval trucks are required?
4. How many delivery forklifts are required?
5. Is one back-of-aisle forklift enough?
6. Can the desired throughput be realized?

UTILIZING THREE-DIMENSIONAL SHADED SURFACE ANIMATION IN PLANNING, DESIGN, AND EVALUATION

Three-dimensional shaded surface animation is a dynamic tool for planning, design, and evaluation of AS/RS, AGV systems, and integrated material-handling and control systems. With color-graphic technology, system layouts have unsurpassed realism which greatly enhances visualization and reduces design time.

With this tool, material-handling simulations are produced for *concept development, detail design,* and *operational change studies.* Thus, an efficient continuity over the full life of a system is provided and in-the-end accountability is assured during acceptance throughput tests. Field results are fed back to the standard simulations models to continually upgrade accuracy.

Color-graphics input is used to define the physical layout of the simulations model easily, quickly, and accurately. Color-graphics output is used to debug and validate simulated movements. Video tape or 16mm film of the animated simulations provide a documentary of proposed systems for management review.

A FIFTEEN-STEP GUIDE FOR PLANNING AN INTEGRATED WAREHOUSE AUTOMATION SYSTEM

Integrating material-handling systems and automation is an objective that most people engaged in warehousing are striving for. What is making this concept a reality is the advent of programmable controllers, computer integrated manufacturing, flexible manufacturing systems, automatic identification systems, and cost-effective distributed data processing, which now makes material-handling integration possible. (See figure 10-16.)

There are many levels of systems integration where different elements are tied together, for example, an AS/R system and an AGV system. Typically, a warehouse's automation program begins with such things as bar code readers in receiving, conveyors, sorters, and automated packaging machines. Over a period of time, it expands to incorporate various automated functions, such as material-handling equipment or entire production lines. As these individual components begin to mesh, it becomes an integrated system.

Integrating disparate types of equipment into a common system is becoming the provence of either a systems integrator, a company that designs and installs integrated "turnkey" systems, or a large equipment supplier who will take on prime systems integration responsibilities. In either of these cases, it would include coordinating the efforts of material-handling systems manufacturers, machine tool manufacturers, robot manufacturers, software companies, automatic identification equipment suppliers, consulting engineers, and systems users. There is certainly common agreement that a team approach is essential to the success of an integration project. It may be an in-house team, a vendor/customer team, or any other combination of people with particular specialized talents, but regardless of its composition a variety of disciplines needs to be involved. Such a project requires a well conceived step-by-step phased implementation approach. Experience appears to indicate that adhering to these steps seems to result in the installation of a successful systems integration project.

1. *Establish a well defined strategy.* The first requirement for any full-blown integration system is that it be consistent with the company's overall strategic plan.
2. Commitment to proceed must come from top management.
3. Objectives, scope, and expectations must be clearly understood by the entire organization.
4. With the long-term goals clearly established, the planning process can begin with the development of an overall system concept.
5. Set up a project team consisting of people with the requisite expertise. A team leader should be designated and the project organization formalized.

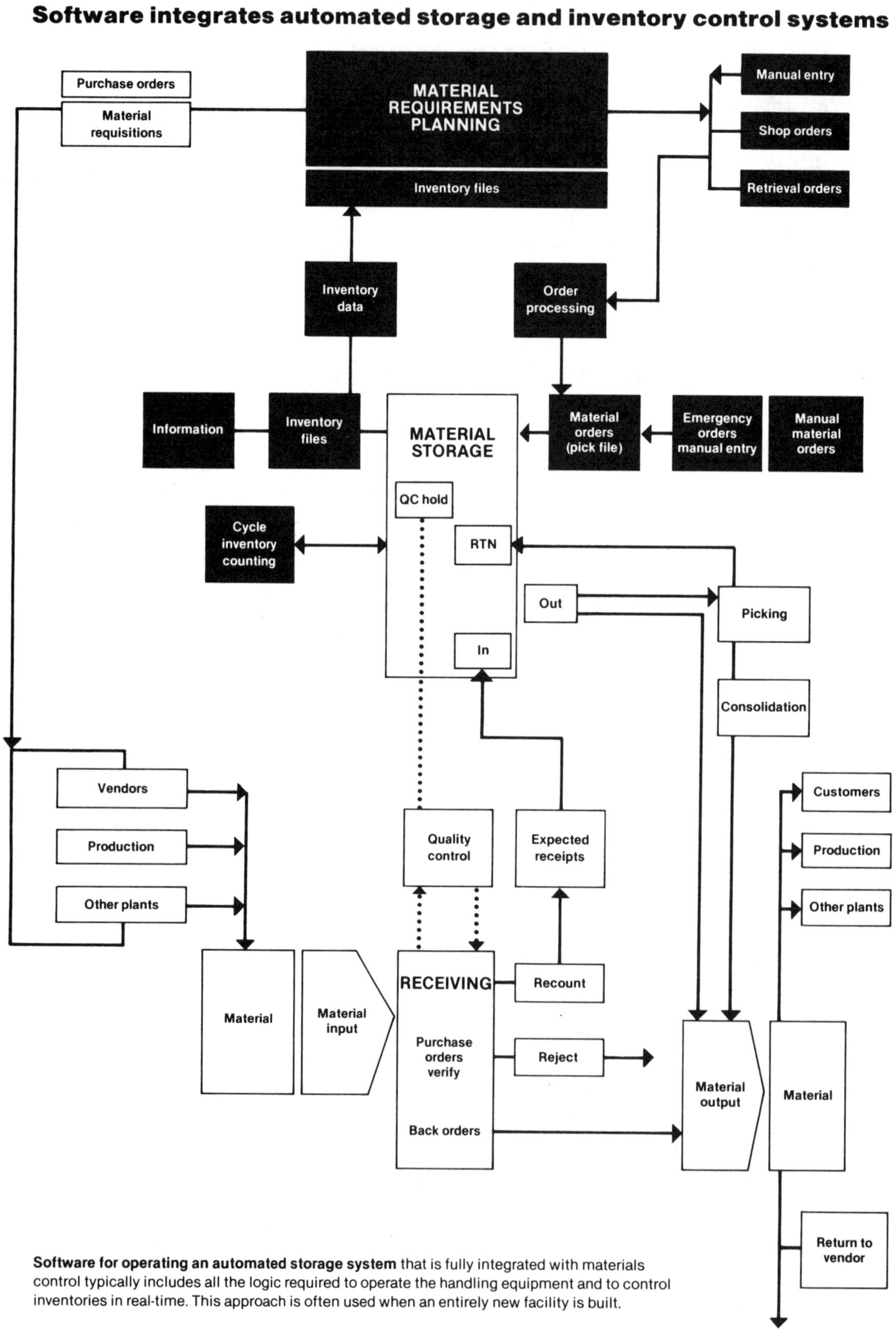

FIGURE 10-16 Software for Operating an Automated Storage System (Courtesy of Integrated Automation, Alameda, California).

6. Begin to compile all relevant data, such as capacities, volumes, quantities, sizes, weights, etc. Establish the data base as early in the project as possible to be in a position to collect, assimilate, store, and retrieve vital information.
7. Define in detail systems requirements of the integrated system using the data on current operations and projections of future needs.
8. Prepare operational descriptions based on the system requirements obtained in Step 7. This will give form and substance to the project by outlining the flow of materials and the equipment to be used in the various subsystems.
9. Establish a benchmark to determine whether the planning so far is on track, and to measure and control further progress. At this point it is advisable to develop a computer model to determine if the system envisioned will be workable.
10. Develop functional specifications that define the system and equipment that will be required. These detailed specifications should describe exactly how the integrated system is expected to operate.
11. At this point the team members should make vendor selections based on the detailed specifications that have been established.
12. The task force members in collaboration with outside vendors, systems integrators, software consultants, or others should complete the detailed system and subsystem design.
13. As the detailed design work nears completion, computer simulations should be used to verify performance of the system.
14. Once the simulation has verified the feasibility of the integrated system, the task force and others can now plan and schedule project implementation.
15. When all these steps are completed, the project is now ready for implementation.

INSTALLING AN INTEGRATED WAREHOUSE AUTOMATION SYSTEM*

As discussed previously, a company may take one of many approaches in organizing a project of this magnitude. Usually, it is done with a combination of in-house personnel and outside specialists and vendors with various expertise. However, since the combinations of equipment, systems, methods, software, and other project components can be so extensive, many companies prefer to rely on a systems integrator house to assume overall project responsibility. One of these firms, Integrated Automation, has become a leader in designing and installing complex warehouse integrated systems. They seek to optimize integration of a warehouse facility by providing systems engineering for automatic identification-based systems which control physical distribution, warehousing, and inventory. Thus, they develop innovative approaches that link material-handling systems, computer systems, and automatic identification methods. These three interrelated components are the key to a high order of integrated automation.

The following case histories are illustrative of installed projects designed by Integrated Automation. These are examples of leading edge technology aimed at fully automating the warehouse facility. Innovation in this functional area is moving ahead rapidly, so that many more changes will occur in the next five to ten years making some of today's advances obsolete. However, limitations in resources, size, and logistical constraints will still inhibit growth to an evolutionary process that results in innovated change developing in stages,

*With permission from Integrated Automation.

rather than in one gigantic thrust. But, to those warehouse operators backed by a supportative management willing to invest substantial sums on the future, the rewards that may accrue through complete integration of automated warehouse operations can result in complete optimization of all activities in this most important support function.

CASE-IN-POINT

COMPLETELY INTEGRATED DISTRIBUTION WAREHOUSE AUTOMATION SYSTEM U.S. SHOE FINISHED GOODS DISTRIBUTION CENTER DESIGNED BY INTEGRATED AUTOMATION/AUTOMATED SYSTEM GROUP

This case history is illustrative of leading-edge technology in the design and development of a completely integrated warehouse automation system. It was installed in a warehouse facility with the following characteristics:

- *Product Type*—finished goods, women's shoes, cases, and pairs
- *Floor Space*—490,000 ft.^{2} on three levels with a 30′ high cube narrow aisle storage area
- *Stockkeeping Units*—250,000 SKUs
- *Storage Capacity*—150,000 cases in high cube; 3,000,000 pairs in active storage
- *Material Handling*—conveyors, hanging basket monorails, tugs and carts, order pickers
- *Scanning Methods*—TEKscan matrix code and interleaved 2 of 5 for portable bar code data entry
- *Computer Hardware*—IBM 4331, IBM Series/1, three integrated automation microprocessors

Situation

At U.S. Shoe's new 500,000 square foot distribution warehouse, the integrated automation system begins with the smooth computer-controlled flow of incoming and outgoing cases, made possible by a high-speed code reader in a central conveyor loop. Cases pass through the loop from receiving docks to two major storage areas, and from storage area to shipping docks.

The dock and storage areas include a high-bay rack area in which workers use order-picking trucks with a nearly 33-foot maximum lift, a two level mezzanine shelving area served by input and output conveyors, and receiving and shipping docks with extendable conveyors that reach into highway trucks.

A corporate-level computer schedules factory shipments to the warehouse, and prints the coded case identification labels that are applied to cases at factories. As a result, the warehouse computer system already knows the expected contents in every incoming case. Receiving is paperless, because the code reader identifies cases, and an adjacent weigh-in-motion scale reports case weights, used to check the amount of contents.

The computer system directs incoming cases to storage zones most convenient for picking. It does this either directly by conveyors or via a sortation system, and transfers to a tractor–trailer train.

The computer system also provides pick lists in location sequence. It also provides the associated case identification labels that are needed for the pick-and-pack operations in the mezzanine storage area. This is where boxed shoes are picked by order and packed into cases.

In addition, the computer system directs outgoing cases from the conveyor loop to a sortation system and accumulation lanes at shipping docks. It prepares documents automatically, including bills of lading and manifests. Furthermore, it accumulates scale weights on

UPS shipments for a direct payment system thus eliminating UPS stickers on individual cases.

A four-level computer hierarchy schedules warehouse receipts and shipments, manages the inventory, and controls or directs all warehouse operations.

Computer-Controlled Input

Conveyors carry incoming cases directly from highway trucks to the central powered belt and roller conveyor loop.

Next, they pass under the high-speed code reader. This is a charged couple device (CCD) that reads "presence–absence" codes printed on labels affixed to the tops of cases. Such a code pattern uses the presence or absence of standard printed slugs or marks to represent a number.

The code reader uses a camera that has a linear array of photosensitive elements. It reads the image of each small code pattern, which is reflected onto the grid.

This equipment reads codes at rates up to 1400 per hour or more. Accurate reading can be achieved at any orientation: a tilt of 35°, a skew of 30°, and a 30″ depth of field.

The number that is read identifies a specific incoming case for the computer system, which already knows the expected contents and destination. This is possible because the corporate-level computer in the U.S. Shoe's hierarchical computer system specifies case numbers and contents in scheduling factory shipments, and provides the case–identification labels that factories use.

These are two-part case labels. One part goes on the top of a case. The second part goes on the side, where it is handy for code scanning with a hand-held light pen.

Portable terminals with light pens can be used in a storage area because the labels also include a bar-coded case number along with human-readable information.

The CCD code reader's input to the computer system is followed immediately by a second input from the automated weigh-in-motion scale. The computer system then compares actual and expected case weights to verify the case contents. If there is a discrepancy, it routes the case to a troubleshooting station.

To do this, the computer hierarchy includes an operations control computer. This is linked to several microcomputers that receive inputs from sensors and operate equipment. The equipment includes the code reader, weigh-motion-scale, receiving and shipping sortation systems, and tracking system motors and diverters.

As a result, the computer system can dispatch incoming cases to specific storage zones via conveyors and other equipment.

One major routing of cases is from the central conveyor loop directly to storage zones in the two-level mezzanine area.

Another major routing is to zones in the high-bay storage area. This routing includes several steps. Cases first go to a sortation conveyor system, which diverts them into accumulation lanes by zone.

The next step is to load cases onto four-wheel carts equipped with trailer couplings. Finally, a battery-powered tractor hauls trains of trailers to the storage zones.

The overall warehouse layout includes two major storage areas. One is the 125,000 square foot high-bay area, with rows of racks up to a 40-foot high ceiling. The second is the two-level mezzanine, filled with rows of 8-foot high shelving.

High-Bay Storage Systems

The high-bay storage area can hold as many as 170,000 cases of shoes—about 3.4 million pairs—in 170-foot long rack aisles. All of this inventory is readily accessible to the operators of several high-lift order-picking trucks.

Thirty-eight storage aisles in the high-bay area are dedicated to "pre-sold" merchandise: full cases of shoes that factories have made to customer order. The warehouse stores these pre-sold cases for scheduled shipments to customers in regional areas.

For convenience, the high-bay area is divided into storage zones assigned to regional areas of the country. Each regional area is served by shipments on a given day of the week. As a result, the computer system can readily sort incoming pre-sold cases to the correct storage zones, each of which is a single-storage aisle. Each day's picking activity in the high-bay area is efficiently concentrated within a few aisles.

Such an approach speeds retrieval and minimizes interference between picking and put-away operations.

Human-readable information on the factory-applied case labels designates the storage zones for putting away pre-sold stock.

Truck operators select any convenient rack openings within zones for random storage. They then use hand-held terminals equipped with light pens to read and record barcoded numbers—both case-identification codes and the storage location codes at the selected rack openings.

The truck operators download this data at intervals, through terminals to the warehouse management computer system. The computer then updates the pre-sold inventory quantities and locations. As a result, the computer also can generate pick lists in location sequence for fast order filling and shipping of orders.

Cases to be shipped from the high-bay storage area are first transferred from four-wheel trailers to a conveyor line that feeds the central conveyor loop.

The loop then diverts outgoing cases to the shipping sortation system. This sorts cases to accumulation spurs, and conveyor extensions can carry them directly into highway trucks. They are floor-loaded inside the trucks.

Mezzanine Storage Systems

The mezzanine storage area can hold as many as 250,000 SKUs—up to one million pairs of shoes. This is an "open-stock" inventory of made-to-stock shoes—forward picking stock, in fixed shelf locations assigned to speed manual handling. Pairs of shoes are stored in individual boxes, unpacked from incoming cases.

All mezzanine shelf locations are readily accessible to manual putaway and picking operations, aided by conveyors. The input arrives on roller conveyor spurs in each of eight storage zones. Picking is done into hanging carrier baskets on loops of a power and free conveyor system. The computer system provides pick lists by customer order, along with coded, case-identification labels. The power and free conveyor then delivers picked shoes to a mezzanine packing station. The packer puts boxed shoes into cases, and applies the computer-generated case-identification labels that match pick lists and orders. The cases advance on roller conveyors to automatic case sealing equipment.

Cases to be shipped from the mezzanine area are first conveyed to the central conveyor loop. They are then dispatched along with other outgoing cases to accumulation lanes at shipping docks.

Flexibility in Dispatching

Having the single permanently positioned code reader in the central conveyor loop gives U.S. Shoe's computer hierarchy great flexibility in dispatching cases through the warehouse.

The computer system can, for example, route some of the incoming cases to "reserve" storage zones in the high-bay area. Later it can recall these cases selectively. It then dispatches them to mezzanine storage zones, where they are needed to replenish the forward picking stock.

Furthermore, the computer system can initiate mezzanine picking and case packing operations in advance of customer's requirements. It then dispatches cases to a "prepack" storage zone in the high-bay area. Later on, it can recall these cases selectively, and dispatch them to shipping docks.

This "prepacking" of customer orders is usually done on days when the routine picking and packing workload falls below average. Such a procedure helps to level the workload in the mezzanine area. It can, for example, eliminate subsequent peaks in activity that might otherwise delay shipments.

System Diagrams

System Layout Diagram. The layout of the U.S. Shoe facility is shown in figure 10–17. Of particular interest is the small amount of space allocated to receiving, the tree structure of the conveyor system, the interleaving of the shipping and receiving sortation systems, and the small amount of laydown space dedicated to shipping.

System Block Diagram. The block diagram of the U.S. Shoe Warehouse Control System is shown in figure 10–18. The arrangement shown is a classic data/control hierarchy

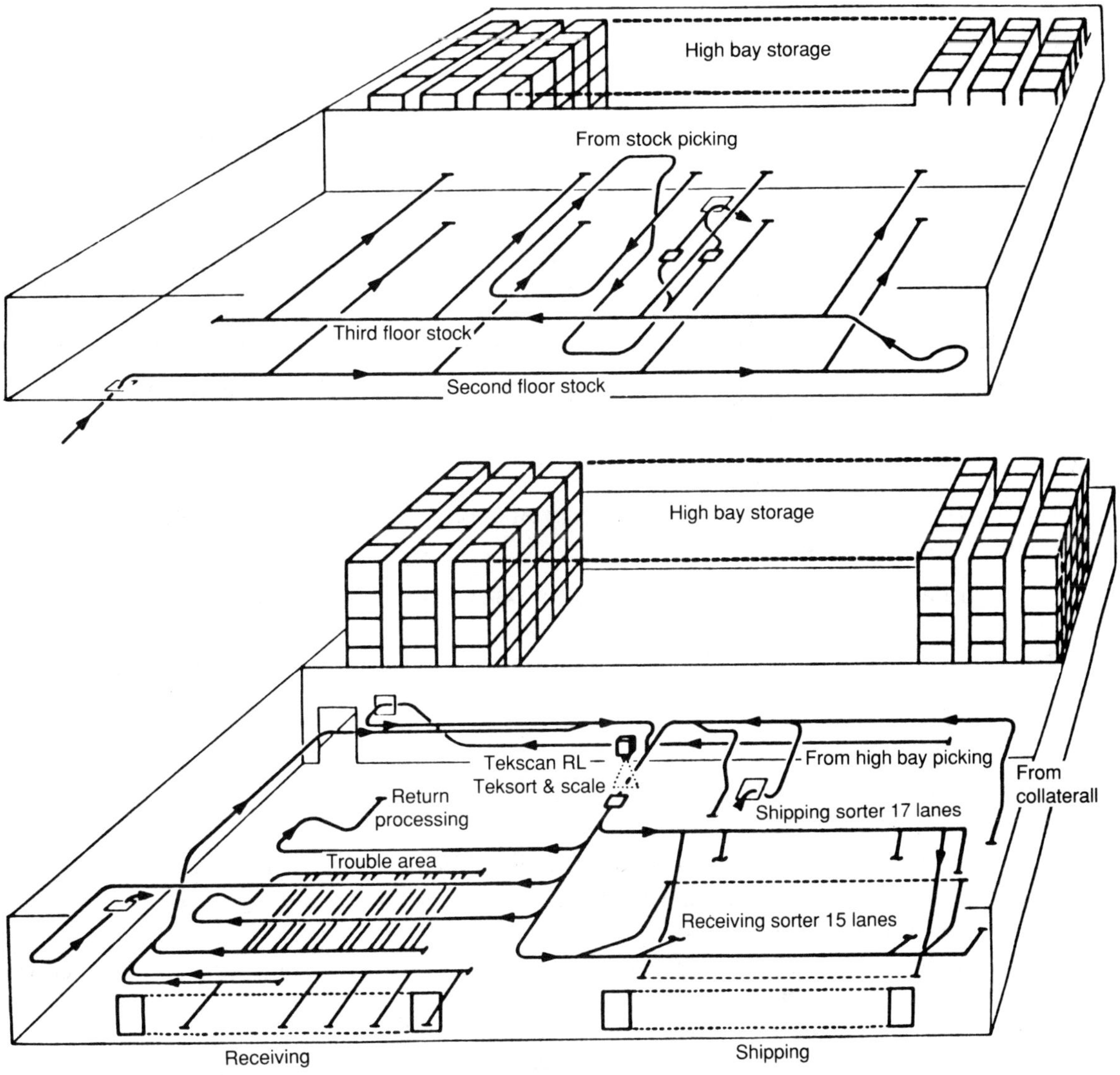

FIGURE 10–17 U.S. Shoe System Configuration (Courtesy of Integrated Automation, Alameda, California).

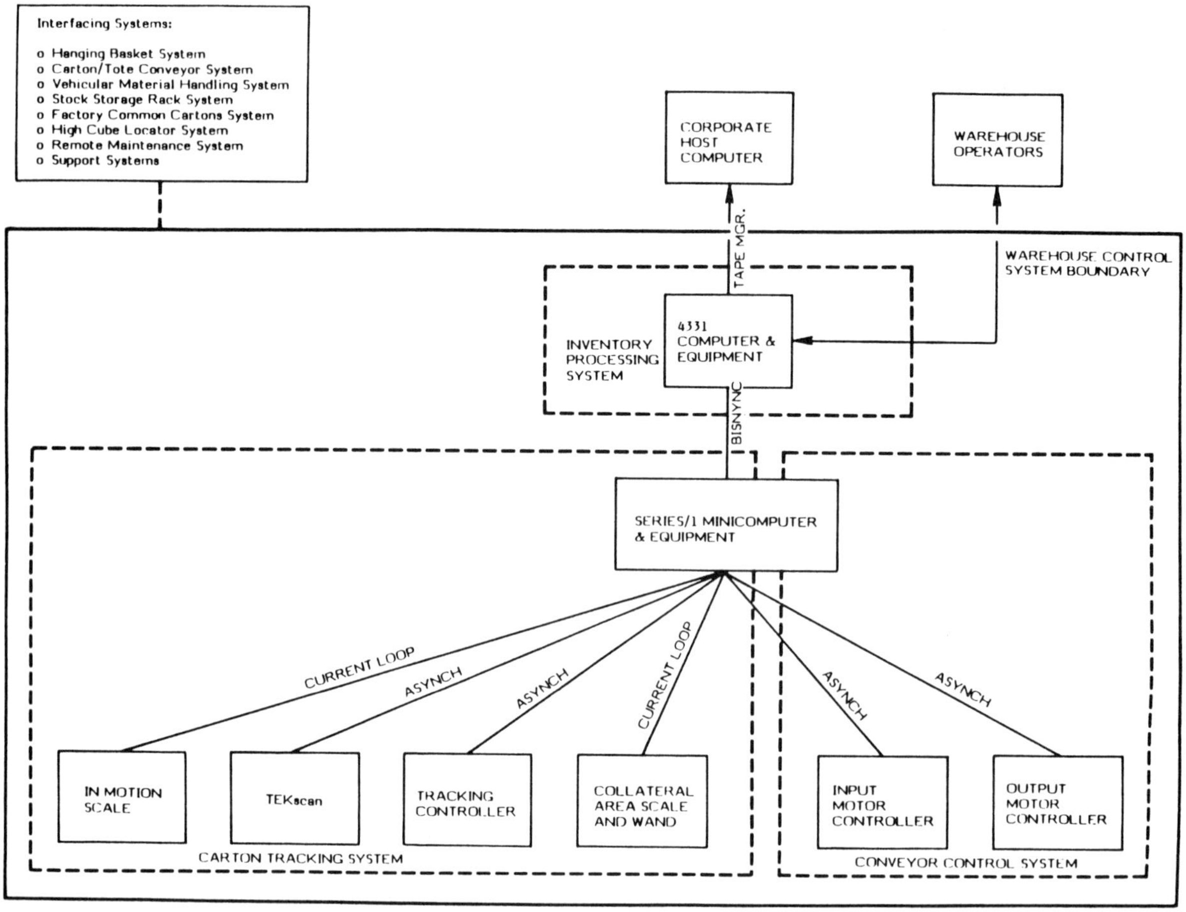

FIGURE 10–18 Warehouse Control System Equipment (Courtesy of Integrated Automation, Alameda, California).

which assigns computations to the appropriate computer based on the timing of response required and the computational complexity.

File and Programs Elementary Diagrams. The U.S. Shoe software consists of 115 programs and 50 files in the 4331, and 20 programs and 6 files in the Series/1. The block diagram shown in figure 10–19 is a summary of the 4331 software architecture.

Unique System Features

- Completely integrated hierarchical control system
- Fluid, paperless, blind receiving
- Complex sortation control scheme minimizes the number of scanners required
- Unique close coupling of raw materials warehouse and finished goods distribution minimizes transportation costs
- Completely automated preparation of UPS manifests, bills of lading, trailer manifests, and other manifests
- Post putaway locating transactions for random high cube storage
- Virtually all cases having exceptions conditions automatically routed to a single trouble area where system support is provided to resolve the exceptions

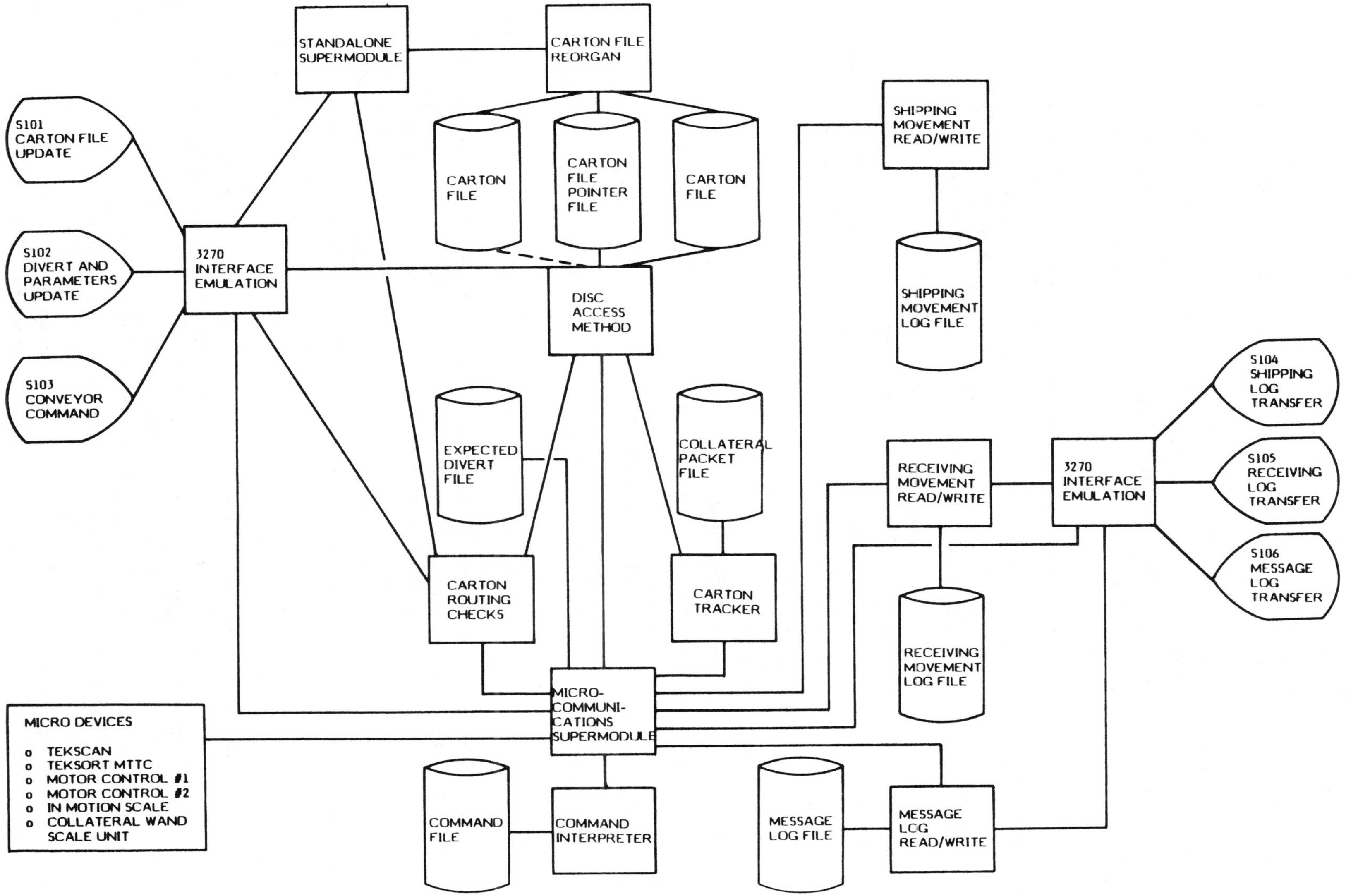

FIGURE 10-19 Carton Tracking and Conveyor Control System (Courtesy of Integrated Automation, Alameda, California).

Results

The following results were achieved by U.S. Shoe upon implementation of this advanced state-of-the-art, completely integrated warehouse automation system:

- Labor force reduced 50 percent, while daily shipping capacity was increased by a factor of three.
- High cube location inventory accuracy is 99.996 percent on a sample of 67,000 cases.
- Customer service is improved by achieving an inventory accuracy of 99.9 percent.
- Computer-integrated systems aid manual handling where and when necessary.
- The computer-integrated system has doubled overall warehouse productivity.

CASE-IN-POINT

COMPLETELY INTEGRATED AUTOMATION SYSTEM
EXCEL CORPORATION BOXED BEEF DISTRIBUTION CENTER
DESIGNED BY INTEGRATED AUTOMATION/AUTOMATED SYSTEMS GROUP

This case history covers the installation of a completely integrated distribution center automation system. This project consisted of the following mix of facilities characteristics:

- *Product type*—boxed beef
- *Floor space*—90,000 $ft.^2$
- *Stock keeping units*—500 SKUs
- *Storage capacity*—3600 pallets in a five-aisle automatic storage and retrieval system
- *Material handling*—carton conveyors, pallet conveyors, automatic palletizers, depalletizers, automatic storage cranes
- *Scanning method*—bar codes on two labels on each carton (one = product code, one = weight) and unique bar coded labels on each slave pallet
- *Computer hardware*—two IBM Series/1, six integrated automation microprocessors

Situation

At Excel's 90,000 square foot distribution center, the integrated automation system begins when cartons of boxed beef are conveyed from the fabrication floor to the box handling floor. As cartons enter the system, a laser scanner identifies the product label before the carton passes over an in-motion scale. A bar coded weight label is printed and applied to the carton while it is in motion. A second scanner is used to verify the proper operation of upstream devices. Defective cartons are diverted from the main line to return to the fabrication floor.

After the cartons are sealed and pass an open flap detector, they are scanned and sorted into one of eighty-one sortation lanes. Products produced in extremely low volumes are diverted instead into an orbiting slow movers loop where they can be recalled at will.

When one of the eighty-one production sortation lanes contains eighteen cartons—that is, a pallet quantity, the lane is released to an automatic palletizing machine. Completed pallets are conveyed to the AS/RS. At the AS/RS the slave pallet label is scanned at the same time that a carton on the pallet is scanned. The control system then directs the putaway of the pallet in the AS/RS to speed products through all five aisles and to place higher velocity products closer to the front of the AS/RS.

Shipping orders are received via telecommunications from a remote host computer. The shipping supervisor assigns orders to trailer loads with system assistance which provides for optimal trailer utilization.

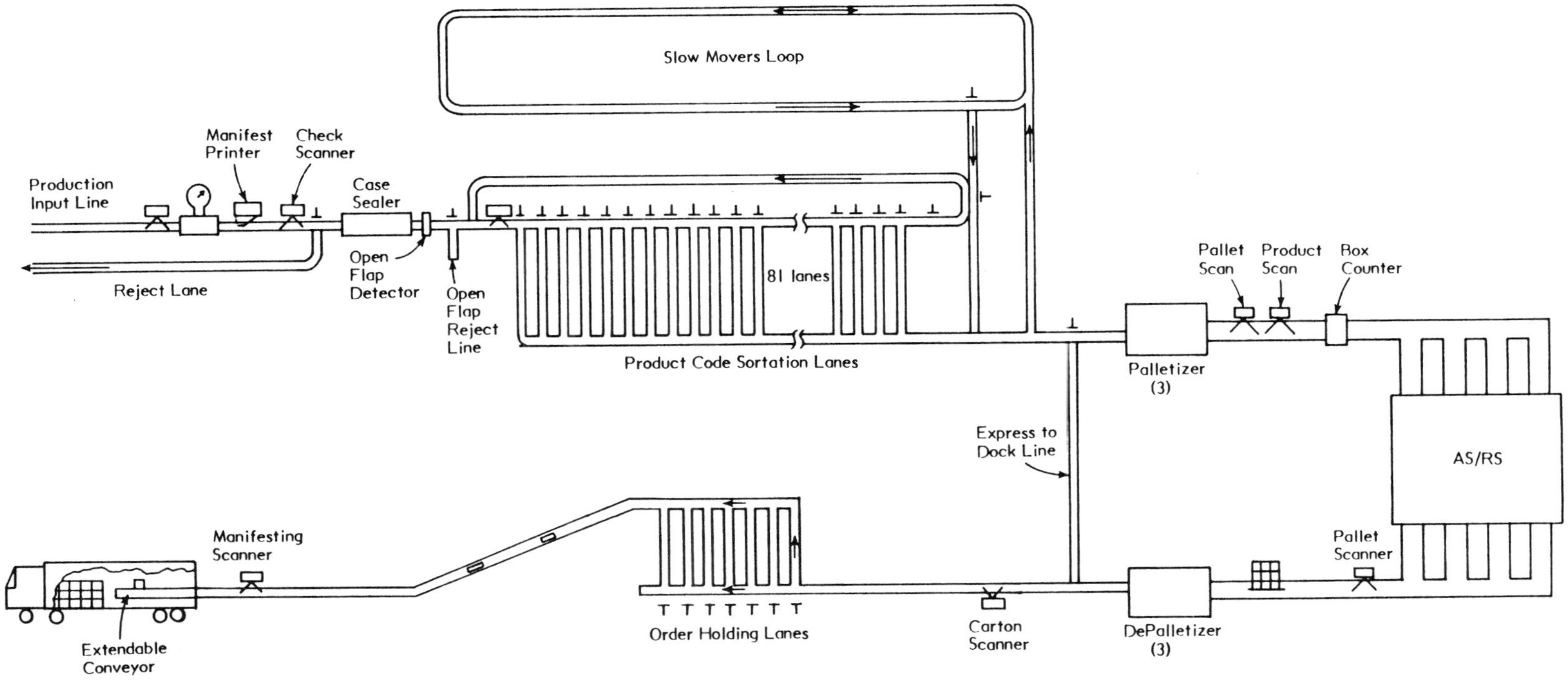

FIGURE 10–20 Layout of the Excel Facility Material Handling System (Courtesy of Integrated Automation, Alameda, California).

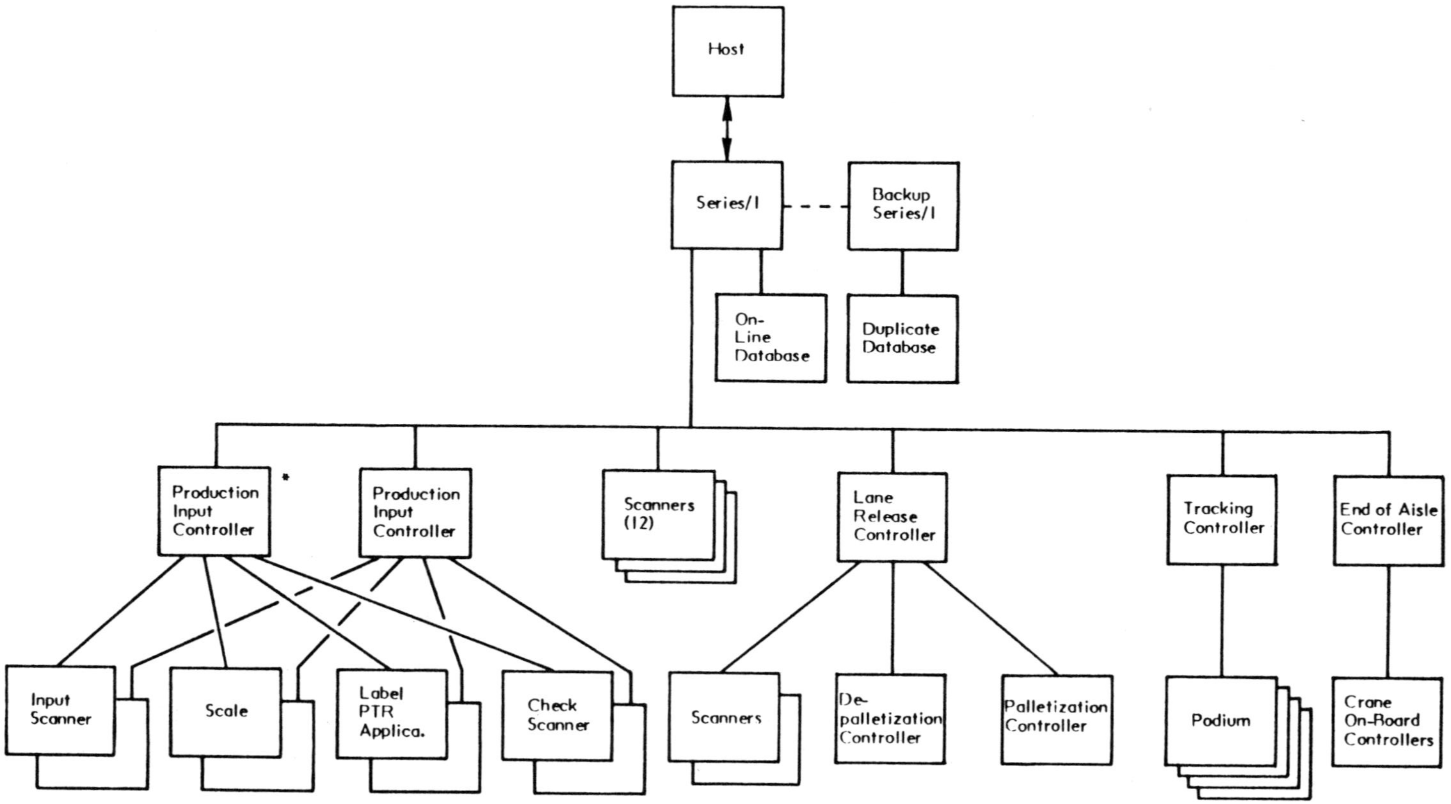

* Redundant at conveyor level. Either production input line has capacity of 100% of production capacity.

FIGURE 10–21 Excel System Block Diagram (Courtesy of Integrated Automation, Alameda, California).

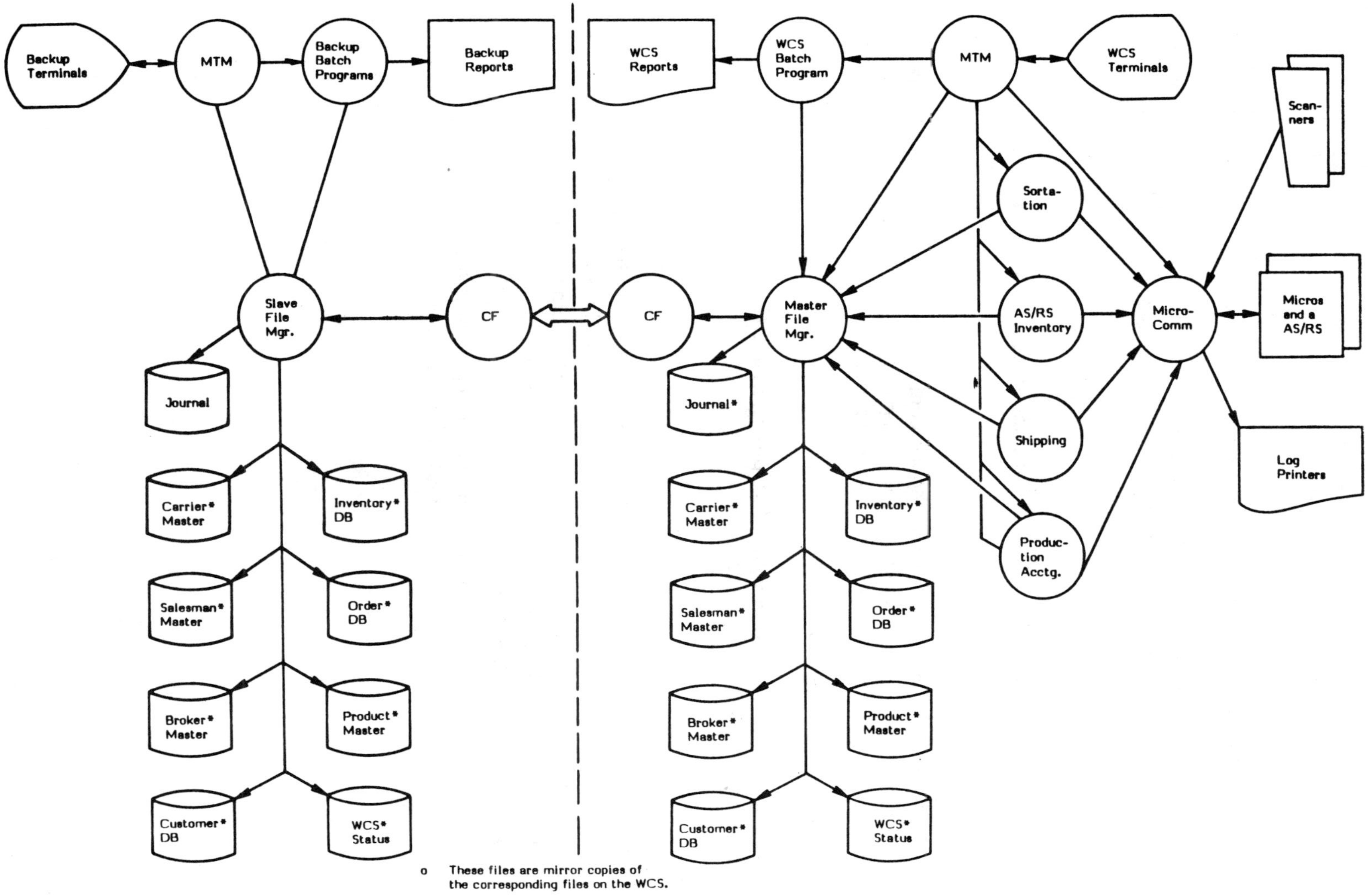

FIGURE 10–22 Excel Software Diagram (Courtesy of Integrated Automation, Alameda, California).

As trailers arrive at the shipping doors, the loads are scheduled for picking. Pallets are then withdrawn from the AS/RS to fill the orders. The pallets are brought to automatic depalletizers and depalletized. Slow movers and express-to-dock cartons are conveyed separately to the order staging conveyors. After the carton labels are scanned, the various orders are automatically assembled onto staging conveyors and released one lane at a time to the shipping dock.

At the dock, the cartons are continuously loaded into trucks with an extendable conveyor. Each carton is scanned as it enters the truck for purposes of manifesting and billing.

As each truck load is completed, the control system prints a manifest for each order.

Periodically, billing and production accounting data is uploaded to the corporate host computer.

System Layout Design

The layout of the Excel facility material-handling system is shown in figure 10–20.

System Block Diagram

The block diagram of the Excel warehouse control system is shown in figure 10–21. Of particular interest are the reliability features that provide for hot backup of the minicomputer and selective redundancy for the controls microprocessor.

File and Programs Elementary Diagram

The Excel software consists of eighty-five programs and thirty files. The software elementary diagram is shown in figure 10–22.

Unique System Features

- No manual material movement required at any point in the system
- Automatic production accounting based on received cases
- Extremely flexible load assignment and load scheduling process
- Use of integrated automatic control storage cranes
- Automatic bill of lading preparation
- Flexible implementation of FIFO product assignment to orders
- Implementation of completely redundant control system

chapter 11

Computer-Integrated Manufacturing: Building the Factory of the Future One Step at a Time

USING COMPUTER-INTEGRATED MANUFACTURING TO STIMULATE PRODUCTIVITY GAINS

Service industries account for approximately two-thirds of the gross national product (GNP), nevertheless it is incumbent upon American industry to improve its capability in the manufacturing sector. Manufacturing still accounts for 19 percent of the GNP. The use of computer-integrated manufacturing (CIM) technology can be of considerable value in stimulating productivity gains in manufacturing, thereby reinvigorating this basic industrial sector.

The objective of CIM is to develop a cohesive digital data base that integrates engineering design, manufacturing, and other related functions. Information is sent on demand and as needed to the largest number of intra- and interdisciplinary groups as possible. A completely operative CIM system designs and manufactures parts without disruption, from raw materials to the finished product. In a CIM controlled factory, machines create products under the hierarchal control of computer networks that optimize production flow and scheduling. After executing their respective tasks, all machines report on work and operating status, current state of being, and other technical and administrative data.

Most manufacturing components in the United States are produced in low-volume lots of fifty pieces or less. This means that a great deal of time is spent altering part routing and reprogramming machine operations, which result in slow and costly production sequences. It is estimated that up to 95 percent of a batch-type metal working shop's production time is spent in simply moving and storing workpieces. To speed up the production process for small-lot batch manufacturing, a company should implement CIM capabilities that allow for the flexibility and changing manufacturing requirements needed for product variations and for controlling material flow. The objective of CIM is to optimize yield while minimizing lead time, by using a network of automated manufacturing operations located at various machining centers or plants. In some instances this may involve linking together several NC machines.

What Is CIM?

CIM is a group of modular subsystems, each of which are controlled by computers that are interconnected to form a distributed manufacturing system. Each subsystem of independent technology interacts to produce a higher level of performance than the sum total of the subsystems acting independently. This automation of information flow includes the collection, storage, modification, analysis, and transmission of data for engineering, production, fabrication, materials, and administration.

Eight Goals of CIM

The goals of a CIM program are:

1. *Reduce production lead time.* Install CAD, CAE, and CADD software programs to approach zero time in design, engineering, and drafting.
2. *Create balanced production lines.* Install balanced production lines to achieve zero work-in-process queues.
3. *Develop capability to produce in a lot size of one.* Develop the system so that when a customer's order is received for one part, the manufacturing facility is able to procure raw materials and labor or produce one part.
4. *Achieve zero setup time.* Develop improved methods that will reduce setup time to an absolute minimum to become an efficient small lot producer.
5. *Maintain high process reliability with zero down time.* By monitoring machine running time, parts produced and other conditions on the shop floor, and integrating this with a preventative maintenance system, seek to achieve zero down time.
6. *High product repeatability with zero defects.* All products produced must be within specifications. Since machines would be expected to operate 24 hours a day, seven days a week, high product reliability would be mandatory since there would be no operators to shut the line down should specifications be exceeded.
7. *Minimum data handling resulting in zero data errors.* Data input by bar code reader is much more reliable than manual key entry of data. Even greater reduction of errors can be achieved where data is collected from sensors mounted on machines, thus enabling data to go directly into the computerized system without any data handling whatsoever.
8. *Development of distributed intelligence that enables zero-manned operations.* This goal will allow machines to operate 24 hours a day, seven days a week, thereby reducing capital investment and increasing machine utilization.

WHY CIM REQUIRES A SHIFT IN CORPORATE STRATEGY

Implementation of CIM requires a completely different corporate strategy from a focus on minimizing cost to one of maximizing competitiveness. Manufacturing will have to shift from being a product fabricator, to providing a fast reactive service. Management will have to develop new skills based on optimizing production processes to provide quick responsive delivery in consonance with customer demands. Product life cycles will be compressed since the flexibility of an automated manufacturing operation will permit fast, new product turnaround. The major strategy in a CIM environment will be to maintain constant production throughout the full seven day work week to drive production costs down and recover capital investment. To install CIM, companies must make long-term commitments based on achieving a significant improvement in overall manufacturing productivity rather than in obtaining greater effectiveness in one particular manufacturing or marketing task.

An Effective Technique for Strategic Planning*

One of the most comprehensive techniques involving strategic planning methodology (SPM) has been developed by BDM Corporation. This technique consists of a generated procedure for the development, evaluation, and selection of alternative processes, and systems and control techniques for factory automation and enhancements. The overall procedure contains a set of individual modules that incorporate the critical parameters of work unit configuration definition, schedule, risk, cost, and levels of automation sophistication (complexity) into a relative evaluation process. The process facilitates the selection of acceptable enhanced/automated manufacturing configurations that can then be developed in detail and implemented in the factory. The SPM procedures have been automated and developed into a computer software package which fully supports all analysis steps. The software contains all the required worksheets, automatically calculates values and transfers data between worksheets when required, and documents the analysis.

This procedure provides the following benefits:

- Results that are traceable, auditable, and repeatable
- Technological advancement of manufacturing processes
- Better defined computer hardware–software information data-base requirements
- Improved long-range asset planning
- Improved inter-organizational communication and understanding
- Reliable sources of information and validated techniques for planning
- Integrated top-level strategic plan development

Without a strategic plan for implementing the factory automation/modernization program, a company risks the following:

1. Inadequate management direction and control
2. Increase in missed profit opportunities due to lack of cost savings or incentive awards
3. Possible implementation of mismatched/uncoordinated equipment, facilities, and computer systems
4. Possible lost opportunity to take the industry lead in developing and implementing advanced technology

The output and effects of the strategic planning process are shown in figure 11–1. The results of the strategic planning process provide the foundation for coordination and development of detailed tactical plans and schedules for implementation of factory automation and enhancement.

*Permission granted by the BDM Corporation, from document entitled "BDM Capabilities to Provide Advanced Manufacturing, Technical Support, Design & Development." BDM/ABO-85-1296-ON-Z, pages III–1 to III–10 including Figures III–1 to III–4.

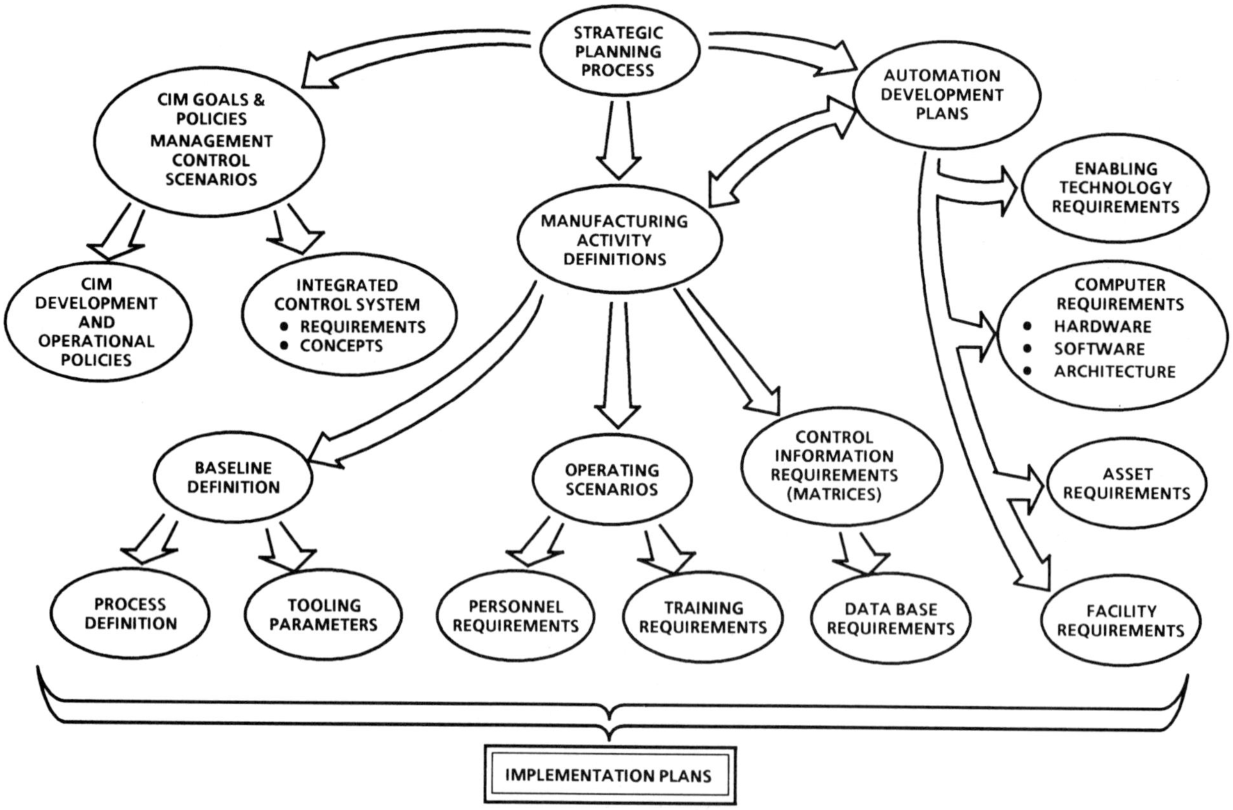

FIGURE 11-1 Outputs and Effects of Strategic Planning (from "BDM Capabilities to Provide Advanced Manufacturing, Technical Support, Design, and Development" BDM/ABQ-85-1296-ON-2, pp III-1–III-10).

This procedure provides a disciplined methodology for a company to use in the development of strategic plans, and for the upgrading/modernization/automation/integration of manufacturing work units. The steps of the methodology are as follows:

1. Develop baseline definition
2. Assess baseline
3. Generate alternatives to enhanced baseline
4. Analyze and compare alternatives
5. Select "best" alternative—new baseline
6. Repeat analysis (by returning to Step 2) until one of the following conditions is satisfied:
   (a) Alternatives are exhausted
   (b) Automation/enhancement goals are satisfied

CIM goals and policies are contained in Steps 3, 4, 5, and 6.

Why Business Plans Are Prerequisites for CIM

Prior to conducting this strategic planning process it is necessary to have a well defined strategic business plan (figure 11–2) which provides guidance and direction for

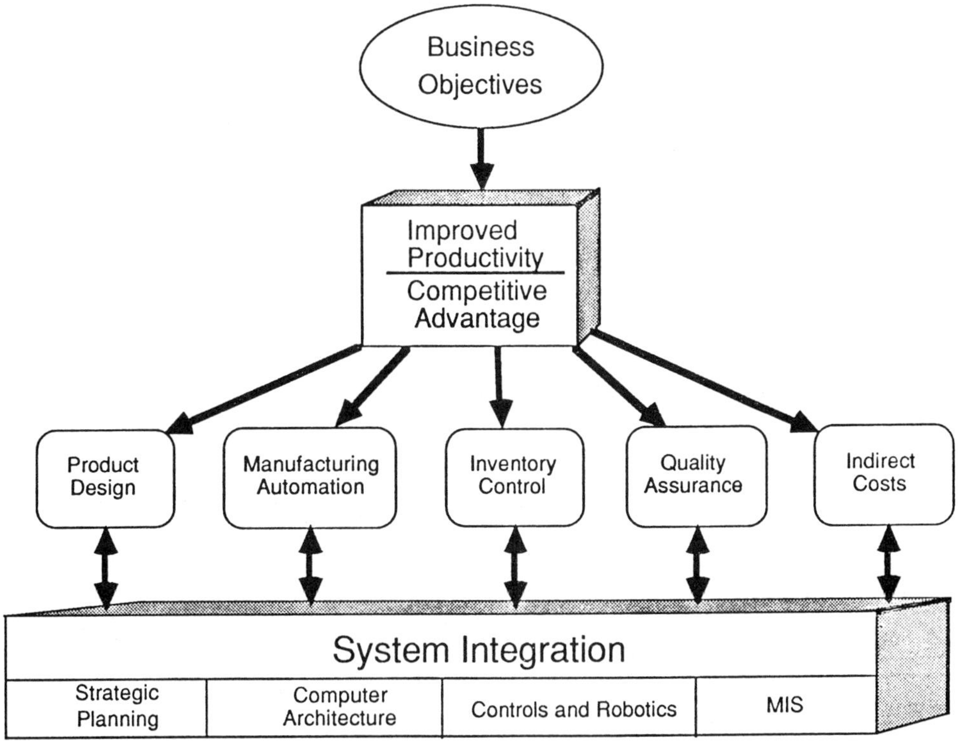

FIGURE 11–2 AMS—Part of an Overall Corporate Strategy (from "BDM Capabilities to Provide Advanced Manufacturing, Technical Support, Design, and Development" BDM/ABQ-85-1296-ON-2, pp III-1–III-10).

this more definitive activity (refer to Chapter 1 for more details on this subject). This executive-level strategic business plan consists of objectives, goals, and policies for the development and establishment of an advanced CIM environment. These goals and policies provide the clear and consistent direction from top levels of management that organizations must use during the development and implementation of the enhanced "factory of the future." It cannot be overemphasized that if this strategic charter (CIM goals and policies statement) does not exist then it must be developed prior to undertaking this far reaching program. The reason for this is because the procedures and steps to follow in effecting the strategic planning methodology depend on the goals and policies of an organization for clear direction in the development and evaluation of alternative systems, equipment, and software.

The procedure requires that the definition of an existing or planned work unit (for example, manufacturing cell) be developed and designated as the original baseline configuration. An assessment of the baseline manufacturing operations is made, and alternatives are developed to upgrade and enhance these operations. Analysis of all costs, implementation schedule, automation level, and risk are conducted for all alternatives developed. A relative comparison is then made among these alternatives. One alternative is selected, and the others are either saved or eliminated, depending on whether or not they offer further automation and/or enhancements to the work unit being studied. Additional alternatives may be developed based on the goals and policies for automation, timing, budget constraints, and so on. Normally, several automation implementation changes occur between the original baseline and the ultimate or final work unit configuration. Each new configuration (change 1, 2,

3, and so on) becomes a new baseline for subsequent analyses that are performed sequentially relative to each automation/enhancement change. The analysis continues until the strategic plan for the automation of the work unit is as complete as possible based on known and forecasted technology development and/or program, product, cost and time constraints (figure 11–3).

The results of using this planning procedure can best be illustrated by graphically depicting the level of automation and integration of a manufacturing work unit in terms of the general type of equipment being used to perform the operations, and how the control information and communications for the processes are generally being accomplished. Figure 11–4 exemplifies a sample of summary results obtained from using this planning methodology on a manufacturing work unit. As shown, the work unit's baseline operations in 1984 are summarized as being generally done by power hand tools and controlled (via information and communications) generally by an on-line computer. The automation/enhancement, changes 1 through 4, drive the work unit to full automation in 1990 by having the operations carried out by automated, mechanized means that incorporate adaptive equipment. The control, information, and communications are generally accomplished directly by means of an integrated computer system.

When analyzing complex manufacturing processes and controls it is advisable to make a simulation analysis in conjunction with the strategic planning methodology to help validate the results of integrating automation and enhancements into the factory (simulation analysis techniques are discussed in a later section of this chapter). The baseline configuration for

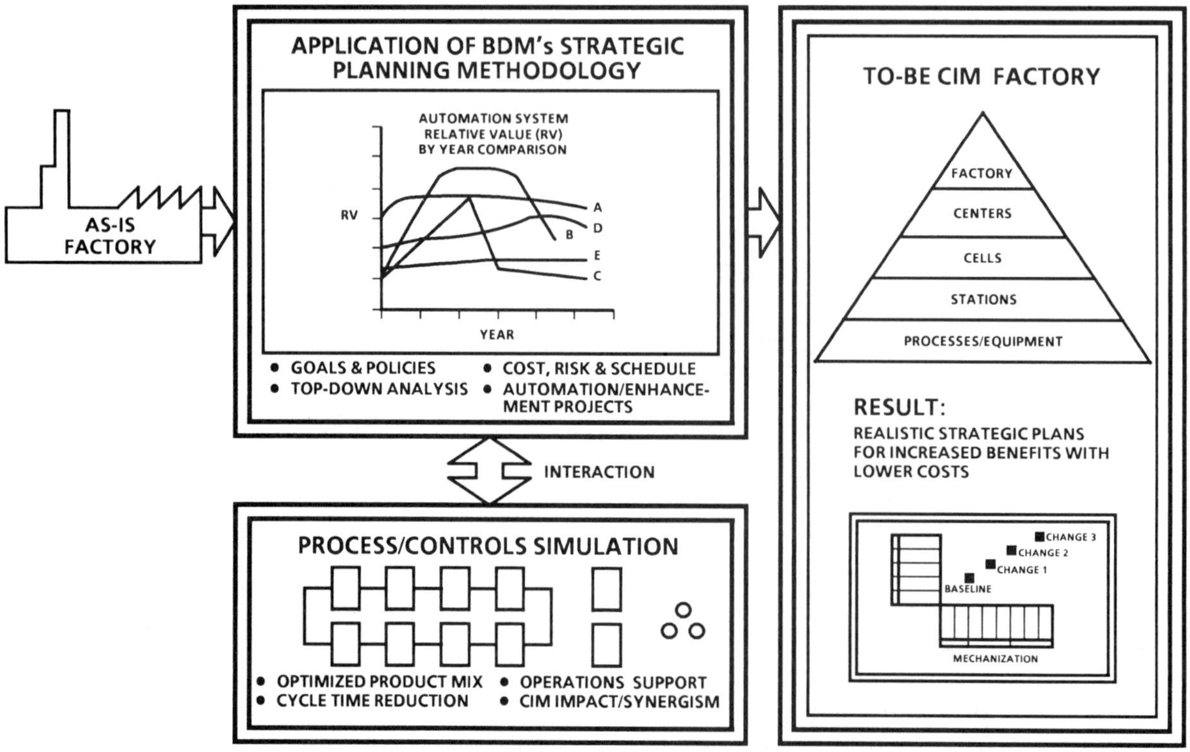

FIGURE 11–3 AMS—Alternatives Based on Relative Value (from "BDM Capabilities to Provide Advanced Manufacturing, Technical Support, Design, and Development" BDM/ABQ-85-1296-ON-2, pp III-1–III-10).

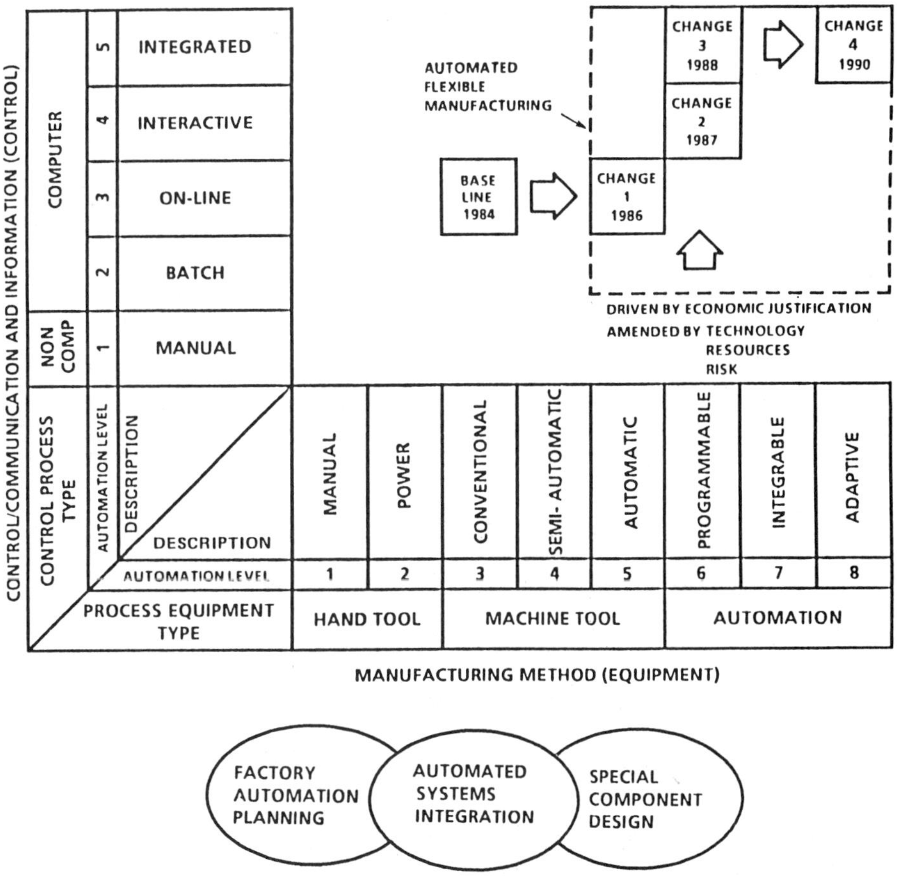

FIGURE 11–4 **Sample Summary Results of the Strategic Planning Process** (from "BDM Capabilities to Provide Advanced Manufacturing, Technical Support, Design, and Development" BDM/ABQ-85-1296-ON-2, pp III-1–III-10).

the manufacturing processes and controls can be modeled and analyzed in conjunction with the first steps of the SPM. SPM analysis results of an automation project yield a selection for implementation based on cost, risk, and implementation timing. The proposed automation/enhancement project or projects can be modeled and analyzed in an integrated fashion by modifying the baseline simulation model, analyzing the results, and identifying any adverse or synergistic effects of multiple automation/enhancement projects. Figure 11–5 illustrates the use of these two powerful tools to develop an integrated strategic factory automation/enhancement plan which will have a very high probability of success when implemented.

The resultant strategic plan is valid only for the time and conditions under which the plan was produced. These long-range, justified, planning results can now be used to develop detailed tactical implementation plans and schedules based on factual data. Variation of the facts or basic assumptions used to develop the plan requires a reassessment of the procedures to assure their validity. Periodic reviews and updates of the strategic plans should be conducted so that as technology, systems, products, and markets change so will the long-range strategic plans.

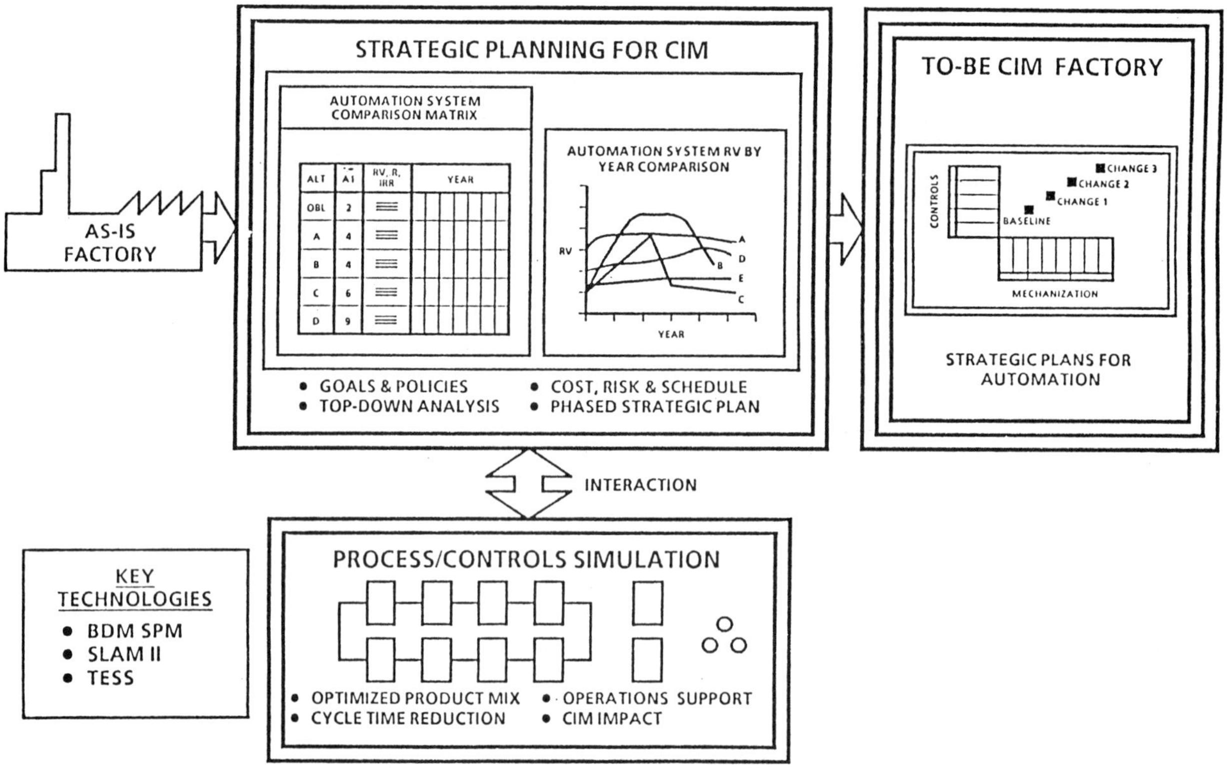

FIGURE 11-5 Simulation Analysis Integrated with Strategic Planning Methodology (from "BDM Capabilities to Provide Advanced Manufacturing, Technical Support, Design, and Development" BDM/ABQ-85-1296-ON-2, pp III-1–III-10).

HOW SIMULATION HELPS MANAGEMENT EVALUATE DECISIONS

Effective factory planning and project/system design requires an understanding of how a particular decision will impact the system environment. This understanding can be quite difficult to grasp due to complexity of most problems in today's factories. Decision consequences can be exposed by the construction of a simulation model that describes the dynamic structure and response of the system to controlled inputs. Useful and effective models depend on sound engineering judgment, experience gained from similar systems, close examination of the system to be modeled, operating constraints, and managerial goals and policies that impact it. Since a simulation model is a representation of the actual system to be studied, it provides an understanding of system behavior while avoiding costly problems associated with building a proposed system, disturbing a current system, destroying a current system.

As a diversified decision support tool, simulation may be employed to accomplish and support activities such as the following:

- Strategic Planning
  - (a) Design new process/system/policies
  - (b) Determine effect of different priorities
  - (c) Forecast production levels/required resources
  - (d) Estimate cost of alternatives

- Management Control
  - (a) Improve throughput
  - (b) Identify effect of changes in resource capacities/equipment failures
  - (c) Identify effect of delays in raw materials
  - (d) Improve system efficiency
- Operational Control
  - (a) Optimize equipment utilization
  - (b) Significantly reduce bottlenecks
  - (c) Define operational requirements
  - (d) Identify critical operation rates
  - (e) Optimize staffing configurations
  - (f) Reduce in-process inventories

There are a number of computer simulation languages currently available, including SLAM II, TESS, SIMAN, GPSS/II and IDSS 20. Due to recent improvements that have resulted in microcomputer software versions and flowchart structures tailored specifically for process-oriented manufacturing environments, these languages have become quite powerful in their ability to support multiple phases of factory planning and operation.

Pritsker & Associates, Inc. has developed TESS and SLAM II, which are highly sophisticated computer simulation languages. TESS is an integrated simulation support system that combines the capabilities of SLAM II with a powerful relational data base management system and state-of-the-art graphics to provide support for all aspects of a simulation project. TESS allows construction of SLAM II network models at a graphics terminal, thus eliminating the preparation of input statements. Actual system data may be stored in the data base with the model. This ability is combined with a data-base orientation that facilitates both simulation analysis and process operational support by permitting file-driven experimentation across multiple scenarios. TESS color graphics provide model building capabilities, animation of model operation (that is, a time-stepped visual display of the manufacturing process), and a wide variety of graphs of key model performance measures computed from TESS's extensive statistical analysis capability. Model outputs across multiple scenarios may be displayed on the same graph automatically. TESS is a powerful tool not only for performing all phases of a simulation analysis, but also for effectively communicating analysis results to both system users and management.

Simulation modeling is predicated on the assumption that a user can describe a system in terms that are acceptable to a computing system. Since a system consists of a set of variables, with each combination of variable values representing a unique state or condition of the system, then manipulation of the variable values simulates movement of the system from state to state. Changes can occur continuously in the state of a system over a period of time or at discrete instants in time. Although the procedures for describing the dynamic behavior of discrete and continuous change models differ, the basic concept of simulating a system by portraying the changes in the state of the system over time remains the same.

Thus, simulation models can be used to:

- Define a system or problem
- Analyze situations in the context of the factory or other environment
- Synthesize all available data and provide the means to evaluate proposed solutions
- Aid in predicting the course of future events

HOW CIM AFFECTS ORGANIZATION STRUCTURE

The organizational structure will experience considerable changes during a complete CIM program implementation since there will be a need for greater team interface, thus breaking down more rigid functional entities. This will result in more dotted-line relationships between manufacturing engineers, foremen, maintenance, quality control, and others. These functions will become interdependent because of the critical need to keep the factory operating on a full-time basis.

Although CIM will require tasks to become integrated and personnel in various functional areas to work together, it also will require more centralized planning. Large capital investments will place greater emphasis on strategic and long-range planning. Planners will have computer-based tools to assist them. Manufacturing process planning will be performed by design engineers. CIM system loading and labor utilization will be based on manufacturing processing logic incorporated into software. Personnel in production planning, maintenance, and quality control will "talk" to one another constantly through a common data base to ensure that the products are manufactured on schedule and within tolerance levels.

WHY LONG-RANGE PLANNING INCREASES THE CHANCE FOR SUCCESSFUL IMPLEMENTATION

Companies need to engage in long-range planning when implementing CIM so that these new capabilities and technologies can be developed with a minimum disruption to existing processes. Experience has shown that CIM programs are most successful when subsystems are incorporated in a modular and phased fashion. Only then can those involved in CIM planning specify data-base interactions between design and manufacturing functions, such as how information is organized, maintained, and transferred from one data base to another. Analysis must be performed that relates to potential integration of engineering design, manufacturing planning and control, shop floor monitoring, and process automation functions. Afterwards, it is possible to define and cost justify relevant CIM components and prepare the modular and phased implementation plan.

COMPUTER-INTEGRATED MANUFACTURING ARCHITECTURE: AN INTEGRATED ARCHITECTURE FOR PRODUCT REALIZATION AT AT&T*

The computer-integrated manufacturing architecture (CIMA) is a planning tool for AT&T's design and manufacturing systems of the near future. It covers a hierarchy of operations, ranging from basic assembly up to production loading and product design, and coordinates these operations within and between hierarchal levels. CIMA is aimed at helping AT&T to ensure fast, efficient introduction of products, low-cost manufacturing operations, fast and complete response to customer orders, and high quality.

The CIMA Project

CIMA has evolved from AT&T's planning and developing of computerized support systems for integrated, efficient local telephone company operations. The operational and

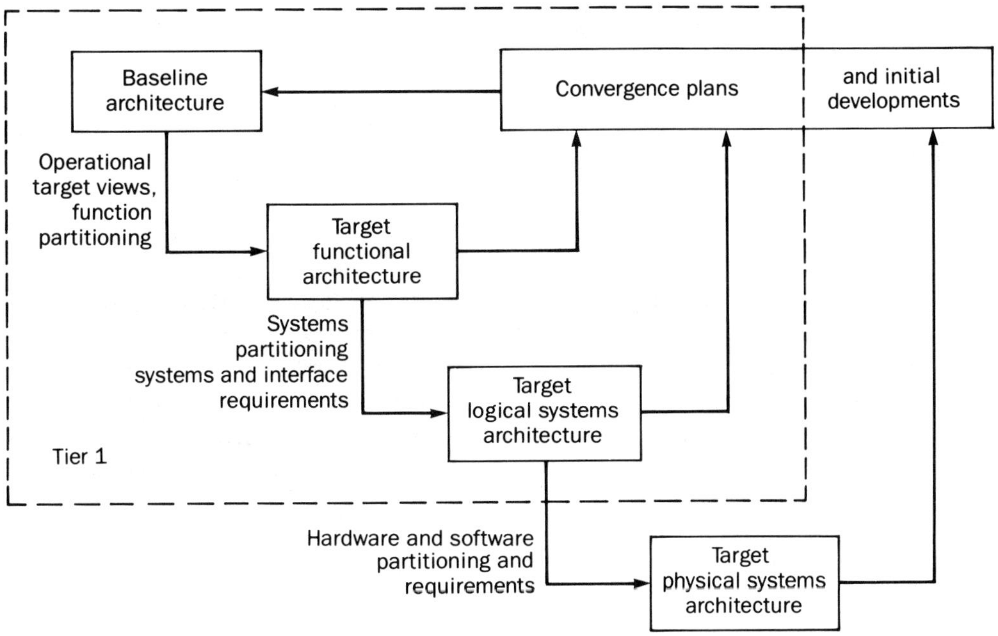

FIGURE 11-6 Computer-Integrated Manufacturing Architecture (CIMA) (reprinted with permission from the *AT&T Technical Journal,* copyright 1986, AT&T).

systems expertise that made this possible is now being applied to CIMA. The approach (figure 11–6) to this new, technically more difficult challenge, is similar to the one used before:

1. Current operational processes have to be analyzed and their limitations clearly understood.
2. Business-related operational target views need to be constructed and accepted by all participants.
3. A target-functional architecture has to be created to describe what functions are required in the future, what their information exchanges should be, and what they will need in terms of systems support, procedures, and measurements.
4. A target logical systems architecture needs to be constructed to cluster the required system functionalities into conceptual systems and to specify the interconnection and environmental needs of such systems.
5. The above architectures need be used as (1) templates for current architectures in design and manufacturing so that the necessary evolution can be planned and implemented efficiently, and (2) as guides for the development of procedures and measurements consistent with the operational target view and for the systems engineering and development of the future operations support systems.

Operational Target View

Customer satisfaction and corporate financial success are the goals that drive the four key targets of CIMA for design and manufacturing operations:

- Fast, efficient introduction of products
- Low-cost manufacturing operations

- Fast, complete response to customer orders
- High quality

These key targets, based on line personnel feedback and process analyses, in turn drive the target architectures.

Fast and efficient product introduction requires the strict enforcement of comprehensive up-front product planning, formation of teams of product and process designers and materials, and deployment planners to provide a multidisciplinary core to the product introduction effort, appointment of product realization team managers with real cross-organizational authority, and detailed methodologies to assure that product and process design adequacy reviews are performed and completed successfully. The target also requires enhanced systems for the paperless creation of design information, improved communication between design and manufacturing engineers, full issue control of information, and storage of the "how-to-build-the-product" information to optimize subsequent factory operations.

Low cost manufacturing operations in assembly factories require low direct and indirect costs of materials. This is critically dependent on product design. Designers must take into account direct and indirect materials costs to factories in all design decisions, from initial product architecture to component selection. On the factory side, this target depends on a resources management discipline that results in stockless ("pull") production and supply operations. It is critical that islands of stockless operations be established wherever possible and that these islands be expanded and linked progressively. This discipline requires that accountability for work-in-progress inventories be pushed down to the lowest organizational levels and that measurements and rewards be consistent with this goal. It also requires new, integrated operations support systems for floor execution, monitoring and control, and flexible operations and materials management systems.

Fast, complete response to customer orders requires modularity and standardization of designs and reduction in the number of product selling configurations. Given these two requirements, the target also requires integrated customer order management, including advanced reservation, sourcing and scheduling, order issue control and tracking, order manufacture, and delivery, so that customers can be adequately served by the sales organizations while manufacturing optimizes its own resource utilization. From a systems perspective, this target requires a very high level of integration of order management, design, and production planning systems. It also requires that adequate support systems be installed to monitor and control final assembly and test operations.

High quality is a target that affects all design and manufacturing operations. It requires clear up-front definition of customer expectations about the product functionality and support, methodologies that stress a zero defect goal, and attention to yield and other manufacturing requirements. The target also implies a culture of preventive attention to performance degradation, and technically strong engineering support aimed at constant process improvements. Systems must support exception reporting of quality degradation and quick follow-up corrective actions.

Functional Architecture

The CIMA functional architecture proposes a structure of functions and information exchanges between functions that supports the achievement of the operational target view. The functional architecture provides the foundation for work center, procedure, and measurement planning. It also is a step toward the CIMA logical systems architecture.

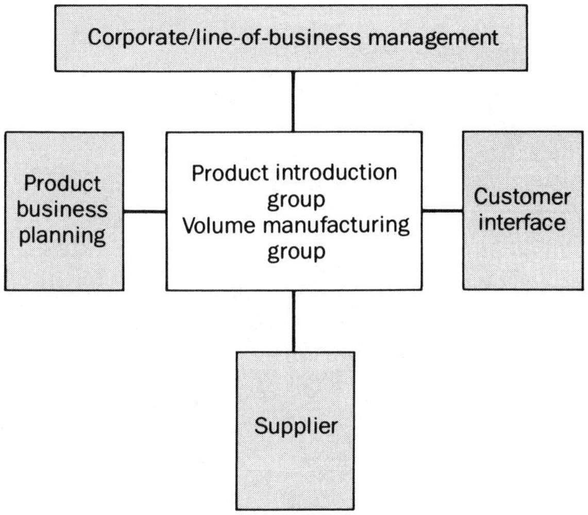

FIGURE 11–7 CIMA Functional Architecture Overview (reprinted with permission from the *AT&T Technical Journal*, copyright 1986, AT&T).

The CIMA target functional architecture abandons the traditional distinction between design and manufacturing and proposes instead that functions be separated into *product introduction* and *volume manufacturing* groups (figure 11–7). This nontraditional functional grouping is necessary if the integration requirements presented in the target view are to be fully realized.

The CIMA product introduction functional architecture is shown in figure 11–8. It contains four parallel functional cycles: (1) hardware and software design and development, (2) manufacturing facilities design and development, (3) materials procurement, and (4) deployment enabling.

These four functional cycles are triggered by the product business plan, controlled by a strong project management function, and brought to a close by a product and process integration verification (PPIV) function. Each one of the four cycles contains early planning, high-level design, detailed design, and verification functions. The information exchanges emphasize product/process communication, closing of information loops, and issue control of information. The product introduction functional architecture provides a formal structure for the evolution of the up-front product business plan information into the full set of information that enables the factory to make the product.

The CIMA volume manufacturing functional architecture is shown in figure 11–9. It contains the functions needed to specify what products to make, when, and in what quantities (production and materials management, manufacturing order service, and order release), the functions that control and execute production (floor execution control, receival control and execution, stores, manufacture and test, and material transport), procurement (procurement control), and manufacturing distribution (distribution control and packing and shipping). Technical support functions, such as engineering, quality, and facility and systems maintenance, as well as manufacturing management and administrative support, also are included. The architecture emphasizes the closing of key information loops in production planning, procurement, production execution, and distribution, as well as the loops with customer orders and product introduction.

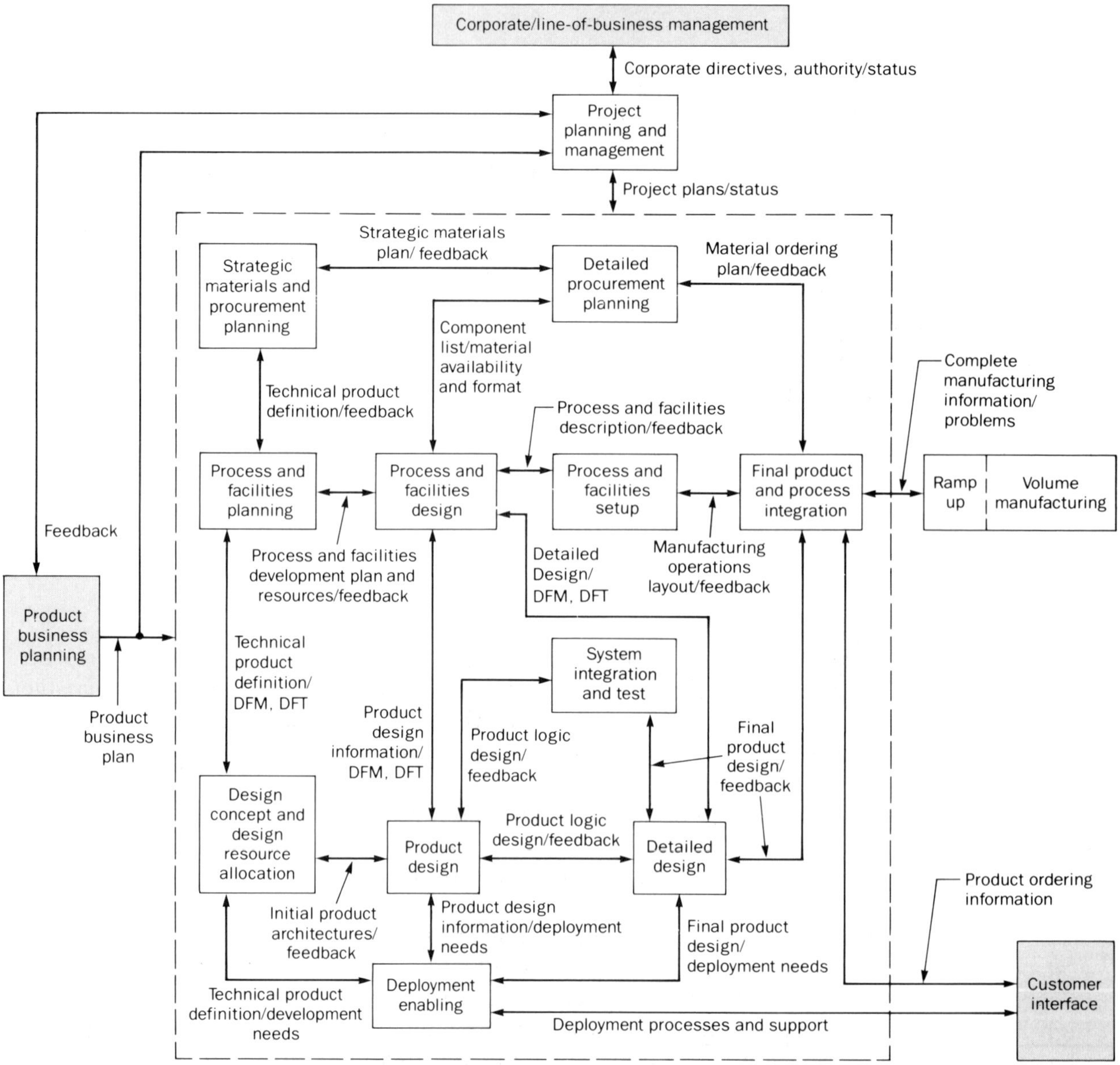

FIGURE 11–8 CIMA Product Introduction Functional Architecture (reprinted with permission from the *AT&T Technical Journal,* copyright 1986, AT&T).

Logical Systems Architecture

The CIMA logical systems architecture proposes a structure of conceptual (logical) systems linked by information exchange networks. Logical systems provide support to one or more functions in the functional architecture.

Distribution and integration are key concepts of the target logical systems architecture. Logical systems are distributed in a CIMA seven-level hierarchy that encompasses design systems outside the factories all the way down to systems that control manufacturing workstations. Integration is emphasized by a single architecture for both product information and volume manufacturing logical systems, with several systems straddling the

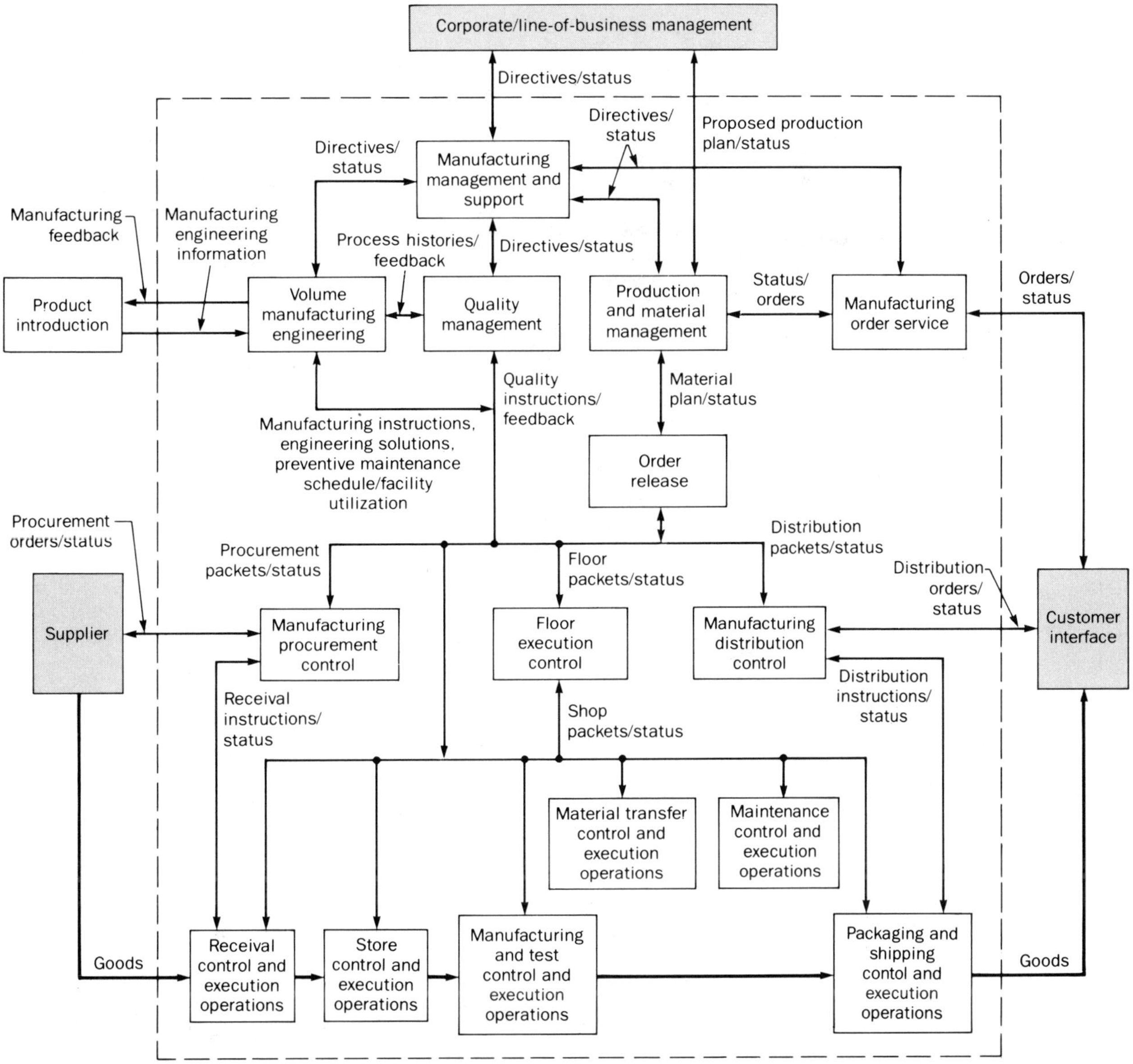

FIGURE 11–9 CIMA Volume Manufacturing Functional Architecture (reprinted with permission from the *AT&T Technical Journal*, copyright 1986, AT&T).

boundary and with an overall information and control management system, which provide a new and critical functionality.

Figure 11–10 presents a high-level view of the logical systems hierarchy. The lowest level of the CIMA hierarchy contains equipment, such as a robot, cart, or bar code reader, which is to be linked to CIMA systems. Equipment performs operations that change the form of, location of, or knowledge about one workpiece. At the next higher level are workstation systems. These control and monitor the operations of the equipment under their supervision. Next are cell systems, not yet broadly implemented in AT&T factories but the subject of much current work. Cells control workstations, and therefore tightly related operations on several workpieces.

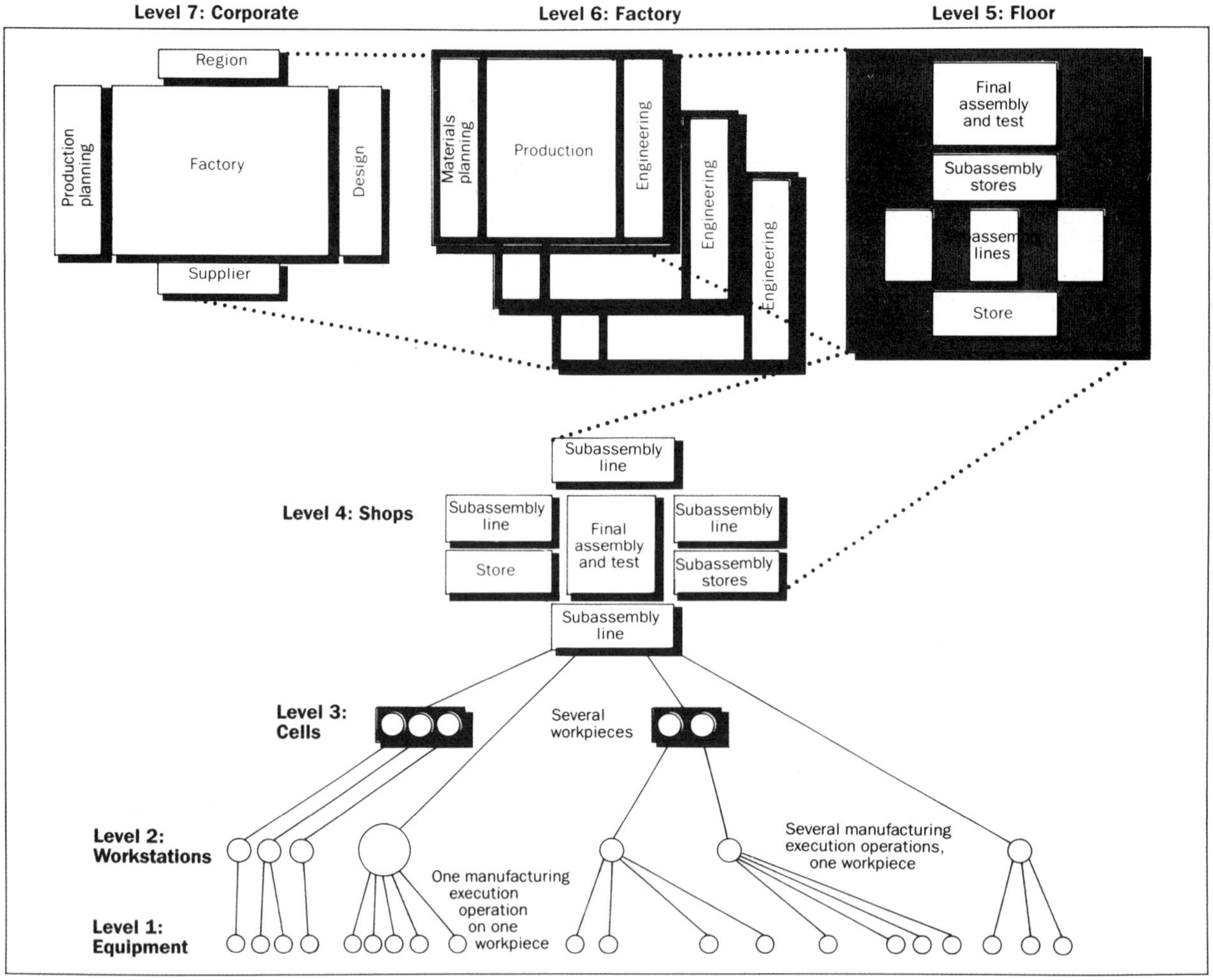

FIGURE 11–10 CIMA Logical Systems Hierarchy (reprinted with permission from the *AT&T Technical Journal,* copyright 1986, AT&T).

At Level 4 are the shop systems that monitor and control natural sets of factory production functions. In CIMA terminology, receiving, shipping, storerooms, subassembly lines, final assembly line, and so forth are "shops." To manage product flows between shops within the time granularity of the instructions from Level 6 systems, CIMA introduces the floor systems. "Floors" tie the shops together and are critical to balancing short-term load among shops and to ensuring pull production.

The next two levels of the CIMA hierarchy plan, support, and enable manufacturing operations. The highest level, corporate systems, acts across factories and results in production loading to factories, designs, order management, and so on. Level 6 occurs in the factory, either at the whole factory or at a factory-within-factory level, and includes such systems as materials planning, engineering, and so on.

The CIMA logical system architecture populates these seven levels with two kinds of logical systems. The operations-support logical systems, as the name indicates, support specific sets of functions in the functional architecture. In addition, these systems interface with other systems and manage their own data. The information and control management system integrates the total information environment.

The CIMA logical systems architecture deals in an integrated way with three main kinds of information flows: engineering information (which tells the factory all details on how to make products), production planning information (which tells the factory what products to make, when, in what quantities, and with what resources), and decision support information (which includes status reports, exception reports, and action-enabling information). The action-enabling information makes possible prompt response to performance degradation and troubles.

The approach to CIMA implementation must rely heavily on standardization of interfaces, reusability of applications software, and provisioning of plug-compatible hardware.

EXAMPLES OF SUCCESSFULLY INSTALLED CIM SYSTEMS

Currently, there are relatively few complete CIM systems installed in the United States. Of these, most have been developed by large industrial firms with the resources and cash flow to invest substantial sums to maintain or expand their high volume of unit sales. As more companies automate either totally or in a step-by-step modular fashion, CIM will undoubtedly become the method of operation for a large segment of American industry. As such, this technology will emerge as the primary force in enabling companies to compete both domestically and in the international marketplace—in quality of product, low-cost operation, and responsive delivery of goods and services.

The author has selected the following award-winning, and outstanding CIM systems as illustrations of work done in this area. Each of these case-in-point examples contain all of the requisite components needed to successfully implement a program of this magnitude.

CASE-IN-POINT

AT&T RICHMOND, VIRGINIA WORKS COMPUTER-INTEGRATED MANUFACTURING SYSTEM

This case covers AT&T's Richmond, Virginia Works CIM System, recipient of the annual LEAD Award, sponsored by the Computer and Automated Systems Association of the Society of Manufacturing Engineers (CASE/SME) (LEAD is an acronym for leadership and excellence in application and development of CIM systems). The award goes to the company or division that, in the LEAD award committee's opinion, demonstrates an outstanding implementation of a computer-integrated manufacturing system.

Situation

The Richmond Works is a high volume manufacturer of *printed wiring boards* (PWBs). It's two plants manufacture a complete range of printed wiring boards, consisting of *multilayer* boards, typically with six or eight layers; *rigid* boards, which have plated through holes; *flexible* boards with the characteristic for bending to meet the contour of its housing; and *connectorized* or *backplane* boards.

The Richmond Works is operated like a job shop, with each production run done in lot form against a specific customer order. Since the job shop environment is highly dynamic, AT&T people recognized early in the plant design process that a flexible computer system would be required to efficiently manage the operation. The PWB is usually the last component designed, yet the first one required for a piece of equipment. It also is the first component changed and the one that is changed most frequently. These considerations require a fast reacting and flexible system. This system also must manage a shop with over a thousand active PWB codes at any given time. The markets for these codes are rapidly changing with an average life expectancy of less than two years. The manufacturing technology is very dynamic because techniques are constantly being improved. These improvements must be implemented quickly and with as little disruption to the manufacturing process as possible.

Another consideration was the short delivery intervals that PWB orders often require, with the attendant need to move lots through more than one operation per day. This requirement motivated a real-time instead of batch product tracking system. Furthermore, these intervals are especially demanding for a new design, so the plant needed the CAM software to quickly generate the required numerical control files and art masters. Competition also had become a greater factor for needing CIM because with the company's divesture, previous AT&T customers could now go elsewhere if the company failed to satisfy their needs. An offsetting gain was that AT&T was now free to compete in the open market. However, this new competitive stance placed a renewed emphasis on CIM at the Richmond Works to enable it to become a successful manufacturer of PWBs.

Solution

The first step was putting together a team to develop and implement the system. This was a multidisciplined group of sixteen professionals, with backgrounds in computer science, engineering, business, mathematics, and management. This team developed the basic modules that evolved into the totally integrated system that is now used to manage the manufacturing facility.

The next step was to select the hardware, software, and data-base requirements. Also, decisions were made on whether to purchase or develop this software in-house. It was decided that the basic modules for the manufacturing information control system and direct numerical control system would be developed by AT&T, as was most of the CAM software.

The DNC system started with a centralized, stand-alone computer system. The experience gained from this system along with the requirement for a long-term growth and support of off-site facilities resulted in the development of a system with the following characteristics:

1. Low-cost, modular distributed system, with local storage for machine data capable of supporting one or more shifts of work if the host computer was out of service.
2. High reliability, durable processor capable of withstanding shop area conditions with minimum maintenance.
3. Capability to interface with the bi-synchronous communications method on the host computer and to support EIA RS232 standard interfacing with shop machines and microprocessors.
4. Ability to support in-plant data communications, as well as off-site communications. Periodic mobility of a processor is required so it can be relocated as shop areas are rearranged or swapped with a spare processor. This must be done with both minimum effort and interference to shop machines.
5. Elimination of dependency on vendors by developing in-house software and interfacing hardware. Support for new machines added on a modular basis with a minimum of new software or hardware development and maximum standardization.

The manufacturing information control system evolved into the current system by enhancing it with needed modules. These modules include a storeroom inventory system, chemical control system, tool control system, product yield system, and a management reporting system.

All the systems are running on an IBM 3083 computer using a MVS/JES2 operating system and CICS for teleprocessing. The following material describes how the processing of a new PWB design is handled using AT&T Technologies' Richmond Works' CIM system functions.

Computer-Aided Manufacturing

New designs are teleprocessed in digital form from remote Bell Laboratories design locations using a corporate wideband transmission network. This data includes a definition of the board's circuitry, drilled hole locations, and test points.

On receipt at the Richmond Data Center, the design is automatically intercepted by a program that checks the characteristics of the PWB (type, size, and so on). Usually, it is found to fit a predefined "family" (developed by using group technology methodology). For these families, the program generates and submits for execution all the *CAM* jobs required. The output of these jobs include: (1) NC files for drilling, routing, and testing, which are added to a central data base, and (2) plots that show drilled hole locations, router cutter paths, and points to be tested. The plotted output is then reviewed by an engineer before the design is released to production. As little as six hours may elapse between the completion of the design at Bell Laboratories and the start of manufacturing at the Richmond Works.

The numerical control data is distributed to a network of 130 NC machines using minicomputers tied into the mainframe. When a machine needs a particular file, it may already reside on the minicomputer disk. If not, the minicomputer will request that the file be downloaded from the central data base. This data base is very large and contains some 20,000 files, totaling about 420 million bytes.

Production Control System

Printed wiring board requirements are received over the same wideband transmission network as design data. These requirements are edited for errors and automatically added to the data base two times per week. At this time, they are analyzed to determine if the requested ship date permits time for the plant's manufacturing interval. If so, a confirmation of the required date is returned to the customer over the network. If the analysis determines that the required date is inside the plant's manufacturing interval, production control negotiates a satisfactory required date with the customer. This date is entered into the data base with a CRT and the confirmation is then transmitted to the customer. For emergency and non-AT&T requirements, a production control clerk enters the requirement into the data base with a CRT and sends the confirmation to the AT&T customers over the network.

These requirements are used to load the shop, reschedule work in process, determine the plant's ahead-or-behind-schedule position at reporting intervals, and produce various management reports. When shipments are made against these requirements, billing data is created and stored, which will be transmitted once a week to AT&T customers and to a corporate billing organization for non-AT&T customers. When a requirement is satisfied, it is purged from the active data base and placed on a history file for use in the company's forecasting systems.

Product Engineering System

When the new design is received, a manufacturing engineer uses a CRT to create a routing for that part number. The routing contains the information the shop requires to manufacture the product, including the sequence of operations, machine parameters at each operation, NC file names, and so on. As an automatic part of the creation process, the expected hourly output rates for each operation also are computed. These rates are important not only in administering a wage incentive program, but in forecasting machine capacity. After the tooling data and routing are available, a customer order for that part number triggers the creation of lots to fill the order.

Shop Status System

The shop status component of the system maintains a queue for each machine, consisting of the lots waiting to be processed by that machine. The queue is ordered into a priority sequence based on the due date of the customer order that each lot is intended to fill. Due dates or quantities are changed in order and this is automatically reflected in the machine queues.

When a machine operator views his or her queue on the CRT, he or she indicates which lot is to be worked on next (normally, the first lot on the screen), and the system prints the manufacturing engineer's instructions for performing that operation using a small printer beside the CRT. The fact that the latest engineering instructions are printed online is a notable feature of the system and makes it much easier to implement manufacturing changes. With a preprinted routing, it would be necessary to physically find all the lots affected by a change so that the instructions could be modified. It is estimated that this task alone would require more than sixty people if performed manually at Richmond Works.

If the operator's machine is numerically controlled, the selected lot number is used by the system to determine (from the engineer's routing) the name of the appropriate NC file to "drive" the machine. This file is then downloaded to the machine.

Once the operation has been completed on the lot, the operator reports the completion to the system using a CRT or a bar code reader. The system then removes the lot from the queue and places it at the proper position (as determined by its priority) in the next machine's queue.

A history of each transaction also is recorded so that batch programs can use it for deriving yield data machine throughputs, operator efficiencies, and so on. Since yield and quality are important issues in PWB manufacture, special attention has been given to developing software to assist manufacturing engineers to find the causes of problems in these areas. For example, it might be found that a group of lots with drilling defects were all processed by one particular drilling machine.

At the final operation in the routing, a "ship" transaction is entered. This transaction removes the lot from the system and updates the appropriate customer order to reflect the shipment. It also results in a delivery document being printed and in the updating of the shipping statistics for the plant.

Tips for Successful Automation: What AT&T Learned from CIM Installation

1. Before undertaking the development of a CIM system it is necessary to have complete management support. If not, the system will never succeed, especially if it becomes necessary to retrain employees who are comfortable with an existing system.
2. A successful CIM program does not allow for complacency since a successful system evolves, and the evolution appears to be a never-ending phenomenon.
3. Get all users of the program, shop operators, production control clerks, engineers, and others committed and involved so that they will willingly assist in the development of enhancements. In this instance that caused the basic software to evolve into a truly integrated system.
4. Without fully automating its manufacturing facility, a company may be unable to continue to operate in a competitive mode.

What AT&T learned from this experience is that it could not operate the Richmond Works efficiently without installing a complete CIM system. With the high volume of PWB's short intervals, design changes, and new competition, AT&T would not be able to meet the service and quality standards that had been its goals even before CIM became a reality.

Benefits of CIM

AT&T achieved numerous benefits as a result of implementing this successful CIM system, such as:

1. Reduction in both tooling and manufacturing intervals which resulted in better customer service
2. Reduction in direct labor, thereby lowering production costs

3. Freeing up users from mundane tasks and allowing them to become more versatile employees
4. Increased job satisfaction and self-esteem of employees through their becoming an integral part of the total system installation
5. Improved quality of product
6. Increased productivity
7. Greater manufacturing flexibility
8. Development of a simplified method of calculating labor rates
9. Development of simplified cost estimating
10. Improved capacity planning techniques
11. Reduced manufacturing cycle time, from receipt of customer order to shipment
12. Reduced overall lead time in half

All of the basic information systems benefits are largely attributable to having all the data now stored on a common data base and therefore available to various groups that interface with one another.

The following listing consists of some system statistics that are indicative of the type support offered by this system:

- 100,000 MICS transactions per day
- 80 percent of the transactions originate from the shop floor
- 40 transactions per employee per day
- 1,100 designs received per month
- 130 DNC facilities
- 600 CRT terminals
- 115 printers
- 55 bar code readers
- 10,000 control files distributed each month

CASE-IN-POINT

*GENERAL ELECTRIC COMPANY'S SCHENECTADY, NEW YORK COMPUTER-INTEGRATED MANUFACTURING SYSTEM**

The General Electric Company is another recipient of the prestigious LEAD Award for an outstanding CIM application. This case history shows in complete, step-by-step detail how the system functions in the company's Schenectady, New York facility—headquarters for the largest manufacturer of steam turbine generators in the world.

An important factor in this effort are the five Data General Eclipse MV/10000 systems and three MV/8000 systems employed in GE's computer-integrated manufacturing system. Total data storage capacity for the Eclipses is approximately 6.4 gigabytes, using mostly 190MB and 354MB disk systems. Additionally, GE employs six Eclipse S/140 systems integrated into its Calma Corporation CAD/CAM systems.

A Typical Cycle of the Award-Winning CIM System

The production cycle of a replacement part—one of the 350,000 studs, bolts, packing rings, spill strips, and other parts ordered each year for turbine generator maintenance—illustrates the operation of the GE system.

*Permission granted by Data General.

A customer order for the part begins the process. The quotation is entered through the direct order entry system and converted, online, into an order. The order is then transmitted to GE's mainframe computer.

The automated process planning system, an expert-type system running on a Data General MV/10000 system, receives the order from the mainframe and matches it to a part recognition code (a group technology code). The system determines the most efficient routing of the part through the shop, selects the proper material, and chooses the correct NC family part program to be used.

This information is automatically downloaded to the NC programming system. The NC programming system processes and post processes the family part program for the part, and passes the planned time to manufacture it (generated by the post processing procedure) back to the automated process planning system.

It also transmits the part program to the small parts shop factory management system, which includes shop floor control, factory communications, and distributed numerical control. The part is scheduled and dispatched to the appropriate workstation by way of shop floor control functions, which also include labor reporting, scheduling, and dispatching.

The NC program received from the NC system is then transmitted to a factory-hardened intelligent terminal at the machine tool, where the part is machined.

As the machine tool finishes the part, it notifies the factory management system, and operators enter labor-vouchering details in real time. The factory communications module then notifies an inspector that the part is awaiting inspection at a specific station.

The inspector, after approving the part, enters certification at the terminal. Factory communications then notify the material handler that the part is ready to be moved from the machine tool area to another machine tool or to be shipped.

When the part is shipped, required information is entered into the appropriate terminals.

Results

The General Electric computer-integrated manufacturing system has achieved a number of improvements, including the following:

- Substantial increase in labor productivity
- Improved customer satisfaction through reduction in manufacturing cycle time
- Ability to provide more timely order deliveries and faster response on emergency shipments
- Integration of parts design, which is assigned a part recognition code (a group technology classification) and entered into the common data base
- Substantial savings in the small parts shop by eliminating the need to create 500,000 punched cards under the previous system to handle a volume of more than 15,000 orders a year
- Reduction of excessive queue time of raw material waiting to be machined
- All information entered automatically into the other elements of the system, which keeps the data base current and precise
- Ability for an operator to step through a menu-type display to generate a service call
- Ability for an operator to use the factory management system for other types of communications such as "talking" to the foreman, methods engineer, quality control inspector, or a host of personnel that would otherwise have to be physically tracked down

CASE-IN-POINT

*GENERAL ELECTRIC COMPANY'S LOUISVILLE, KENTUCKY COMPUTER-INTEGRATED MANUFACTURING SYSTEM**

This case is a good example of how important some of the intangible benefits can be in a CIM system. Before it was installed, GE had a small market share of the dishwasher market. Because their new CIM line produces such high quality dishwashers, the company has become number one in market share. GE's increased market share has been gradual and was initially intangible. But since then the result has been higher profits—which are, of course, very tangible. They invested much effort designing a product that was easy to manufacture consistently. In current vernacular, this approach is referred to as "design for manufacturability."

Situation

The consumer was the driving force behind the redesign of the GE dishwasher product line. General Electric's major appliance business group undertook the most comprehensive program of market research in the history of its dishwasher business. A major part of the research focused on "consumer cues"—the product attributes that consumers rate as the most important to them. Actual dishwasher users identified and ranked product characteristics that mattered most. These "consumer cues" were the basis for planning and designing the company's new Permatuf® C dishwashers. The result is a product line that responds to the customer's demand for performance, reliability, durability, capacity, convenience, and quiet operation.

Solution

The investment that was required to develop General Electric's new line of dishwashers and its supporting manufacturing system was sixty million, with some of this sum due to increased product volume. The manufacturing system was designed to produce dishwashers of unparalleled quality with a degree of consistency and efficiency that was unattainable with earlier generation production methods. The system incorporates the use of some exotic technology and modern manufacturing approaches—technology that includes the use of automation at selected critical stages in the production process to ensure high levels of consistency and precision. Manufacturing approaches like just-in-time inventory control and "management by sight" also help to assure that the proper part is in the right place at the right time—and that it's right in every respect. The result has been to achieve a high standard of quality.

Computerized parts processing and assembly operations are the foundation of General Electric's new dishwasher manufacturing system. Computer control begins with an electronic order entry system that integrates input from purchasing, cost accounting, production planning and shop floor control. Computers also monitor all key functions including injection molding, steel fabrication, and quality control.

The manufacturing process includes other advanced features, such as:

- Laser readers
- Robot handlers
- Nonsynchronous conveyor systems
- Vibration and ultrasonic welding
- "Fast-stitching" and "pierce riveting" assembly

*Permission granted to use General Electric Company's Louisville, KY, computer-integrated manufacturing system as a case-in-point.

The CIM system's manufacturing process follows this cycle of events:

1. *Metal frame and leg fabrication* begins with the blanking and trimming of coil steel. "Point-of-use manufacturing" provides instant availability of parts as they are needed. These components are measured electronically for quality before reaching assembly and finishing operations. Parts that do not meet strict specifications are rejected.

2. *Permatuf® tub and door liner fabrication* is accomplished through high-pressure injection molding of a unique GE formulated material. Programmable controllers provide instant pressure changes when required, producing the Permatuf® tub and door liner to assure conformance to specifications.

3. *Tub structure assembly system* weds the Permatuf® tub and the steel structure members. Since both tub and frame have been previously checked electronically, a thorough seal with the door is assured. The possibility of leakage has been virtually eliminated.

4. *Inner door assembly system* automatically installs accessories, detergent cups, dispensers, gaskets to the Permatuf® door liner. Torques are measured to a tenth of an inch-pound, dimensions to the thousandth of an inch—(levels of tolerance that help assure top performance and reliability).

5. *Nonsynchronous conveyor system* frees the final assembly operator from time as the controlling factor because assembly operations are performed with quality, not speed as the highest priority, thus enabling the company to produce products that are reliable and durable.

6. *Product test system* allows a test operator to evaluate every dishwasher produced. Each is put through an operational cycle; each is subjected to a final examination that takes half an hour. As a result, the consumer is assured of receiving a product that has been thoroughly tested.

7. *Sound room* subjects each finished product to both human and electronic analysis because the computer alone cannot always discern harmonics that are irritating to the ear (a response to the important customer demand for quiet operation).

8. *Computer control center* monitors the manufacturing processes and work-in-process inventory. The control center specialist interfaces with each station in real time, monitoring the functioning of the system and making adjustments to schedules when necessary.

Results

Within the manufacturing environment, this advanced CIM system achieved the following:

- Reduced the manual manipulations—the "handlings"—of tubs and door liners during the manufacturing process by 85 percent.
- Point-of-use manufacturing did away with long-term storage and inefficient delivery of structural components.
- Computerized, on-the-spot checking of every automated operation substantially reduced the on-line repair of finished goods and eliminated incomplete units from progressing on through the factory.
- A "worker-paced" assembly line allowed the product to move on only if the worker was satisfied that the work has been done properly.
- Assembly-line changes and automation together reduced total manufacturing time—from raw materials to finished product—to less than eight hours, from three days.
- Productivity was improved 18 percent.
- A computer-based plant quality information system collected test and inspection measurement data for each unit at key points in real time, signaling process control engineers if unfavorable trends developed.
- Prior to installing this computer-integrated manufacturing system, market share was about 28 percent. By 1986 GE was in "first place," with an *increase* of more than 33 percent in market share.

Future Plans

General Electric plans to invest about one billion dollars over a five-year period in upgraded product designs and facilities.

The Louisville, Kentucky dishwasher project was the prototype for such major investments to upgrade product designs and to improve facilities as well.

The same treatment is now taking place in GE's refrigerator facilities, to be followed by appliance park plants in Louisville that manufacture home laundry (washers and dryers) and ranges.

USING MODULAR INSTALLATION OF CIM TO BUILD THE FACTORY OF THE FUTURE

The ultimate goal of CIM is to enable the manufacturing facility to become the factory of the future. In the final analysis, the key to the factory of the future is management, not computers and not technology.

Although much has been written about the benefits and capabilities of computer-integrated manufacturing systems available today—they can be large and expensive—and when a CIM system is purchased and installed as a complete system in a new plant or as a replacement for an existing plant operation, it still can be an overwhelming project when implemented. For small companies in particular, a project of this scope is not feasible. The capital required and the disruption of existing operations are prohibitive. Instead, the most appropriate approach is to install CIM using a modular approach. A modular installation reduces the risk, lowers initial investment, and does not overwhelm the company making the installation. As each module is implemented, the company should be able to obtain noticeable productivity gains without the prohibitive costs of the larger-scale projects.

Installation of CIM using a modular approach still requires management to have a clear understanding of the final product of this work effort. This may involve such advances as:

- Customer orders received via modems and placed directly into a data base
- Production schedules automatically created from data stored in the data base
- Parts specifications and process parameters loaded directly into numerical controlled machines
- Raw materials automatically ordered via modem from vendors
- Invoicing and order releasing done by computer-to-computer communications with reports created indicating backorder status, shipping discrepancies, etc.
- Machines monitored for production counts and machine run time
- Product quality and process parameters monitored by means of statistical analysis
- Machines requiring no direct labor and operating unattended for 24 hours a day, seven days a week
- Use of CAD and CAE to reduce product lead times to less than production times
- In-process inventory maintained only with buffers within work centers
- Preventative maintenance elevated to a level where down time approaches zero
- All systems communicate by means of a common language or protocols, such as MAP RS232
- Real-time status of production, quality, order backlog, maintenance, and other critical factors available to management on real-time displays or in printed reports

All of these advances must be incorporated into the company's strategic plans so that new machines and processes have computer interfaces and automation capability to mesh with the ultimate goal of the factory of the future. This point cannot be emphasized too strongly, because if it is not you will have merely created isolated "islands of technology."

When viewed in its totality, all the technologies are not yet in place to install a fully automated CIM program. Of those required, the following components are available today.

- *MRP Systems*—identify materials and their requirements, production schedules, machine loads, and other factors that are necessary to run an efficient factory.
- *Cost Accounting*—computerized cost accounting systems to control and measure performance standards against operating results.
- *Factory Monitoring to Close the MRP Loop*—monitoring of machines and processes, counting parts as they are manufactured, and relieving raw material inventory as it is used.
- *Factory Floor Data Collection Systems to Close the MRP Loop*—obtain data on time and plant activity, work-in-process inventory, lot tracking, and traceability.
- *Statistical Control*—real-time statistical process control and real-time statistical analysis of product quality.
- *Work Cell Automation*—FMS using robots, vision technology, numerical controlled machines, and other advanced methods.
- *Electronic Communications*—between off-site and other computers within a manufacturing facility employing specialty software packages.

Where to Start Implementing a CIM System

The best place to start is with the area or product or process, or whatever is most important to your particular situation.

For example, demand has increased for one of your lower margin products, and to meet this demand, machine time must be increased by 50 percent. Your options are to purchase another machine, add another shift, increase overtime, add more supervision and support personnel, or decline to accept the business.

A better alternative solution in this case might be to automate that particular product so production can run three shifts or seven days unattended without direct labor. This approach would allow acceptance of more business and reduction of costs by spreading capital cost over a larger number of parts.

A Modular Approach to Installing a CIM System

A modular approach, particularly for smaller firms, can enable a company to begin automating portions of their factories instead of tackling the entire plant at once. Smaller modular projects require less resources and have more intermediate project deliverables. This differs from islands of technology because such islands form when inadequate planning leads to isolated equipment that cannot communicate with the rest of the factory, whereas having an automation plan that incorporates *temporary* "islands" can be regarded as an acceptable approach.

The starting point in proceeding with a modular CIM installation is to focus on functional modules. These are defined by the functions they perform, such as:

- Real-time statistical process control
- Machine and process monitoring
- Shop floor data collection
- Employee time and activity
- Preventative maintenance

These are all functional modules rather than vertical or product modules of a CIM system since they apply to all products or a class of products.

Types of Modular Subsystems

In a comprehensive program of this type, there will be a number of modular subsystems that will ultimately be required to achieve CIM goals. These modular subsystems should be distinguished from workstations or functional modules. It also is important to note that you must be careful when implementing subsystems so that they do not become islands of technology.

These subsystems may include the following:

Material Handling. This is the ability to move raw materials to the workstation; move finished product from workstation to customer's location; and move materials from workstation to workstation between operations.

Electronic Sensors, Scales, and Controls. Electronic sensors are mandatory to identify what is happening on the factory floor, to monitor performance, and make adjustments.

Vision Systems. Vision systems enable machines to "see" what is going on.

Programmable Controllers. Programmable controllers provide the factory floor with the ability to react quickly and to control hundreds of parameters, such as temperature, pressure, etc.

Material Handling Robots. The use of robots in material handling in the transfer of products and in the unloading of machines are key modular subsystems because of their flexibility and programming capabilities.

Installing CIM by Vertical Approach

Another method for installing CIM is by product or class of product using a vertical approach. This is achieved by dividing the manufacturing facility into a hierarchy of equipment and operations. From this hierarchy it is then possible to identify and focus on the crucial elements.

The application of group technology can be helpful in logically defining the hierarchy. Group technology defines an end product as variations on current designs of manufactured components. It is a method of grouping similar parts, rather than similar machines. As such, it is used to form cell modules or work cells.

A workstation is any machine that performs a single operation on a part without requiring additional part handling or operator assistance.

A cell is one or more stations that is served by the same material buffer.

In the modular approach to installing a CIM system, a workstation has the lowest level of complexity and is at the bottom of the hierarchy. The cell or work center is the next level of the hierarchy. The modular approach involves integration from the bottom up. That is, it integrates the elements of the workstation, followed by integration of the cell module, then the system, and finally the factory.

Figure 11–11 consists of two workstations that form a cell and are connected through an event/status controller, which in turn, is interfaced to a host system—usually a micro- or a super-microcomputer. At the workstation level, there are sensors to evaluate, monitor, and control the station. At the event/status controller level, decisions are made on controlling machines in the cells. Messages are relayed through to the host system and messages from the host system back to the workstations. At the host level, a disk drive is incorporated, along with terminals, printers, and other output devices to advise machine operators and management of the status of the cell.

The modular approach incorporates a greater degree of local control. A host system can support a single cell, a single workstation, or multiple cells depending on the requirements of the application.

This local control allows greater flexibility in manufacturing operations. Thus, a company can apply any degree of integration desired within a module. It may decide to automate and integrate only one machine, or it may automate and integrate all machines and processes affecting one product.

To determine whether it is feasible to integrate an existing machine, an evaluation must be made of the machine's present state. If the machine does not have capabilities for communicating with a computer, those interfaces must be installed, or the machine must be

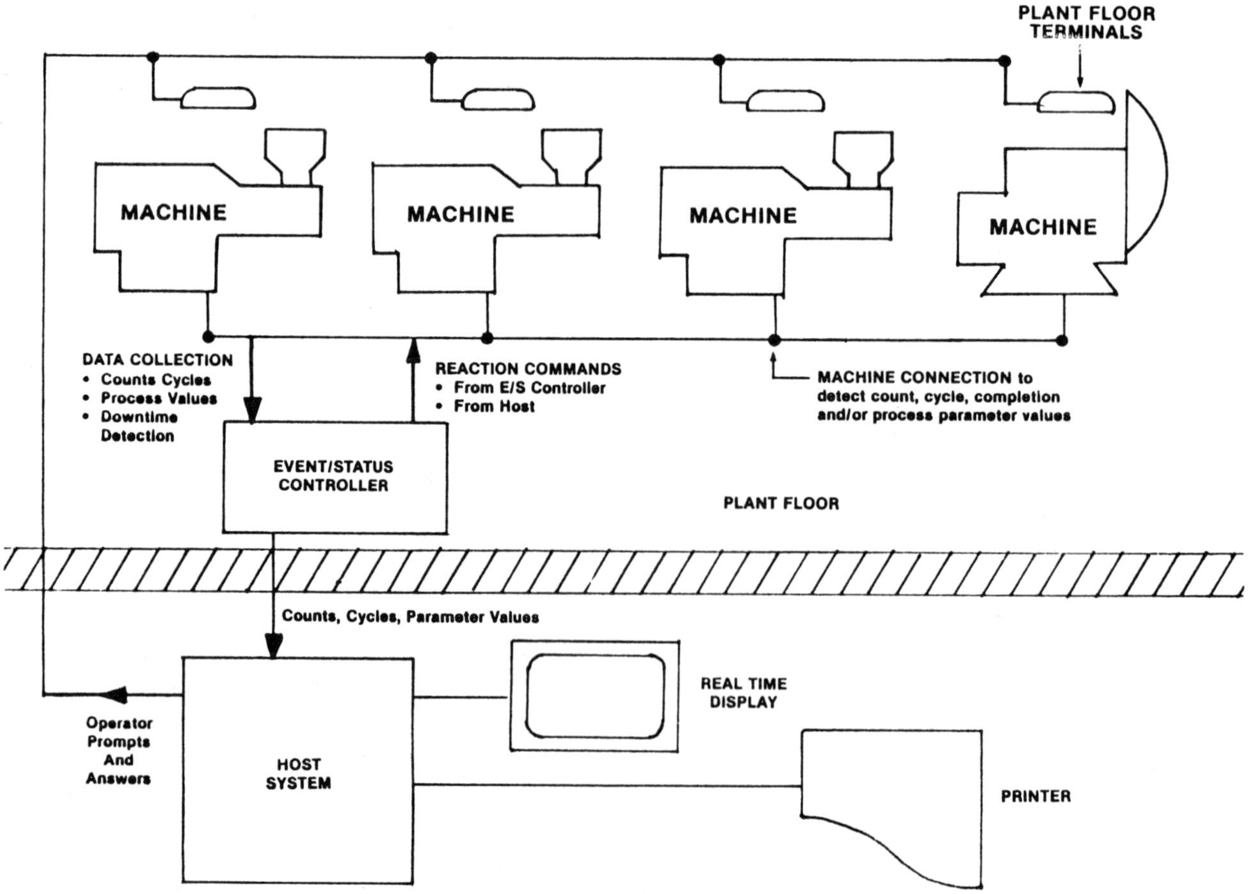

FIGURE 11–11 Integration of Workstation, Cell Module, System, and Factory (Courtesy of LDJ Computer, Inc. and Systems Specialist, Inc.).

replaced before integration can occur. In addition, machines must be equipped with sensory switches for the detection of machine performance and machine run time. The host system with the disk drive is used to store and download individual, NC part instructions to the machine controller. Therefore, the machine controller must have a communications port to accept the instructions. The status controller contains the digital I/O capability to communicate with the machines.

The host computer must manage several sets of data. It must contain a file for each part and station routing named within the cells. If a robot is used for material handling, the host system must contain movement instructions for each task required of the robot. The host computer collects and summarizes data needed for numerous reports and analysis.

All data generated within the cells is stored on the host computer disk. This can be addressed for further analysis. The status controller monitors the number and types of parts, the time the machine spends to manufacture each part, the utilization time for each station, the number of defective parts, the time and date for each operation, the station and tool maintenance, and any other data that is of interest to the manufacturer of the part.

Statistical analysis can be performed on the process parameters and the part characteristics and attributes to determine how to improve the product repeatability.

Machine operators use the data to access what went wrong when the machine shut down, or to evaluate the idle time for a particular part or workstation. Management can use the data to access the machine utilization. The maintenance department uses the data to identify preventative maintenance time intervals.

The mainframe computer which is handling basic data processing activities is a modular subsystem. It integrates part of the overall system by two-way data transfer between the shop floor host computers and the mainframe. The CAD system and CAE system are modular subsystems.

Numerical controlled machines that are part of workstations or cell modules become a modular subsystem.

Other subsystems include quality control computers or shop floor data control systems. Shop floor terminals, bar code printers, and other peripherals are subsystems of a CIM system.

As discussed previously, with computer integrated manufacturing systems a company must have developed a strategic plan indicating where it is headed, and what the totally integrated system will look like when it gets there. This strategy may change as technology produces new products and new developments, so that it may require refinement from time to time.

To attempt to implement the Factory of the Future completely at one time would be impractical for most companies. Therefore, the alternative is to modulize the system by creating a hierarchy of machines and process in the factory as well as using group technology to define similar parts and the manufacturing process to produce them. Or, CIM can be implemented by a functional modular subsystem approach.

Many technologies currently exist to enable companies to start installing CIM systems. These advanced technologies include sophisticated numerical controlled machines, vision systems, monitoring and control systems, real-time statistical process analysis systems, shop floor data collection systems, and so on.

Combined with these "hard" technologies are many innovative software programs that allow companies to fully utilize the capabilities of these methods and techniques. Thus, the total impact of CIM can be achieved through proper planning, coordination, and execution of a well conceived program.

UNDERSTANDING THE COMPONENTS OF A CIM SYSTEM

The core building block that provides the linkage of a computer-integrated manufacturing system is computer technology. It is a computer hardware and software that allows workers in all areas of plant activity to have access to information that drive CIM applications. The various components that interface in a CIM system are shown below:

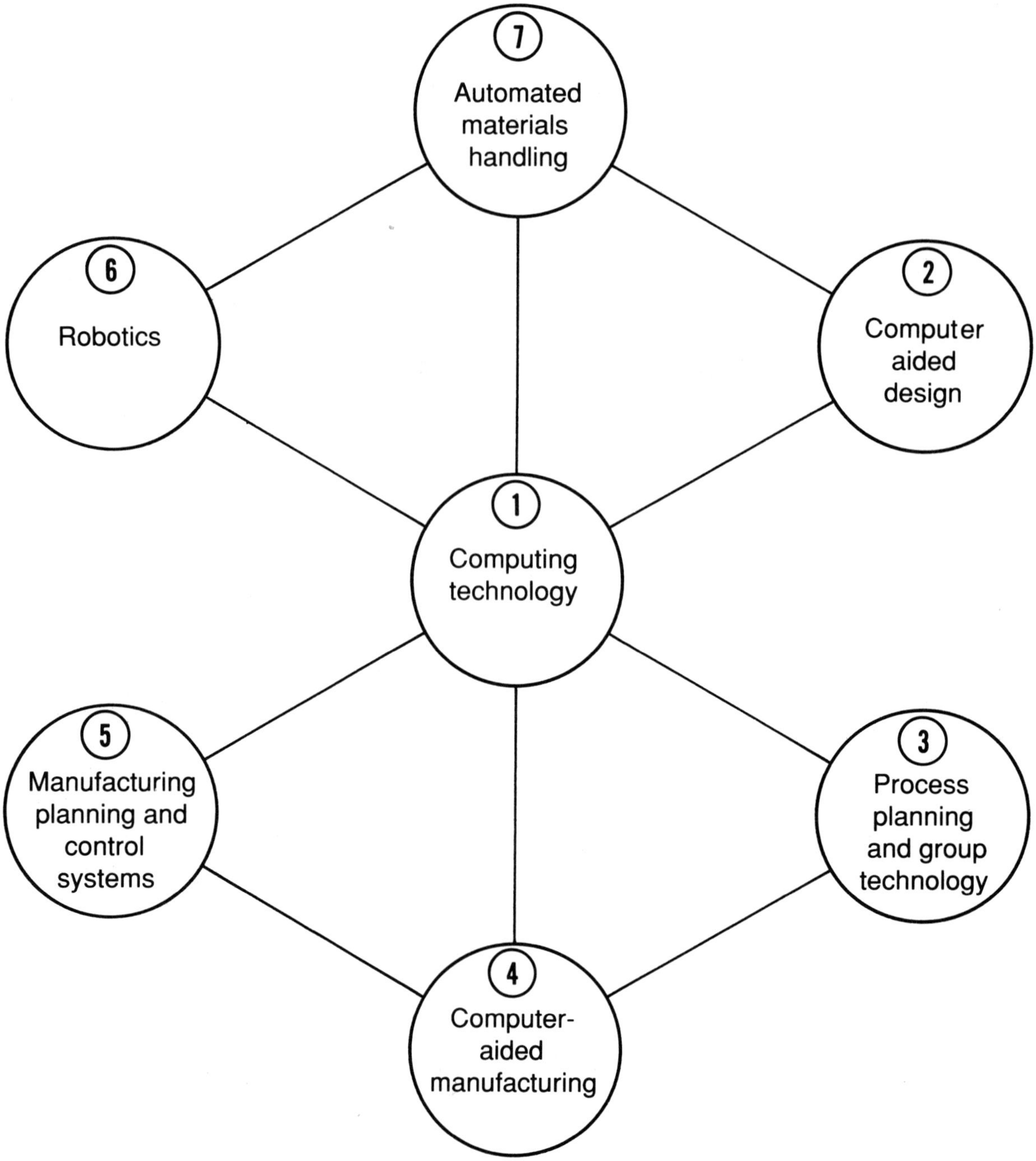

Each of these CIM components consist of major activities that must be coordinated and implemented with extreme precision to achieve a fully automated plant operation. The key elements of this CIM network are:

1. *Computer Technology*
   (a) Computer hardware
   (b) Software
   (c) Data-base management systems
   (d) Computer network
   (e) Telecommunications
   (f) Graphics, peripherals, etc.
2. *Computer-Aided Design/Process*
   (a) CAD
   (b) CAE
   (c) CAPP
   (d) NC part programming
3. *Group Technology*
   (a) Coding and classification structure
   (b) Family grouping of common parts
   (c) Process standardization and simplification
   (d) Standardization of nomenclature
4. *Computer-Aided Manufacturing*
   (a) Machine tools
   (b) Process control devices
   (c) Computer-aided test and inspection
   (d) Electronic assembly
   (e) Factory data collection/data acquisition and control
   (f) Programmable controllers
5. *Manufacturing Planning and Control Systems*
   (a) MRP
   (b) Forecasting
   (c) Master scheduling
   (d) Capacity planning
   (e) Shop floor control
6. *Robotics*
   (a) Fixed sequence robots
   (b) Variable sequence robots
   (c) NC robots
   (d) Intelligent robots
   (e) End effectors and sensors

7. *Automated materials handling*
   (a) Automated storage/retrieval systems
   (b) AGV systems
   (c) Miniloaders
   (d) Conventional material movement and storage systems

Elements of CIM's Computing Technology

A fully integrated system consists of a mainframe computer, real-time microcomputer and work cell controller supported by software for work cell–control applications.

Work cell control is needed to handle the considerable amount of automation action that takes place between the MRP system and PLCs, which are used for direct control of manufacturing machines and equipment. By providing the vital components to link plant management systems with dedicated equipment controllers on the factory floor, a company now can implement a CIM strategy.

The work cell controller serves as a communications hub, enabling a manufacturer to have a two-way flow of information over a plantwide factory information network stretching from the factory floor to the plant management computer system.

The primary function of a work cell controller is to coordinate and direct the activity of a logical grouping (or cell) of factory devices.

A work cell controller also must be able to connect to a wide variety of factory floor devices, including programmable logic controllers, robots, bar code readers, material-handling systems, vision systems, and machine tools.

In addition to overseeing or giving instructions to these factory-floor devices, a work cell controller collects production status information and other data as rapidly as events occur.

The work cell control then must send selected production data upward to a mainframe computer. Work cell controllers must be able to "talk" to each other to implement distributed intelligence systems.

Work Cell Controllers for Harsh Environment

Because a work cell controller usually is located on the factory floor, it must be rugged enough to survive in harsh manufacturing environments. Its hardware must endure extremes in temperature, shock, vibration, and electrical noise common to most factories.

Collecting data from factory devices can be done automatically by a work cell controller, or by workers using bar code wands, badge readers, industrial terminal keyboards, or other manual entry devices. A combination of automatic and manual data collection often is the best solution.

Once data is collected, the work cell controller must manipulate it for a variety of purposes. One of these is to generate management reports on the status of work being done at the particular work cell.

Work cell controllers can be a valuable part of management decision–support systems, by enabling managers to get immediate answers to unscheduled or other unexplainable situations on the shop floor.

Need for Real-Time Computers

If the work cell application is fairly simple, a personal computer (PC) may be adequate as the controller. However, if the application requires the controller to define, identify, and prioritize physical events, and take action on those events within a required period of time, then a real-time minicomputer should be used as the controller.

If the events assigned to the controller must be measured, controlled, or reported in fractions of a second, a real-time minicomputer should be used as the controller. If events of interest take place more slowly—from seconds to minutes apart or on an hourly or end-of-shift basis—then a PC may be adequate.

Historical Archiving Capabilities

The historical archiving capabilities in a work cell controller can be used to generate reports according to a schedule, event, or on demand.

Work cell controllers need operator interfaces within the shop for displaying layouts or other drawings in color graphics or on less expensive monochrome graphic terminals. Other requirements could be display of alarms while running another application on a terminal; display of a process screen within two seconds from a request; and display of CAD drawings on an industrialized terminal.

Because every manufacturing process is different, system customization is necessary. Software and tools customization is required to configure as much functionality and performance on a node as needed.

Work cell controllers can support a wide variety of protocols for factory device and computer-to-computer communication. Industry standard protocols such as parallel, RS232-C, IEEE488X.25, IBM protocols, and IEEE802.3 must be supported by work cell controllers. Manufacturing automation protocol (MAP) links to unit controllers need to be in place to conform to baseband or carrier-band MAP (refer to Chapter 4 for additional discussion).

Controllers Must Provide Growth Path

To provide for future expansion it is essential that any acquisition of work cell controllers have provision for a growth path. This necessitates having product growth that ranges from entry-level control all the way up to powerful real-time cell controllers.

A wide variety of work cell–control software applications that can be used in discrete and process industries is steadily reaching the market.

Software

Improved software packages are constantly being developed for work cell control applications. Most are menu-driven packages that can be tailored to specific applications by manufacturing or production engineers without programming expertise, enabling users to:

- Monitor production activities and capture production data
- Archive historical data in a variety of formats
- Generate reports to document the production process
- Support flexible manufacturing through recipe management
- Create operator interfaces that contain real-time process data
- Monitor alarm conditions and react to those by informing the operator, scheduling a task to handle the alarm, or documenting the alarm condition
- Coordinate the operation of several manufacturing cells without the use of a higher-level supervisor

How CIM Utilizes CAD

Product design data from remote CAD/CAM systems can be teleprocessed to a central facility where numerical control tapes can be produced for various manufactured parts. Individual manufacturing units are connected to a single CPU, supplying all production instructions using DNC techniques.

The Roles of Process Planning and Group Technology

Process planning is concerned with the total manufacturing cycle. Computer software that optimizes this process must take into account such factors as the status of the workpiece at each workstation, as well as the routing of the part around a shop floor.

As a company manufactures its products, even in low-volume lots, there inevitably will be underlying similarities among the parts, despite the fact that these parts change over time. Recently, FMS have been designed based on these part similarities. CIM process planning systems are used to retrieve relevant part and process information using group technology whereby families of parts are organized, based on either shape similarity or common manufacturing operations (discussed in Chapter 2).

Some CAD/CAM vendors actually enable users to store standard plans and their associated operational procedures for part fabrications for each part family. Users can customize by simply editing the generic part family to produce operational manufacturing codes for each new part as it is designed.

Eventually the computer will generate its own data, internal computations, and independent decision making. In these generative systems, process plans will be produced directly from the geometric (CAD/CAM) data base with almost no human assistance.

In interactive graphic NC machining, the programmer defines tool motion directly from the design data base, which in turn, is graphically displayed on a CRT while the tool path is created. Design changes that might be necessary because of collisions are made in the data base and reflected in all associated drawings and part information. Although a generic NC code generator is under development, specific code for various NC machines is generated by one of several hundred post processors.

Off-line robotic programming means programmers create and verify programs without using the traditional "teach" method whereby robots are taken out of production during retraining cycles. The data base includes models for describing robotic movements that are, in turn, graphically simulated.

Group technology programs allow parts with similar geometric features and manufacturing characteristics to be classified by code numbers. The code numbers are used to retrieve existing designs from the data base for revision and customization. Added software facilities generate process plans (including graphics and text), rather than parts, and retrieve and format similar plans resident in the data base using group technology techniques.

Tooling and fixture design software performs shrink and bend analyses on geometric models and by interfacing other computer equipment to coordinate measurement machines. This allows input of part information to a common data base from the actual part model, pattern, or prototype. Software can unfold a 3D design into corresponding flat patterns. This information is used for NC tool path generation for punch presses, welders, saws, or laser cutters.

Engineers currently create products using a series of part drawings. The management of information for part drawings for a complex assembly, which can consist of thousands of such drawings, is a difficult task for most CAD/CAM systems. For CIM to be effective, CAD/CAM systems must handle both increased storage requirements and software facilities to allow integration of all drawings for subassemblies and other components employing group technology procedures. Additionally, CAD/CAM data bases must be integrated and linked to traditional corporate data-processing functions such as forecasting, production planning, scheduling, materials management, process inspection, and other related processes.

In the future, manufacturing data bases will expand many times over current data processing data bases, and be composed of process plans, fixture and tooling design, NC programs, quality control and inspection programs, robot programs, and process programs.

Although CIM techniques will increase productivity, lower costs for materials, labor and energy, and improved product quality, its implementation on a broad scale will entail considerable effort. Because of its integrative nature, CIM technology will have a profound effect on a company's organizational structure and labor relations. The historical approach of the American business community to concentrate on short-term profits rather than long-range strategies has inhibited the growth rate of CIM installations. If this approach can be altered, then CIM technology can become an effective tool toward making companies in the United States more competitive in the international marketplace.

Why Coupling Computer-Aided Manufacturing with CIM Lowers Costs

The basic objectives of CIM are to increase corporate profits, improve on-time delivery performance, and maintain a high standard of product quality.

Within CIM, the technologies of CAD/CAM can enable a company to design and produce a product at a significantly lower cost. However, one of the major problems confronting companies in proceeding toward even higher levels of automation is the integration of CAD/CAM and other computer-based support systems with the machine tools on the factory floor.

Most NC part programs are created using computer-assisted systems. A primary bottleneck that has slowed implementation of CAD/CAM systems has been the post processor. Post processing is one of the most fundamental components of integration of CAD/CAM with plant equipment. These software programs translate the data that comes from the CAD/CAM system into a format that is compatible with CNC and specific machine tool combinations. However, there are a number of problems associated with post processing, such as:

- Cost of purchasing or developing internally post processors
- Time element waiting for a deliverable system, which lengthens the effective implementation cycle of new equipment
- Quality of output is often less than desired
- No migration path from one generation of computer assist equipment to the next
- Lack of program portability in moving parts programmed using a post processor from one machine to another without reprogramming and reprocessing

*Binary Cutter Location Exchange Input Format for Numerically Controlled Machines**

To overcome these problems an advanced technology has been developed that provides for more effective integration of CAD/CAM. It is binary cutter location (BCL). It comes from the title of Electronic Industries Association (EIA) Standard RS-494, "Binary Cutting

*Permission granted by Charles Rosbeck, Fairchild Republic Corporation, and Michael Hitchcock of the Manufacturing Technology Division of The ATR Force Wright Aeronautical Laboratories, Wright-Patterson AFB.

Location Data Exchange Format for Numerically Controlled Machine Tools." BCL was developed to aid NC users in reducing the cost of maintaining many different post processors for programming and the resultant incompatibility of NC programs. Instead, BCL provides compatibility to move jobs from one machine tool to another or from one machining facility to another without reprocessing the part program.

This advancement has overcome the previous concern that the lack of NC program compatibility could become an obstacle to integrated and automated manufacturing, especially since CAD/CAM systems are often the source of geometric data for NC programs. Conversely, the importance of manufacturing standards in building the factory of the future is becoming more evident, and BCL may very likely become a cornerstone in this advanced technology.

RS-494 is a standard that specifies a particular binary format for data used as input to a machine tool. If NC programs are produced in this format, the same programs can be exchanged between machine tools even though the machine tools or control units are not the same. Both machines must have CNCs designed for BCL and the machine tools must be appropriate for the work, have sufficient capacity, and be loaded with the right tools and fixtures. But the program does not have to be reprocessed to make the switch. Most importantly, the traditional kind of post processor is not needed for the original programming.

Creation of the BCL File

The part programmer, working from the part drawing, writes a source program describing the part geometry and the tool path. Instructions are included as to tool size and shape, spindle speeds, feed rates, and so forth. This program is written on a coding sheet to simplify the data entry operation. The program is transferred from the coding sheet by a data entry operation. The part program source code is fed to a large data processing program. (It should be noted that many CAD systems available presently can create such a program automatically based on the part geometry stored in a data base. These CAD systems eliminate the human interface and can transmit the part programs electronically to the NC machine.) The APT program processes the part program and produces a BCL file which it stores on its disk memory. This file consists of cutter location data. Cutter location data is the computer-generated information that defines where and how the tip of the cutting tool is going to move as it machines the workpiece, and it is the heart of any NC program.

The Post Processor Operation

The post processor program in the data processing computer uses the BCL file stored in the computer's disk memory as its input. It processes the BCL file data and punches the post-processed data on a mylar tape, magnetic tape, or to a disk as required by the particular control unit, which is then transferred to the particular machine tool/control unit combination to be used. The post processor program also generates a human-readable printed listing of the post processor output.

Each machine tool has its own individual characteristics, whether or not it has an automatic tool changer, the number of tools in the changer, where the origin of the coordinate system is located, the location of the home position, what is the clearance plane for tool changing and axis travel, maximum travel rates, coolant control variations, and so on. All of this information is stored by the post processor, which uses it to create a part program tailored for use on only one machine tool/control unit combination.

Since part programs must be tailored for every different combination of machine tool and control unit, shops with more than a few CNC machine tools will have several post processors. The more machine tools and the greater their variety, the larger the number of post processors needed. Acquiring and maintaining them becomes extremely costly.

In addition, once an NC program has been post processed, it usually cannot be used at any machine tool except the one for which it is tailored. If that machine is backlogged, down, or unavailable for some other reason, reprocessing the program makes it difficult and time-consuming to switch to another machine. The post processing program also is of little use to other shops if they have different machine tools and control units.

BCL gets around these obstacles by taking the CL-file output of the APT processor, putting it into a simplified format, and using it as the machine input. Post processing capability is embedded in the computerized numerical control unit by the control builder in the same way added functions would be built into a personal computer.

This is a key point concerning BCL—namely, that the control unit must have the computer processing capability and the right software to process the BCL data internally. Stated another way, the last step to processing an NC program is left for the BCL control unit. A BCL program is not "tailored" for a specific machine tool/control unit combination until it is being executed by the control unit to machine a workpiece. *That is why BCL programs are transportable.*

The Cutter Location Converter Operation

To complete this approach, there has to be agreement about what the cutter location (CL) data will look like before it goes to the control unit and there has to be agreement about exactly what the control unit will do with the program once it is there. For this reason, BCL requires a "CL converter" to give CL data the correct look. The BCL converter is a program that runs after the CL data is generated. It converts CL data into a form that can be interpreted by the BCL processor in the control unit.

Essentially, the converter reads the numerical representations of point coordinates, arc centers, end points, and so on as well as the numerical codes for machine–function commands, and translates them into a 32-bit binary integer format. This format allows the BCL processor to run more efficiently in real time, while the machine tool is cutting metal. The BCL data can then be recorded on a floppy disk or any suitable medium.

This disk containing the BCL program can be loaded in any BCL control unit to machine the workpiece, provided that the machine has the configuration, capacity, tools, and fixtures to do the job. These disks can be used on various machine tools with BCL control units to move a job from one machine tool to another. The CL converter program also creates a program listing on a line printer.

CAD/CAM Integration

BCL can be adapted to facilitate the integration of CAD with CAM. Most CAD/CAM systems generate some kind of CL file in the process of creating tool-path data for NC programming. This CL file may be post processed within the system to yield programs that run a machine tool or be processed by other software programs to drive plotters and create dynamic displays of tool paths on a graphics monitor.

Historical Background Leading to Adoption of RS494 by Fairchild Republic Corp.

Post processing represents the conversion of a standard APT part program output format to a data format acceptable as input to a machine tool controller. Historically, every different type of machine required its own post processing program with attendant storage and maintenance requirements in the mainframe computer. With the advent of the microprocessor, machine tool manufacturers began installing computer "intelligence" into individual machine tools to enhance their capability and reliability.

At this point, Fairchild Republic distributed RFQ's specification that CNCs assume the task of post processing so that similarly configured machines could use a standard, common format of part program data.

Vega Servo Controls responded to the RFQ and stated that it could meet the Fairchild specification, and that a very similar request had been received by Vega from another source just prior to meeting with FRC Management. The other source was Rockwell International. Thus, Fairchild retrofitted seventeen NC profilers and thirteen ATCs with a common CNC capable of using a single format of input data: 32-bit integer and binary CL data, now referred to as "BCL data." All of the mainframe resident post processors were removed and replaced by a lone "CL converter" program that converts APT output to a common BCL data format.

All of the BCL programs at FRC are stored on a minicomputer that is linked to each CNC in the NC shop. This enables a part program to be downloaded to a CNC via a communications link. Different machine tools can then readily call and load any BCL program in the data library without the delay and cost associated with reprocessing and reproving.

The use of BCL data part programs permits the programming of parts without reference to the specific tool on which the part will be machined. A number of benefits accrue from the resultant part program:

1. Shop loading can be established without regard to part program processing.
2. The part can be run on any available machine that has adequate capacity.
3. If the machine for which the part was scheduled is busy or broken down, the part can be run on any available machine of adequate capacity without delay.
4. If no machine is available within the plant, the part can be off-loaded to an outside vendor who has a BCL data format machine.
5. The part program can be stored for future use without requiring reprocessing even if the machine that originally made the part has been replaced.
6. Part programs for military equipment could be furnished as part of the equipment procurement so that replacement parts could be manufactured at any number of facilities without delays due to reprogramming and reprocessing.
7. In an emergency, part programs can be transmitted by data link from one facility to another. Part programs can be stored in central locations and transmitted to a manufacturing or repair facility.

Additional benefits are realized at the part programming and data processing end of the manufacturing process:

1. Only one post processor, called a *BCL converter,* resides in the data processing computer.
2. No longer must a new post processor be installed each time a new machine is installed.
3. If the data processing computer is replaced, only one post processor program must be re-installed if all machines run on BCL data.
4. The usual delay in installing and debugging a new post processor is eliminated.
5. Post processing accounts for a significant portion of program processing cost. This cost is virtually eliminated.

Since Fairchild Republic now regards this as the most advanced state-of-the-art technology for post processing, all NC and CNC systems and machine tools, including FMS, are now designed to accept BCL data (RS-494) part program format.

Evolutionary Stages of Distributed Numerical Control

Distributed numerical control (DNC) represents an integral part of an overall CAM system. DNC performs a wide variety of communication tasks and functions centered around the transmission of information between engineering and manufacturing CAD/CAM systems and remote production machines in the factory. The information consists of input data to production machines, instructions for machine operators, status information going back to the host system, and eventually all the shop paperwork that is now being handled manually.

At this point, it may be beneficial to reflect on the evolutionary development of this advanced software technology since its previous four generations of upgrade have set the stage for its use in such applications as FMS.

The first generation consisted of stand-alone machine tools where tape had to be hand carried to each machine.

The second generation was BTR (behind-the-tape reader) mode, in which the host computer, by way of parallel 8-bit transmission lines, provided a substitute for the tape reader and passed one block of information at a time in the same fashion as a tape reader.

The third generation evolved during the period when minicomputers were still very expensive but where some of the benefits of computer numerical control could be obtained by using a medium sized process control computer. The power of the process control computer would permit the bulk of the interpolation to be done in the computer with a resultant substantial savings in control hardware at the machine tool.

The rapid decrease in the cost of minicomputers quickly eliminated the third generation DNC as a competitive mode of operation.

The fourth generation is the buffered mode of real-time DNC in which data terminals are connected to the hard-wire control through a BTR hookup, or a computer numerical control having storage of several blocks of information permitted the transfer of several blocks of information at a time. This approach resulted in a reduction in CPU load in the host computer.

The first four generations of DNC were deficient for two reasons. First, if the host computer goes down, all machines connected to it are down. Work cannot be transferred from one machine to another, nor can operators be transferred to other work. Also, backup by means of conventional tape required the maintenance of tape libraries and a backup tape reader as an input device. Tape also must be punched just as it would have been without DNC.

Second, because of the limit of the storage capacity, every machine has to be serviced at very short time intervals. Thus, all machines in the system are being serviced at the same time. The CPU load on the host computer is therefore very high and requires a comparatively expensive host computer for any reasonable number of machines.

Both these deficiencies are removed by the fifth generation DNC which is host-free DNC. In this mode of operation, the DNC system makes use of the part program storage capability of current CNC controllers which have the capability of storing in random access memory or in bulk storage media such as flexible disks one or more entire part programs. By storing the entire part program, the CNC becomes host independent during the entire machining operation. If the host goes down, the only machines affected are those that are in the process of loading a part program or waiting to load a new part program.

Host failure for an hour might affect only 5 percent of the machines in the system, as compared with 100 percent in the earlier generations.

A typical hardware configuration that supports the system is *part program data flow.* It would use BCL exchange input format for data used as input to a machine tool. The use of BCL standard in conjunction with the hardware already in place affords a company a logical transition to a completely automated FMS, and will become mandatory as work progresses toward the fully automated factory of the future.

Integrated Computer-Aided Manufacturing

Integrated computer-aided manufacturing (ICAM) is an important application area. It consists of four distinct levels:

Level I—Automated Products. The robot controller is based on a 32-bit computer and seven other microprocessors, and is responsible for coordinating motion control and teaching and programming of the arm. Programming for new parts or handling procedures can be done quickly with the teach pendant and joystick or off-line with a CRT.

Level II—Automated Systems. This level incorporates a compact cluster controller, a 32-bit computer system with optional high-resolution color graphics. This controller supports multiple robot controllers and can issue operational commands and status inquiries to the robot controllers. The cluster controller is used for off-line programming, robot simulation, performance monitoring, and production reporting. Programs or data also may be downloaded from the controller to any robot controller. One cluster controller can be used to support one or more complete FMS work cells.

Level III—Computer Assisted Preproduction. At this level, a host computer can provide hierarchal control for multiple cluster controllers. This level executes factory-level application programs and data retrieval reporting functions. It also integrates many separate FMS work cells into a complete FMS.

Level IV—Computer Integrated Manufacturing. This level points toward tomorrow's fully automated factory. Computer hardware for this level is usually available in the company. The robot manufacturer provides the communications protocols and tools to interface its robotics FMS into the factory mainframe computer. FMS information is used by the mainframe for materials requirements, planning, production scheduling, inventory management, and quality reporting.

Products for CIM

Products developed by robot vendors to handle intelligent robots for CIM have been designed to be compatible with each other to provide users with the ability to increase the level of automation in a factory at their own rate. Thus, the line of intelligent robots can include gray scale, high-speed vision systems, universal controllers, and user-friendly programming languages.

A universal controller allows a user to control robots, vision systems, sensors, and other factory automation equipment simultaneously. It allows the direct integration of a vision system with robots and other automation systems if the architecture is a compatible real-time multitasking operating system.

Vision Systems

Vision is needed in applications like glueing, parts positioning, and seam tracking. A gray-scale vision system provides real-time feedback to the robot. This feature is particularly important in the automotive and electronics industries where robots must be highly flexible to reduce the amount of expensive fixed tooling.

A vision system can make precision measurements of part dimensions, orientations, and surface quality. This allows the system to include complex surface details and various component shapes in any analysis. By creating a data base from a physical model, the system uses search and processing algorithms to focus on important image regions while ignoring changes in color and the general background.

Extended Arm, Overhead Versions

The ability of a robot to perform applications in environments with limited floor space or other constraints is extremely important. An overhead version that can handle payloads up to 50 pounds and an arm that extends the robot's reach to 60 inches enhance its ability to handle a wider range of applications in factories.

Programming Language

Another item of importance is the development by the robot manufacturer of programming language specifically for CIM applications to allow a wide variety of factory automation equipment to be programmed from one common interface.

Mini-Intelligent Assembly Robots

Intelligent assembly robots and systems provide versatile solutions to the problems of flexible assembly and computer-integrated manufacturing. These are three- or four-axes robots offering high speed, high precision, and selective compliance with payloads and reach. These horizontally articulated robots provide capabilities to meet user needs for applications such as:

- Precision electronic assembly
- Precision mechanical assembly
- High-speed material handling
- Four-axes arc welding

Manufacturing Planning and Control Systems

Most companies still have rather rudimentary manufacturing-related systems. Many companies that have installed MRP have had mixed results. The drive now is to install MRP II, which includes MRP and contains other elements of resource utilization strategies and management controls. These advanced systems are more highly integrated, user friendly, and more directed toward performing specific applications. Latest computer-based technologies include a wide range of applications including group technology, factory automation and robotics, computer-aided manufacturing, flexible manufacturing systems, automated in-process inspection, JIT manufacturing, artificial intelligence, expert systems, and eventually as most of these advances are implemented, the unmanned factory of the future.

To achieve some or all of these objectives, management must be willing to accept the fact that this will represent a major change in the manufacturing environment. Hardware and software systems are changing so rapidly that previous systems—even those of relatively recent origin—are rapidly becoming obsolete. Computers are becoming faster and easier to use, thereby enhancing the ability to store, manipulate, and retrieve data.

The goal of these computerized systems is to reduce labor costs, slash inventory investment, improve delivery performance, and increase product reliability. These goals can be met in part or in whole by improving management efficiency through tighter process control,

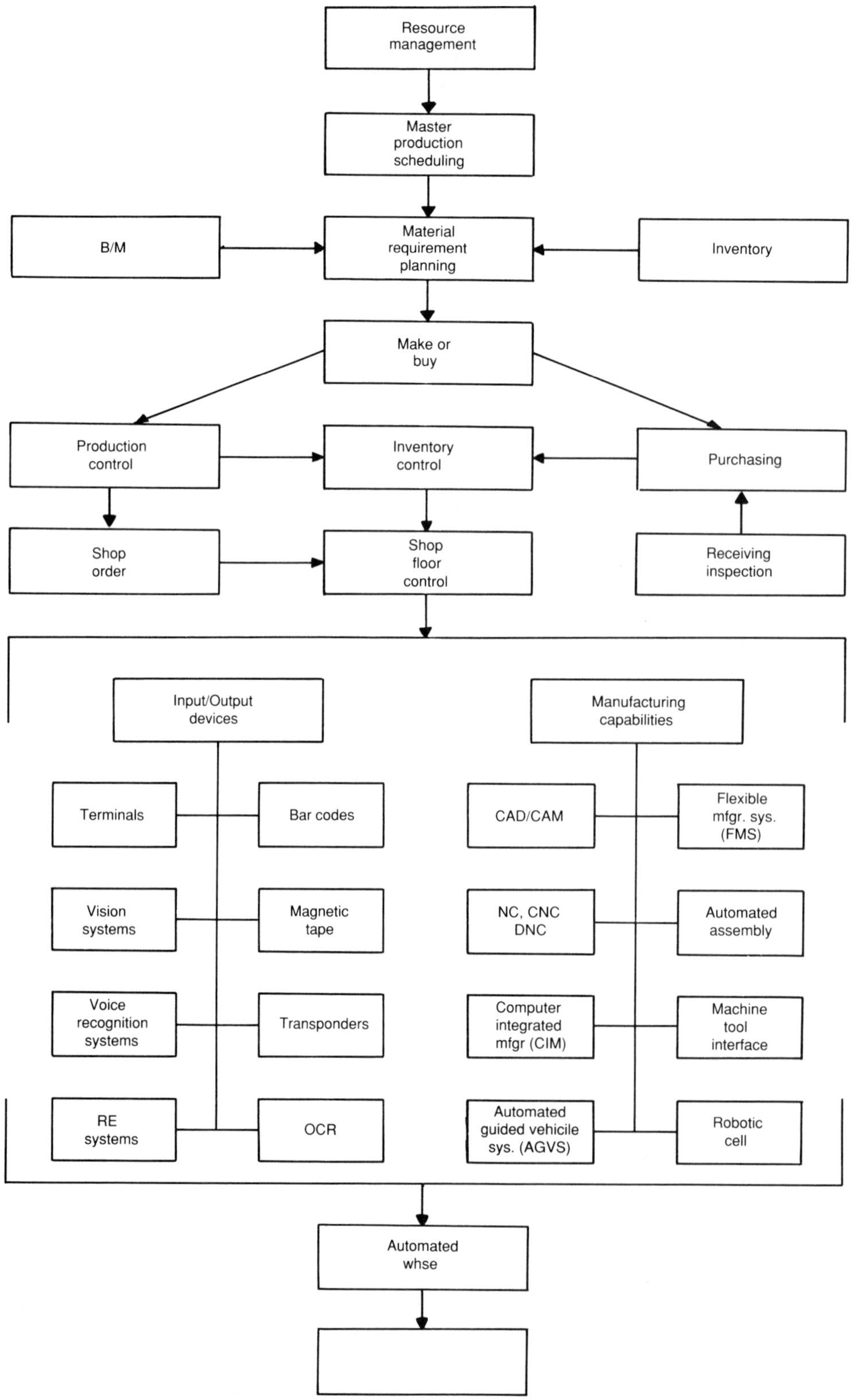

FIGURE 11–12 Integrated Manufacturing System

higher quality standards, sharp reduction in vendor lead times, reduced levels of inventory, and improved production efficiencies.

Achievement of these goals must begin with development of a completely integrated system. Since CAE and CAD systems are used heavily in industry today, and NC, CNC, and DNC software have had sufficient industrial applicational exposure, users can now move toward a computer-integrated manufacturing environment with a high probability of success.

Computer-integrated manufacturing can be viewed as part of an overall strategy for linking existing technologies and people to manage previously independent activities. From this basic framework a company can achieve the integration of each component into a cohesive structure. Figure 11–12 contains all elements of a completely integrated manufacturing system.

Figure 11–12 contains the total integration of all systems required to implement a complete CIM program. In conjunction with the introduction of these application systems, there is an equal need to develop the applicable systems architecture. The following points contain CIM architectural characteristics:

- Uniform operation system and programming language
- Integrated hardware and applications
- Logical communications paths
- High degree of vertical integration
- Communication addressed explicitly
- Simple and planned for growth and change

How Robotics Fit into CIM

The Japanese Industrial Robot Association (JIRA) defines six distinct categories of robots. The simplest are manual devices with a manipulator worked by an operator; the most complex are intelligent robots, which contain sensory perceptors (visual and/or tactile) that detect changes in working environments and employ automatic decision-making facilities to regulate operations. In general, all six categories of robots involve the concept of programmability and are best employed in the performance of material-handling functions and repetitive tasks.

Using robots in an integrated CIM environment is an important factor in performing low-cost, flexible manufacturing operations (that is, where the machines can be configured to perform a variety of functions on a family of parts). Commercial CAD/CAM vendors offer data bases to provide machine and tool information, and offer the capability of off-line programming of robots. Such programming is important because it enables robot simulation to allow for procedures testing without tying up or taking the work cell out of production. Additional software developments permit the CAD/CAM data base to be used to check clearances and interferences of robotic components and assemblies; verify robot paths; and include information on robot, task, and work cell variables related to specific robotic tasks.

At present, there is no standard robot language. Industrial vendors are developing proprietary languages that enable robots to work interactively with a person (rather than being a blind slave) and allow for the automatic teaching and reprogramming of robots—currently the slowest and most error-prone aspect in robotics.

Automated Materials Handling

Chapter 7 contains a complete discussion on automated guided vehicle systems (AGVS). The following case-in-point illustration shows an application of AGVS to CIM.

CASE-IN-POINT

*MANUFACTURING VERY LARGE SCALE INTEGRATION CIRCUITS FLEXIBLE MANUFACTURING SYSTEMS, INC.**

Situation

To remain competitive and realize the benefits of very large scale integrated (VLSI) technology, it had become apparent to companies in this field that they would have to automate the manufacturing process. The demand for improved yields and lower manufacturing costs made this a necessity.

Previously, efforts to automate the manufacturing line had concentrated on wafer handling within individual pieces of equipment with some extension to transporting wafers from one machine to another. These automation methods consisted of rigid, interconnecting links of islands of automation. Efficient operation of these systems required that production tools be sequenced or online. Other mobile systems were restricted to a predetermined path. The inflexibility of these systems limited their implementation into multiproduct manufacturing lines. Furthermore, complex controls must be provided to compensate for equipment failure or alternate process sequences.

Solution

The solution is to design a new generation CIM system that offers operating flexibility and is easily integrated into existing production lines. These systems permit evolutionary change from manual handling procedures to fully automated wafer manufacturing. With these objectives, Flexible Manufacturing Systems, Inc. developed an intelligent vehicle that transports wafers in an ultra-clean environment. Coupled with the transport system is a minicomputer-based control system that manages material flow and work-in-process inventory.

These components are contained in the entire FMS CIM system which consists of five interrelated subsystems: a mobile transport unit, docking module with tool interface, system controller, intelligent Work-in-Process (WIP) station, and integration software (figure 11–13).

Mobile Transport Unit. The mobile transport unit (MTU) is an autonomously guided vehicle that transports wafer cassettes, boxes, or other clean-room containers from one workstation to another. Containers are transferred between the MTU and production tools by a robotic arm. Self-contained and battery powered, a single MTU can efficiently service ten or more production tools. Obstacle detection allows the MTU to avoid all obstructions in its path, allowing operation amongst workers.

Docking Module with Tool Interface. The docking module (DM) may be mounted on a variety of production tools or WIP stations, providing a docking point for the MTU. The tool interface (TI) is a display terminal that enables personnel to communicate with the systems controller. It may be installed on a production tool or WIP station, providing a convenient means of entering instructions or verifying data relevant to process sequencing, wafer lot location, formation of new wafer lots, and other production variables. It also provides for a bar code reader and hard-wired data link between the production tool and the systems controller.

Systems Controller. The systems controller (SC) controls the entire CIM system. It manages all MTU guidance and location monitoring functions as well as local interfacing between the MTU and production tools being serviced. A single SC, operating in a high-volume fabrication line, can control as many as five MTUs and a combined total of fifty

*Based on information from Flexible Manufacturing Systems, Inc. Used with permission.

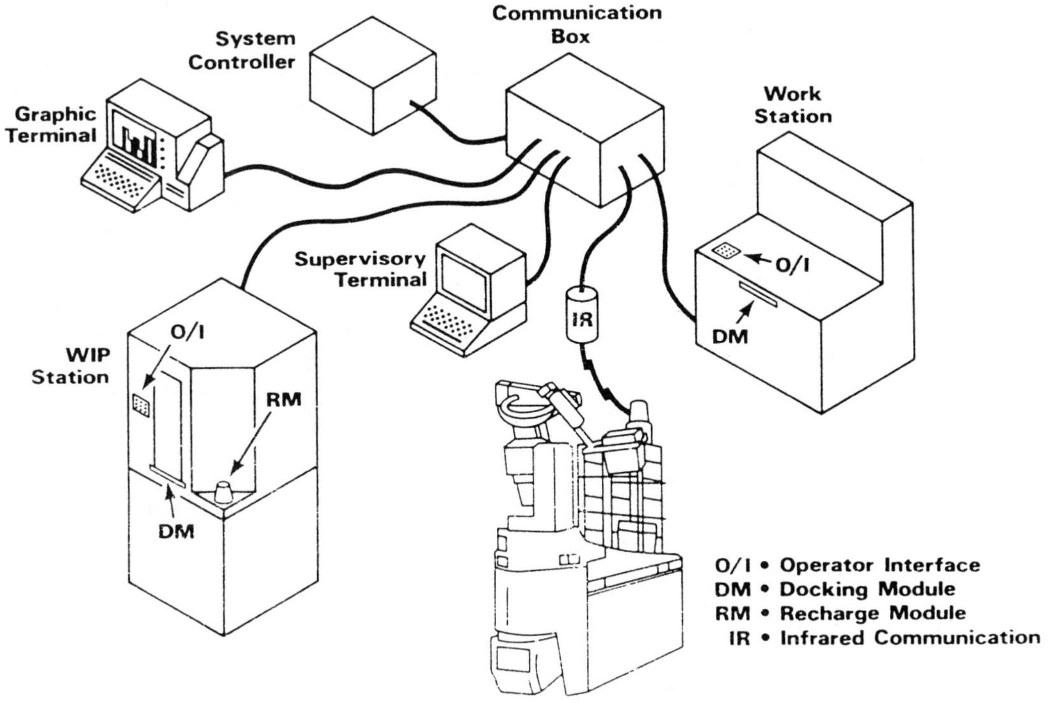

FIGURE 11-13 CIM System (Courtesy of Flexible Manufacturing Systems Inc.).

production tools and intelligent WIP stations. The SC also collects real-time data relevant to product lot location, CIM system fault analysis, and the operating status of production tools. System capabilities permit introduction of new tools to the production line, new vehicle routes for alternate process sequences, alteration of process parameters, and formation of new wafer lots to initiate rework cycles. A SECS II (semiconductor equipment communications standard) link is provided for data exchange with a host computer.

Intelligent Work-in-Process Station. Strategically placed intelligent WIP stations provide local storage and inventory control for wafer cassettes or cassette boxes. Several stations may be interconnected to provide expandable storage of WIP. Various configurations provide storage from 16 containers to 100.

Integrating Software. A comprehensive software package provides material tracking, system and production tool status, and system control. When teaching the vehicle new transfer functions, software formulates these into routes for subsequent processing of material. A security system allows only qualified personnel to access the system at various management, operation, and teaching and maintenance levels.

Flexibility of Motion for Practical Manufacturing Needs. The MTU employs an inertial guidance system for autonomous guidance. Moving about the production area without the aid of rails, rods, embedded wires, painted pathways on the floor, the vehicle moves along previously taught paths. Because the MTU has its own power and guidance system, it moves with absolute freedom—unheard of in other automation systems.

Each time the MTU interfaces with a DM, its location is recalibrated by the SC to eliminate cumulative positioning errors. The SC confirms that the MTU is addressing the correct DM, that the production tool is operational, and that the wafer lot to be processed on that tool is correct. If a designated process tool is not functioning, the MTU can automatically be routed to a backup tool providing the same process.

Results

Improved Yields. The FMS CIM system increases fabrication line yield two ways. First, it eliminates misprocessed wafer lots resulting from operator error, and second, it reduces wafer contamination due to operator handling. The resultant elimination of contamination improves wafer probe yield. By limiting or eliminating the number of people in the fabrication area, particle and chemical contamination also are reduced.

Lowered Manufacturing Costs. In this important area, the FMS CIM system offers substantial long-range benefits. Chief among them are:

Reduced Cycle Time and Material Costs. It is estimated that full automation can shorten by 70 percent the throughput production cycle of today's complex integrated circuits. Because wafers are processed faster with higher yields, rework is eliminated thereby reducing material costs.

Better Equipment Utilization. Reduced process cycle time and labor requirements increase equipment usage. Since fewer people are required, three production shifts can be scheduled daily without increasing labor overhead. Furthermore, improved equipment utilization will reduce the amount of equipment and floor space required to achieve optimum productivity.

Reduced Product Development Time. The CIM system, by optimizing work flow and materials handling, will minimize the effect of process variations through shortened process cycles. As a result, product development time will be reduced significantly.

Labor Reduction. Automation can reduce labor by as much as 50 percent. In addition, reduced labor and benefit costs with other cost savings resulting from less required operator training, less overtime, and less supervision also can be achieved.

Consistent Quality. The CIM system combines automated materials handling and data-base management tools to ensure process repeatability and consistent quality.

Accelerated Return on Investment. The automated system provides higher returns on investment. It is projected that these significant savings will provide a return on investment for automated equipment in less than one year.

Innovations to Watch: Reaching for the Factory of the Future

A long overdue change is taking place in the United States with the accelerated pace of collaborative R&D ventures among the government, industry, and academia. In particular, company scientists are working in concert with academic researchers. Competitors in the same industry are forging alliances with one another, and R&D budgets are dramatically increasing. A recognition is emerging that industrial research has become the driving force behind the process of technological change.

In 1986 alone industry spent $2.7 million on basic research, and more than forty research consortiums were formed to find solutions to common problems.

Some large industrial companies have financed scores of research projects in many universities. As a result of this flow of money into both internal company R&D and external investment in consortium or single contract R&D projects, this concerted surge of activity is resulting in numerous applications achieving fruition. In short, business is getting results from its R&D investment dollar. For instance, GE's R&D personnel in collaboration with Rensselaer Polytechnic have developed computer software to automate industrial design and production. Ford Motor Company recently purchased 10 percent of Carnegie Group, Inc., a spin-off from work done at Carnegie-Mellon University, that is heavily involved in robotic and artificial intelligence research and development. The CMU Robotics Institute has become the world's largest industry-financed center for research on robotics and manufacturing technologies, engaging in developmental work for over thirty industry sponsors. Companies participating in this arrangement can draw on the expertise of the center's 200 scientists and engineers to design manufacturing systems using robotics, computer vision, and other artificial intelligence technologies. As an example, the center designed a fully automated system to forge turbine-engine blades for Westinghouse Electric Corp. Other areas of CMU effort are directed toward speech recognition systems, expert systems, and flexible factory-automation systems.

The relaxation of Federal anti-trust laws has encouraged companies in the same industry to form strategic alliances. Also, the National Science Foundation (NSF) is establishing cooperative research centers to provide partial funding with industry picking up the major part of the cost. In 1985, a dozen companies including Control Data, Burroughs, Cray Research, and IBM invested heavily to become participants in five NSF-funded supercomputer research centers chartered to conduct research in such fields as composites manufacturing and biotechnology process engineering.

In another far-reaching development, Microelectronics and Computer Technology Corporation (MCC) was formed by ten otherwise competing U.S. computer and semiconductor companies to develop advanced VLSI technology. This effort also involves the United States government, and is America's answer to Japan's push into fifth generation computing (artificial intelligence realized in hardware).

MCC's goal is to develop seven areas of attainable computer technology and linking the results. These areas of concentration are as follows:

- New semiconductor packaging to increase the density and decrease the size of systems containing ICs
- More efficient software design by large software teams
- More efficient IC design through computer-aided design
- Parallel processing machines for faster computation
- Artificial intelligence and knowledge-based systems for computers that can "reason"
- Large, complex data bases to store the mass of information required by knowledge-based systems
- Better interfaces between computers and people so users can work more effectively

Recently, the original areas of research have been expanded so that the program in advanced computer architectures (ACA), for example, has been split into four full programs: parallel processing, artificial intelligence and knowledge-based systems (AI-KBS), data bases, and human-factors technology. These four programs remain linked, and these programs plus packaging, software, and CAD give MCC its seven programs.

The ultimate goal of MCC is to go beyond the present thinking of segmentation—by computer industry, semiconductor industry, or telecommunications industry—into a unified information-handling industry. Thus, the main thrust of MCC is to evolve that move toward a systems approach to information handling in programs. The ultimate success of this venture will depend upon MCC's ability to transfer developed technology to the participating companies and the United States government.

Industrial firms are exploring new ways to get technology onto the production line.

One approach used by General Motors Corp. is to take people who worked in R&D during development of critical technology projects out of the laboratory when a project is ready, and move them into operating divisions to implement the technology.

This accelerated pace of R&D effort by the government, industry, and academia will have a decided impact on advances in technological innovation, eventually leading to a coalescence of scientific and engineering innovation referred to as the factory of the future.

In conjunction with these joint ventures, certain notable industrial concerns have forged ahead on their own. Of these, the author has selected the work done by IBM's Lexington, Kentucky plant as illustrative of a completely automated manufacturing facility.

CASE-IN-POINT

IBM LEXINGTON, KENTUCKY
*THE FACTORY OF THE FUTURE FROM PELLETS TO PROPRINTERS**

Situation

The IBM Lexington, Kentucky manufacturing facility is one of the most advanced automated manufacturing facilities of its kind in the world. The heart of the automated plant is the final assembly system.

The 63,000 square foot assembly area contains a highly flexible assembly line which currently produces twenty IBM products. Supporting the final assembly area are six automated subassembly areas, each designed to act as stand-alone CIM factories.

The key to the efficiency of the plant lies in the flexible automated systems that control and monitor the movement and assemble and test the IBM products.

The Lexington facility includes the following:

- More than 180 IBM Series/1 computers maintaining the system
- 140 IBM robotic manufacturing systems
- 86 injection molding presses, ranging in size from 200 to 1500 tons, used to manufacture many of the printer's plastic parts
- 7 1/2 miles of conveyors

The automated systems assemble, cover, test, and package the printer for shipment.

About 70 percent of the IBM Proprinter's parts are made of plastic and are molded in presses in the plant. The plastic parts snap together tightly, which offers two advantages. The robotic manufacturing assembly process is able to handle the situation with greater ease, and with fewer parts both the reliability and quality are improved.

The remainder of the parts—there are only sixty numbered parts in the IBM Proprinter, another design element that eases the assembly process—are shipped just in time from local vendors when needed by manufacturing. This results in a considerable savings in cost of inventory and stock handling. Parts not made in the plant include transformers, motors, shafts, and acoustic foam. Parts are purchased when buying is the most cost-effective approach.

How the Process of Building an IBM Proprinter by Full Automation Works (figure 11-14)

The production process begins with the arrival of plastic pellets (raw material) by rail car or truck. The pellets are stored in bins until they are pumped directly into the injection molding machines where they are melted and fabricated into parts, including covers, bases, and paper guides.

Totes containing parts move via overhead conveyors from the fabrication areas and the loading docks to temporary storage in automated stackers and vertical carrousels which

*Used with permission of IBM Corporation.

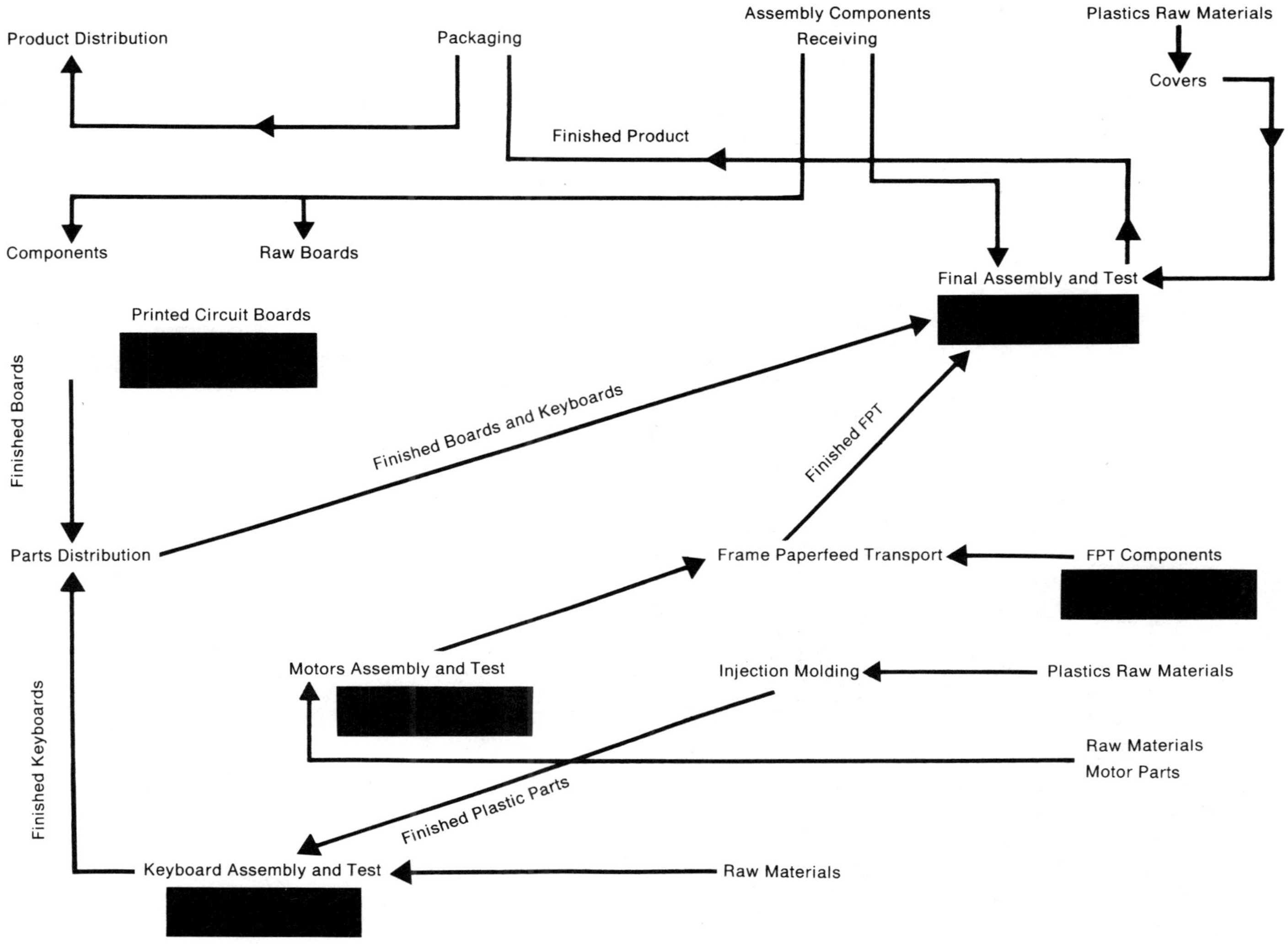

FIGURE 11–14 Factory of the Future from Pellets to Proprinters (Courtesy of IBM Corporation).

operate under the control of IBM Series/1 computers. From the stackers, conveyors deliver the parts to the assembly stations as needed. Each IBM Proprinter moves through the series of automated stations, while robotic manufacturing systems add parts from the base up in an efficient "layered" building process.

Each printer is automatically tested and then given an extended manual test to ensure quality products leave the plant. The printer is then shuttled by conveyor to the packaging area. Foam end caps are placed on the printer, the unit then is boxed, labeled, and loaded on a truck for shipment. The plastic pellets that earlier were being funneled into the facility have become the major part of the IBM Proprinter.

The automation control system, a series of programs developed specifically for the Lexington facility and running on IBM Series/1 processors, control the final assembly operations. The programs act as traffic controllers, ensuring that parts and finished products move—at 60 feet a minute—over the plant's more than 7 1/2 miles of conveyors.

Even the logic card in the IBM Proprinter is assembled by an automated process designed by Lexington's electronic card assemblies area. The card, which drives the IBM Proprinter and provides its many capabilities, is assembled in another subassembly area at Lexington.

The IBM Proprinter was developed by a team of product designers and manufacturing engineers so that it could be easily assembled by automation at low cost. It therefore had to be designed for the most efficient use of the robotic manufacturing systems. For example, the product designers and engineers eliminated all screws, belts, pulleys, and springs from the IBM Proprinter assembly because they are difficult for robotic manufacturing systems to handle. It also is easier for a robotic system to snap parts together, which is one of the reasons plastic was chosen. Also, the fewer the parts handled, the more efficient the assembly.

But the automated plant is not designed just to assemble the IBM Proprinters. It is a generic plant because almost any product within a certain size and number of parts can be assembled in the plant at the same time.

chapter 12

Training, Retraining, and Measuring the Productivity of People in the Automated Factory

PART A: TECHNIQUES FOR TRAINING AND RETRAINING PEOPLE IN THE AUTOMATED FACTORY

TRAINING: WHAT'S HAPPENING NOW AND WHY MORE EFFORT IS NECESSARY

The previous chapters of this book covered in detail all of the various components—the parts of which comprise the *total integration of all manufacturing elements* (TIME®). When taken in their totality, these advances represent an extraordinary assemblage of technological progress unsurpassed by any other period in the history of the United States. Yet, these gains will appear modest in comparison with what will be developed during the next decade. You must also factor in to this growth equation the potential rewards of esoteric sciences still in their embryonic state such as artificial intelligence, expert systems, and parallel distributed processing as a framework for explorations in the microstructure of cognition.

What this means is that by harnessing the awesome power of the computer with present or near term, sophisticated high technology, the factory of the future will soon become a generic reality in American industry. As this transition occurs, the remaining labor force will require many different skills and more intensive training than at present. Without question, this critical need will have to be met by the *coordinated effort* of government agencies, industrial concerns, and academia.

Although it is difficult to obtain data on the various aspects of government training or those provided by academia, current figures are available in *Training* magazine's industry report for 1986.* With their permission, this data is presented as a most valuable means of establishing a base point for further discussion on the changing role of human resources in industry.

*Permission granted by Lakewood Publications, Inc. of Minneapolis to use material and exhibits from *TRAINING Magazine's Industry Report, 1986.*

Training magazine's industry report for 1986 is an excellent source of statistical data and other relevant information for quantifying the extent and nature of employer-sponsored training in the United States at this time. The following material will contain answers to:

1. How much does corporate America spend on employee training?
2. What subjects are taught?
3. What types of employees get the most training?
4. How is the instruction delivered?

Highlights of the 1986 Training Magazine Report

These are the highlights derived from this comprehensive survey:

- Total dollars budgeted for formal training in 1986 by U.S. organizations with fifty or more employees totaled $29 billion.
- Total number of individuals who will receive formal, employer-sponsored training totaled 36.5 million.
- Total man-hours of training those individuals will receive totaled 1.3 billion.
- Total dollars budgeted by training departments for outside expenditures (i.e., money spent outside the organization on audiovisual equipment, seminars, computers, packaged-training programs, etc.) totaled $6.2 billion.
- Percentage increase in outside expenditures over 1985 totaled 11.7%.
- Industrial sectors with the highest average (per organization) outside expenditures for employee training were transportation, communications, utilities.
- Lowest average was public administration.
- Type of employee likely to get the greatest amount of formal training was middle management.
- Least amount was clerical workers.
- Most prevalent type of formal training in U.S. organizations was new employee orientation.

It should be noted that the estimate for formal training of $29 billion includes the salaries paid to trainers and the overhead expenses charged to training departments, neither of which were studied or reported in previous industry reports.

The other significant change in the 1986 survey, also reflected in the budget section, was the addition of bigger budget categories for outside expenditures on seminars, hardware, packaged training programs, and so on.

How Large Are Training Budgets?

Figure 12-1 consists of a pie-chart presentation of the various components that make up the dollars budgeted for formal training by U.S. organizations with fifty or more employees. Of the $29 billion total dollars expended, $20.87 billion (72 percent) is for training staff salaries, with all but $2.07 billion (7 percent) devoted to outside expenditures (21 percent). These outside expenditures break down as follows:

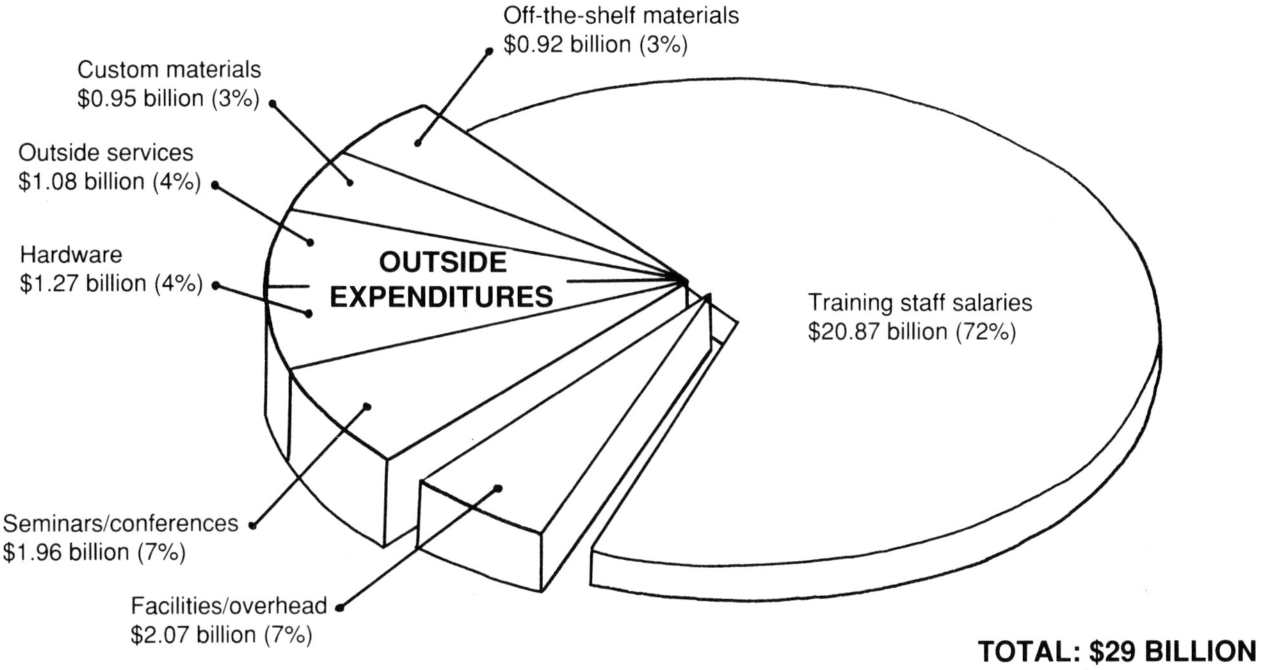

Definitions
Facilities/Overhead--charges to the training department for buildings, classrooms or other space to be used, remodeled or built during 1986, and for utilities, administrative support from other departments, etc.
Outside Expenditures--dollars budgeted for the following five categories:
Seminars/Conferences--training by outside providers conducted either as the respondent's location or off-site, including public seminars but *not* trainee travel and per diem costs.
Hardware--audiovisual and video equipment, computers, teleconferencing equipment, etc.
Outside Services--consultants (not acting as seminar leaders), printing, material production costs, etc.
Custom Materials--audiovisual, video, printed material, computer courseware, etc., tailored to meet respondent's needs or designed specifically for respondent's organization.
Off-the-Shelf Materials--prepackaged in any format: books, films, computer courseware, structured classroom packages, etc.

FIGURE 12–1 Dollars Budgeted for Formal Training [by U.S. Organizations with 50 or more employees] (Courtesy of Lakewood Publications).

- Seminars/conferences total $1.96 billion (7 percent)
- Hardware total $1.27 billion (4 percent)
- Outside services total $1.08 billion (4 percent)
- Custom materials total $950 million (3 percent)
- Off-the-shelf material total $920 million (3 percent)

Together, the sum of all outside expenditures amounts to almost $6.2 billion.

Where the Training Goes

Based on the findings of this survey, 82 percent of all formal training conducted by companies occurred on paid time. That is the "average" expressed as a mean. The median is

95 percent since in the model (most typical) organization, employees are paid for 100 percent of all time they spend in training. From this data, the estimate of 36.5 million people receiving formal training and total labor hours of training of 1.3 billion were arrived at.

As a related cost factor, the American Society for Training and Development estimates that on-the-job training may run as much as about $180 billion a year.

In regard to the 1.3 billion labor hours of training mentioned previously, most is for training production workers. Collectively, production workers have spent more than 300 million labor hours in formal training programs in 1986.

Computers in Training

Of all organizations with fifty or more employees, 47.9 percent are using computers in connection with their training efforts. This involves such things as using computers for word processing (revising training manuals), data processing (keeping records on trainees, trainers, course offerings, and so on), or computer-based training (CBT).

Organizations that use CBT—that is, those that employ computers to deliver actual instruction to trainees, the subject they most frequently use it to teach is, by far, how to use computers. Of all respondents who employ CBT to teach anything 91.7 percent use it to teach computer-related skills such as programming, word processing, how to use applications software, and so on, 29.8 percent use it to teach technical skills not related to computer use, and 23.4 percent us it to teach nontechnical subjects.

Following is the total 1986 compilation of all sizes of organizations with 100 or more employees that use computers in training, broken down by various categories:

Category	Percentage
Using computers in training	53.1
Using computers for computer-based training	32.1
Using computers for data management	43.1
Using computers for word processing/graphics	43.9

Who Uses Interactive Video and How

The 1986 survey defined interactive video as the use of a computer, in conjunction with videotape or videodisk, as a delivery system for training.

Based on this definition, the survey findings are as follows:

- Of all U.S. organizations with fifty or more employees in 1986, 15.5 percent used interactive video to deliver job-related training
- 20.2 percent of respondents who did not use interactive video expected to in 1987
- 73 percent using interactive video did it with videotape; 18.3 with videodisk
- 86.4 percent using videotape used "packaged" programs (i.e., bought ready-made from vendors)
- 20.1 percent used programs custom designed by or for their organizations
- 67.2 percent of those currently using videodisk, used packaged programs

- 42.1 percent used custom-designed programs
- 49.2 percent of the 15.5 percent of respondents using interactive video in some form (tape and/or disk), used it in an individual-instruction format (i.e., for self-paced training)
- 60.2 percent used it in a group or classroom format
- 36.3 percent of all organizations with 10,000 or more employees used interactive video in some form; while only 7.4 percent used it in organizations with 50–99 employees

Options for Classroom Training

From a broad perspective, there are three ways to develop and deliver training:

1. Build a program and teach it yourself
2. Buy a program
3. Build some segments and buy others

Within these general categories, an organization may use either seminar/workshop or self-instruction. The following table is broken down by type of training:

Type of Training	Developed By	Delivered By	Percentage of Organization's Total Training Effort (Mean) 1986
Seminars/workshops (in-house)	In-house training staff	In-house training staff	30.4
Seminars/workshops (off-shelf)	Vendor	In-house training staff	14.8
Seminars/workshops (public)	Various external sources	Various external sources	21.5
Seminars/workshops (closed i.e., for organization's employees only)	Vendor/ consultant	Vendor/ consultant	10.2
Self-instruction	In-house training staff	N/A	12.0
Self-instruction	Off-shelf	N/A	8.6
Self-instruction	Custom-built by vendor	N/A	2.5

Which Instructional Methods Are Most Popular

Following is the percent using these methods for employee training of all organizations with fifty or more employees:

Method	Percentage
Lectures	81.8
Videotapes	80.5
One-on-one instruction	62.2
Role plays	45.1
Case studies	41.5
Films	47.4
Self assessment/self-testing instruments	35.8
Noncomputerized self-study programs	27.6
Teleconferencing (audio/video)	23.9
Teleconferencing (audio only)	10.8
Computer conferencing	3.4

General Types of Employee Training in Descending Order of Use

1. Management skills/development
2. Technical skills/knowledge
3. Supervisory skills
4. Communication skills
5. New methods/procedures
6. Clerical/secretarial skills
7. Executive development
8. Computer literacy/basic computer skills
9. Customer relations/services
10. Personal growth
11. Sales skills
12. Employee/labor relations
13. Disease prevention/health promotion
14. Customer education
15. Remedial basic education

A TEN-STEP PLAN FOR EFFECTIVE EMPLOYEE TRAINING

The statistical data covered previously can be of considerable value in providing a macro overview of what industry is doing in general in the area of employee training. For example, if certain innovative training methods have progressed past the developmental stage or new motivational techniques are proving to be successful, then you may elect to evaluate these approaches to determine whether they are a viable product for employment. Thus, this information may indicate new or alternative directions to take to satisfy your own particular training needs. Although in the final analysis each company must tailor its training program to satisfy its unique requirements, much can be learned from the experience of others.

Each company strives for an investment in human resource training, which over the long term, will improve worker productivity and increase company profitability. To achieve these objectives a company must clearly define the goals that it seeks to accomplish with its training program so that every member of the organization is conversant with exactly what is to be accomplished from this effort. The following steps are provided as a guide to be used in formulating training goals:

1. Start at the top by developing an understanding at the executive and managerial levels what the purpose and value is to the company with better trained, broad-gauged, motivated workers.
2. Provide management with a well defined training program that indicates that this important activity is in the hands of competent professionals.
3. Prioritize your training needs by focusing on organizational segments, functions, or systems that will have the best prospects for short- and long-term payback.
4. Focus training efforts on a few key concepts so that employees will understand the approach being used.
5. Develop training techniques that are user friendly and workable.
6. Provide as much as possible in the way of participation action programs to broaden the base of people involved in implementing benefits derived from the training.
7. Promote the results of the training program through examples of specific accomplishments.
8. Continuously follow up after people have been trained to continually upgrade, motivate, support, and monitor performance on the job.
9. Constantly create a desire on the part of employees to obtain additional skills and greater job knowledge for self-fulfillment.
10. Evaluate and assess the impact of training on subsequent job performance and as a means of relating it to overall company goals and objectives.

TYPES OF TRAINING PROGRAMS

Using Small Semi-Autonomous Teams to Counter Monotony

Today's employees are better educated, more individualistic, and less inclined to accept repetitive tasks. In many respects automation runs counter to the desires of this changing work force.

To counter this concern, many companies have reacted in a positive manner by setting up small semi-autonomous teams. This technique aims to change a stiltified, monotonous work system by grouping people into small semi-autonomous teams. In this environment, each team performs all the functions that were previously done in separate centralized repetitive-oriented work centers. Team members are given additional training to learn new skills, thereby enabling them to perform more varied jobs. This has enhanced the worker's job satisfaction, and usually gained the worker a higher pay scale. As a result, workers are more highly motivated; supervision is self-disciplined by team members; worker morale is substantially improved, and productivity usually improves dramatically.

The productivity gains that this approach represents are due to the synergism that accrues by pairing people with automation. This approach represents a recognition on the part of corporate America that technological innovation by itself is not the answer to increased productivity. Instead, to realize the full potential of automation, workers and technology must be integrated in socio-technical systems that completely restructure the way work is organized and managed.

This is a significant change from past company procedures—one that is producing a new model of job design and work relations that will shape the work place for years to come. Although some companies are adapting its work force to a team approach, the primary use of this motivational technique will undoubtedly parallel growth in automation in general, as the full momentum of this movement becomes more evident.

Contrary to the engineers' vision of factories run by robots, the high-technological work place depends more than ever on people. Although factory automation will require fewer people, those that remain will play a critical role in maintaining an unimpeded operation. Thus, it is evident that advanced technology calls for a radical change in traditional work practices. The old "scientific management" method of dividing work into discrete tasks that require little skill or training becomes obsolete in a computerized work place where many functions are integrated by the computer. This integration makes it more difficult to define individual jobs or measure individual worker performance. For these reasons, as well as in recognition of the "quality-of-work-life" movement, companies are installing work systems that emphasize broader-based jobs, teamwork, participative managers, and multiskilled workers.

Another important factor that has force-fed this change, is global competition, which has required American industry to seek a high rate of product innovation. Although the United States may not be able to compete with countries that turn out standardized products and parts at low wages, it can create new market niches for customized products. This not only involves the use of fast reacting FMS, but also highly skilled broadbased workers capable of performing in this new production mode. Thus, the old system characterized by authoritarian management and an extreme division of labor epitomized by the assembly line will be replaced by integrating the psychological and social needs of workers into semi-autonomous teams that are proving to be as much as 30 to 50 percent more productive than their conventional counterparts.

Why Companies Use Productivity Action Team Program

The productivity active team (PAT) program is somewhat of a hybrid in that aspects of a variety of techniques, programs, processes, and methodologies have been integrated into the design program. The PAT program provides management with a dynamic way in which to:

- More fully utilize the skills and talents of employees.
- Create improved goal congruity between management and employees.
- Enhance opportunities for productivity gains by instilling a sense of motivation that can engender improved work-oriented results.
- Provide an environment conducive to employee participation in organizational problem solving.
- Enhance the decision-making process based on group consensus rather than based on autocratic or consultative style of decision making.
- Encourage participants to take a fresh approach to problems and seek to resolve previously untenable productivity constraints.

As you can readily see, the methodology employed in the PAT program is similar to a quality control program. However, its distinguishing characteristic is that it integrates certain critical features that have been designed into the process to enhance its effectiveness and efficiency in American organizations. Thus, the primary thrust of the program focuses on basic management and employee development; pragmatic problem-solving techniques; interrelationship cooperation; quality as a function of productivity; shop floor problem solving methods; developing qualitative and quantitative approaches to labor productivity measurement, evaluating employee data feedback; and most importantly, developing an ability to take action by changing or modifying human behavior.

In general, the relationships between changes that occur in the areas of participation and job enrichment as a result of a PAT program or any other involvement strategy and improved job performance, are not as identifiable as the relationship between money and goal setting and performance improvement on the job. However, although short-term productivity improvement resulting from participative techniques applications may be difficult to measure, experience tends to confirm that considerable long-term productivity improvement does accrue from employee participation in a PAT program.

Evaluating Productivity Improvement

Productivity improvement comes about when the relationship between results or output to inputs for a given system increase. More specifically, you may determine that involvement strategies will cause some or all of the following outcomes:

1. Increase motivation thereby increasing effort which increases the quantity and/or quality of work done by employees.
2. Promote employees to work toward developing greater skills or in more diversified jobs.
3. Increase problem-solving abilities on the part of employees so that problems are identified and dealt with on an ongoing basis.

Productivity Action Team Program

Much work has been done in this area to arrive at a soundly constructed methodology for implementing this developmental technique. The following steps consist of the PAT program:

1. *Necessary Preconditions*
   - (a) Production management planning
   - (b) Management style/productivity improvement
     - Technique
     - Compatibility analysis
   - (c) Productivity measurement technique
     - Productivity improvement
     - Technique survey and analysis
   - (d) Base period measurement
   - (e) Management briefings
   - (f) Employee briefings

2. *PAT Program Steps*

Step 1: Productivity action group (PAG)

Ideas for Productivity Improvement: Nominal Group Technique (NGT)—1 1/2–2 hours

Output: Prioritized/Consensus List of Ideas for Productivity Improvement

Step 2: Productivity action group (PAG)

Implementation Issues: Structured by form session—1 1/2–2 hours

Output: Productivity action team (PAT) Formation

Implementation cost forecasts

Step 3: Productivity action team problem solving

Work related problems

Problem analysis (PA)

Decision analysis (DA)

Implementation analysis (IA)

Issues

Issue analysis

Structured by problem solving process—1 hr. per week, one cycle runs approximately 3–4 months through recommendations to management. Issue analysis may only run 3–4 weeks.

Step 4: PAT review sessions to PAG

Results of PA and DA from PATs are reported to PAG. These are essentially dry runs to ensure all in group are in accordance with work done by PATs—2–4 hours over several weeks.

Step 5: Management review session

Results of PA and DA are presented to a management "steering committee"—2 hours.

Step 6: Management feedback session

Decisions and action planning laid out for PAG assignments and targets set—1 hour.

Step 7: Implementation

Step 8: Evaluation/productivity measurement

Step 9: Knowledge of results/feedback/charting

Step 10: "Incentives" system design

Step 11: Recycle/maintenance process (a typical cycle, Steps 1–6 will take between 3–6 months)

What an Effective Participant Action Planning Program Can Achieve

The approach used by exponents of participant action planning (PAPA) differs somewhat from the methodology employed by PAT, but the end result is similar. Both programs are action-oriented, problem-solving programs that rely heavily on the individual analyzing a problem developing a plan to solve the problem, implementing the change, and then following up the results of the improvement.

In each instance, the ultimate objective is to achieve a targeted goal and to train the participant so that the person's behavior will be modified constructively toward the way his or her job is performed.

PAPA seeks to transfer classroom learning to the actual working situation. This is done by requiring each person to develop specific items that will be concentrated on when back on the job once the training program is completed. This detailed plan of action for implementing new behaviors defines in concrete terms not only what tasks will be undertaken, but when these action-oriented tasks will begin and how final results will be evaluated.

The PAPA approach relies on follow-up of participants by the trainer, normally at intermittent periods over a 3 to 6 month period. Figure 12-2 shows the sequence of events in the training evaluation model. It includes the three key areas that comprise the evaluation process: learning, behavior, and productivity.

A well defined, properly executed, PAPA program can enable a company to achieve the following:

1. Train the participant in defining a task that requires corrective action; developing a specific plan to accomplish this task; and striving to implement a satisfactory resultant solution.
2. By being trained as part of a small group, the participant tends to accept commitment to accomplish certain goals more readily.
3. As part of a small group it is easier for the participant to discuss the problem and simulate what the condition will be like in the real-life situation back on the job.
4. The participant tends to be more committed than under less formal ad hoc problem-solving endeavors because the person is aware that the program includes follow-up to the actual training.
5. When the follow-up occurs, the trainer is supportive in evaluating the results, determining whether there were any constraints in the work place that mitigated

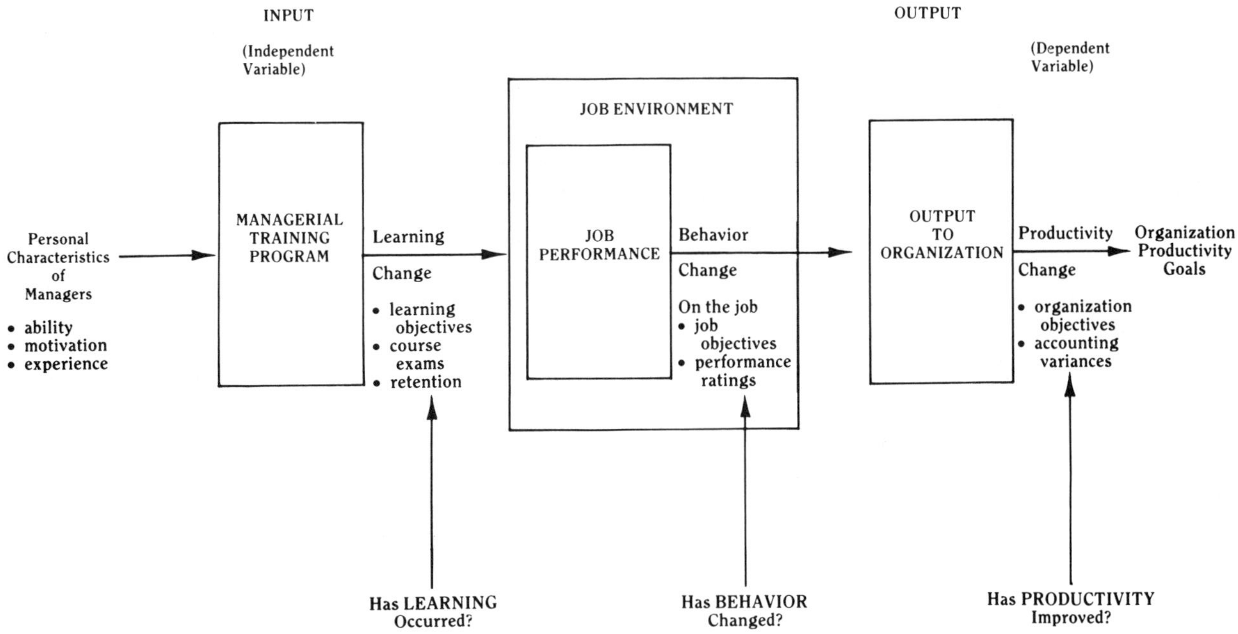

FIGURE 12–2 Training Evaluation Model (Courtesy Lakewood Publications).

successful accomplishment, or whether the person's behavior had not been modified sufficiently to achieve the intended goals.

6. The trainer can evaluate the degree to which the participant's superior has become more supportive in the post-training environment.

In summary, PAPA provides all the attributes for improving the "people element" of an organization. Participants are trained in small groups that provide each person with a feeling of belonging. Through the methodology of the program, a participant's behavior may be modified constructively toward planning, goal setting, and action-oriented accomplishment. The program completes the cycle by evaluating targeted results, determining and seeking to resolve any failed accomplishments, and maintaining a continuity of purpose presence through periodic follow-up and critique.

How Participant Management Influences Success

As a corollary to the approach used in a PAT program, if PAPA is to be successful it is of equal importance that management participant attitudes be improved, as well as seeking to modify worker behavior. This involves indoctrination of complete organizational participatory concepts with everyone becoming attuned to problem identification and problem solving.

What makes participant management so difficult to institute in an organization is that it is perhaps the most difficult form of management because it is highly disciplined, controlled, and goal oriented. More fundamentally, participant management is an organized set of ideas about *how* to manage a modern and progressive business enterprise. It is a set of concepts predicated on a philosophical conviction that the vast majority of employees can make significant contributions to the business, human, and social objectives if they are provided the opportunity, knowledge, support, and reinforcement to do so. Given this philosophy, the term "management" must be treated as a set of concepts, not as a group of people. Such concepts as planning, coordination, control, and decision making can be utilized and applied by every employee.

Participant management is grounded in the value premise that all people possess leadership ability—the capacity to influence the attitudes and behavior of others toward the accomplishment of common goals and objectives. The essence of participatory organizations is that they are not human systems where people are mere executors of work, but are instead systems where people are the organization's architects and engineers. They help design work systems, and having designed them, make them perform.

Every human system needs people who have a vision about what the system should and can be. Every human system needs a plan—a process for transforming the system from its present state of effectiveness to qualitatively and quantitatively higher states of effectiveness.

In authoritarian systems, only a limited number of people are permitted to engage in this process, whereas in participatory systems, every employee is encouraged to be involved creatively to the limits of his or her interests and capabilities. This process has also been referred to as quality of work life (QWL), employee involvement (EI), and other similar descriptive terms. But regardless of what it is called, in the final analysis, participant management may be the most flexible and committed method of maintaining a highly adaptive organization in today's complex industrial environment.

EFFECTIVE TRAINING TECHNIQUES*

Why Many Companies Prefer Custom Training

From all indications, a good percentage of companies prefer to develop or obtain a customized training program, rather than an off-the-shelf training package. What distinguishes custom training materials, from the others, is that it tends to reflect actual work life in the company, and thus has a greater chance of being accepted by the trainees. This involves close liaison and coordination with the people to be trained to identify and convey the priorities of these individuals.

This in-depth analysis of people-training needs requires considerable skill and understanding on the part of the custom trainer because most people are not particularly interested in the more mundane development of course outlines or design documentation. Basically, busy people want to see the finished product before reacting to its training value.

Another consideration in this process is that much of the customized developmental work must, of necessity, be disruptive to the future trainees. This in turn, is a by-product of how much support management is willing to allow for this time consuming task in terms of lost worker productivity. To partially offset this concern, it is advisable to obtain an up-front commitment from management that a customized training program will have certain benefits above and beyond a packaged training approach. To gain this commitment it is usually necessary to determine what measurable results will accrue from this training. This means identifying what changes are desired after the customized training program has been implemented, and when to expect these improvements to occur.

After approval to proceed, the trainer should include in the design, industry-specific or company-specific information and situations. Obtaining this information will take longer to develop and be more expensive than a program that incorporates some generic material. On balance, cost may preclude less than complete customization.

As the custom training material progresses to completion, care must be taken to adapt it to the particular organization. For example, if many of the trainees read at less than a high school level, plans to use computers or videotape may be downgraded to more basic methods. Again, the key to customization is being able to tailor the training program to the exact requirements of the organization, within an acceptable cost parameter.

How to Establish a Cost-Effective, Hands-On Training Program

Whenever the particular training requirement permits, companies prefer to use hands-on classroom training. However, this option requires an evaluation to decide how many training systems (computers, videodisk players, simulators, and so on) to buy or rent so that the cost of the training is as cost effective as possible.

In each case, the organization needs to know how many systems to acquire to minimize the ultimate expense of the training program. In addition to system or equipment costs, there are other variables to consider, including instructor cost, course length, and total number of trainees.

*Permission granted by Lakewood Publications, Inc. of Minneapolis to use material and exhibits from *TRAINING Magazine's Industry-Report* and from "Hands-On Training: How Much Equipment Do You Need?" by Mark D. Shanks (*TRAINING Magazine*, August, 1985).

By applying the appropriate data to a relatively simple formula, it is possible to determine the optimum number of training systems. The same equation can be used to figure out the number of times the course must be taught to be as cost effective as possible.

For example, a company is installing a new minicomputer system in all its office locations nationwide. The director of training judges that two trainees can learn on a single system about as effectively as one can. In a different situation, the appropriate trainee-to-system ratio might be 3:1 or 1:1, depending on the particular piece of equipment.

Given a specified trainee-to-system ratio, only two cost variables are affected by varying the number of training systems. One is the total cost of the equipment itself; the other is the instructor's total fee. As for equipment cost, two systems clearly cost more than one. In terms of instructor cost, the more systems there are, the more trainees can be taught in one class. In turn, the more trainees per class, the fewer total classes the instructor will have to teach. It is assumed that class length, whether it be hours, days, or weeks, remains the same for each class, and that the trainee to system ratio also is unchanged throughout the training.

Other training costs incurred by the trainees, such as wages, transportation, temporary housing, and meals stay the same even if the number of training systems changes. The costs of course development remains constant as well. Because these costs are fixed whether two or twenty systems are used, they do not enter into the analysis.

To see how equipment and instructor costs vary according to the number of systems used, look at figure 12-3. Generally speaking, the relationship between total equipment cost and number of individual systems is a linear one. As shown in the figure, the new minicomputers cost $30,000 apiece. Therefore, two systems cost $60,000, three cost $90,000 and so on. (Note: In some cases, there may be volume discounts for large orders of equipment. In other instances, however, more systems may mean more maintenance and repair bills. Therefore, the system-cost line shown on the graph actually may curve slightly in either direction.)

Now, to see how instructor costs vary according to the number of systems used, suppose the company must train 2,000 employees on the new minicomputers over five years. The course lasts five days, and the instructor's charge is $200 per day, which gives us an instructor fee of $1,000 per course. Remember, two trainees share one system in this example.

If one system is used, there can be only two trainees per course, and the instructor will have to teach the course 1,000 times. At a rate of $1,000 per course, the total instructor cost will be $1 million. If two systems are used, four trainees can attend one course and the instructor will only have to teach the course 500 times to accommodate all 2,000 trainees. In this case, instructor costs comes to $500,000. With eight systems, the instructor cost for 125 courses is $125,000.

As figure 12-3 illustrates, using two training systems is half as expensive (in terms of instructor costs) as using one. Using three systems is one-third as expensive, and using eight systems is one-eighth as costly. Notice that as the number of training systems increases, the reduction in instruction cost becomes marginally less. Eventually the reduction becomes insignificant. The most dramatic reductions in instructor cost result from increasing the number of systems from one to ten rather than from ten to fifteen.

The sum of the equipment and instructor costs is the lowest when the costs are equal—that is, where the two cost lines intersect in figure 12-3. Total costs are at a minimum at this point, and this is the number of training systems that will minimize total training expenses.

It is not necessary to plot a similar graph for your own classroom training situation to find the optimal number of training systems. Simply plugging the appropriate numbers into the mathematical formula corresponding to the graph will do. Figure 12-4 contains definitions for each of the variables in the equation and their specific numerical values for our example.

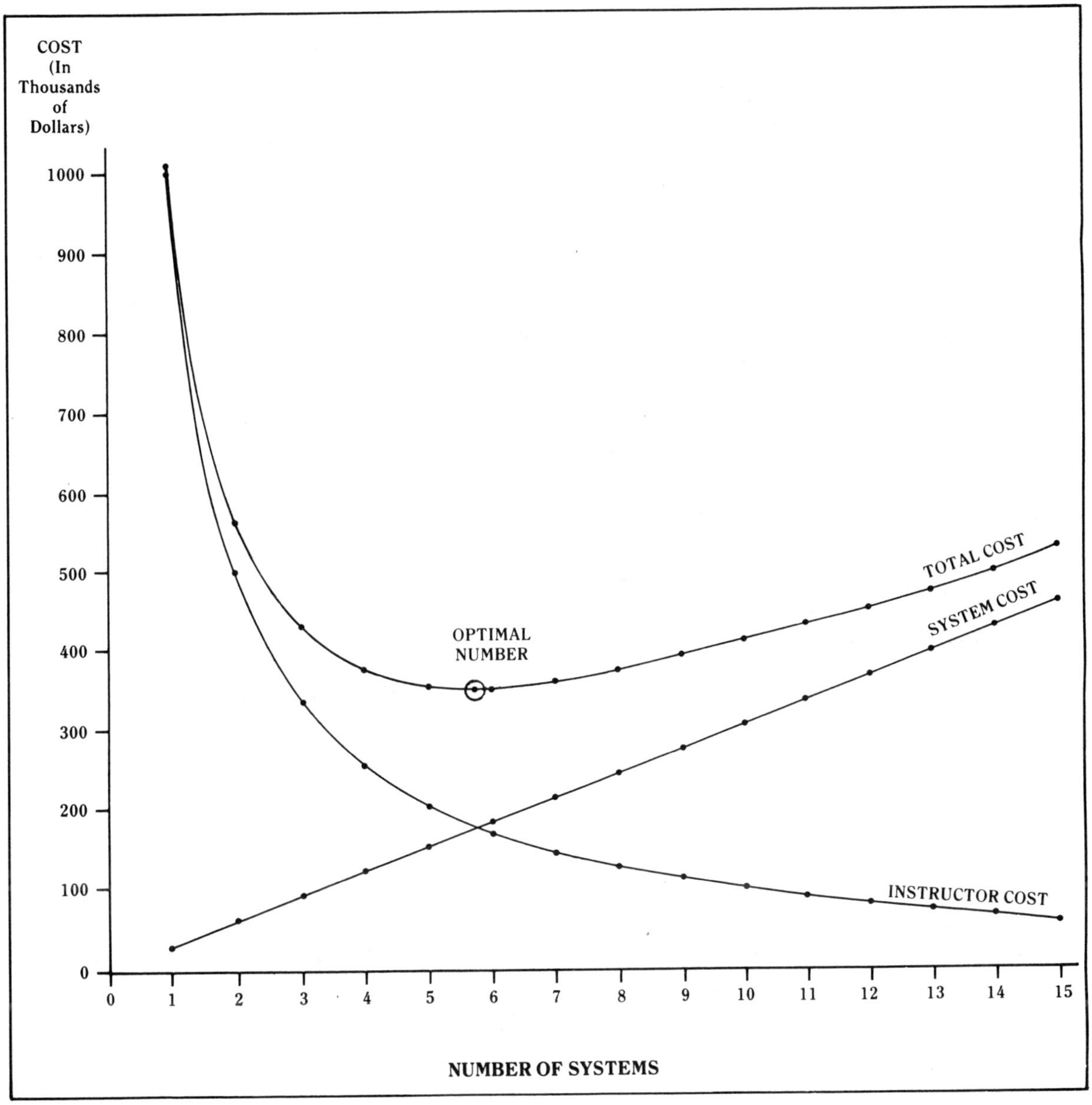

FIGURE 12-3 Relationship between Total Equipment Cost and Number of Individual Systems (Courtesy of Lakewood Publications).

The total costs are expressed mathematically as

$$A = TIL/RN + NC$$

where

A = total costs
T = total number of trainees
I = instructor cost per day
L = length of course per day
R = ratio of trainees to systems
C = total cost of system (including taxes, freight, other additional charges)

The first term to the right of the equal sign (TIL/RN) is the total instruction cost. The second term (NC) is the total equipment or training system cost.

Example: A Company Installs New Minicomputer Systems in all its office locations

Total Cost Formula: $A = \frac{TIL}{RN} + NC$

$$\left(\text{Total Cost} = \frac{\text{Trainees} \times \text{Instructor Cost} \times \text{Course Length}}{\text{Trainees Per System} \times \text{Number of Systems}} + \text{Number of Systems} \times \text{System Cost}\right)$$

Definition of Variables:
- A — Total Costs
- T — Total Number of Trainees
- I — Instructor Cost Per Day
- L — Course Length in Days
- R — Ratio of Trainees to Systems (Number of Trainees Per System)
- N — Number of Training Systems
- C — Total Cost Per System (including taxes, freight, maintenance, etc., over the life of the system)

Data for Company Example:
- T = 2,000
- I = $200 Per Day
- L = 5 Days
- R = 2 Trainees Per System
- C = $30,000

Computation: $N = \sqrt{\frac{TIL}{RC}} = \sqrt{\frac{2{,}000 \times 200 \times 5}{2 \times 30{,}000}} = \sqrt{33.33} = 5.77$ or 6 Systems

FIGURE 12-4 Method for Computing How Instructor Costs Vary According to the Number of Systems Used

You want to find out the optimal number of training systems (N) that will minimize your total training cost (A). Through use of calculus, you can do this by taking the first derivative (solved for N as a minimum) of the previous equation, and then taking the square root of the following term: TIL/RC (see figure 12-5).

Year	Instructor Cost (I)	No. of Trainees (T)	Computation
1	$180 per day	250	180 × 250 = 45,000
2	$190 per day	500	190 × 500 = 95,000
3	$200 per day	500	200 × 500 = 100,000
4	$210 per day	500	210 × 500 = 105,000
5	$220 per day	250	220 × 250 = 55,000
			45,000 + 95,000 + 100,000 + 105,000 + 55,000 = 400,000
			400,000 ÷ 2,000 (T) = $200 (Weighted Average Instructor Cost)

FIGURE 12-5 Computations for Determining the Optimal Number of Training Systems That Will Minimize Total Training Cost

Tips for Conducting Successful Teletraining

Teletraining is a method of training where the trainer is communicating to dispersed groups of people via a teleconferencing system. There are a number of important pointers that you should be aware of to run a successful teletraining program. They are:

1. Make certain that the audio quality is good. If there is static, faint reception, or other interference flaws in the telephone communication it can affect the level of interest of the people being trained. The best approach is to use half-duplex or a "voice-switched" bridge.
2. Use trained facilitators for each group of trainees since unlike classroom training, teletrainers cannot see the participants and directly supervise their activities.
3. The teletrainer must establish credibility and authority without the psychological advantage of occupying the strategic location at the front of the classroom.
4. Modulate your voice. Vary your tone, volume, and pitch.
5. Keep the momentum going. Use team teaching, prerecorded audio- or videotapes, examples, and stories. Avoid any monologue that runs longer than eight minutes.
6. Make a point of maintaining clear communication.
7. Encourage interaction both within and among the various groups of trainees by providing thought-provoking material and creating a nonthreatening environment in which comments and questions are reinforced and treated with dignity.
8. Build a library of slides, tapes, and handouts that can be reused.
9. Ask students for feedback on the training program.

How to Implement and Use Computer-Based Training

Computer-based training (CBT) is still in its infancy in terms of industry usage, with many knowledgeable people expecting virtually everybody to be using CBT in some form or other within the next ten years. The areas in which CBT seem to offer the greatest growth potential are:

- Teaching the use of computer hardware or applications software
- Generic business and management topics (decision making, negotiating, selling, stress management, etc.)
- Simulations, either equipment-related or topic-related
- Information transfer (teaching employees about new products or services to be sold)
- Delivering specific operational and procedural information about the business

Within these various categories, the largest growth potential appears to be CBT for people learning to use computer systems, disk tutorials, and simulation/modeling training. This is closely followed by training people in basic business skills.

Tips for Successful Automation: How to Choose the Right CBT Components

Today, CBT is thought of as being performed on the personal computer, a minicomputer, or a mainframe. Actually, the personal computer and the larger computers can complement each other. Virtually any personal computer can be used for the remote delivery of instruction, but a larger and more powerful host computer (either a large microcomputer, minicomputer, or mainframe could be the preferred computer for the management of CBT). In addition, the host computer might be used as a centralized location to store student records, and as a central library from which courses can be downloaded to the user's personal computer.

For management of computer-based training the usual configuration is to link the delivery computer (microcomputer) to some kind of more powerful host computer. Thus, after a lesson is taken on the microcomputer, an evaluation of the student's performance can be sent as output to the host computer, where the student's records would be accessed by the training department. Alternatively, students' results can be stored on the microcomputer and retrieved upon being polled by the host according to a preset schedule. These possibilities allow for decentralized course delivery, but centralized recordkeeping. At the user location, a printer can be used to create a permanent record of the trainee's performance and/or any diagnostic assessment.

In addition to the usual keyboard, CRT, and printer, microcomputers support a number of I/O devices that can be used in conjunction with computer-based training such as:

- *Voice Synthesizers.* Allows the computer to "speak" to the user with preprogrammed words and phrases on the screen.
- *Voice Recognition.* Allows the computer to recognize voice responses.
- *Light Pens.* Allows the user to point to a particular object on the CRT.
- *Pressure Sensitive or Touch Screen.* Allows the user to simply touch the screen with his or her fingers.
- *Mouse (or Joystick).* "Pointing" and drawing tools that allow easy use of many software packages.

The following section on the uses of CBT has been excerpted from *The Design and Development of Computer Based Training* by Technical Educational Consultants.*

Computer-Based Training is Useful in a Number of Situations

Testing

Testing is potentially the easiest and fastest form of CBT to produce. From drill and practice, to a comprehensive testing program, the computer can administer, grade, diagnose, and manage based on the input of the student. Naturally, as the program becomes more sophisticated, so does the programming effort. Before designing a test to use on the computer, two questions should be considered:

1. What type of response is expected?
2. What, if anything, will be done with this response?

The easiest response to handle is the one that must be a unique answer. Typically, this is an answer to a quantitative or mathematical problem.

The computer is a very precise machine so that although it can be programmed to accept all the variations that can be thought of, a logical time-saving device would be to offer

*Used with permission of Technical Educational Consultants.

the user a choice of a limited number of responses. If, for example, a multiple choice format is used, the answer then consists of a unique single letter or number response.

A wide spectrum of computer response and management may be incorporated into the testing program to process a given user input. On the simplest level the computer would provide any or all of the following services:

- Determine if the response is correct
- Provide the correct answer or a hint
- Determine how many of the questions are correct and give a percentage score

Diagnostics

The inclusion of diagnostics, while always a desirable feature, is inherently less cost effective than the creation of the testing program itself. Successful diagnostics rely on the ability of the designer to test rudimentary skills through the response to one or more questions. Thus, the writing of good, targeted questions is the first prerequisite to sound diagnostics. Second, a decision process must be outlined which will tell the user what to do if he or she gets a certain combination of questions incorrect. The programming of these diagnostics is generally only as complex as the decision process itself. By necessity it parallels and is interconnected with the testing program, thus increasing its complexity. Since the user should only be taking a test if he or she expects to answer the majority of the questions correctly, most of a diagnostic program would not be utilized. This results in lower cost effectiveness. Consequently, the decision to include diagnostics should result from the judgment that they would be used repeatedly—and taken seriously—so as to make the commitment of development time worthwhile.

Error Analysis

The next level of sophistication is the inclusion of an error analysis in the diagnostic routine. Here the user is told not only what area he or she does not understand, but also what algorithm or process is being performed incorrectly. Now the computer must record which questions the user got wrong and what incorrect answers were entered. Patterns must be discerned involving several incorrectly answered questions for a successful error analysis. The computer is ideally suited to do such work considering its speed and storage capacity. However, the amount of human effort required to do this programming reduces its cost effectiveness so dramatically that only large-scale use would make the development of an error analysis program economically feasible.

Tutorial

As an instructional aid, the proper way to view a computer is that it is one among several options.

A typical application of the tutorial is as a self-contained module, where the program first presents text, examples and then drill and practice. Unlike testing, the tutorial does not lend itself to "mass production," as the formatting is typically different for each part of the lesson. Thus, the decision to use the computer as a tutor should be carefully weighed as the time commitment can be substantial.

When designing a tutorial program, you must look to use the computer to its full advantage. This means drawing upon its speed, interactive nature, and graphics capabilities.

It also means avoiding its limitations, such as limited display area, and when using graphics, limited picture resolution. For example, we would not wish to display a table of information unless we wanted to highlight parts of the table as part of a problem solving process. Similarly, a line graph looks much better on a printed page. However, this viewpoint will change if the computer is to draw the graph based on information the user inputs, or specific points are to be emphasized during the construction process. Even in this latter case, the point should be made dynamically, such as with flashing, rather than with words. Words are the general province of the printed page; and if much information need be presented, it is best to use a textbook that provides a ready reference. The computer should only be used when it can do a better job than the textbook. Thus if the tutorial lends itself to both CBT and text, then both should be used. We should draw upon the strengths of each, and have them complement each other. In such cases, to provide maximum continuity and benefit, the instructional material should be organized so that the user spends sufficient time with each medium—that is, the learning experience must be cohesive.

Simulation

In a simulation the computer is programmed to assume the characteristics of either a person or a machine. There are, for example, several "touch typing" packages that are available commercially. Here the computer aids the user in becoming a better typist by indicating errors, calculating typing speed, and noting error frequency. In effect the computer becomes an electric typewriter that "talks back" by comparing the actual input from the user with the theoretical data stored in its memory. A much more elaborate project is one where the computer is programmed to respond like the console of an airplane or a nuclear reactor. Here we are not referring to an actual simulator (which has mechanical manifestations), but to an electronic simulation. As a simulator might be very expensive, a typical use of the larger microcomputer is to simulate a simulator electronically. The user thereby learns how to operate a simulator before actually training on it.

A very good area for CBT is to teach how to use various application programs such as a data base, order entry systems, or spreadsheets by simulating these programs. These simulations, however, are usually complicated, and throw the development project outside the realm of the company training department.

A third form of simulation is that in which the computer plays the role of an individual. This approach could be used in a management or sales training program. The computer, for example, can play the customer who will make a purchase if the user's sales pitch is effective. As with testing, response recognition is crucial in defining the complexity of the project. Again a multiple choice format is simplest for a user response. Generally it is not necessary to rigidly define each of the branching routines, rather the program design should include a probability function that picks the particular computer response from among those stored. In addition, key words entered by the user can appear in the output by embedding variables at the appropriate places.

Programs that affect role playing should be extremely well thought out to provide for a meaningful dialogue. It is very easy for the training objectives to get lost as the user/computer exchange reduces to mere gibberish. Voice synthesizers, while generally of poor voice quality, offer one option for adding more realism to the simulation. With additional hardware purchases, you also can use interactive audio and interactive video where the computer, based on user input, controls the next audio or video response.

CASE-IN-POINT

COMPUTER-BASED TRAINING
MAJOR INDUSTRIAL COMPANY
DEVELOPED BY TECHNICAL EDUCATIONAL CONSULTANTS
HICKSVILLE, NEW YORK

Situation

A major industrial company with hundreds of retail outlets throughout the United States was experiencing very high turnover in field personnel, sometimes in excess of 100 percent per year. As a result constantly training new hires became a burdensome task for the company. Management felt that one of the major reasons for this inordinantly high turnover rate was the difficult and time-consuming task required to prepare numerous handwritten forms which were mailed to the home office on a daily basis. However, this information was extremely important for inventory control, cash control, and sales statistics.

At the home office, a staff of clerical personnel laboriously compiled and processed this information, and key entered the data into the mainframe computer.

Partial Solution—Phase 1

After a detailed analysis of the paperwork flow, the Management Information Systems (MIS) Department designed a computerized data collection system, and installed the system in the retail outlets. It consisted of PCs and storage devices to enter and process data, and modems to transmit data via telephone lines to a mainframe computer at the home office. It was felt that this new computerized system would offer better inventory control and help solve the problem of data collection since field people could now enter data on the PC screen, rather than by the previous method. However, even with entry screen display formats, the logistics of the business still required manual preparation of the form before screen entry. Although still not completely satisfactory, the new system did speed up data processing since the mainframe computer could now "telephone" the retail outlets' computer, collect the data from the computers' hard disk storage, and automatically collate the data and generate management reports. This systems improvement enabled the company to eliminate most of the clerical work performed previously in the home office. More importantly, it provided management with a faster method of obtaining inventory, cash, and sales information.

In the initial field test implementation, management became aware of some important problems that required immediate attention. Although the major thrust of the system was to control field operations through quicker data access, too much information was inaccurate, and therefore of limited value. Another fault was that the system had been designed by MIS people without sufficient concern for the end users in the retail outlets. To operators in the field, the system seemed very confusing and appeared too complicated. Although the system contained all required information, through lack of adequate training, the operators did not know how to use it.

Complete Solution—Phase 2

Although the application program enabled the company to establish the mechanism for a computerized control system, it did not take into account the training required to teach the operators how to use the PCs and the system. Once this critical need was recognized, the company decided to hire technical educational consultants to develop a computer-based training program for its field personnel.

The following steps cover the approach used to achieve the level of competence desired by the company:

Step 1—Tutorial. This step entailed getting the people comfortable with the computer, which involved teaching them about its parts and how to enter data and become familiar with its basic functions.

Step 2—Tutorial. This step involved training the operators on the data collection system. The application system was broken up into a number of data entry screens that required daily input. A CBT module was designed and developed for each data entry screen. These were accessed by the user through a series of menu choices. Each CBT module consisted of the following parts:

- An introduction to the data collection screen and the purpose of the screen—telling them where they can get the data for the screen—and indicating any special qualities about the screen.
- An illustration of the data entry using a split screen approach:
  (a) The top half created a facsimile of the data entry screen.
  (b) The bottom half highlights fields or groups of fields and shows the user what they were expected to enter, and where to get the information. This is referred to as *illustrative data entry.*

Step 3—Simulation. The third step of the CBT program is to have the operator practice entering data (simulation). This involved calling up a facsimile of the data entry screen and having the user refer to a printed training guide exercise for sample data. Using this data, the person fills in the fields on the screen as if in the live application by using the cursor keys to locate the desired field. The objective of this exercise was to simulate the data entry screens and follow the exact procedures that would be required in a live mode. The user was given two chances to complete the data entry screen correctly, after which incorrectly entered fields were displayed.

The entire computer-based training program took 4–8 hours for the trainee to compete, and some 1000 hours to produce.

Results Achieved

Although it was difficult to measure the effect that this CBT program had on improved inventory control, cash control, and sales reporting, it did result in MIS designing subsequent application programs to be more user friendly, with screens less crowded and easier to use.

Overall, user responses indicated a 98 percent acceptance level for this training program. This compares very favorably with industrywide standards for computer-based training which typically yield acceptance levels ranging between 75 to 85 percent.

Why Interactive Videodisk Training May become the Most Popular Technique

Of all the technological products developed to aid people involved in training, the one that may prove to be the most preferred device over the long term is the computer-controlled interactive video using a videodisk instead of videotape. Videodisk technology continues to improve, costs are coming down, and new applications are becoming more evident as its use increases. Its primary value lies in its extraordinarily versatile interactive feature. Almost any course material can be converted to an interactive videodisk mode, such as a "hands-on" simulation application.

Nearly every training method of teaching is linear with instruction proceeding in a straight line from beginning to end (slides, videotape, flip charts, textbooks, and so on). Since learning is related to the pace of a particular medium, some trainees are unable to keep up with the course content, while others lose interest because the pace is too slow. The most productive method for training people in nearly any type of instruction is face-to-face, one-on-one, individually paced teaching, where the student can interact freely with the teacher. That is essentially what happens in computer-based training. But, even though CBT is a dynamic advancement, combining the features of interactive and self-paced learning, its disadvantage is that trainees have to read a lot of text, and replacing text-based instruction with video imagery is limited by the speed with which a person can move back and forth among images. Interactive videodisk technology solves both problems. Fast random access enables the trainer to jump any segment of a disk in less than a second, and a videodisk holds up to 54,000 single video frames per side, or up to 30 minutes of moving imagery. Controlling this access is a microcomputer which also displays text or graphics, commands video segments and determines how the program will "branch." Depending on how the viewer responds to questions, simulations or other variables, the trainee may be advanced to the next section, quizzed, returned to view a segment over again, shown further clarification, or asked where the trainee would like to go next. The options are virtually unlimited. Although, as mentioned previously, CBT is also interactive (and soundly positioned as an extremely cost-effective training device), videodisk's main advantage is its ability to simulate reality. Thus, the joining of interactive CBT and the videodisk has created a new geometry of thinking, involving simulation, decision making, realism, personalized feedback, and facets that border on artificial intelligence.

The major drawback to videodisk is still cost. Particularly for smaller organizations, interactive videodisk programs, with their requisite computer design time and video production costs, can be prohibitively expensive. At present only large organizations with a sizable student population can cost-justify the development of applications. The other option for the smaller companies is to buy off-the-shelf videodisk training, but unfortunately, this approach loses the advantage of customized courseware. Or, as with factory automation, less tangible long-term benefits may have to be built into the economic cost-justification equation when judging the value of the videodisk option. Therefore when considering this innovative state-of-the-art training technology, determine if your situation fits at least several of the following criteria:

- There will be many users of the videodisk training program.
- The users are dispersed geographically.
- The need is for fast, visual, interactive training material.
- Live demonstration is inappropriate for training your people in required applications.
- The programs need to be extensively variable.
- Users vary in experience or skill.
- The program is relatively stable or vital.
- The program can be used repeatedly.

Four Levels of Interactive Video Instruction

In addition to making the determination of whether or not to go the route of videodisk, another factor to consider if the decision is positive, is which of the following four levels of interactive video instruction to select.

Level One uses a consumer videodisk player with very limited memory capability and no computing ability. It is essentially a playback machine for the videodisk with which the user can control the point of entry into the disk.

Level Two uses an industrial player with approximately 1,000 bytes of memory, which contains a simple program with branching. Computer code is actually contained on the videodisk. Level Two systems are faster than Level One systems. They are more flexible, more rugged, and more easily controlled by the user.

Level Three combines all the capabilities of the video player and the microcomputer for a broad range of instruction possibilities. Sophisticated branching is possible for learning complex concepts and tasks.

Level Four uses two videodisk players and allows almost instantaneous feedback and more complex branching. These advanced capabilities are needed in training people in complicated occupations.

How Videodisk Compares with Videotape

The type of training program that is planned for use should determine whether to use videodisk or videotape since both methods contain advantages and disadvantages.

In the most common videodisk format, a laser stylus reads an electronically coded signal that resides just below the disk's surface. The signal is separated into audio and video, and these signals are sent from the disk player to the television monitor.

The videocassette tape recorder also reads signals, but the technology is quite different. The main difference between videotape and videodisk is that on tape, signals must be sequentially accessed rather than randomly accessed as on disk.

Interactive videotape has two major disadvantages: The seek time or shuttle-time is longer, and if an individual wants to move back and forth within a program there can be a significant wait while the tape rewinds. The second major disadvantage is that the tape cannot give clear resolution to a single frame. If you want to leave a single picture or diagram on the screen for a period of time, that image must be shot to run continuously as the tape continues to run. However, recently developed tape interfaces allow exact location of a single frame, but single freeze-frame playback still could have better resolution.

Tape tends to wear out, while disks are nearly indestructible. Videotape also requires sequential access—that is, each succeeding frame must be viewed or passed over before getting to the next frame. Trainers who choose tape as a medium often use linear instructional designs to avoid the long shuttle-time while the recorder seeks the portion of the tape to be viewed next.

Tape also has some significant advantages. A tape is less expensive to reproduce in small quantities than a disk. A program is relatively easy to change even after the mater is made. Shuttle-time can be significantly reduced by using two tape decks and alternating between them, and tapes can be designed and edited to minimize shuttle-time.

Videodisks allow quick random access to any portion of the video program. Disks are better for single-frame viewing. They allow very good resolution, precise location, and prolonged viewing. On the negative side, disks are expensive and relatively inflexible.

One of the technological problems in combining video with the computer is locating the beginning and end of a video segment so that it can be called up at any time during the program. Each frame on a videodisk is addressed with a unique frame number, so any segment is relatively easy to locate on a videodisk.

The best method for keeping track of frame locations on videotape is a special time code laid down on an unused audio channel. The most accurate time code is the SMPTE code developed by the Society of Motion Pictures and Television Engineers. The SMPTE code counts exactly thirty pulses per second—one for each frame of a standard videocassette.

In addition, there is a "universal video interface," which consists of a circuit board that fits inside the computer or a separate chassis that can attach to any computer with an RS232 C port (most microcomputers have one) and a videodisk or videotape player.

What to Expect in the Future

Although both videotape and videodisk are fine training devices, there is general agreement that videodisk has the greater potential. And yet, ironically, even though videodisk is still a number of years away from becoming a major player in the field of computer-based training devices, some consider it an intermediate step toward development of optical storage technology. Every indication is that when perfected, full read-write-erasable capability on optical disk will be a cost-effective technology. And, at that point in time—when cost, writing capability, and improved quality coalesce—there will be a decided shift in focus to courseware rather than hardware.

RETRAINING: WHO IS RESPONSIBLE FOR DISPLACED WORKERS?

The critical question that must be answered in regard to displaced workers is the public policy issue raised by automation: Who is responsible for retraining workers whose jobs are obsoleted by new technology? In the past, our society tended to forfeit this responsibility to government agencies. Now, with belt-tightening in Washington, industry is being compelled to pick up the slack. As *Training* magazine's industry report on training indicated, although considerable sums are being invested in this area, it may not solve the problem simply because most displaced workers lack the ability to become trained for more technically oriented jobs being created by automation. This is because the new jobs require much greater literacy and skill than they, as a group, possess.

According to the National Commission on Employment Policy, it is estimated that through the remainder of the 1980s, 400,000 workers a year may need extensive retraining to find new jobs. This includes people who must acquire higher technical skills, as well as those who will need remedial training in basic English and mathematics, whether they are active employees or just entering the work force.

At present, those companies that are retraining workers are doing it both to provide access to another position, and to broaden the worker's knowledge base so that he or she can perform several jobs outside the company if none is available inside. At times, these programs are jointly sponsored by the company and its union.

Who Risks becoming Displaced?

Training programs that teach students new or upgraded skills encompass both existing employees and those that have become, or will become, displaced workers. Although certain of these job losses are due to foreign competition and economic employee dislocations, the vast majority of these new unemployed workers will be the result of our country's technological revolution. The advanced technologies discussed in this handbook will create new jobs but the net result will be greater unemployment. On the plus side, over the next decade there will be a substantial need for scientists, engineers, and computer specialists. However, technology will take its toll with industrial robots which means fewer jobs for welders, painters, and assembly line workers; forklift operators, stockroom pickers and packers being eliminated with AGVs and other automated methods; draftsmen losing out to computer-driven design systems; and work processors replacing typists.

In addition, substantial numbers of people will become victims of structural unemployment. The changing dynamics of factory automation will not only displace people in specific job categories, but also will have a pronounced effect on those that are economically and educationally disadvantaged, and those whose skills do not fit the demands of America's new industrial composition. In the case of the disadvantaged, most are minorities living in central cities, with substantial numbers of unemployed even today. But in the latter instance, this loss of skill transfer represents the middle-class suburban worker who has always been regarded as the backbone of our society.

Need for Immediate Action

A concerted retraining effort must be initiated *now* so that as this impending shift in job skills and worker displacement gains momentum—and begins to accelerate—the United States will be able to cope with this problem in an orderly manner. Although the expenditure by industry of $29 billion in 1986 to train employees is certainly a large sum, it is unfortunately insufficient to meet the total demands built into the structure of our emerging technological economy.

The most realistic appraisal of this situation that seems to be based on a common consensus, is that a bold initiative must be forthcoming from various levels of government, industry, and labor to meet this challenge head-on. If the immense potential of computer-based factory automation is to be realized, then its immeasurable benefits must not only include increased product output, but an acceptable level of employment as well. This involves addressing the human relations aspect in consonance with expending our energies to develop new technologies.

A number of publications have addressed this problem by offering specific action plans. Without attempting to evaluate the merits of each of these proposals, *Industry Week's* plan appears to be soundly conceived, and therefore as good a point of departure as any. It consists of the following:

1. Fully implement Title II of the 1982 Job Training Partnership Act, which although primarily targeted at the economically and culturally disadvantaged, also includes provision for retraining displaced workers. *At the same time, consider a complementary "human capital" income-tax credit for companies that hire and train disadvantaged youths and adults.*
2. *Create an Individual Training Account (TA) program* to replace Title III of the 1982 Job Training Partnership Act (which provides federal assistance to dislocated workers).
3. *Encourage states to customize entry-level job-training programs to the needs of employers* and spread the word about successful programs in place in North and South Carolina, Georgia, and Oklahoma.
4. *Formalize systems by which business people and vocational and technical schools can coordinate their educational programs to match local needs.* Executives and union

officials should seek out school officials to let them know of their work place needs.

5. *Link the fifty states and the District of Columbia into a reliable, up-to-date, computerized jobs-and-skills matching system.*
6. *Promote greater CEO involvement in human resources.*
7. *Create a national commission to examine existing training and retraining programs and mechanisms, and to recommend ways that labor and management can resolve conflicts that hinder training efforts.* The goal should be to stress long-term employment security over protection of specific jobs.

How Extensive Is Corporate and Union Involvement?

For many years, large companies have provided employees with tuition reimbursement programs as an inducement to acquire new related job skills. However, these programs make no provision for displaced workers. Although corporate interest still remains more attuned to established training programs, for a number of reasons there is a discernible shift toward greater recognition for instituting comprehensive retraining programs. For one thing, companies are making use of the funds provided by the Job Training Partnership Act. Another reason is that firms are finding it cost effective to retrain workers to fill the new jobs that technology is creating. Perhaps most importantly, negotiated union contracts with some of the major corporations include clauses that stipulate some degree of worker retraining. In some instances, provisions of these agreements include arrangements for career counseling, job-search training, and placement assistance for laid-off employees as well as retraining. The development of local and national programs of this type is being coordinated by the UAW-Ford National Development and Training Center in Dearborn, Michigan. This center is directed and guided by a joint governing board of union and management representatives. It is funded by withholding 5 cents per employee hour worked. The center runs general and targeted retraining courses, acts as a clearinghouse for retraining information, and coordinates lower-level programs established nationwide.

In General Motors retraining is decentralized. At the Buick and Oldsmobile assembly plant in Wentzville, Missouri, which became highly automated, GM established a new technology training program. Many of the initial participants were displaced workers from a nearby GM plant, who were given 1,000-hour basic skills training courses. The basic training involved such things as math and physics, electronics, programmable controllers, robotics, data communications, and an introduction to minicomputers. The purpose of this program is to provide a worker with enough of a foundation in basic skills so that the company can build on this body of knowledge as more technology is brought into the work place. Initially, only laid-off workers were permitted to participate in the program, but it has been so well received that the company is now including employed workers who want to upgrade their job classifications.

Despite these inroads, many people feel that as important as counseling and retraining are for dislocated workers, such programs only treat the symptoms rather than the disease. Therefore, many are concerned that the national focus should be on job preservation and creation, not retraining. In essence, it is their opinion that it is better to train and upgrade employed people within the system than to lay them off and retrain them. Certainly, it can be argued that retraining is not a substitute for job creation, and unless new jobs are created in society, retraining just takes unemployed workers and makes them higher-skilled unemployed workers.

PART B: HOW TO MEASURE HUMAN PRODUCTIVITY IN THE MODERN FACTORY

HOW TO DEVELOP SHORT- AND LONG-TERM HUMAN RESOURCES STRATEGY

When expending considerable sums on personnel training and/or retraining, a company should develop these programs in consonance with its short- and long-term human resources strategy. To implement this strategy the initial step should be to identify the strengths of its most successful competitors so that this can become the basis upon which to measure your own performance.

Once these standards are in place, long-term goals can be devised to include developing a flexible response to any changes in the external business environment. Also, training/retraining plans can be devised to provide optimum use of learner-controlled instruction while maintaining close liaison with internal and external groups to ensure early awareness of new products or services where training/retraining needs are evident. The company also can remain current with industry trends and technological advances and maintain overall budgetary control of training/retraining expenditures.

Once overall planning and control steps have been taken, the next phase should be quantitative labor productivity measurements of employed personnel for which accurate and meaningful data can be gathered. There are four basic measurements that can be used for this purpose: (1) total cost per student hour of training, (2) hours of training delivered to company employees per trainer, (3) training hours per company employee, and (4) the ratio of total company employees to training employees.

In addition, it is advisable to agree upon standard definitions for four "critical" terms to be used in the measurement program. These are as follows:

- *Benchmark*—a standard against which performance is measured.
- *Gap* (either positive or negative performance)—level differences between your company and competitors based on analysis of internal and competitor data.
- *Objective*—a short-term goal established to reduce or eliminate gaps.
- *Target*—the "ultimate" benchmark, which signifies that the company's performance had met or exceeded the expected performance of its competitor.

Understanding Workers' Concerns about Productivity Measurement Programs

The advent of factory automation has made it incumbent upon management to develop technical training and retraining programs for both new hires and employees displaced through technological change.

This in turn, has placed demands on management to institute productivity measurement programs to evaluate benefits expected to accrue from automation. These measurements provide management with an objective appraisal of results achieved and productivity gains of the work force.

Employees should be motivated to meet desired objectives not only through incentive compensation but by job satisfaction derived from sociological and psychological means of fulfilling their inner needs. When all of these motivational techniques are synchronized properly, a company can expect to increase labor productivity, which translates into increased profits and decreased unit costs, that can then be reflected in a gain in sales volume and market share.

When introducing methods of measuring productivity involving factory automation, concerns of both management and direct labor must be addressed in a straightforward, positive approach. Should workers be unionized, then union representatives also must be involved in this discussion process.

Studies have found that these concerns typically include the following six factors:

Are Measures Being Misunderstood or Misused?

Even when time study was used as the primary method of measuring worker productivity, the leveling factor applied by the industrial engineer was subjective. Historical standard data also often became obsolete due to subtle changes on the shop floor or through negotiations with worker representatives. Often peer pressure limited the amount produced to only enough to achieve a "reasonable" incentive pay without necessitating a methods change or review of the existing standards.

In most past instances, workers accepted the standard without challenging the methodology that was used. In today's environment, workers are better educated and perhaps more threatened by the advent of the automated factory of the future designed to operate with robots instead of people. To allay this fear, it is necessary to clearly communicate how the measurements were developed, and what criteria will be used in interpreting production data. Both lower-level management and hourly employees must be assured that senior management is cognizant of any factors beyond their control so that proper analysis can be made that is objective and constructive.

Will Inadequate Performance Be Exposed?

As productivity measurements expand beyond the direct labor work force to also embrace white collar workers, concern over inadequate performance can become endemic throughout the entire organization. Hence, management must ensure all employees that the program is designed to improve the company's posture, vis-a-vis competition and profitability, and not as a means of ferreting out sub-par performance. Although performance measurement does undoubtedly result in a "weeding process" in terms of maintaining a work force of people meeting established production norms, this should not be regarded as the primary reason for instituting these programs.

Will the Program Create Additional Time and Reporting Demands?

Just as short-interval scheduling and other techniques developed in recent years in the office environment created additional paperwork controls that annoyed managers responsible for maintaining these reports, some managers will consider newly developed productivity measurement devices in the same light. It is therefore imperative that any program of this nature be designed to minimize paperwork chores, and be more than offset by benefits to be derived.

Will the Staff Be Reduced?

This is the most delicate item to be addressed. Certainly you must expect that some correlation exists between productivity measures and staffing levels. Obviously, a company must have alternative plans for displaced workers through retraining, attrition, or early retirement or workers will not accept the program fully. Without employee cooperation, any expectation of long-term benefits will surely be muted by employee resistance. To offset this concern, a company must demonstrate that displacements will be temporary and offset with some positive means such as increased sales or greater rewards to remaining personnel.

Is Distortion of Performance Possible?

It is necessary to develop productivity measures only after specific objectives and goals have been defined. When this is achieved measurement of important actions will take precedence over those of less importance and will provide a means for senior management to track tangible gains along a predetermined path.

Will Individuals Lose Autonomy?

In today's manufacturing environment the direct labor work force is shrinking while the indirect number of employees is growing. These latter groups are more technically skilled and have a higher academic background, which tends to make it more difficult to institute productivity measures. Also, whereas much of the direct labor measurements were machine paced (based on machine speed and feed), office work consists almost entirely of human-paced jobs. Thus, measurement is entirely dependent on the person involved, and since tasks may vary considerably from position to position—and even within a skill classification—measurement that can be monitored with any degree of accuracy becomes more involved.

FOUR PRIMARY TECHNIQUES FOR MEASURING PRODUCTIVITY

There are a number of reasons for developing and instituting productivity measurement techniques. One of the primary reasons is to arrive at a sound method of determining worker productivity output to control unit labor cost. This in turn, can enable a company to develop accurate cost data for product pricing. Of equal importance is application of a properly conceived productivity measurement device that can instill a positive attitude throughout the organization resulting in a more highly motivated work force. Specifically, introduction of productivity measurement techniques aims to improve *total* employee performance by satisfying these seven criteria: (1) effectiveness, (2) efficiency, (3) quality, (4) productivity, (5) quality of work life, (6) innovation, and (7) profitability/budgetability.

There are four generic and primary techniques used in industry for measuring productivity.* They are:

1. *Multi-Factor Productivity Measurement Model* (MFPMM). (Also called the total productivity measurement model, total factor productivity measurement by product in other related forms and variations.)
2. *Normative Productivity Measurement Methodology* (NPMM)
3. *Multi-Criteria Performance/Productivity Measurement Technique* (MCP/PMT). (Also called the objectives matrix.)

*Dr. D. Scott Sink, "Productivity Measurement Techniques," © *National Productivity Review*, 22 West 21 Street, New York, NY 10010. Reprinted with permission of the publisher. For further information on this subject see also *Productivity Management: Planning, Measurement and Evaluation, Control and Improvement* by Dr. D. Scott Sink , Ph.D., P.E., Associate Professor, Department of Industrial Engineering and Operations Research, Director, Virginia Productivity Center, Virginia Polytechnic Institute and State University (New York: John Wiley and Sons).

4. *Surrogate Productivity Measurement Techniques*
   (a) Common Staffing Study
   (b) Checklists, Audits, etc.

Multi-Factor Productivity Measurement Model

The multi-factor productivity measurement model (MFPMM) is a dynamic, aggregated, indexed, and computerized approach of measuring productivity. It is capable of blending the major inputs of a particular organizational system together and relate the resultant aggregate input to the total output of that same system. MFPMM can be utilized to measure productivity change in labor, materials, energy, and capital. It also measures the corresponding effect each one has on profitability. With essentially the same accounting data used to track revenues and costs, the MFPMM can provide additional insight into the individual factors that are most significantly affecting profits.

The MFPMM is based on the premise that profitability is a function of productivity and price recovery—that is, an organizational system can generate profit growth from productivity improvement and/or from price recovery. Productivity relates to quantities of output and quantities of inputs, while price recovery relates to prices of output and costs of inputs. Price recovery can be thought of as the degree to which input cost increases are passed on to the customers in the form of higher output prices.

The data required for the MFPMM is periodic data for quantity, price, and value of each output and input of the organizational system being analyzed. Since value equals quantity times price, having two of the quantity, price, and value variables obviously yields the third algebraically. Quantity, price, and/or value of the various outputs produced and most of the inputs consumed are straightforward and should be provided by most basic accounting systems.

The MFPMM compares data from one period (base period) with data from a second period (current month). This comparison forms the basis of the productivity/price recovery/profitability analysis. The choice of a base period is a critical decision since it establishes the period against which the current period will be compared. Therefore, the base period should be as representative of normal business conditions as possible. If the data exists, the budget or "standards" could be used as the base period data. Depending on the needs of the user, the availability of data, product cycle time, and so on, period length could be a week, quarter, year, or any other period for which input data can be matched to output data.

From the base and current period data, the MFPMM generates a series of ratios and indexes, each communicating different information about the system under study. The ratios and indexes, can include weighted change ratios, cost/revenue ratios, productivity ratios, weighted performance indexes, and total dollar effects on profits. Weighted change ratios depict the percentage increase (or decrease) of an item from the base to current period. Both price and quantity weighted change ratios are generated by the model to show the percentage changes from period to period. Cost/revenue ratios reflect the percentages of reported revenue consumed by a particular input in a given period. This information provides the user with insights as to where leverage exists. The most common method of productivity improvement is cost reduction and these ratios show exactly where cost reductions will pay the biggest dividends. Productivity ratios depict absolute productivity values in the base and current period. These absolute values are used in calculating the price weighted productivity indexes which show increases or decreases in productivity for the overall system and each component. Weighted performances indexes are actually output over input change ratios from Period 1 to Period 2. The final set of indexes are the dollar effects on profits.

These indexes indicate what impact (in dollars) is caused by changes in productivity, price recovery, and profitability. The ratios and indexes, alone or together, provide management with information on their systems. The ratios and indexes identify areas that need improvement and they also identify areas that are operating at an acceptable level. If the information is used correctly, productivity can improve, which in turn should increase profits.

The MFPMM is most appropriate at the firm and plant level and would be most useful to senior management. It could be used at the cost center level as a separate accounting system for an assembly line, individual product line, and so on, however at lower levels of organizations, managers do not normally need the kind of detail offered by the model. The MFPMM has been most often applied in manufacturing settings but it can be used anywhere the necessary data exists.

Normative Productivity Measurement Methodology

The normative productivity measurement methodology (NPMM) incorporates structural group processes to identify appropriate productivity measures for such work groups as engineering, marketing, and personnel. The structured group processes are used as mechanisms for shaping consensus and for developing a commitment for further follow-up. Once the productivity measures are identified, it becomes the task of the group to operationalize and implement the productivity measurement system. The final and perhaps most important task of the NPMM is to provide feedback to the workers in hopes of identifying productivity improvement opportunities.

The NPMM, as a component of a productivity measurement system, is not implemented until several necessary preconditions are fulfilled. Without support from all levels of management and labor, the NPMM cannot be successful. Once it is determined that NPMM is to be used in the productivity measurement system, there are five distinct stages to be followed.

Stage 1 of the NPMM involves execution of a structured group process such as the *nominal group technique* (NGT) or the *delphi technique* (DT). This process is used as a mechanism for generating a prioritized list of productivity measures, ratios, and/or indexes for each specified unit of analysis. The basic reason for using a "normative" or participative approach is to ensure adequate motivation commitment and accountability on the part of key participants for implementation and acceptance of the resultant productivity measurement system.

Stage 2, the design of the productivity measurement system, requires intervention from one or more productivity analysts, who works at operationalizing measures obtained from the structured group process. The analysts also must determine how and where to collect the data for the productivity measures previously suggested by the structured group processes. Eventually it becomes the responsibility of the analysts and the group members to shape a workable productivity measurement system.

Stage 3 of the NPMM process requires a briefing, review, discussion, potential revision, and eventual approval of the draft operating system of the productivity measurement program. This also is the stage that provides feedback to the participants in the organizational system for which the productivity measurement system is being developed. This feedback is important in keeping commitment from all participants involved in the process.

Stage 4 is the integration and implementation of the productivity measurement system with existing performance measurement systems within the organization. This integration and implementation requires consultative or managerial backing for the resultant productivity measurement system to be fully implemented.

Stage 5 occurs once the measurement program has been implemented. It involves continuous monitoring of the productivity measurement system after implementation is recommended. This monitoring will provide feedback and validity for the program. The monitoring also provides flexibility to the method. If at any time the program developed by the NPMM does not accomplish its objectives, corrections can be made by repeating the NPMM process.

The NPMM can be implemented in any organization that is interested in pursuing a participative method of developing productivity measures. By tapping the resources of its employees, an organization can capture a group wisdom that is most likely to identify the real problems and opportunities involved in productivity improvement.

It has been determined that the NPMM process functions best when the measurements focus is not on the corporation as a whole or the individual employee, but rather on the individual department. The productivity emphasis, therefore, would be placed on the "producing" unit. The NPMM is most often implemented in organizations where white collar employees prevail.

Multi-Criteria Performance Productivity Measurement Technique

The multi-criteria performance productivity measurement technique (MCP/PMT) is an innovative, widely applicable, and reasonably simple approach to measuring performance or productivity. The technique can be nicely integrated with NPMM to facilitate more effective use of the tremendous number of measures and ratios of productivity or performance that can be attained from groups in organizations.

The value of a participative approach lies in the creation of an "ownership" for the resultant measures. Successful implementation of ensuring measurement, evaluation, and control systems is more assured with effective participative approaches.

Although difficult, evaluating performance against a list of measures that is often highly heterogeneous is critical in using MCP/PMT. This can be done by developing a prioritized set of productivity/performance measures utilizing the NGT. A "utility" curve for each of the priority (top eight to ten) measures can be developed, with a ranking and rating assigned based on relative importance of each productivity/performance measure. The utility curve is then utilized to transform actual performance against each specific measure or criteria to a common 0 to 1.0, then multiplied by the relative value. The various performance values for each of the top priority measures are then added together to obtain a productivity/performance index.

To implement MCP/PMT assume you have just generated a consensus and prioritized list of productivity/performance measures for a given organizational system utilizing the NGT. You have a list of heterogeneous measures. You are interested in aggregating or evaluating performance against these criteria in an integrated fashion.

A common performance scale or utility scale needs to be developed that converts all the uncommon measures into some common denominator. The performance scale commonly is rather arbitrarily allowed to range from either 0 to 1.0, 0 to 10.0, or 0 to 100.

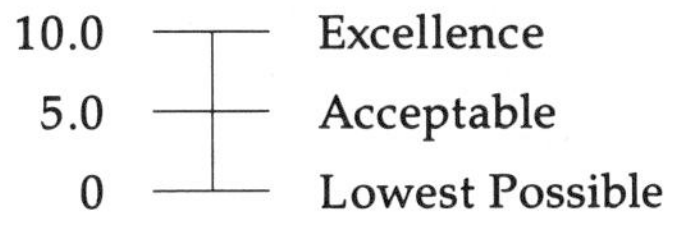

Common Performance Scale for each Productivity/Performance Measure

Level 0 represents the lowest level of performance possible for a given measure. Level 5 represents a minimally acceptable performance level (MAPL). Level 10 represents the perception of best performance or excellence. Levels 0, 5, and 10 should be clearly defined and accepted benchmarks.

Each productivity/performance measure or criterion has at least one "natural" scale that it can be or is measured with. Often this "natural" scale is simply an industry consensus or norm. For example, measuring liquids in gallons or liters, measuring coal in tons, measuring profitability performance in terms of ROI, and so on.

The objective in the MCP/PMT is to develop a valid set of natural scales used to measure performance against a given criteria and to match levels of performance on that scale to levels of performance on the common utility scale.

A utility curve, (such as curves a, b, or c) which can, and often will be subjective, is developed and used to transform performance on one scale to a common scale.

The M in the MCP/PMT stands for multi, which signifies that there are many criteria or measures of productivity/performance against which we are attempting to perform. Therefore, there often will be many utility function graphs as discussed previously. The question becomes one of how to aggregate performance scores. Since these scores are all from a common scale, we might be tempted to just add them up. However, the question of relative impact of performance against various criteria on overall performance becomes critical.

The NGT will have given you a ranked list of productivity/performance measures. The first step, once the criteria have been ranked, is to arbitrarily assign 100 points to the top priority measures or criteria. Next, the relative importance of the second most important measure is assessed. In performing this process, you have a sense for the relative importance or contribution of each measure or criterion to overall performance/productivity. There are some critical nuances to this procedure that are not described here, however the approach is basically as straightforward as it appears. These weights can be determined unilaterally by the manager of the group or by an analyst, or participatively by the same people who identified the criteria and their rankings.

The next step in the MCP/PMT is to integrate the performance (utility) graphs (scales and curves) with the criteria weightings. This will allow the development of one performance/productivity indicator which will indicate the overall performance of the organizational system. The final computational step is to add all the weighted scores together.

From this exercise, a general matrix format is developed which represents the X-axis and Y-axis of the utility curve. It is possible to have sub-criteria for a given measure. In other words, there could be more than one way to operationalize a given performance/productivity measure. In that case, the weighting for the given measure or criteria is simply divided up among the sub-criteria.

Surrogate Productivity Measurement Techniques

There also are a few distinct techniques that are not specifically for productivity measurement. A surrogate productivity measurement approach is one that does not measure productivity directly but measures something that is highly correlated with productivity. The implicit assumption behind these approaches is that if we measure, evaluate, control, and improve upon these factors then productivity also will be managed.

Any measurement, evaluation, control, and improvement technique that focuses on the broader issue of performance could therefore be labeled a surrogate productivity measurement approach since efficiency, effectiveness, quality, quality of working life, and innovation are highly interdependent with productivity. In this respect, the MFPMM, NPMM, and the MCP/PMM are partially surrogate approaches. More specifically, however, the following techniques typically do not measure productivity directly but have productivity management implications.

The *common staffing* technique is designed as an attempt to measure, plan, and improve productivity in the indirect labor areas. These areas tend to encompass work that is complex, nonrepetitive, irregular in character, and often unpredictable. The common staffing approach is based on the assumption that it is not feasible or economical to measure most "indirect" manufacturing jobs in the sense of determining the minimum requirements to accomplish a task with 100 percent performance. Instead, the objective is to describe the level of productivity, whatever it is, at one point in time and then to strive to continuously improve that productivity in all areas through future points in time.

Productivity audits and checklists can serve as starting points toward the design of productivity measurement systems or be used to evaluate the effectiveness of existing productivity efforts. They fit the definition of surrogate measurement techniques and assist in the productivity management process.

Some Conclusions About Productivity Measurement Techniques

To summarize, productivity measurement techniques have to differentiate between effectiveness, efficiency, quality, productivity, quality of work life, innovation, and profitability. It also is axiomatic that you cannot measure what cannot be defined.

Basically, there are four generic productivity measurement techniques. You must recognize the distinctions among operational differences, the phrases "measurement of productivity or productivity measurement," and "measurement, evaluation, and verification, of productivity improvement." There are a limited number of ways to measure productivity (input/output), although there are a great number of ways to measure whether productivity has improved and by how much. Productivity measurement can tell you this but so can a number of other approaches. Any measurement and evaluation approach that will indicate that input resource consumption changed favorably relative to outputs from the system can be utilized to verify productivity improvement. Cost/benefit analysis, engineering economic analysis evaluation, and so on, all are mechanisms that could be utilized to verify productivity improvement. A critical factor, however, is the ability of these techniques to measure in constant value terms. An advantage to utilizing the MFPMM is that it automatically indexes prices and costs to some established base period value.

Finally, you must distinguish between a measurement system designed to control and verify and a measurement system designed to improve. There are productivity measurement approaches that are effective at control, diagnosis, and evaluation such as MFPMM. There also are productivity measurement approaches that are effective at evaluation and improvement (NPMM and MCP/PMT). However, there are no techniques that are effective at doing both. Also, there are certain techniques that are not accurately labeled as productivity measurement approaches but that can measure, evaluate, and verify the existence and magnitude of productivity improvement (cost benefit analysis and tracking, engineering economic analysis, and evaluation approach). These approaches can be effective at control and verification but often are less effective at successfully linking measurement to improvement. Thus, an organization may need to develop a variety of productivity measurement approaches and develop mechanisms for integrating these systems together.

THE FIVE PHASES OF PRODUCTIVITY IMPROVEMENT

Worker productivity is one of the key variables that creates additional wealth in a society, thereby leading to a higher standard of living. Although there are many approaches to developing, installing, and monitoring a productivity improvement effort, basically there are five phases that define the productivity improvement process. These are:

Awareness. An understanding on the part of the organization that there is a definite need to increase labor productivity to remain a viable entity. This awareness must be recognized at the executive level since these executives are ultimately responsible for implementing any actions deemed necessary to improve human resources efficiency.

Planning. It is of critical importance that a well conceived plan be developed before embarking on a program of this type. In addition to all the normal planning elements, it is necessary in the planning phase to have reliable and accurate financial data for use in establishing a base point of departure and to develop the financial goal parameters that must be achieved.

Assessment. This phase is concerned with an audit of results of the program. It requires specific details in terms of how and where productivity improvements are to be made, such as divisions, plants, shifts, and so on. This phase should establish certain quantitative values to audit, based on a determination of which corporate areas seem to possess the most potential for productivity gains and which approaches may be most cost effective.

Implementation. Actual implementation of a productivity improvement effort should be within the jurisdiction of line management. They may receive guidance and some degree of audit control from various staff departments, but accountability for success of this undertaking is vested in the direct line organization chain of command.

Evaluation. The final phase should consist of an analysis of productivity improvements that have been made throughout the company. These changes should be compared with the financial data developed in the assessment phase to validate the findings in quantitative terms. Some discretion is required in this stage to allow for qualitative or less definable gains to be considered as well.

PROCEDURES FOR MEASURING PRODUCTIVITY

Basically, productivity is determined by how efficiently a company uses its resources. This means of measurement is concerned with how much output is produced from a given amount of input. This ratio can be calculated at any level of the business process. However, to arrive at this ratio properly, it is necessary that the output divided by an input is from the same process.

A Formula for Finding Total Productivity*

In the past, labor represented the largest factor of production, and hence was afforded the primary means of measuring company productivity. This has changed drastically, so that today direct labor may only account for 10 to 15 percent of the cost of goods sold. Because of this, the attributes of production have been broadened to consist of four key factors for analyzing productivity: (1) labor, (2) material, (3) energy, and (4) capital.

When a single input or factor of production is used in an input-output ratio, this is called a *partial productivity ratio.* The theoretical ideal, however, is the *total productivity ratio,* which is calculated by dividing the output of the process by the sum of all the factors of

*Material taken with permission from the American Productivity Center.

production: labor, capital, energy, and materials. The formula for arriving at total productivity is:

$$\text{Total Productivity} = \frac{O}{L + C + E + M}$$

A similar type of productivity measure is *total factor productivity,* which is defined as output minus purchased goods and services (or "value added") divided by the sum of labor and capital. Since "purchased goods and services" are mainly material and energy, total factor productivity ends up involving the same four factors of production as total productivity, assembled in a different manner, as follows:

$$\text{Total Factor Productivity} = \frac{O - (M + E)}{L + C}$$

> ***Key Point:*** A productivity ratio that merely arrives at volume per employee is of no value as an analytical tool unless it is put in a specific context such as trend analysis, level comparisons, or standard performance.

Four Types of Measures

Within a corporation, there are basically four types of measures that make up a complete measuring system. These are described below:

1. National or Industry Measures.

These provide executive management with realistic benchmarks against which to gauge the overall performance of the organization.

Bureau of Labor Statistics Division. The Bureau of Labor Statistics (BLS) is still regarded as the main source of this country's productivity by using output per employee hours paid for as its base for a measurement device. The output data utilized for the BLS productivity series is based on the Commerce Department's national income and product accounts series that uses a "value added" approach to each of the three segments: private business economy, manufacturing, and nonmanufacturing. For its "industry productivity" series, BLS calculates an output in physical terms, so far as possible, with a heavy reliance on the Federal Reserve Board's indexes of industrial production and some associated data.

Census Bureau Measures. The Census Bureau produces a census of manufacturers and of other segments of the economy every five years. These are supplemented by an annual sample survey that appears in various annual publications. This data is gathered for very broad purposes, and productivity measures can be derived from them only indirectly. The output of an industry (as expressed in "value of shipments") can be divided by the labor input and a level of labor productivity can be arrived at that can be compared with the same ratio from the previous census to get an approximate growth rate.

American Productivity Center Measures. To fill the obvious gaps in BLS and census bureau productivity data, the American Productivity Center has created and maintains on a regular basis a series of productivity measures for the economy as a whole, for major segments of the economy, and for the twenty major manufacturing industries. This series provides not only labor productivity ratios but also capital productivity ratios and a combined labor and capital (total factor) productivity ratio, supplemented in some cases by total productivity indexes.

Industry Interplant Comparisons. In some cases, reliable and useful industry productivity statistics can be developed only by the industry itself. A relatively small number of industries are covered by the BLS, and the census data are released too late to be very useful.

2. *Operational Control Measures*

These monitor the productivity of a company's production process on a regular basis for the guidance of line managers.

Partial Measures. Partial measures are used for operational control in the typical factory, sales region, or distribution center. They form the simplest element in a productivity measurement program in that they include only one (usually physical) output and an equally simple input. Where the measure can be calculated immediately and explained easily, the partial measure provides the basic control requirement.

In many operations, however, not only is more than one factor of production important, but there are various outputs at various stages of the process to consider. Once an organization has used a family of simple measures, it can proceed with calculations that combine unlike outputs or inputs, or compare unlike whole productivity measures of different parts of the process.

Physical Weighting. Since most business processes are rarely so simple that they can be fully described using either one output or one input, most analyses require taking into account different factors. Whether the analysis makes level comparisons with another organization, or trend comparisons with their own performance, "weighting" mechanisms for combining unlikes must be developed.

Trend Analysis. Trend analysis is used to compare a "base" period against present actuals and future projections.

Adjustments for Inflation. In developing financial and operating data it is necessary to also include an adjustment for inflation in any calculations.

Analyzing Input. Most companies still use labor as the main input in productivity analysis, and labor is still considered to be the most important factor of production because it is the most costly for the "average" industry. The simplest way to quantify labor input is to count the number of employees. A more precise approach is to use employee hours.

Other factors to consider in analyzing input are:

- Outside contract work
- Employee benefits
- Indirect labor
- Materials input
- Energy input
- Capital input

3. *Professional Staff Measures*

These must be treated differently than more routine occupations because of the more elusive outputs and special human elements involved.

Measures of professional work groups are becoming increasingly important in productivity analysis as more of an organization's work is being performed in the professional departments rather than on the line.

The nature of professional work that requires analysis is:

- Nonrepetitive work
- Collective efforts
- Shifting objectives
- Productivity distinctions
- Staff participation
- Employee involvement in nominal group technique

4. Total Productivity Measures

These capture and analyze historical performance levels and are useful in the long-range and strategic planning process.

Measures of productivity provide a company with a means of evaluating its business. Thus, productivity forecasts and goals should be a key part of a long-range planning process. Although "bottom-line" profit is still the primary yardstick, productivity improvement is an important factor in improving productivity.

There are several valid approaches to total productivity analysis such as:

- Combination of physical partial measurements
- "Frozen" cost standards for each factor of production
- American productivity center performance measurement system which links the key elements of profitability, productivity, and price recovery

Index

A

B

D

E

F

G

J

L

M

N

O

P

Q

R

S

T